Studies
in the History of Mathematics and
Physical Sciences

5

Herman H. Goldstine

A History of the Calculus of Variations from the 17th through the 19th Century

Springer-Verlag
New York Heidelberg Berlin

Herman H. Goldstine
IBM Research, Yorktown Heights, New York 10598/USA
and
Institute for Advanced Study, Princeton, New Jersey 08540/USA

AMS Subject Classifications (1980): 01A45, 01A50, 01A55, 49–03

With 66 illustrations

Library of Congress Cataloging in Publication Data
Goldstine, Herman Heine, 1913–
A history of the calculus of variations from the seventeenth through the nineteenth century.
(Studies in the history of mathematics and physical sciences; 5)
Bibliography: p.
Includes index.
1. Calculus of variations—History. I. Title.
II. Series
QA315.G58 515′.64′0903 80-16228

Printed in the United States of America.

9 8 7 6 5 4 3 2 1

ISBN 0-387-90521-9 Springer-Verlag New York Heidelberg Berlin
ISBN 3-540-90521-9 Springer-Verlag Berlin Heidelberg New York

To Mady and Jon

ἀγαθαὶ δὲ πέλονται ἐν χειμερίᾳ
νυκτὶ θοᾶς ἐκ ναὸς ἀπεσκίμφθαι δύ' ἄγκυραι

Very good they are
In night of storm, two anchors firmly fixed from the swift ship.

Pindar, *Olympian*
VI, 100-101

Preface

The calculus of variations is a subject whose beginning can be precisely dated. It might be said to begin at the moment that Euler coined the name calculus of variations but this is, of course, not the true moment of inception of the subject. It would not have been unreasonable if I had gone back to the set of isoperimetric problems considered by Greek mathematicians such as Zenodorus (c. 200 B.C.) and preserved by Pappus (c. 300 A.D.). I have not done this since these problems were solved by geometric means. Instead I have arbitrarily chosen to begin with Fermat's elegant principle of least time. He used this principle in 1662 to show how a light ray was refracted at the interface between two optical media of different densities. This analysis of Fermat seems to me especially appropriate as a starting point: He used the methods of the calculus to minimize the time of passage of a light ray through the two media, and his method was adapted by John Bernoulli to solve the brachystochrone problem.

There have been several other histories of the subject, but they are now hopelessly archaic. One by Robert Woodhouse appeared in 1810 and another by Isaac Todhunter in 1861. In addition to these there are good historical accounts by Lecat in the *Encyclopédie des sciences mathématiques*, II 31, 1913 and 1916, by Ernesto Pascal in his 1897 book on the calculus of variations as well as accounts by Kneser, and Zermelo and Hahn in the *Encyclopädie der mathematischen Wissenschaften*, II A8, 1900 and II A8a, 1904. The reader might also wish to look at Lecat's excellent bibliography from the beginnings of the subject through 1913.

In light of the fact that these previous accounts are either out-of-date or inadequate I have in this volume attempted to select those papers and authors whose works have played key roles in the classical calculus of variations as we understand the subject today. In doing this I have excluded some otherwise excellent papers and very many that seemed to me to be pedestrian. Other readers may not agree with this selection process. I must bear sole responsibility.

It is with pleasure that I acknowledge my deep appreciation to the International Business Machines Corporation, of which I have the honor of being a Fellow, and especially to Dr. Ralph E. Gomory, Vice-President for Research, for the encouragement and support that have made this book possible. I would like to express my gratitude for the patience and helpfulness of Professors Marshall Clagett and Harold Cherniss of the

Institute for Advanced Study in discussing with me many points that arose in the course of writing this book. It would be remiss of me if I did not also acknowledge the helpfulness of the Institute for Advanced Study in sustaining me intellectually and for providing me with its facilities. My colleagues both at IBM and at the Institute for Advanced Study have served me as examplars of the highest standards of science and scholarship. In particular I wish to single out Professor Otto Neugebauer who has been a great inspiration to me.

In closing I extend my thanks to Mrs. Irene Gaskill, my secretary, for her tireless, good-natured and patient help in preparing the manuscript and in reading proofs, and to Springer-Verlag for its excellent help and for the high quality of its work on this volume.

Fall 1980 HERMAN H. GOLDSTINE

Table of Contents

Introduction

The ancients were certainly well aware of isoperimetric problems, and their results were preserved for us by Pappus (c. 300 A.D.). Their methods were, of course, geometrical, and I have accordingly ignored this work preferring to limit the scope of this book to problems solved analytically. I have also given almost no account of Galileo's analysis of the brachystochrone and the heavy chain problems for more or less the same reason. Instead this volume begins with Fermat's analysis of the passage of a light ray from one optical medium to another. He postulated that light moves in such a way that it will traverse the media in the least possible time.

Fermat's analysis does make use of the calculus and hence seemed to me to be a reasonable starting point for this book (see Section 1.1). It is also important because John Bernoulli took Fermat's ideas and used them to solve the brachystochrone problem by an adroit use of Fermat's principle (see Section 1.7): He divided the space between the two given points into narrow horizontal strata and supposed that the descending particle moved uniformly in each stratum and was refracted at each interface as if it were a quantum of light so that the total time of descent was a minimum.

After Fermat's papers of 1662 there was a lull in the subject until 1685 when Newton solved what can be viewed as the first really deep problem of the calculus of variations. In the course of his researches he investigated the motions of bodies moving through a fluid of "a rare medium of very small quiescent particles . . . ." He led up to the general problem of the shape of the body encountering the least resistance upon its nose by first calculating the resistances encountered by several special ones. He thus discussed a hemisphere, a cylinder, and the frustum of a cone. After this he was able to proceed to the general case (see Section 1.2). This result of Newton appeared in his *Principia* without any details being given. In 1691 Huygens studied this problem of Newton without, however, getting all the way through. He was, however, more successful than anyone else.

David Gregory persuaded Newton in 1694 to write an analysis of the problem, which made Newton's ideas available to the mathematical community. Newton's technique is very interesting and was duplicated by subsequent authors until Lagrange developed an elegant and superior analytical apparatus. This study did not exhaust Newton's interest in the problem: He proceeded to study a similar problem in which one end-point is allowed to vary on a vertical line. He succeeded in solving this problem and found the transversality condition for this case. For some reason this work of Newton on a variable end-point problem seems to be unknown in the mathematical literature. In his analysis Newton exhibited the family of extremals transversal to the given end-curve.

In 1699 John Bernoulli challenged the mathematical world to solve the brachystochrone problem: The problem first formulated in 1638 by Galileo. The problem was solved not only by him (see Section 1.7) but also by Leibniz (Section 1.6), his brother James (Section 1.8), and anonymously by Newton (Section 1.5). Bernoulli wrote of Newton's solution that one could tell "the lion from his touch."

There were in fact two quite different solutions given by John Bernoulli (see Sections 1.7 and 1.11). The second of these seems not to have received attention. In this solution Bernoulli gave a very elegant sufficiency proof, probably the first in history.

The brachystochrone problem served as a stimulus to the brothers Bernoulli to formulate and solve a number of other more general problems and thereby to establish a mathematical field. Indeed Euler was probably led into the subject by John Bernoulli, and he certainly gave the new subject a beautiful framework in his 1744 opus *The Method of Finding Curves that Show Some Property of Maximum or Minimum . . .* (see Chapter II).

In this book Euler treated 100 special problems and not only solved them but also set up the beginnings of a real general theory. His *modus operandi* was still very like that of Newton's, but he made it systematic. It also served to influence Lagrange, then a quite young man, to seek and find a very elegant apparatus for solving problems (see Sections 3.1–3.3). This new tool of Lagrange's was, of course, his method of variations and caused Euler to name the subject appropriately the calculus of variations. Lagrange, however, did much more than replace Euler's archaic method by a much better one: He moved ahead to variable end-point problems—only with difficulty—and he explicitly formulated the Euler-Lagrange multiplier rule even though he did not prove it. This rule became a sovereign tool in his hands for discussing analytical mechanics.

In an effort to discuss sufficient conditions Legendre in 1786 broke new ground by extending the calculus of variations from a study of the first variation to a study of the second variation as well. Legendre's analysis was not error-free and incurred harsh criticism from Lagrange. The essence of the difficulty was not truly appreciated by either man, however; and it was not until 1836 when Jacobi wrote a remarkable paper on the second variation that the root of the matter was recognized. Among other things he showed that the partial derivatives with respect to each parameter of a family of extremals satisfy the Jacobi differential equation. He then proceeded to discuss the relationship between solutions of that differential equation and the zeros of the second variation. With the help of this he put Legendre's transformation on a rigorous basis and discovered the fundamental concept of conjugate points (see Section 4.2). None of Jacobi's results was proved in his paper. As a result a large number of commentaries were published, mainly to establish an elegant result of his on exact differentials. All that work on exact differentials was rendered obsolete by Clebsch and then Mayer. Jacobi viewed the problem based on the integrand $f(x, y, y', y'', \ldots, y^{(n)})$ as a more general problem than that based on $f(x, y, y')$, which in a sense it is. He did not, however, see how the former problem could be reduced to the latter by the addition of very simple side-conditions.

At about the same time that Jacobi wrote his classic paper just mentioned, Hamilton published two remarkable papers on mechanics. In these papers he showed that the motion of a particle in space, acted on by various forces, could be comprehended in a single function which satisfies *two* partial differential equations of the first order. In 1838 Jacobi criticised this result, pointing out that only one partial differential equation, the Hamilton-Jacobi equation, was needed. Jacobi then showed, in effect, the converse of Hamilton's result (see Section 4.5). This work of Hamilton and Jacobi underlies some of the most profound and elegant results not only of the calculus of variations but also of mechanics, both classical and modern. Indeed it is really here that we see the deep significance of the calculus of variations and the reason why it is basic to the physical as well as the mathematical sciences.

At this point there was a lull in the theory for about twenty years. The calculus of variations as a subject was now at a fork in its intellectual path: The entire subject was in need of a rigorous re-analysis; the theory of weak and strong extrema was not yet understood or indeed even articulated; and general problems had yet to be enunciated. Two quite different directions were now taken by the nineteenth century analysts: Weierstrass went back to first principles and not only placed the subject on a rigorous basis using the techniques of complex-variable theory but, perhaps more importantly, discovered the Weierstrass condition, fields of extremals, and sufficient conditions for weak and strong minima. Clebsch tentatively and A. Mayer decisively moved on quite another route: They succeeded in establishing the usual conditions for ever more general classes of problems.

In Chapter 5 we consider in some detail Weierstrass's researches on the calculus of variations. This work was elegantly organized by Rothe in a volume of Weierstrass's collected works and constituted a real milestone in the subject. Here we find yet another example of Weierstrass's rigor, clarity, and depth of understanding. He mainly studied only the simplest problem in parametric form; but for it he found his necessary condition, discovered fields of extremals, and made sufficiency proofs. He made several of these sufficiency proofs using not only series expansions but also fields. In about this same period—1886—Scheeffer also gave a sufficiency proof for a weak minimum in ignorance of Weierstrass's results.

Clebsch worked in a quite different direction from Weierstrass and considerably before him—1857. He concentrated upon more general problems of the calculus of variations—ones having differential equations appended as side-conditions. By means of these he was able to render obsolete some of the complex but ancillary results that Jacobi and his school had succeeded in establishing—in particular all their studies of integrands containing derivatives higher than the first. Clebsch, working primarily with the second variation, succeeded in generalizing Legendre's form of the second variation to his general problem. As a by-product he also found the Clebsch condition. His work is, however, quite opposite to Weierstrass's in rigor, clarity and elegance (see Section 6.2).

It was not until 1868 that Mayer reworked Clebsch's material into reasonable form and proceeded in a long series of papers more or less in parallel with Weierstrass to discuss very general problems of the calculus of variations. He described the problems we now call the problems of Lagrange and of Mayer in 1878 and 1895. He also gave an elegant treatment of isoperimetric problems in which he formulated his well-known reciprocity theorem. In a paper in 1886 Mayer took up the problem of establishing the famous multiplier rule, which had never been established rigorously for a general problem. His proof had a fundamental gap in it, which was later filled in by Kneser and also by Hilbert at the turn of the century. In the course of his analysis Mayer discovered abnormality although it was Hahn who in 1904 invented the name for this case (see Section 6.5). At the turn of the century Mayer systematically treated the problem of Lagrange with the help of a Mayer family of extremals and also gave an elegant treatment of the Hamilton–Jacobi theory for this case.

At the international mathematical congress of 1900 Hilbert gave a beautiful discussion of the calculus of variations summarizing his lectures at Göttingen on the subject. In the course of this work Hilbert put the subject pretty much into its final classical shape in the sense that later extensions made by Morse and his school use new ideas arising from deeper geometrical understandings and are not analytical generalizations but have moved the field into entirely new realms. Hilbert's greatest contributions were perhaps his discovery of his invariant integral together with the elegant results that stem from it; his perception of the second variation as a quadratic functional with a complete set of eigenvalues and eigenfunctions; and his examination of existence theorems, i.e., whether there actually exist minimizing arcs for given problems (see Sections 7.1 and 7.2).

Osgood and Bolza each investigated yet another kind of existence theorem: When do systems of implicit equations have solutions and what continuity and differentiability conditions do their solutions have? In particular they were able to give proofs that a field can be constructed about a given extremal (see Section 7.2). Thus under quite reasonable hypotheses any given extremal can be viewed as a limit point of a suitable region of arcs.

Hilbert also gave the first completely rigorous proof of the multiplier rule for the problem of Mayer in 1906, and showed the relation between his independence theorem and the Hamilton-Jacobi theory. In the course of this discussion he formulated the modern definition of a field (see Section 7.3).

A new approach to the calculus of variations was provided by Kneser in a paper in 1898 and in two editions of his *Lehrbuch*. He started from an elegant result of Gauss on geodesics: Given a curve on a surface draw through each point of the curve the geodesic through it and mark off a fixed length on each. The end-points of these arcs form a curve that cuts the geodesics orthogonally and conversely. Kneser generalized this result by his discovery of the more general concept, transversality, which includes orthogonality as a special case. Out of this notion he was able to establish the envelope theorem, first discovered by Zermelo in 1894. Using the result, Kneser was able to give a geometrical interpretation of conjugate points. This interpretation is not different from Jacobi's, but his analysis deepens and enriches our understanding of what is meant.

Kneser proceeded in his *Lehrbuch* to examine variable end-point problems, and as a result he discovered focal points. His work was followed by Bliss and Mason who continued the discussion in depth. He also introduced what he termed normal coordinates of a field in an elegant way which permitted both himself and Bolza to give sufficiency proofs by showing that the Weierstrass $\mathcal{E}$-function is invariant under coordinate transformation (see Section 7.6).

In the course of extending results by Bliss and Mason in 1908 and 1910, Hahn in 1911 established a theorem of great importance for the treatment of sufficiency theorems for variable end-point problems (see Section 7.7).

At about the same time W. Cairns, Mason and Richardson, carrying out Hilbert's ideas, introduced boundary-value methods into studies of the second variation viewed as a quadratic functional (see Section 7.8). In the course of this work Richardson gave several interesting oscillation theorems. As mentioned earlier Hilbert investigated under what conditions a nonnegative integral of the calculus of variations attains its lowest bound over a given class of arcs (see Section 7.9).

In 1913 Bolza formulated what Bliss called the problem of Bolza as a generalization of the problems of Lagrange and Mayer. (Bliss showed that in fact all three problems are equivalent.) He proceeded to give a seminal and penetrating first discussion of the problem (see Section 7.10).

In 1904 Carathéodory took up John Bernoulli's second method for handling the brachystochrone problem. This enabled him to define at each point what he called the direction of steepest descent and the notion that a family of curves are geodesically equivalent. With the help of these concepts Carathéodory developed the usual results of the calculus of variations with considerable elegance.

Upon this note the present volume ends. It does not contain any discussion of the classical theory of multiple-integral problems because the great insights into this subject did not occur until this century, at which time Courant, Douglas, Morrey and others made their discoveries. For similar reasons Bliss's masterful improvements and summations of the classical theory, Morse theory, the elegant discussions of abnormality by Graves, McShane, Morse, Reid and others, and the generalizations of the subject to arbitrary functionals by Goldstine and others are all omitted as is any mention of control theory.

1. Fermat, Newton, Leibniz, and the Bernoullis

1.1. Fermat's Principle of Least Time

Fermat wrote nine interesting and important papers on the method of maxima and minima which are grouped together in his collected works. The last two in this set were sent by him in 1662 as attachments to a letter to a colleague, Marin Cureau de la Chambre.[1] As their titles, "The analysis of refractions" and "The synthesis of refractions," imply, they are companion derivations of the law of refraction, now usually known as Snell's law. These papers are fundamental for us because Fermat enunciates in them his principle that "nature operates by means and ways that are 'easiest and fastest.'" He goes on to state that it is not generally true that "nature always acts along shortest paths" (this was the assumption of de la Chambre). Indeed, he cites the example of Galileo that when particles move under the action of gravity they proceed along paths that take the least time to traverse, and not along ones that are of the least length.[2] This enunciation by Fermat is, as far as I am aware, the first one to appear in correct form and to be used properly.

By means of his principle and of his method of maxima and minima, Fermat was able to give a very succinct and completely clear demonstration that Snell's or Descartes's law holds for a refracted ray.[3] An interesting thing about Fermat's conclusion is that it is based on correct physical laws which were contradicted by Descartes: Fermat assumed that light moved with finite velocity and that this velocity was slower in denser media. The Descartian model for light was quite the opposite: Descartes

---

[1] Fermat, *OP*, Vol. I, pp. 132–179. The last papers appear on pp. 170ff. They were appended to a letter of 1 January 1662.

[2] Ibid., p. 173. The reference to Galileo is to his attempt to show that an arc of a circle is the solution of the brachystochrone problem. This appears in Galileo, *TWO*, pp. 97, 212–213. See also Section 1.3 below.

[3] See Sabra, *TOL*, pp. 99ff. This is a good discussion of the physics of light in this period. We will see in Section 2.8 below how Maupertuis subsumed Fermat's principle under the least-action one.

assumed that light moved faster in a denser medium (cf. Sabra, *TOL*, p. 105n). It was therefore a source of astonishment to Fermat that he reached by precise means apparently the same conclusion about refraction that Descartes had.

The main reason I have started a history of the calculus of variations at this point is that Fermat's work seemed to be clearly the first real contribution to the field and certainly served as the inspiration for the imaginative proposal and solution of the brachystochrone problem by John Bernoulli in 1696/97.[4] Indeed, Bernoulli's solution depends directly on Fermat's principle. As we shall see in Section 1.4, Bernoulli replaced his material particle moving under the action of gravity by a ray of light moving through a series of optically transparent media of properly varying densities.

It might be argued with some reason that Galileo's treatment of the brachystochrone problem should be taken as the proper starting point, but I felt that since Galileo is incorrect in his argument and does not make use of the calculus, it is more fair to choose Fermat as the legitimate sire of the field.

In Figure 1.1 Fermat considers a circle $ACBI$ with center at D and made up of two media ACB and AIB of different optical densities. He supposes that a ray of light starts from the point C in the upper and rarer medium and moves along the broken ray COI across the interface ADB to the point I in the denser medium. As the reader can see, there is also an external line segment M in the figure. Fermat assumes, in effect, that the length of M is a measure of the resistance of light in the rarer medium and that the length of DF is a measure of the resistance in the denser one. At least he assumes that they are proportional to these with a common factor of proportionality.[5] Moreover, he uses the term "resistance" to mean the

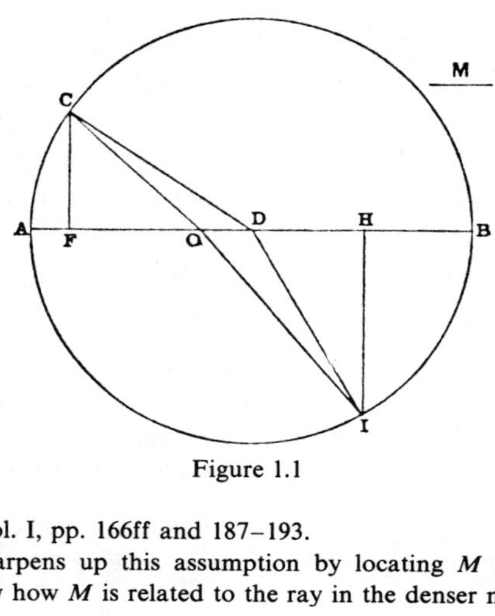

Figure 1.1

[4] John Bernoulli, *OO*, Vol. I, pp. 166ff and 187–193.

[5] In his synthesis he sharpens up this assumption by locating M in the figure itself and specifying more precisely how M is related to the ray in the denser medium.

reciprocal of the velocity. The problem he proposes is then, given these things, to locate the point O so that the time for a ray to move from C to I via O is a minimum. To do this, he introduces some notations to simplify matters: let the radius CD be called N; the segment DF, B; and DH, A (F and H are the feet of perpendiculars from C and I onto AB). Then the minimum value for the time is, in Fermat's terms (p. 171), "N in $M + N$ in B" or, in ours, $N \cdot M + N \cdot B$ by his assumption about velocities in the two media, since in a homogeneous medium time varies directly as the distance traversed and inversely as the velocity. To show this, he makes use of the law of cosines to express things in terms of $DO = E$. Thus he has, in modern terms,[6]

$$CO^2 = N^2 + E^2 - 2BE, \qquad OI^2 = N^2 + E^2 + 2AE;$$

and hence the quantity to be minimized can be written as

$$CO \cdot M + IO \cdot B = M\sqrt{N^2 + E^2 - 2BE} + B\sqrt{N^2 + E^2 + 2AE}\ .$$

Fermat says that the expression above is to be handled by "my method of maxima and minima." (Fermat developed this technique around 1629.) That is, the point O is to be so located on the line AB that the expression $CO \cdot M + IO \cdot B$ for the time is a minimum.

Let us see what his method was. Fermat indicates it in his paper, "Methodus ad disquirendam maximam et minimam,"[7] where he gives as an example the problem of dividing a line segment AC by a point E so that the rectangle AEC—the product $AE \cdot EC$—is a maximum. His procedure is to let B be the segment AC and A be AE; then $BA - A^2$ is to be a maximum. To find its value, suppose A is replaced by $A + E$. Then the expression

$$B(A + E) - (A + E)^2 = BA - A^2 + BE - 2AE - E^2 \tag{1.1}$$

viewed as a function of E is to be a maximum. Now Fermat notes for E small that

$$BA - A^2 + BE - 2AE - E^2 = BA - A^2;$$

and so

$$BE \sim 2AE + E^2 \quad \text{or} \quad B \sim 2A + E. \tag{1.2}$$

Thus

$$B = 2A \tag{1.3}$$

is the desired necessary condition. [Essentially what Fermat does when he wishes to maximize or minimize a function f of E is to calculate $f'(0)$ and set this value to zero.]

[6] Fermat, *OP*, Vol. I, pp. 171–172. This is in his "Analysis."

[7] Fermat, *OP*, Vol. I, pp. 132–136.

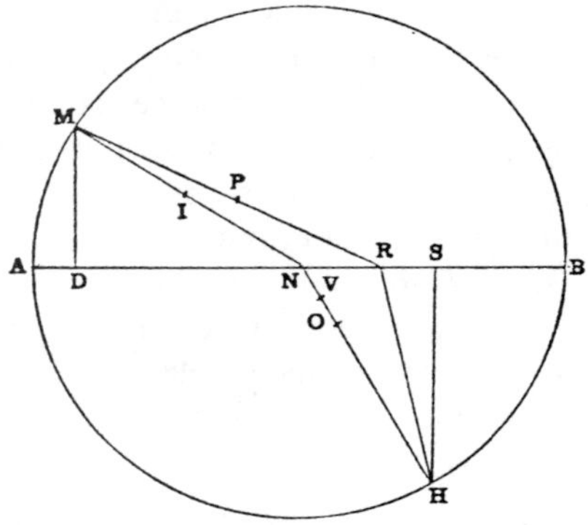

Figure 1.2

Now in the case of the time for the light ray, Fermat sets

$$CO \cdot M + IO \cdot B = N \cdot M + N \cdot B,$$

does the algebraic manipulation, and finds the elegant result $A = M$; that is, the segment DH is in fact the segment M. By his original assumption that the denser the medium, the more slowly light proceeds, he then has the fundamental relation

$$\frac{DF}{DH} = \text{const.} > 1,$$

and this leads directly to the familiar law of refraction

$$\frac{\sin FCD}{\sin HID} = \frac{DF}{DH} = \text{const.} > 1. \tag{1.4}$$

This derivation of the necessary condition (1.4) is the substance of Fermat's "Analysis ad refractiones." He says (p. 172) that his formula (1.4) "agrees in every respect with the theorem discovered by Descartes; the above analysis, derived from our principle, therefore gives a rigorously exact demonstration of this theorem."[8] In his next paper, "Synthesis ad refractiones," Fermat takes up the postulate of Descartes that light moves more rapidly in a dense medium than in a rare one. He notes that he has assumed quite the contrary and asks whether it is possible to reach the same conclusion from contrary assumptions. He does not pursue this philosophical question here but goes on to show that given the law (1.4), a ray moving (see Figure 1.2) from the point M in the upper medium to H in the lower one along the path MNH, with N the center of the circle, will take the least time. Thus in his "Analysis" paper Fermat establishes a necessary condition, and in his "Synthesis" paper he shows it to be sufficient.[9]

[8] Sabra, *TOL*, p. 147 or Fermat, *OP*, Vol. I, p. 172 and Vol. III, p. 151.

[9] The former paper was sent to de la Chambre on 1 January 1662 and the latter, in the following month (cf. Fermat, *OP*, Vol. I, pp. 170 and 173). The former paper is on pp. 170–172 and the latter, on pp. 173–179.

Sabra does, however, examine in detail the relation of Descartes's to Fermat's rule. The interested reader will see that Descartes's result is that the sine of the incident angle is to the sine of the refracted one as the velocity in the second medium is to that in the first.[10] On the contrary, Fermat's result is that the sine of the incident angle is to the sine of the refracted one as the velocity in the first is to the velocity in the second. (This is not the exact way Fermat states it; he uses the segments DF and DH instead of the sines.) In "Synthesis" Fermat starts with the result about the ratio c of the velocity in the upper medium to that in the lower one

$$\frac{DN}{NS} = c > 1 \tag{1.4'}$$

and goes on to show that a ray moving from point M in the upper medium through the center N of the circle to H in the lower one will arrive in the least time. The situation is as shown in Figure 1.2. Fermat explicitly assumes that the velocity in the upper (rarer) medium is the faster; thus the ratio c in equation (1.4′) above is greater than one. In Figure 1.2, D is the foot of the perpendicular from M onto the interface ANB between the media, and S is the foot of the perpendicular from H onto that line. The point R is an arbitrary point on the interface; and I, P are points on MN, MR, respectively, so chosen that the relations

$$c = \frac{DN}{NS} = \frac{MR}{RP} = \frac{MN}{NI} \tag{1.5}$$

obtain.

Fermat next chooses two more points O and V on the line RH so that

$$\frac{MN}{DN} = \frac{RN}{NO}, \qquad \frac{DN}{NS} = \frac{NO}{NV}. \tag{1.6}$$

He then wishes to show that the time for a ray to move on the broken line MNH is less than on the line MRH. His proof is quite simple and direct. For notational purposes, let t_{XY} be the time required by light to move along a line segment XY in a homogeneous medium. Then since velocity in such a medium varies directly with distance and inversely with time and since the ratio of the velocity in the upper to that in the lower medium is c [by (1.5)], Fermat has

$$\frac{t_{MN}}{t_{NH}} = \frac{MN}{NH} \cdot \frac{1}{c} = \frac{NI}{NH}, \qquad \frac{t_{MR}}{t_{RH}} = \frac{MR}{RH} \cdot \frac{1}{c} = \frac{RP}{RH}.$$

He concludes from these relations that

$$\frac{t_{MNH}}{t_{MRH}} = \frac{t_{MN} + t_{NH}}{t_{MR} + t_{RH}} = \frac{NI + NH}{RP + RH}$$

---

[10] Sabra, *TOL*, pp. 111, 149.

since the velocities along NH and RH are equal, and now he needs to show that $RP + RH > NI + NH$. To see why this is so, recall the relations (1.6). We know that $DN < MN$ and $NS < DN$ by (1.4′) and thus

$$NO < RN, \qquad NV < NO. \tag{1.7}$$

By an application of the law of cosines, Fermat is able to express the relation[11]

$$MR > MN + NO.$$

But he clearly has the relations

$$\frac{DN}{NS} = \frac{MN}{NI} = \frac{NO}{NV} = \frac{MN + NO}{NI + NV} = \frac{MR}{RP},$$

and he therefore concludes that $RP > NI + NV$.

He next shows that $RH > HV$ by having recourse to the law of cosines as applied to the triangle NHR and the second inequality (1.7). He then has $RP + RH > NI + NV + HV$; that is, he has his desired relation $RP + RH > NI + NH$, and so the time along the broken ray MNH is less than that along any broken ray MRH, at least when R is as shown in Figure 1.2. If R lies on the other side of the center N, Fermat gives another proof to show his relation is still valid.[12]

By way of conclusion to the subject, we should perhaps read what Fermat wrote Clerselier, a defender of Descartes's views, on 21 May 1662[13]:

> I believe that I have often said both to M. de la Chambre and to you that I do not pretend, nor have I ever pretended to be in the inner confidence of Nature. She has obscure and hidden ways which I have never undertaken to penetrate. I would have only offered her a little geometrical aid on the subject of refraction, should she have been in need of it. But since you assure me, Sir, that she can manage her affairs without it, and that she is content to follow the way that has been prescribed to her by M. Descartes, I willingly hand over to you my alleged conquest of physics; and I am satisfied that you allow me to keep my geometrical problem—pure and *in abstracto*, by means of which one can find the path of a thing moving through two different media and seeking to complete its movement as soon as it can.

---

[11] This follows by expressing the side MR in triangle MNR in terms of the other sides. Thus he has, by the first of the ratios (1.6) and of the inequalities (1.7),

$$MR^2 = MN^2 + NR^2 + 2MN \cdot NR \cdot \cos DNM = MN^2 + NR^2 + 2NR \cdot DN$$
$$= MN^2 + NR^2 + 2MN \cdot NO > MN^2 + NO^2 + 2MN \cdot NO = (MN + NO)^2.$$

[12] Fermat, *OP*, pp. 177–178.

[13] Sabra, *TOL*, p. 154. In this book the interested reader will find a very complete discussion of the whole problem.

1.2. Newton's Problem of Motion in a Resisting Medium

The first genuine problem of the calculus of variations was formulated and solved by Newton in late 1685.[14] In Cajori's edition of Newton's *Principia* (this is a revision of Motte's 1729 translation) the theorem in question is in the scholium to Proposition XXXIV.[15] In this part of Book II Newton investigates the motion of bodies moving under rather specialized and restrictive assumptions through an inviscid and incompressible fluid, or as Newton says, "a rare medium of very small quiescent particles of equal magnitude and freely disposed at equal distances from one other." Newton says, "This Proposition I conceive may be of use in the building of ships."[16] He actually takes up two closely related problems in the scholium referred to above. The first concerns the motion of a frustum of a cone moving in the direction of its axis through the medium and the second a considerably more general body. The former is solved by Newton using the ordinary theory of maxima and minima; it is in the latter that he introduces to the world the concept of the calculus of variations.

This latter problem involves finding the shape of the solid of revolution which moves through a resisting medium in the direction of its axis of revolution with the least possible resistance. (We will make this more precise shortly when we take up Newton's solution.) This problem is of cardinal importance to the calculus of variations for a number of reasons. First, it is the first problem in the field to be both formulated and correctly solved, and the techniques used in that solution by Newton were those later adopted with suitable modifications by both Newton and James Bernoulli in the solution of the problem of the so-called brachystochrone; these ideas were still later systematized by Euler. In truth the Newtonian method prevailed in one form or another until Lagrange's superior variational method of 1755 swept the older ones away. Second, the problem itself has an inherent interest since it is an unusual one which can possess a solution curve having a corner—a discontinuous slope—and which may have no solution in the ordinary sense if the problem is formulated parametrically (see p. 16). Third, it is one of the most complex special cases of the entire theory, as will be seen in what follows. Minimizing arcs can be made up of three quite dissimilar types of curves in not altogether obvious ways. (See Bolza, *VOR*, pp. 412–413.)

Newton's solution of the problem of the body moving in a resistive medium appeared in the *Principia* with no suggestion of his method of

---

[14] The reader should consult Whiteside's elegant and invaluable edition of Newton's mathematical papers. See Newton, *PAPERS*, Vol. VI, pp. 456–480. In particular, see the wealth of illuminating and perceptive footnotes there of Whiteside.

[15] Newton, *PRIN*, pp. 333–334 and pp. 657–661.

[16] Ibid., p. 333.

derivation and mystified the mathematical community. It was apparently not understood by Newton's contemporaries, with the possible exception of Huygens, who studied the problem in his notes dated 22 and 25 April 1691.[17] Indeed, the first person who discussed the matter with Newton seems to have been David Gregory, a nephew of James Gregory and a Savilian professor of astronomy at Oxford; he was unable to solve the problem and persuaded Newton to write out an analysis of it for him. He then lectured on the material in the fall of that year at Oxford and made it available to his students and peers.[18] Possibly it was via this route that James Bernoulli and Leibniz became aware of how to handle such problems, although I have seen no evidence to verify this point. It could also have come to one or the other via Fatio de Duillier, directly via David Gregory, or perhaps via Huygens, who was a close friend of Leibniz. In this connection Whiteside tells us that Leibniz noted in the margin of his *Principia* an ambiguous remark which suggests that he did not see how to carry out the analysis. [Nicholas Fatio de Duillier was a peripatetic figure, who lived at various times in Geneva, The Hague, Paris, and London; he was a friend of many of the great figures of his time, including Cassini, Huygens, Leibniz, and Newton. His book on the brachystochrone problem (Fatio [1699]) helped initiate the quarrel between the latter two men.] It is therefore not certain whether the contributions of James Bernoulli and Leibniz are derivative from Newton or were recreations by them of similar processes, independently rediscovered.

Newton imagines that his body moves through the resisting medium in such a way that no forces act on its tail (there is a vacuum behind it), that it is frictionless (there are no forces tangential to it), and that particles of the fluid which it meets rebound from it along the surface-normals at the points of impact. (Moulton tells us this is approximately the case for bodies moving at velocities considerably above the velocity of sound for the fluid.[19])

Let us examine Newton's problems and their solutions. In Proposition XXIV, Theorem XXVIII, he undertakes to derive the result[20]:

---

[17] Newton, *PAPERS*, Vol. VI, p. 466n. Huygens's studies are in Huygens, *OC*, Vol. 22, pp. 335–341, where (pp. 327–332) there is a discussion by the editors which Whiteside terms "impercipient." I am not convinced that it is. Perhaps it is best to review the evidence; this is done briefly at the end of the present section.

[18] D. Gregory, *NMF*. This compendium, which Whiteside tells us was widely circulated, appears in Motte's 1729 translation of the *Principia*. It is in Volume II, Appendix, pp. v–viii. Motte says he received the analysis from a friend (see, e.g., Newton, *PRIN*, pp. 657–658 and *PAPERS*, Vol. VI, pp. 466n). Newton's problem was also brought to the public's attention by Fatio de Duillier, in his *Investigatio* [1699]. This appears in the latter's *Lineae* [1699], where he derived the differential equation for the brachystochrone problem. There is further English literature on the calculus of variations in Taylor, *MET* and Maclaurin, *FLUX*, as well as in books of Emerson and Simpson. It is not impressive by comparison to the continental literature of the same period, and I will say little more about it.

[19] Moulton, *BAL*, p. 37.

[20] Newton, *PRIN*, p. 331.

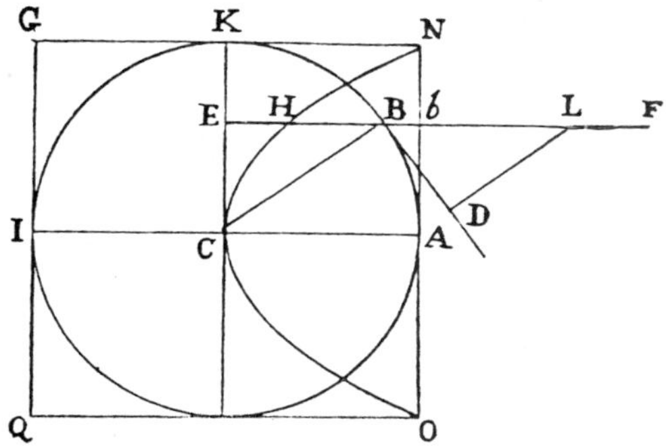

Figure 1.3

If in a rare medium, consisting of equal particles freely disposed at equal distances from each other, a globe and a cylinder described on equal diameters move with equal velocities in the direction of the axis of the cylinder, the resistance of the globe will be but half as great as that of the cylinder.

To follow his demonstration consider Figure 1.3, in which a section of the cylinder is $GNOQ$ and of the sphere is $ABKI$ with center at C. Suppose that both bodies move in the direction of the axis ICA of the cylinder, that the line EBF is parallel to ICA, and that LD and $BC = LB$ are perpendiculars onto the tangent BD to the circle.

Newton then remarks that according to his third law he can hold his bodies, the cylinder and the sphere, fixed and allow the fluid to flow past them with uniform velocity in the direction AC, the axis of the cylinder. At an arbitrary point B on the sphere the resistive force is along the normal to the surface—Newton assumes tacitly that there are no frictional forces present. If b is the point on the cylinder corresponding to B on the line EF parallel to CA, Newton says that "the force with which a particle of the medium, impinging on the globe obliquely in the direction FB, would strike the globe in B, will be to the force with which the same particle, meeting the cylinder . . . , would strike it perpendicularly in b, as LD is to LB, or BE to BC." This follows from the fact that $LD = LB \cos BLD = BC \cos EBC = BE$ since $LB = BC$, by construction. Thus if f is the amplitude of the force on the cyclinder at b in the direction bF, $f \cdot \cos\theta$ is the force on the sphere at B in the direction CB extended, where $\theta = \operatorname{arc}\cos(BE/BC)$.

The component of this force on the sphere in the direction BF is called by Newton its *efficacy* in the direction BF and is evidently $f\cos^2\theta$ (By symmetry, the sum of all components orthogonal to the direction BF vanishes.) It now remains for Newton to evaluate the integrated effect of this force over the entire surface. To handle this, he introduces a point H on BF extended to E so chosen that

$$bH = \frac{BE^2}{CB} \quad (= CB \cdot \cos^2\theta). \tag{1.8}$$

He then remarks that "bH will be to bE as the effect of the particle upon the globe to the effect of the particle upon the cylinder. And therefore the solid which is formed by all the right lines bH will be to the solid formed by all the right lines bE as the effect of all the particles upon the globe to the effect of all the particles upon the cylinder."

Newton then needs to calculate the volumes of these two solids. To this end he notes that the solid formed by "the right lines bH" is "a paraboloid whose vertex is C, its axis CA, and latus rectum CA."[21] The other solid is "a cylinder circumscribing the paraboloid." That the former solid is as described can be seen from equation (1.8) by noting that

$$bH = \frac{CB^2 - CE^2}{CB} = \frac{CA^2 - CE^2}{CA};$$

and hence if $x = HE = bE - bH = CA - bH$ and $y = CE$, there results $y^2 = CA \cdot x$, as was to be shown. As Whiteside points out, Archimedes, in his *On Conoids and Spheroids*, shows "that the volume of the frustrum of a paraboloid of revolution is half that of the circumscribing cylinder."[22] It follows that "if the particles of the medium are at rest and the cylinder and the globe move with equal velocities, the resistance of the globe will be half the resistance of the cylinder, Q.E.D."

For completeness, let us look at the problem from the point of view of analysis rather than geometry and calculate the resistances of the globe and cylinder. At the arbitrary point B the force of a particle rebounding from the surface along the normal to the sphere will evidently be $f \cdot \cos\theta$, where $\theta = \arccos(BE/BC)$ in Figure 1.3 and f is the magnitude of the force with which the particle strikes. It may be resolved into a component $f \cdot \cos^2\theta$ in the direction EF and one, $f \cdot \cos\theta \cdot \sin\theta$, normal to that direction. But by symmetry there is a point B' (not shown) symmetrically placed to B with respect to the axis of rotation CA for which this normal component is $-f \cdot \cos\theta \cdot \sin\theta$, and they just balance each other. There are therefore no forces in the direction normal to EF, and the net force on a segment of the generating curve KBA is $f\cos^2\theta\, ds$, where s measures arc-length along that curve.

Let us choose C as our origin, CA as the direction of the x-axis and CK of the y-axis. Then the number of particles striking the element ds is evidently proportional to $ds \cdot \cos\theta = dy$, and so the total resistance of the surface is proportional to

$$R = \int_0^{2\pi} d\varphi \int_{s_1}^{s_2} \cos^3\theta \cdot y\, ds = 2\pi \int_{y_1}^{y_2} \cos^2\theta \cdot y\, dy, \tag{1.9}$$

and if we choose units properly, we may regard this as the value of the resistance.

---

[21] Newton, *PRIN*, p. 332.

[22] Newton, *PAPERS*, pp. 469n–470n; Heath, *ARCH*, pp. 129–131.

It is clear that this expression for the resistance is valid for surfaces of revolution in general and not just for a sphere of radius c. In that case, however, we have $y_1 = 0$, $y_2 = c$, and

$$\cos^2\theta = \frac{x^2}{c^2} = 1 - \frac{y^2}{c^2},$$

and hence R is $\pi c^2/2$; in the case of the cylinder $\cos^2\theta = 1$ and R is πc^2.

In his Scholium on pp. 333–334 Newton goes forward with his analysis. He has just considered the bow resistance of a ship or a projectile when it is blunt or hemispherical. He now goes to the case where it is a frustum of a right-circular cone and finds, for a given altitude and base, the best shape. From this he proceeds to the general question of the best-shaped bow, as we shall see.

In Figure 1.4 the circle $CEBH$ with radius b is the base of the cone, S is its vertex, and the altitude of the frustum is $OD = a$. The radius of the circular cap $FD = c$, and Q is the midpoint of the altitude OD. For convenience, let $z = DS$. Then by the value of R in (1.9), the resistance of the curved surface of the frustum is

$$R_1 = 2\pi\int_c^b y\cos^2\theta\cdot dy = \pi(b^2 - c^2)\cos^2\theta,$$

since θ is a constant; in fact,

$$\cos^2\theta = \frac{b^2}{b^2 + (a+z)^2}.$$

The total resistance is, of course, the sum of R_1 and $R_2 = \pi c^2$, the resistance of the cap of the frustum, since the frustum moves in the direction from O to D; that is,

$$\begin{aligned} R = R_1 + R_2 &= \pi(b^2\cos^2\theta + c^2\sin^2\theta) \\ &= \pi\cdot\frac{b^4 + c^2(a+z)^2}{b^2 + (a+z)^2} = \pi b^2\frac{b^2 + z^2}{b^2 + (a+z)^2}, \end{aligned}$$

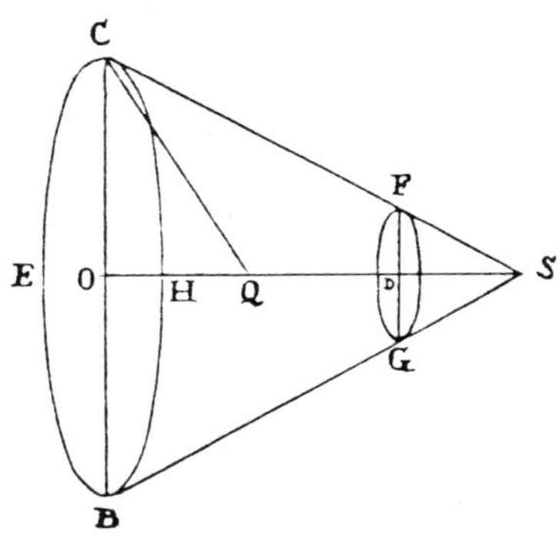

Figure 1.4

since $b/c = (a+z)/z$. From this it is clear that the minimum value with respect to z occurs at

$$z = -\frac{a}{2} + \sqrt{\frac{a^2}{4} + b^2}\ ; \tag{1.10}$$

and so the height of the cone OS is $(a/2) + ((a/2)^2 + b^2)^{1/2} = OQ + QC$, as Newton asserts. He says "bisect the altitude OD in Q, and produce OQ to S, so that QS may be equal to QC, and S will be the vertex of the cone whose frustum is sought."[23]

Whiteside gives us two versions of Newton's analysis. The first, "The solid of revolution of least resistance to motion in a uniform fluid," was carried out in late 1685 and the second, "Recomputation of surfaces of least resistance," in the summer of 1694. They both appear in Whiteside's Vol. VI.[24] Both versions contain the frustum problem as well as the general surface of least resistance. The former is essentially the same in both. Newton says that the force on the surface of the cone is to that on the cap centered at D as $OC^2 - FD^2$ into $(OC - FD)^2$ is to FD^2 into CF^2—note that $\cos\theta = (OC - FD)/CF$. The total resistance is then proportional to

$$\frac{OC^2 \cdot (OC - FD)^2 - FD^2 \cdot (OC - FD)^2 + FD^2 \cdot CF^2}{CF^2}$$

$$= \frac{OC^2 \cdot (OC - FD)^2 + FD^2 \cdot OD^2}{CF^2}.$$

To simplify the notation, Newton now sets $2OQ = OD = a$, $OC = b$, and $OC - FD = x$ and notes that $CF = (a^2 + x^2)^{1/2}$, $FD = b - x$. Finally he calls the resistance y and observes that

$$y = \frac{a^2b^2 - 2a^2bx + a^2x^2 + b^2x^2}{a^2 + x^2}.$$

He notes that this is equivalent to

$$a^2b^2 - 2a^2bx + a^2x^2 + b^2x^2 = a^2y + x^2y$$

and differentiates both sides as to "time," finding

$$-2aab\dot{x} + 2aax\dot{x} + 2bbx\dot{x} = 2x\dot{x}y + aa\dot{y} + xx\dot{y}$$

in his notation, where $\dot{x} = dx/dt$ and $\dot{y} = dy/dt$. He then observes that at

[23] Newton, *PRIN*, p. 333.

[24] Newton, *SOL* and *APP 2*. This latter reference is to his 1694 (summer) recomputation for the frustum of a cone and the general solid of revolution written out, as mentioned above, for David Gregory (see *APP 2*, pp. 470n).

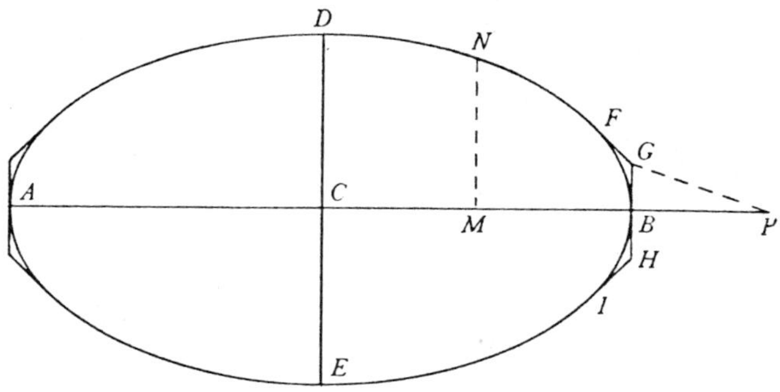

Figure 1.5

a minimum $\dot{y} = 0$; and after a small calculation he finds that

$$x = \frac{-aa + a\sqrt{aa + 4bb}}{2b}$$

and concludes that $QS = CQ$.[25]

From the frustum in his *Principia* Newton moves on to the oval figure $ADBE$ in Figure 1.5 and remarks[26]

> Whence, by the way, since the angle $\widehat{CSB}$ (in Figure 1.4) is always acute, the consequence is: if the solid $ADBE$ (in Figure 1.5) be generated by rotating the elliptical or oval figure $ADBE$ round its axis AB, and the generating figure be touched at the points F, B and I by three straight lines FG, GH, HI, where GH is restricted to be perpendicular to the axis at the contact point B, while FG, HI contain with GH angles $\widehat{FGB}$, $\widehat{BHI}$ of 135°, then the solid generated by the rotation of the figure $ADFGHIE$ round the axis will be resisted less than the previous solid, provided that each moves forward along the line of its axis AB, the end-point B leading in each case. This proposition will, I reckon, be not without application in the building of ships.

To understand this result of Newton we first note that his theorem on frustums of cones holds true even when the altitude $OD = a$ approaches zero. In this case the minimum value of z, as given in (1.10), becomes b, and the height of the cone also becomes b, which is the radius of the base. Thus the sides of the cone meet in a right angle at the vertex. Newton now needs to show that the frustum of the cone, $FGHI$, will encounter less resistance than will the portion FBI of the oval. He does not indicate how

[25] See Newton, *APP 2*, pp. 470n–471n for an interesting discussion by Whiteside of the material and its relation to Gregory and others. See also Newton, *SOL*, for his 1685 analysis of the problem.

[26] Newton, *PRIN*, p. 333 or *SOL*, p. 463 and *APP 2*, pp. 477ff; in particular, see Whiteside's note 42 on p. 478. (The quotation above is from p. 463.) It is also of interest to read on p. 473n that Newton understood about towing tanks and observed that the best shape for a ship could be determined at little cost by the use of such a tank.

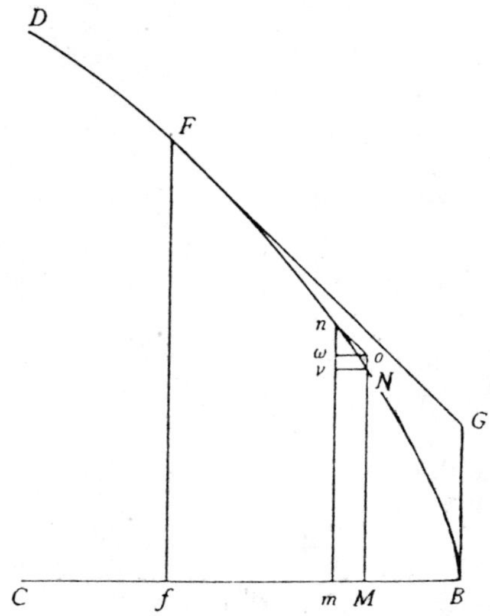

Figure 1.6

he showed this, but Bolza and Whiteside have each given possible reconstructions which are almost surely quite similar to Newton's method (see Newton, *PAPERS*, Vol. VI, pp. 462nff).

In Figure 1.6 (which is Whiteside's), let FG be inclined at 45° to the x-axis which extends in the direction from B toward C; MN is an ordinate of the curve FB at an arbitrary point M, mn is the corresponding ordinate at m, a point *near* to M, and no is drawn parallel to FG, that is, inclined at 135° to the axis CB. Since, as we just remarked, the cone frustum having this slope for its generator experiences minimal resistance among cone frustums with common bases, the resistance felt by the frustum $noMm$ is less than that felt by $nNMm$, provided that m is near enough to M, as we can see by continuity considerations. Thus we see that the resistance felt by the line Nn rotated about CB is greater than that felt by the figure generated by rotating the broken line segment $No + on$ about CB. Since oN is $dy - dx$, this latter resistance is expressible as $2\pi y(dy - dx) + \pi y\,dx = 2\pi y\,dy - \pi y\,dx$, since $x + dx = Bm$, $y + dy = mn$, and $dx = Mm = o\omega = \omega n$. It follows directly by integration that the resistance R experienced by rotating the arc BF is greater than

$$\pi(Ff^2 - BG^2) - \pi\int_0^{Bf} y\,dx \geqslant \frac{\pi}{2}(Ff^2 - BG^2).$$

(This integral above, the area under the curve BF, is less than the area under the line FG.) But since FG is inclined at 135° to CB, $Ff - BG = Bf$, and hence

$$\tfrac{1}{2}(Ff^2 - BG^2) = \tfrac{1}{2}(Ff + BG)\cdot Bf,$$

which is the area of the trapezoid $BGFf$ encompassing the area under Bf. Thus the total resistance R experienced by the solid of revolution generated by BF is greater than that of the frustum generated by $FGBf$, which is what Newton asserted.

Bolza summarizes the situation in an exercise in his *Vorlesungen über Variationsrechnung*: "If PQ is an arc whose slope is always [numerically] > 1, then the surface generated by rotating it about the x-axis will experience greater resistance than will the surface generated by rotating the broken line segment PRQ, where PR is parallel to the y-axis and where RQ has slope [numerically] $= 1$."[27] Notice that the broken segment does not necessarily generate the surface of least resistance. In fact, we show below that it does not.

Newton then proceeded to the general theorem. Whiteside's translation of the text refers to Figure 1.5 and is this:[28]

> But should the figure $DNFG$ be so shaped that, if from any point N in it the perpendicular NM be let fall to the axis AB, and the straight line GP be drawn parallel to the tangent to the figure at N, cutting the axis (extended) in P, there shall be MN to GP as GP^3 to $4BP \times GB^2$, then will the solid described by this figure's rotation round the axis AB be resisted less than all others of the same length and breadth.

Before proceeding to Newton's derivation of this result let us first examine in somewhat more modern notation the problem as posed. We saw earlier in equation (1.9) that the resistance on the surface formed by rotating the curve DNG in Figure 1.8 about the axis CB is, in proper units,

$$\int_{BG}^{CD} y\cos^2\theta \cdot dy = \int_{BG}^{CD} y\left(\frac{dy}{ds}\right)^2 dy = \int_{BG}^{CD} \frac{y\,dy}{1+(dx/dy)^2} = \int_{t_1}^{t_2} \frac{y\dot{y}^3\,dt}{\dot{x}^2+\dot{y}^2}, \tag{1.11}$$

where θ is the angle the normal makes at the point (x, y) with the x-axis, s measures arc-length, and $\dot{x}$, $\dot{y}$ are derivatives with respect to a parameter t. In Newton's Figure 1.8 N is an arbitrary point (x, y) on the curve so that $MN = y$, $CM = x$, and GR has been drawn parallel to the tangent to the curve at N. Since GR is parallel to the tangent at N,

$$\frac{BR}{GR} = \cos GRB = \sin\theta \qquad \frac{GB}{GR} = \sin GRB = \cos\theta;$$

---

[27] Bolza, *VOR*, p. 455, exercise 43. He correctly attributes the result to Newton. In his next exercise he gives the interesting datum that the surface of least resistance of given base and height experiences a resistance of 0.3748 times that of a cylinder of the same dimensions.

[28] Newton, *SOL*, pp. 463–465. This is the heart of Newton's contribution. It is his determination of a first necessary condition for a minimum in this problem. As we shall see, it is a first integral of the Euler differential equation. (Newton's Figure 1.5 is not quite right, since it portrays the minimizing arc as made of DNF and the line segments FG and GB. What he clearly intended is to have the configuration in Figure 1.7, depending upon the slope of the minimizing curve at G. (See Bolza, *VOR*, p. 414.)

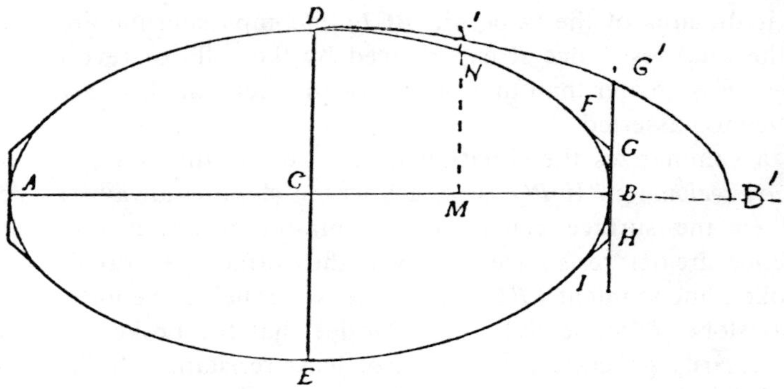

Figure 1.7

and Newton's proportion is equivalent to the statement that

$$\frac{y}{GR} = \frac{1}{4} \sec GRB \csc^2 GRB;$$

or if $q = \cot GRB$ and $GB = y_0$, to

$$\frac{yq}{(1+q^2)^2} = \frac{y_0}{4}, \tag{1.12}$$

since $GR = y_0 \csc GRB$.

We need to make the assumption that our arcs are representable parametrically as $x = x(t)$, $y = y(t)$ $(t_1 \leqslant t \leqslant t_2)$ with $\dot{x} \geqslant 0$, $\dot{y} \geqslant 0$ at each point. Newton realized this and made provision for it. Otherwise, as was first shown by Legendre, Newton's problem has for a solution a saw-toothed curve giving a total resistance of zero. We discuss this later (pp. 144). Under the assumptions above we can show that all extremals are representable as $x = x(y)$ with x single-valued, continuous, and having a piecewise continuous derivative.

The nonparametric form of integrand function $f(y, x, x')$ in (1.11) is clearly $y/(1 + x'^2)$, where $x' = dx/dy$; and the problem to be solved is to find among all admissible arcs of the form $x = x(y)$ joining the points (D) and (G) in Figure 1.8 the one which renders the resistance on the

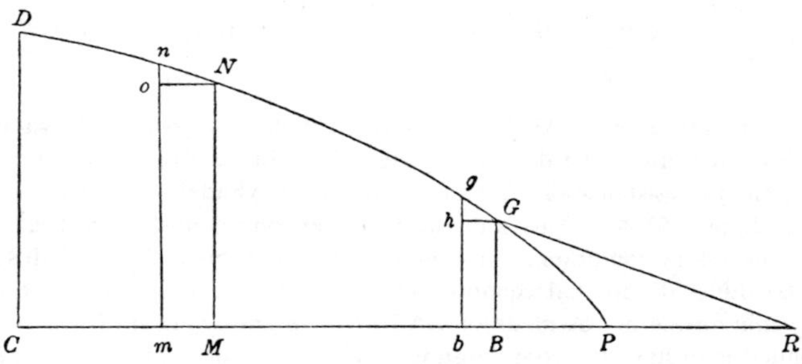

Figure 1.8

corresponding surface of rotation

$$\int_{BG}^{CD} f(y, x, x')\, dy$$

a minimum. (Note that the integrand function does not contain x explicitly.) The parametric form of the integrand function $F(x, \dot{x}, y, \dot{y})$ is $y\dot{y}^3/(\dot{x}^2 + \dot{y}^2)$, and the problem in that form is to find in the class of admissible arcs, as represented parametrically above, one which renders

$$\int_{t_1}^{t_2} F(x, \dot{x}, y, \dot{y})\, dt$$

a minimum. It is clear that

$$F_{\dot{x}} \equiv \frac{\partial F}{\partial \dot{x}} = \frac{-2\dot{x}y\dot{y}^3}{(\dot{x}^2 + \dot{y}^2)^2}, \qquad F_{\dot{y}} \equiv \frac{\partial F}{\partial \dot{y}} = \frac{3\dot{x}^2 y\dot{y}^2 + \dot{y}^4 y}{(\dot{x}^2 + \dot{y}^2)^2}$$

and that the Euler equations for the nonparametric and parametric forms of the problem are, respectively,

$$\frac{d}{dy}\left(\frac{\partial f}{\partial x'}\right) = \frac{d}{dy}\left[\frac{-2x'y}{(1 + x'^2)^2}\right] = 0 \tag{1.12$'$}$$

and

$$\frac{d}{dt}\left(\frac{\partial F}{\partial \dot{x}}\right) = \frac{d}{dt}\left[\frac{-2\dot{x}y\dot{y}^3}{(\dot{x}^2 + \dot{y}^2)^2}\right] = 0,$$

$$\frac{d}{dt}\left(\frac{\partial F}{\partial \dot{y}}\right) = \frac{d}{dt}\left[\frac{(3\dot{x}^2 + \dot{y}^2)y\dot{y}^2}{(\dot{x}^2 + \dot{y}^2)^2}\right] = \frac{\dot{y}^3}{\dot{x}^2 + \dot{y}^2}. \tag{1.12$''$}$$

(Note in (1.12′) that the variable y is the independent one, and not x.) If q is taken to be $-x' = -dx/dy$, then relation (1.12′) becomes, on integration,

$$\frac{yq}{(1 + q^2)^2} = \frac{y_0 q_0}{(1 + q_0^2)^2}, \tag{1.12$'''$}$$

where (x_0, y_0) is some fixed point on the minimizing arc. (Notice that this q is the slope of the normal. It is nonnegative for oval curves such as those Newton discusses.) If the slope of the curve is -1 at this point, then $q_0 = +1$, and (1.12‴) becomes Newton's relation (1.12).

We show below that there is always such a point (x_0, y_0) on a minimizing arc. Before doing this, let us first look at some properties of minimizing arcs. We see that curves satisfying Newton's relation (1.12) include both horizontal and vertical line segments. It is, however, clear that lines with finite and nonzero slope cannot satisfy the relation (1.12′) or (1.12″). Thus frustums of cones such as that generated in Figure 1.5 by rotating *FGB* cannot be part of a surface of least resistance, as we mentioned above. Furthermore, we see below that all extremals—solutions of (1.12′)—have $y > 0$ and, in fact, $y \to +\infty$ as $x \to +\infty$. This is why Newton in Figure 1.8 picks the points D and G, through which the minimizing curve passes, above the axis of revolution.

It is not hard to find a parametric representation for a minimizing arc in terms of the quantity $q = -dx/dy$. If we call the constant right-hand member of Newton's relation (1.12) a, then

$$y = \frac{a(1+q^2)^2}{q} = a\eta(q), \qquad \eta(q) = \frac{(1+q^2)^2}{q} \tag{1.13}$$

and

$$\frac{dx}{dq} = -q\frac{dy}{dq} = -a\left(-\frac{1}{q} + 2q + 3q^3\right)$$

if we regard x, y as functions of the parameter q. Then integrating as to that variable, we find

$$x = -a\left(q^2 + \tfrac{3}{4}q^4 - \log q\right) + b \equiv -a\xi(q) + b. \tag{1.13′}$$

Relations (1.13) and (1.13′) define a two-parameter family of extremals parametrically. It is then easy to calculate the slope of one of these arcs in terms of q. We have the derivatives

$$\frac{d\xi}{dq} = \frac{(3q^2-1)(q^2+1)}{q}, \qquad \frac{d\eta}{dq} = \frac{(3q^2-1)(q^2+1)}{q^2};$$

and the radius of curvature is seen to be

$$\rho = +\frac{(3q^2-1)(q^2+1)^{5/2}}{q^2}.$$

Moreover, for $q = +1$, we find $y = 4a$, $x = -7a/4 + b$ for each a, b and conversely. Thus every solution of the Euler equation, apart from a line segment, has on it a point at which $q = +1$ (see p. 27 for a detailed analysis).

From these expressions we see that as q varies from 0 to $(1/3)^{1/2}$ the curve $x = \xi(q)$, $y = \eta(q)$ is convex to the x-axis, at $q = (1/3)^{1/2}$ there is a cusp with slope of $3^{1/2}$ (the cuspidal line is inclined at an angle of 60° to the x-axis), and then as q increases there is another branch of the curve concave to the x-axis.[29] It is this latter branch that furnishes minimizing arcs. This follows from the Legendre condition.

There is a result in the calculus of variations known as the *Weierstrass–Erdmann corner condition* (see p. 153) which tells us that at each point including corners of a minimizing arc—points at which the slope of the arc is discontinuous—the partial derivatives of the integrand function F with

[29] Bolza, *VOR*, p. 411.

respect to $\dot{x}$ and $\dot{y}$ must remain continuous. Thus at the point defined by the value y we have for $\partial F/\partial\dot{y}$ the condition that

$$\frac{\partial F}{\partial \dot{y}}\bigg|^{q_L} = \frac{\partial F}{\partial \dot{y}}\bigg|^{q_R}.$$

In other words, since $q = -dx/dy = -x'$,

$$\frac{(3q_{\mathrm{L}}^2+1)y}{(q_{\mathrm{L}}^2+1)^2} = \frac{(3q_{\mathrm{R}}^2+1)y}{(q_{\mathrm{R}}^2+1)^2},$$

where q_{L} and q_{R} are the values of q on the left and right sides of a corner, respectively. From this we see that at a corner such as the point G' in Figure 1.7 where we have $q_{\mathrm{R}} = 0$,

$$\frac{(3q_{\mathrm{L}}^2+1)y}{(q_{\mathrm{L}}^2+1)^2} = y;$$

and hence since $y \neq 0$, we have $q_{\mathrm{L}}^2 = 1$. We see from this that at G a minimizing curve must cut the vertical line GB at an angle of 45° or 135°.

We now go back to Newton's discussion of both relation (1.12) and his conclusion that $q_{\mathrm{L}}^2 = 1$. To this end we first give his 1694 version and only later his 1685 one. Newton wrote out for Gregory the details of his demonstration of his assertion in the *Principia* scholium, indicated earlier (p. 12), in a letter dated 14 July 1694. Whiteside tells us that the original document was damaged in a number of places but was first restored by John Couch Adams, the astronomer, in 1888 in an unsatisfactory way, and then later by Bolza in 1912/13. Both restorations have been reexamined by Whiteside and a more definitive one prepared by him.[30] To this end we again make use of Newton's Figure 1.8, where $BGhb$ and $MNom$ are narrow parallelograms with their distance apart Mb and altitudes MN, BG assumed to be given; $s = (Mm + Bb)/2$, $x = (Mm - Bb)/2$, and "the infinitely little lines *on* and *hg* be equal to one another and called *c*." Newton then imagines that the figure generated by rotating the segment $mnNgGB$ of the curve about the axis BM is moved "uniformly in water from M to B according to the direction of its axis BM." He then states his conclusion that "the sum of the resistance of the two surfaces generated by the infinitely little lines Gg Nn shall be least when gG^{qq} is to nN^{qq} as $BG \times Bb$ to $MN \times Mn$," that is, when, in modern notation,

$$\frac{gG^4}{nN^4} = \frac{BG \times Bb}{MN \times Mm}. \tag{1.14}$$

He says that the resistances of the surfaces generated by revolving the small lines Gg and Nn are proportional to BG/Gg^2 and MN/Nn^2, as we

[30] See the preface by Adams, *CAT*, pp. xxi–xxiii; Bolza [1912/13], pp. 146–149; and Newton, *APP 2*, pp. 475n–476n.

may see by recalling that the resistance is proportional to $y\cos^2\theta$, where y is the altitude BG or MN and θ is the angle made with the line CB by the normal to the curve. Hence the resistances are proportional to $BG\cos^2\theta = BG\sin^2 gGh = BG \cdot hg^2/Gg^2$ and to $MN\sin^2 nNo = MN \cdot on^2/Nn^2 = MN \cdot hg^2/Nn^2$ since $on = hg$. The resistance of the two figures together is then proportional to

$$\frac{BG}{p} + \frac{MN}{q},$$

where Newton has set $p = Gg^2$, $q = Nn^2$; and it will be a minimum "when the fluxion thereof $-(BG \times \dot{p})/pp - (MN \times \dot{q})/qq$ is nothing." He then calculates that $p = Gg^2 = Bb^2 + gh^2 = s^2 - 2sx + x^2 + c^2$, $q = nN^2 = on^2 + oN^2 = c^2 + s^2 + 2sx + x^2$ and that $\dot{p} = -2s\dot{x} + 2x\dot{x}$, $\dot{q} = 2s\dot{x} + 2x\dot{x}$. From these facts he finds that

$$\frac{p^2}{q^2} = \frac{BG(s-x)}{MN(s+x)} = \frac{BG \times Bb}{MN \times Mm},$$

which was his assertion (1.14).

He now needs to discuss what happens to the intermediate solid $bgNM$ and says this[31]:

> 2. If the curve line $DnNgG$ be such that the surface of the solid generated by [its] revolution feels y^e^ least resistance of any solid with the same top & bottom BG & CD, then the resistance of the two narrow annular surfaces generated by y^e^ revolution of the [arches Nn &] $G[g$ is] less than if the intermediate solid $bgNM$ be removed [along BM a little from or] to BG & by consequence it is the least that can be & therefore gG^{qq} is to nN^{qq} as $BG \times Bb$ [to $MN \times Mm$.]
>
> [3. If GR be drawn parallel to the tangent at N &] gh be equal to hG so that y^e^ angle [BGg is 135^{degr}, & thus $gG^q = 2Bb^q$, t]hen will [$4B$]b^{qq} be [to Nn^{qq} as $BG \times Bb$ to] $MN \times Mm$, & by consequence $4BG^{qq}$ [to GR^{qq} as $BG \times BP$ to $MN \times BR$ & therefore] $4BG^q \times BR$ to GR^{cub} [as GR to MN.]

To discuss the behavior of the entire minimizing arc $DnNgG$, Newton translates the arc Ng in the direction MB a short distance. Thus N goes to a new point N' so that the line NN' is parallel to MB and g to g' so that gg' is parallel to MB. Now if the arc $DnNgG$ corresponds to the surface of minimum resistance, then the new arc $DnN'g'G$ must correspond to a surface of greater resistance. But the resistance experienced by the solid formed from rotating the arc Ng is identical to that experienced by the solid formed by rotating $N'g'$, since they are the same. Thus the resistance experienced by "the two narrow annular surfaces . . . Nn and Gg" is less than that experienced by nN' and $g'G$, and consequently relation (1.14) of

[31] Newton, *APP 2*, p. 477. Notice that expressions such as GR^{cub} and Nn^{qq} mean GR^3 and Nn^4, respectively.

Newton holds at each point N on the minimizing curve. This relation gives us the condition

$$\frac{BG \times hG/hg}{gG^4/hg^4} = \frac{MN \times oN/on}{nN^4/on^4},$$

since $hg = on$; or equivalently

$$\frac{yq}{(1+q^2)^2} = \frac{y_0 q_0}{(1+q_0^2)^2}, \tag{1.15}$$

where $q = \cot gGh$, which is precisely our previous relation, (1.12‴).

In item 3 above, Newton proceeds to complete his analysis by fixing the constant value of the right-hand member of (1.15). He does this with the help of his condition that the frustum of zero altitude makes a 135° angle with the vertical; he has $q_0 = 1$ and so

$$\frac{yq}{(1+q^2)^2} = \frac{y_0}{4},$$

which corresponds exactly, as we saw above, to Newton's assertion that $4BG^2 \times BR$ is to GR^3 as GR is to MN.

His 1685 version is somewhat different from that of 1694, at least in detail, and is worth describing since it was to become the standard tool of John and James Bernoulli and of Euler until Lagrange's superior one came along. The basic idea is that at a minimum a functional is flat in the sense that in a small neighborhood of the "point" which gives the least value, the values of the functional are nearly the same as the minimum value. The reasoning is this: if φ is a twice differentiable functional on a region of some suitable space R and at $r = r^0$ the functional φ is a minimum, then there is a value $\bar{r}^0$ near r^0 such that

$$\varphi(r) = \varphi(r^0) + \tfrac{1}{2} d^2(\bar{r}^0; dr),$$

where the second differential $d^2\varphi$ is quadratic in dr.

There is a second specialized idea that enters also[32]: if a given arc renders the functional a minimum, then any subarc must also. (This is clearly not true for all problems of the calculus of variations, but certainly for a wide class of them.) With the help of these concepts Newton gave in his 1685 notes a very terse analysis suggesting to me, at least, his complete familarity with the ideas involved. This proof is given in a few lines near the second illustration in Figure 1.9. In Figures 1.10 and 1.11 the point B is a general point on the minimizing curve, and D is near to B; AD is perpendicular to BE and $BA = a$, $AD = x$, $CD = o$, where C is near to D. (Thus a is what we call dy and AD, dx.) Newton then says that the difference of the resistances on the surface formed by rotating BC and BD

[32] Newton, *SOL*, p. 459.

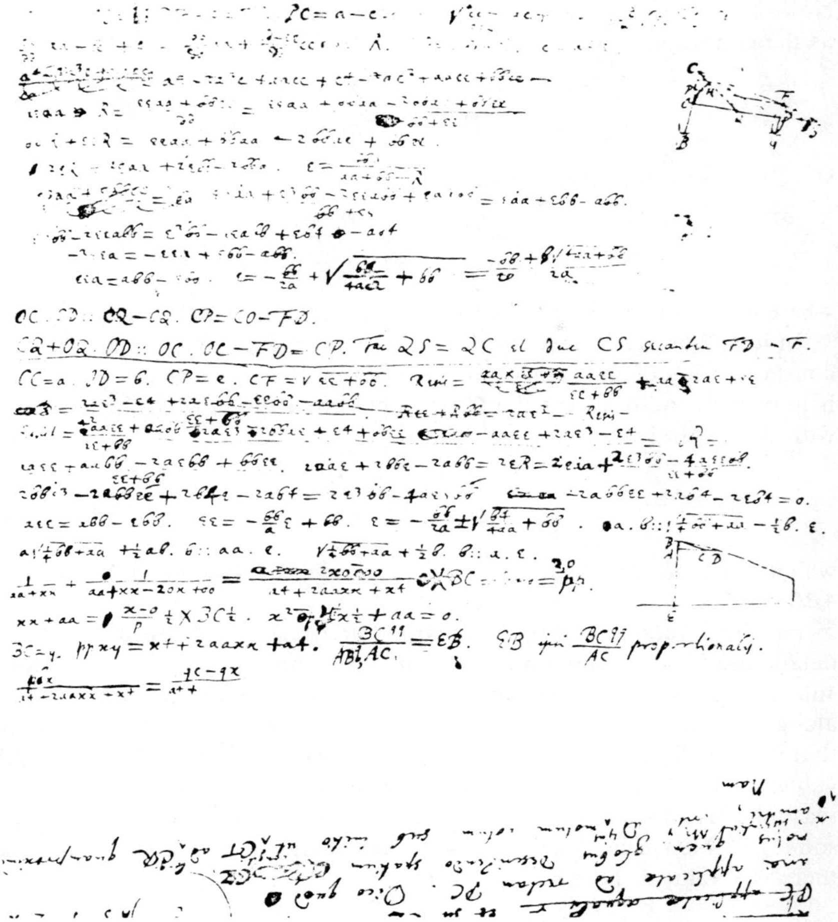

Computation of the cone frustum and solid of revolution of least resistance to translation along its axis

Figure 1.9

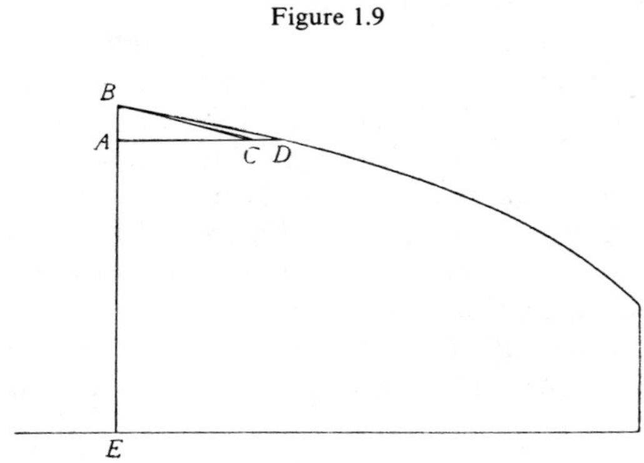

Figure 1.10

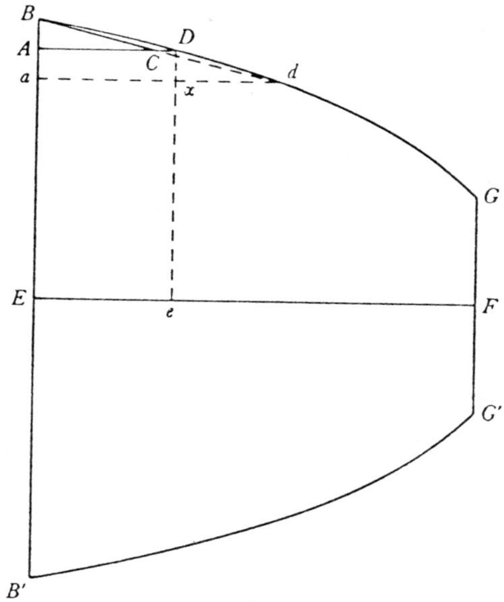

Figure 1.11

about the axis is proportional to

$$\left(-\frac{1}{aa+xx}+\frac{1}{aa+xx-2ox+oo}\right)\times BE=\frac{2xo-oo}{a^4+2aaxx+x^4}\times BE$$

$$=dato=\frac{2,o}{pp}, \qquad (1.16)$$

where p is a constant.[33] (We will explain why Newton took this difference to be a constant in a moment.) From this it follows by properly neglecting "infinitesimals" that

$$ppxy=a^4+2aaxx+x^4,$$

where $y=BE$, since Newton neglects terms of the order of 2 in o and divides out $2o$ as a common factor; in other words, that

$$\frac{\left(a^2+x^2\right)^2}{x}=p^2y.$$

[33] Note that the resistance on the surface formed by rotating BD, for example, is apart from a proportionality factor,

$$2\pi\int_{AE}^{BE}y\,dy\sin^2 ADB=\pi(BE^2-AE^2)\times\frac{AB^2}{AD^2+AB^2}$$

$$=\frac{\pi(BE+AE)a^3}{a^2+x^2}=\frac{\pi a^3(2BE-a)}{a^2+x^2}\sim\frac{2\pi a^3\cdot BE}{a^2+x^2}.$$

A similar expression obtains for the resistance on the surface formed by rotating BC, except that x is now replaced by $x-o$. Also note that $2,o$ means twice o.

In modern terms this last relation tells us that

$$y = \beta \frac{(1 + \xi'^2)^2}{\xi'}$$

where ξ' is what Newton wrote as x/a and where β is a constant; we recognize this as being essentially equation (1.12).

Let us examine why Newton was justified in setting his expressions (1.16) above equal to a constant. This point is very important since, as I mentioned above, it is the device whereby the Euler equation was derived by the Bernoullis and Euler before Lagrange's discovery or invention of the "variation." In Figure 1.11 Whiteside has chosen the point d so that $BA = Aa = D\alpha = a$. Then let us set $\alpha d = \xi$, $\alpha e = \eta$. The resistance on the surface generated by the rotation of the small arc BDd must be nearly the same as that on the surface generated by the rotation of BCd. (This is because BDd gives the least value and the functional remains flat nearby.) Thus we have apart from a factor $2\pi a^3$, approximately

$$\frac{y}{a^2 + x^2} + \frac{\eta}{a^2 + \xi^2} = \frac{y}{a^2 + (x - o)^2} + \frac{\eta}{a^2 + (\xi + o)^2},$$

where $AD = x$, $BE = y$, or

$$\left[-\frac{1}{a^2 + x^2} + \frac{1}{a^2 + (x - o)^2} \right] y = \left[\frac{1}{a^2 + (\xi + o)^2} - \frac{1}{a^2 + \xi^2} \right] \eta;$$

i.e.,

$$\frac{2xy}{(a^2 + x^2)^2} = \frac{2\xi\eta}{(a^2 + \xi^2)^2}.$$

Since we could iterate this process as often as we wish, we ultimately can conclude that the value of the right-hand member of this last relation can be evaluated at any point d on the curve we wish, let us say at G, which we may view as a fixed point. Thus the left-hand member of this relation is a constant; i.e., in terms of the nonparametric integrand function in (1.11),

$$\frac{\partial f}{\partial x'} = \frac{\partial}{\partial x'} \left[\frac{y}{(1 + x'^2)} \right] = \text{const.},$$

which is relation (1.12). Notice that Newton does not yet show how he determined the value of the constant above as he later did in the *Principia*.

So far Newton has confined himself to the simplest problem of the calculus of variations: namely, to find among all admissible curves passing through two fixed points the one that renders a given functional a minimum. But he did more. In a fragment that was never published, but which Whiteside has fortunately brought to light, Newton considers a more complex problem, whose introduction into the literature is usually ascribed to later mathematicians. The significance of this work of Newton

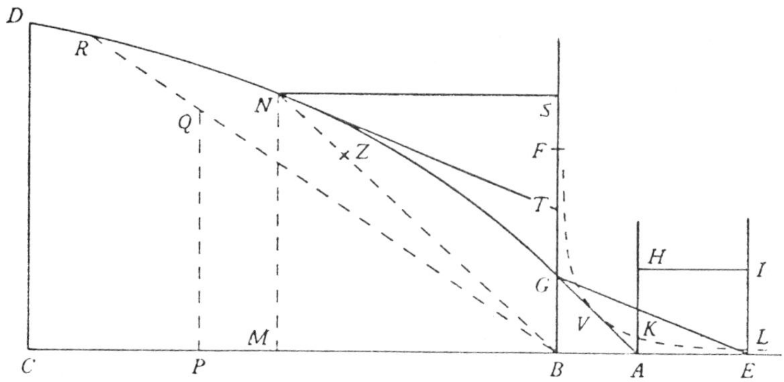

Figure 1.12

on a variable end-point problem has not been noted by Whiteside or others. It is therefore important to emphasize the grasp Newton had of the subject.

What Newton now proposed is explicable in terms of Figure 1.12. He considered how to find among all curves through the point D: (x_2, y_2) that one which intersects the vertical line BS—$x = x_1$—in such a point G: (x_1, y_1) that the functional

$$y_1^2 + 2\int_{y_1}^{y_2} \frac{y\,dy}{1 + x'^2} \tag{1.17}$$

is a minimum. Notice that the point G is no longer fixed but is free to move on the line BS, and its location is to be determined so that the functional (1.17) has its least value. Newton says that the figure of revolution produced by rotating $DNGB$ about the axis CB will experience less resistance in moving through a "rare and elastic medium" from B to A than will any other figure produced by rotating a curve with "the same longitude BC and latitude $2CD$," provided that

$$\frac{BS}{BG} = \frac{NT^{qq}}{4NS \times ST^{\text{cub}}}.$$

In this figure N is an arbitrary point on the minimizing curve, NS is drawn through N parallel to CB, and NT is the tangent to the curve at N.[34] [In modern terms, this says that $BS/BG = NT^4/(4NS \times ST^3)$. Just prior to stating this result, Newton had recorded a somewhat confused statement in which he said that among all curves made up of an arc of an ellipse DNF of latitude (minor axis) DE, center at C, and longitude (major axis) AB in Figure 1.7 and of the frustum of a cone FGB, the one of least resistance is the ellipse which meets BG—extended if need be—at an angle of 135°.

[34] Newton, *APP 2*, VI, p. 479. It is of interest to note a paper by Armanini ([1900], pp. 134–135), who formulated the variable end-point problem posed above and discussed aspects of its solution without knowing of Newton's work.

Very possibly the theorem above represented Newton's final and precise solution of the problem.]

First let us look at Newton's variable end-point problem as formulated in (1.17). The transversality condition in the present case then says that at the point the relation

$$y_1\, dy_1 - \left[(f - x' f_{x'})\, dy + f_{x'}\, dx \right]^{y=y_1, x=x_1, x'=x_1'} = 0 \tag{1.18}$$

holds for $dy_1 = 1$, $dx_1 = 0$, $f(y, x, x') = y/(1 + x'^2)$; i.e.,

$$0 = y_1 - \left[\frac{y_1}{1 + x_1'^2} + \frac{2y_1 x_1'^2}{(1 + x_1'^2)^2} \right] = y_1 \left[\frac{x_1'^2(x_1'^2 - 1)}{(x_1'^2 + 1)^2} \right]. \tag{1.18'}$$

It follows from this that either $y_1 = 0$, $q_1 = x_1' = 0$, or $q_1^2 = x_1'^2 = 1$.

The Euler equation for this problem is (1.12′), the same as for the earlier one. Since $y = a(1 + q^2)^2/q$, $y_1 = 0$ would imply that $a = 0$ and the minimizing curve would be the line CB, but this does not pass through point D; $x_1' = 0$ would again imply $a = 0$. Thus $x_1'^2$ must be 1; i.e., the curve $x = x(y)$ intersects the line $x = x_1$ at an angle of 45° or 135°, as Newton asserted.

In his next paragraph Newton takes up the question of finding a parametric representation for the extremals in terms of the parameter $q = BE/BG$. He considers that the problem is to locate the point N in Figure 1.12 in terms of this quantity q. To do this, he draws GE parallel to NT and notes that $BS/GE = GE^3/(4BE \times BG^2)$ since $BS = MN$, the triangles NST and EGB are similar, and $MN/BG = GE^4/(4BE \times BG^3)$; he then lays off the point F on BS so that

$$\frac{GF}{BG + BE} = \frac{5BG^2 + BE^2}{16BG^2}. \tag{1.19}$$

Given this, he chooses A so that $BA = BG$ and erects $AH = BG + FS$ perpendicular to BE.[35] He then draws the hyperbola VKL with center at B, BE, and BS asymptotes, and which is tangent to the line AG. Finally he fixes N so that SN is the sum of the length GS and the length determined by the ratio of the area $HKLI$ to the length BG; i.e., if the area under the hyperbola between A and E is designated by T, then

$$SN = GS + \frac{AH \times AE - T}{BG}. \tag{1.20}$$

Given these facts Newton can now find his representations in terms of $q = BE/BG$. Since $q^2 + 1 = (BE^2 + BG^2)/BG^2 = GE^2/BG^2$, it is clear

[35] Whiteside here corrects an error in Newton's text. See Newton, *APP 2*, p. 479n. (Note the hyperbola VKL is the dotted curve.)

that

$$y = MN = BS = \frac{GE^4}{4BE \times BG^2} = \frac{BG}{4} \times \frac{(q^2+1)^2}{q} = \alpha \frac{(q^2+1)^2}{q}, \tag{1.21}$$

where $\alpha = BG/4$. This is, of course, Newton's first integral of the Euler equation (1.12′). Next, following Whiteside, we see how Newton found the representation for x using the hyperbola described above.

With the help of the relation (1.20), it follows that

$$SN = GS + \frac{1}{BG} \times (BG + FS) \times AE - \frac{1}{BG} \times \frac{BG^2}{4} \times \log \frac{BE}{BA},$$

since $AH = BG + FS$ and the equation of the hyperbola relative to the axes BS and BE is

$$XY = \frac{BG^2}{4}.$$

Thus $T = (BG^2/4)\log BE/BG$ since $BA = BG$. But $FS = BS - BF = BS - (BG + GF)$ and by (1.19),

$$\begin{aligned} GF &= \frac{(BG + BE)(5BG^2 + BE^2)}{16BG^2} = \frac{BG}{16}(q+1)(q^2+5) \\ &= \frac{\alpha}{4}(q+1)(q^2+5) \end{aligned}$$

so that

$$BG + FS = BS - GF = \alpha \frac{(q^2+1)^2}{q} - \frac{\alpha}{4}(q+1)(q^2+5).$$

From these relations it follows that

$$\begin{aligned} -x = SN &= (BS - BG) + \frac{1}{BG} \times (BS - GF) \times (BE - BG) \\ &\quad - \frac{BG}{4} \times \log \frac{BE}{BG} \\ &= \frac{\alpha(q^2+1)^2}{q} - 4\alpha + \alpha(q-1)\left[\frac{(q^2+1)^2}{q} - \frac{1}{4}(q+1)(q^2+5)\right] \\ &\quad - \alpha \log q \\ &= \alpha\left(\frac{3}{4}q^4 + q^2 - \frac{7}{4} - \log q\right). \end{aligned} \tag{1.21′}$$

(It should be noted that this representation puts the origin at B.)

It is not difficult to show that through any point (x_2, y_2) with $y_2 > 0$ there is one curve of the family of extremals (1.13), (1.13′) which cuts the line $x = x_1 > x_2$ at an angle of 135°. To do this, we note first that for $q = 1$, $x_1 = -(7/4)a + b$, which fixes b as a function of a. Next we go to show that the equation

$$\frac{x_2 - x_1}{y_2} = \frac{-q^2 - \frac{3}{4}q^4 + \log q + \frac{7}{4}}{\eta} \equiv \varphi(q) \tag{1.22}$$

has a unique solution $q = \bar{q}$ given the value $(x_2 - x_1)/y_2 < 0$ and the value of η in (1.13). To this end, note that $\eta = \eta' q(q^2 + 1)/(3q^2 - 1)$ and hence that

$$\varphi'(q) = -\frac{\eta'}{\eta^2}\left[\frac{1}{4}q^4 + q^2 + \log q + \frac{11}{4}\right] < 0 \qquad (3q^2 > 1).$$

We also know that $\varphi(1) = 0$, and hence φ proceeds monotonously to $-\infty$ as q goes to $+\infty$. It must therefore have exactly one value $\bar{q}$ at which relation (1.22) is satisfied.

With this value $\bar{q}$ the parameter a is determined so that

$$y_2 = \frac{a(1 + \bar{q}^2)^2}{\bar{q}}$$

and hence

$$x_2 = x_1 + y_2 \cdot \varphi(\bar{q}).$$

In Newton's final paragraph he examines how the family of extremals [(1.21) and (1.21′)] varies as G moves. He notices that MN/BG and BM/BG depend only on the parameter q and not on G; i.e., not on α. He picks in Figure 1.12 a new width BP, a new height PQ and draws the ray BQ cutting the original curve DNG in R. Then on BN he marks off BZ so that $BZ/BN = BQ/BR$. Now the point Z describes the new curve of least resistance through Q which cuts BS in a new point G' such that $BG'/BG = BQ/BR$. This changes the parameter $\alpha = BG/4$ in (1.21) and (1.21′) into $\alpha' = BG'/4 = (BQ/BR)\alpha$. It also changes BM/BG and MN/BG, as can be seen with the help of the relation (1.21). It is evident that $MN/BG = (q^2 + 1)^2/q$ and by similar triangles that $BM = q \cdot MN$; as a result, $BM/BG = (q^2 + 1)^2$. Newton shows then how to find the family of extremals cutting the line BS transversally (see Newton, *APP 2*, pp. 479–480). We see that Newton has in effect shown the existence of a field of extremals transversal to an end-curve; we will see this again when we discuss John Bernoullis's synchrone on p. 42 below.

In closing this discussion of Newton's contributions we should look at Huygens's notes (see footnote 17, above). In his 22 April 1691 notes he first worked through the case of the frustum of the cone without difficulty. He then took up what may be the general problem of finding the curve, which

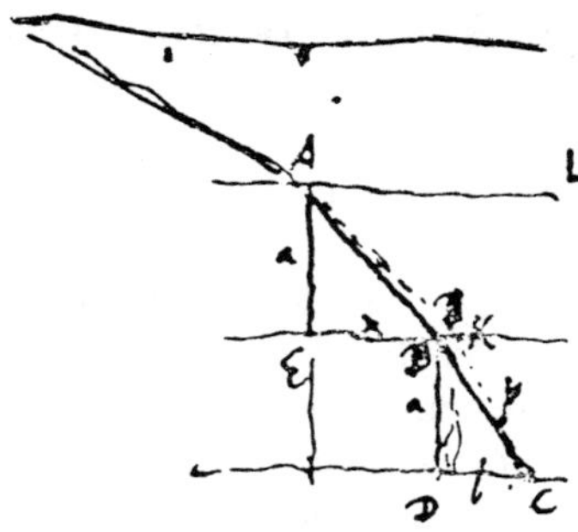

Figure 1.13

when rotated about the axis of motion through the fluid, experiences the least resistance. His figure is the present Figure 1.13. Note that $AE = BD = a$, $EB = x$, and $DC = b$. He allows B to move to K along the line EB. He asserts that the resistance on the figure made up of AB and BC is proportional to

$$\frac{a^4}{a^2 + x^2} + \frac{a^4}{a^2 + b^2} \tag{1.23}$$

and is to be a minimum; this he does by replacing x by $x + e$ and b by $b - e$ and neglecting quadratic terms in e.

This seems not too different from Newton's 1685 solution if we think of a, b, x as differentials. However, if this is the case, his formula (1.23) is not correct. The resistance on the figure formed by the two segments is, apart from a factor,

$$\left[(y + a)^2 - y^2\right] \cdot \frac{a^2}{a^2 + x^2} + \left[y^2 - (y - a)^2\right] \cdot \frac{a^2}{a^2 + b^2}$$

$$\sim \frac{2a^3y}{a^2 + x^2} + \frac{2a^3y}{a^2 + b^2};$$

where y is the height of the point B above the axis of rotation, which is presumably not shown in Huygens's drawing. Note that Huygens has a^4 in his numerator in (1.23); this may mean that he has chosen $y = a$, in which case his analysis will not lead him to the correct Euler equation found by Newton (see p. 23).

However, Huygens goes on in his 25 April 1691 notes to attempt to establish the theorem of Bolza mentioned in footnote 27 above. His proof is incomplete but shows clearly how he was proceeding. In fact, he remarked in a marginal comment that no ship's prow of least resistance can be formed out of a rotund or rounded curve. He clearly went quite far but, I think, perhaps not as far as Whiteside feels he did.

1.3. The Brachystochrone Problem

The problem of finding the curve joining two points in a vertical plane along which a frictionless bead will descend in the least possible time was apparently first considered by Galileo in his 1638 work, *Two New Sciences*.[36] His conclusion is not completely correct since he decided that "one can deduce that the swiftest movement of all from one terminus to the other is not through the shortest line of all, which is the straight line [AC], but through the circular arc." Actually, what he shows is that (Figure 1.14) the time for a bead to go along the inscribed regular polygon $ADEFGC$ is less than along AC, and "Hence motion between two selected points, A and C, is finished the more quickly, the more closely we approach the circumference through inscribed polygons."

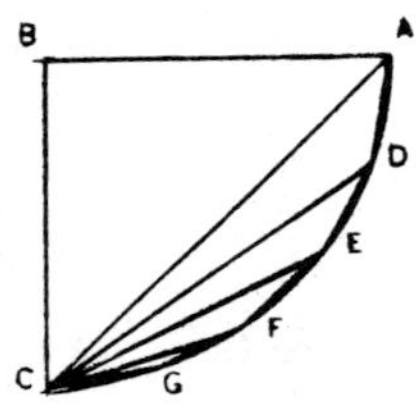

Figure 1.14

Galileo's solution was based not on the methods of the calculus of variations, but his problem was to serve as the inspiration for much work in the field. This work was initiated by John (Johann or Jean) Bernoulli, who challenged the mathematical world in June 1696 to solve the following problem:

> Given points A and B in a vertical plane to find the path AMB down which a movable point M must, by virtue of its weight, proceed from A to B in the shortest possible time.[37]

The statement of the problem is followed by a paragraph in which John Bernoulli reassures his readers that the problem is very useful in mechanics, that it is not the straight line AB, and that the solution curve AMB is very well known to geometers. He says he will show that this is so at the end of the year if no one else does. Then in December 1696 in the *Acta Eruditorum* and again in January 1697 he published in Groningen, Holland, where he had been professor since 1695, an extension of his previous time limit and a further explanation of his mechanicogeometrical prob-

---

[36] Galileo, *TWO*, pp. 97 and 212–213.

[37] John Bernoulli, *PN*, p. 161. See Figure 1.15.

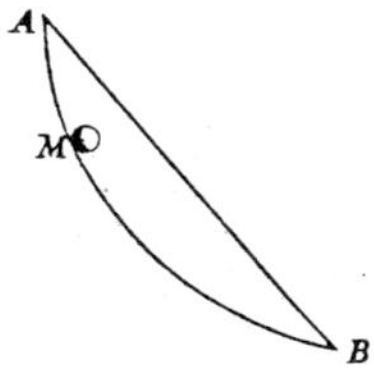

Figure 1.15

lem.[38] He also included in the latter publication another problem of a purely geometrical sort which does not concern us. His new deadline was set as Easter of 1697. However, if no one had succeeded by then in solving the problem, Bernoulli said that he would make not only his solution, but also Leibniz's, available to the world.

He had this latter one of Leibniz's at hand for some time. In fact, John had written Leibniz in Hanover on 9 June 1696 posing his problem privately and had received Leibniz's answer from Hanover dated 16 June 1696![39]

The postponement by Bernoulli was effected at Leibniz's suggestion to allow foreigners, particularly the French and Italian mathematicians, time to receive the *Acta Eruditorum* since its delivery outside of Germany was apparently slow. Leibniz also suggested the problem be called the problem of swiftest descent, "Tachystoptotam" (from *tachystos*, swiftest, and *piptein*, to fall). Bernoulli, however, preferred to name it the problem of shortest time, the "Brachystochrone" problem (from *brachystos*, shortest, and *chronos*, time).[40]

Finally the May 1697 issue of the *Acta Eruditorum* appeared with John Bernoulli's solution on pp. 206–211. This issue also contained the solution of his older brother James (Jacques or Jacobus) on pp. 211–218 and a brief note by Leibniz saying that he had also solved the problem but since his solution was similar to that of the brothers Bernoulli, he would not reproduce it. He also noted that "l'Hôpital, Huygens were he alive, Hudde if he had not given up such pursuits, Newton if he would take the trouble"

---

[38] John Bernoulli, *LB* and *PN*, pp. 165–169. This appears in German translation by P. Stäckel in Ostwald's *Klassiker*, No. 46, Leipzig, 1894.

[39] Leibniz, *LMS*, Part I, Vol. III, Nos. XXVIII and XXIX, pp. 283–290; see also pp. 290–295 for Leibniz's addendum of mathematical details. It is not completely clear from the text exactly when Leibniz produced that addendum, but it seems likely it was written by him before he sent off his letter to Bernoulli (see p. 35).

[40] Leibniz, *LMS*, Part I, Vol. III, p. 298 and p. 312. The former reference is in a letter from Bernoulli to Leibniz on 21 July 1696; it "I have given the curve the name 'Brachystochrone' for the reason you see therein; but if the name 'Tachystoptote' pleases you more, I will let this be submitted in its place." In response on 31 July 1696 Leibniz says "The name of Brachystochrone pleases me for its general significance; when it is a question of the fall of a heavy particle the name Tachystoptote possesses special significance." However, he did not press his point further.

could also have solved the problem. In fact, Newton did in an anonymous paper in the January 1696 issue of the *Philosophical Transactions*, which was reprinted in the *Acta Eruditorum* (see Section 1.5, below). The issue also includes discussions of the problem by L'Hôpital and Tschirnhaus.

In a letter to a M. Basnage, doctor of law and author of the *History of Works of Scientists*,[41] John Bernoulli discussed a number of aspects of the problem and interestingly notes that "I proposed the problem of swiftest descent in the *Leipsic Acts*, as being completely novel, not knowing that it had been attempted previously by Galileo" (p. 194). Again (p. 199) he says "M. Leibniz noted two remarkable things about Galileo: it is that this man who was, without contradiction, the most clairvoyant person of his times . . . , lacking, however our new analysis, could conjecture that the curve of the catenary is a parabola and that the curve of swiftest descent is a circle"[42]

Perhaps we should now turn to the actual solutions themselves to see the state of the mathematical techniques of the time and then study the wealth of problems that grew out of these beginnings.

1.4. The Problem Itself

It is not difficult to set up Bernoulli's problem analytically. Let us choose the x-axis to be vertical and the y-axis horizontal, following Bernoulli. Then what John Bernoulli calls Galileo's hypothesis is that the velocity at the height x is given by the relation

$$v^2 - v_1^2 = 2g(x - x_1),$$

where g is the constant of gravity and the particle starts from the point A: $(x, y) = (x_1, y_1)$ with the initial velocity v_1. (We assume that the positive x-axis points downward.) If the velocity is expressed as ds/dt with s arc-length, the time T to fall from the point A: (x_1, y_1) to B: (x_2, y_2) is given in parametric form by the functional relation

$$\sqrt{2g}\, T = \int_{t_1}^{t_2} \frac{\sqrt{x'^2 + y'^2}}{\sqrt{x - x_1 + k}}\, dt, \tag{1.24}$$

---

[41] John Bernoulli, *LE*.

[42] In his *Two New Sciences* of 1638 Galileo had raised a very interesting physical problem. It was to find the shape of a heavy and inelastic curve hanging between two fixed points. Galileo incorrectly concluded the desired curve was a parabola. Then in a series of papers published in the *Acta Eruditorum* from 1690 through 1692 the Bernoulli brothers, Huygens, and Leibniz independently gave the correct solution. The problem can be formulated as one in the calculus of variations: to find, among all heavy and inelastic curves of a given length joining two fixed points, the one whose center of gravity is the lowest.

where $x' = dx/dt, y' = dy/dt$, and $k = v_1^2/2g$. The problem is then to find among all admissible curves joining the points A and B in a vertical plane the one which renders this integral a minimum. Notice that the integrand function

$$f(x, x', y') = \frac{\sqrt{x'^2 + y'^2}}{\sqrt{x - x_1 + k}} \tag{1.25}$$

in this case does not contain y explicitly. In this sense it is quite analogous to the one in Newton's problem. (Note that there the roles of x and y are interchanged, but this is mere notation.)

One of the Euler equations for this problem is therefore

$$0 = \frac{d}{dt} f_{y'} = \frac{d}{dt}\left[\frac{y'}{\sqrt{(x - x_1 + k)(x'^2 + y'^2)}}\right]; \tag{1.26}$$

or, equivalently, $f_{y'} = \text{const.} = 1/(2a)^{1/2}$ is a first integral, much as before. After a little manipulation this condition reduces to the nonparametric equation

$$\frac{dy}{dx} = \sqrt{\frac{x - \alpha}{\beta - x}}\,, \tag{1.27}$$

where $\alpha = x_1 - k$ and $\beta = 2a + \alpha$. Experience has shown it is convenient to introduce as a parameter the angle that the tangent to the curve makes with the y-axis; thus

$$\frac{y'}{\sqrt{x'^2 + y'^2}} = \cos\tau; \tag{1.26$'$}$$

hence relation (1.26) is equivalent to

$$x - x_1 + k = 2a\cos^2\tau = a(1 + \cos 2\tau), \tag{1.28}$$

that is, $x' = -2a\sin 2\tau$. It then follows from (1.26′) that $y' = \pm 4a\cos^2\tau$ or, equivalently,

$$y - y_1 + c = \pm a(2\tau + \sin 2\tau). \tag{1.28$'$}$$

Relations (1.28) and (1.28′) are thus a parametric form for the extremals, with a and c as arbitrary parameters. (Note that τ is not time but an angle.) It is convenient to replace τ by a related angle $\varphi = 2\tau + \pi$, and these equations reduce to the form

$$x - x_1 + k = r(1 - \cos\varphi), \tag{1.28$''$}$$

$$y - y_1 + b = \pm r(\varphi - \sin\varphi), \tag{1.28$'''$}$$

where r and b are arbitrary constants. These curves are the well-known cycloids. Such a curve is generated by a point fixed on a wheel of radius r that rolls along the line $x = x_1 - k = \alpha$. This curve has been known and of interest since about 1500. One of its most remarkable properties, that of tautochronism, was discovered in 1673 by Huygens. It is known that for very small oscillations the period of a simple pendulum moving on an arc of a circle is nearly independent of its amplitude; that is, it is almost in simple harmonic motion, but for a finite-sized oscillation this is no longer so. However, Huygens showed that when the pendulum bob moves along an arc of a cycloid, its period is precisely independent of the amplitude of its excursion (i.e., the motion is exactly simple harmonic). In John Bernoulli's paper on his solution he aptly remarks that "one will be astounded when I say that the Cycloid, the tautochrone of Huygens, is the sought Brachystochrone."

1.5. Newton's Solution of the Brachystochrone Problem

The story is told by John Conduitt, the husband of Newton's niece, that Newton one day read John Bernoulli's challenge and wrote out his solutions before going to bed that night.[43] Whatever the truth of this tale, it is clear that on 30 January 1697 Newton sent to his friend Sir Charles Montagu, Secretary of the Royal Society, his solution of Bernoulli's two problems.[44]

In the first four and a quarter pages of his paper Newton reproduced verbatim Bernoulli's *Programma* of January 1697 and then said "Thus far Bernoulli. But the solutions of the problems are these."

Problem I

To be found is the curve *ADB* along which a heavy particle will fall under the action of gravity from any given point *A* to any given point *B*.

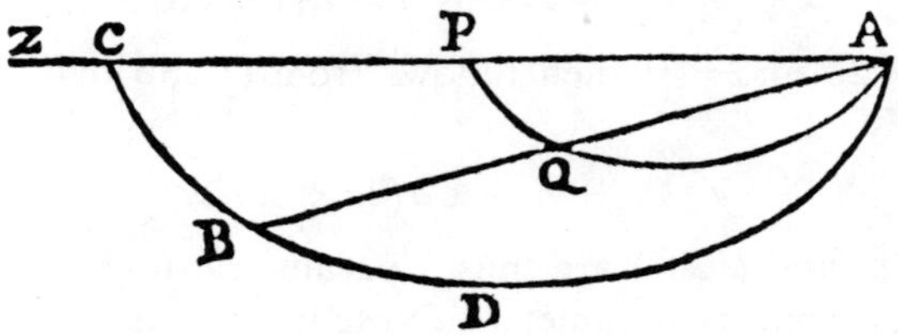

Solution

Through the given point *A* draw the horizontal line *APCZ* and on it first describe any cycloid *AQP* cutting the line *AB* (produced if need be)

[43] Andrade [1958], p. 100.

[44] Newton, *PT*, pp. 384–389. The paper appeared anonymously, presumably as Newton desired. (It is not relevant but of note that Montagu secured for him the two posts he later held at the Mint: the wardenship and then the mastership.)

in the point Q and second another cycloid ABC whose base and altitude are, respectively, as AB is to AQ. This last cycloid passes through point B and is the curve along which a heavy particle will descend most quickly from the point A to the point B. QEI.

This concludes Newton's demonstration. However, it is reasonable that he must have seen at once the decisive similarity between the integrand functions for his surface of revolution problem and this one. That is, he certainly must have been aware of the fact that both lack the derivative of the dependent variable, and hence he certainly knew very quickly a necessary condition for a minimum, namely, that $\partial f/\partial y' = \text{const.}$ or, equivalently, that (1.27) holds in this situation. As we saw earlier, once this first-order differential equation (1.27) is obtained, it is easy to see that the solution is a cycloid. It is, however, interesting to speculate on why Newton gave no clue in his paper as to how he knew the curve was cycloidal and why he bothered to write down only the rather simple existence proof. It is also fascinating to read in Bernoulli's account to Basnage that even though Newton sent in his paper anonymously he, Bernoulli, knew *ex ungue leonem*—knew the lion from his touch.[45]

1.6. Leibniz's Solution of the Brachystochrone Problem

As we saw above, the letter from John Bernoulli to Leibniz brought a reply dated just one week later containing at least a sketch of the latter's solution. Moreover, it probably also contained the extensive addendum cited in footnote 39. There is unfortunately nothing in Gerhardt's edition to indicate whether that material headed *Beilage* was sent then, later, or never, or whether it was merely Leibniz's note for himself. However, let us first see what he did in his letter dated 16 June 1696. Here he writes down the differential equation (1.27) without any indication of how he finds it.

In Figure 1.16 he chooses the x-axis to be vertical and the y-axis to be horizontal. He then asserts without explanation (it occurs in his addendum), "Whence it follows that an element of the curve in this manner is in

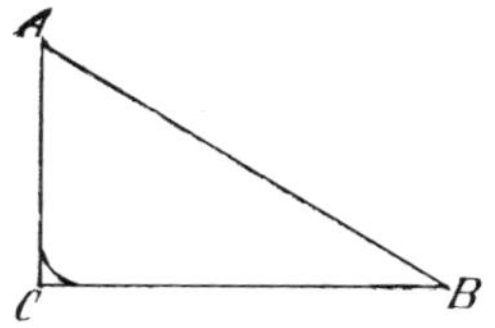

Figure 1.16

[45] John Bernoulli, *LE*, p. 196.

the ratio composed directly as the [corresponding] element of the latitude and inversely as the square root of the altitude itself." This means that if AC is the x-axis and quantities measured along it are called *altitudes* and BC is the y-axis and quantities measured along it are called *latitudes*, and if s is arc-length along the curve in question, then

$$ds = \frac{k}{\sqrt{x}}\, dy. \tag{1.29}$$

Since $ds^2 = dx^2 + dy^2$, it follows directly that

$$\frac{dy}{dx} = \sqrt{\frac{x}{2b - x}}\,, \tag{1.30}$$

where Leibniz has set $2b = k^2$. (He writes this last differential equation in the form

$$dy : dx = \sqrt{,,x : ,2b - x}$$

and then introduces a variable v so that $v : b = dy : dx$. From this he reasons that $y = \int v\, dx : b$.)

It is remarkable that nowhere in his letter or its addendum does he mention that the curve is a cycloid. He presumably knew in 1686 that his differential equation defines the cycloid but makes no mention of it until after Bernoulli sent him his solution on 21 July 1696.

Be that as it may, let us see how Leibniz deduced his differential equation (1.29) and hence also (1.30).[46] His analysis, unlike Newton's rambles over unnecessary paths before getting to its goal. Let us, therefore, extract merely the essence of his proof. In effect, he makes use of Galileo's hypothesis that the time t_{AB} of fall down the incline AB in Figure 1.16 is given by the relation

$$AB = \frac{1}{2}\, g t_{AB}^2 \cdot \sin ABC = \frac{1}{2}\, g t_{AB}^2 \cdot \frac{AC}{AB}\,,$$

and the time t_{AC} down the vertical by

$$AC = \frac{1}{2}\, g t_{AC}^2.$$

Thus

$$t_{AB} = \frac{AB}{AC}\, t_{AC}. \tag{1.31}$$

Next in Figure 1.17 Leibniz has the points A, B, C, and E fixed but wishes to allow D to move on the line through E parallel to the base so that the time of descent along the path made up of AD and DB will be a minimum. To this end he first finds that

$$r \equiv t_{AE} = \sqrt{\frac{AE}{AC}}\; t_{AC}, \qquad n \equiv t_{EC} = \left(1 - \sqrt{\frac{AE}{AC}}\,\right) t_{AC}$$

[46] Leibniz, *LMS*, Part I, Vol. III, pp. 290–295.

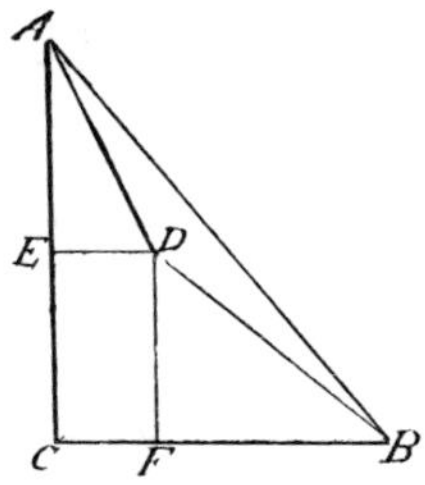

Figure 1.17

and hence by relation (1.31) that

$$t_{AD} = \frac{AD}{AE} \cdot r, \qquad t_{DB} = \frac{DB}{EC} \cdot n.$$

This enables him to calculate $t_{ADB} = t_{AD} + t_{DB}$ and to minimize it subject to the conditions that $AE = $ const., $EC = $ const.,

$$AD^2 = AE^2 + ED^2, \qquad DB^2 = EC^2 + FB^2$$
$$= EC^2 + (CB - ED)^2.$$

From these it follows easily that

$$r \cdot \frac{ED}{AD \cdot AE} = n \cdot \frac{FB}{DB \cdot EC}$$

and thus that

$$t_{AD} \cdot \frac{ED}{AD^2} = t_{DB} \frac{FB}{DB^2}. \tag{1.32}$$

Given this result, Leibniz makes use of Figure 1.18, in which he has plotted a parabola AE with vertex at A and axis AB so that a particle which falls from A to B vertically arrives there in the time BE. The curve AC is Leibniz's tachystoptote, and B_1, B_2, and B_3 are near to each other and equally spaced. By relation (1.32), he has

$$t_{C_1C_2} \cdot \frac{D_1C_2}{(C_1C_2)^2} = t_{C_2C_3} \cdot \frac{D_2C_3}{(C_2C_3)^2},$$

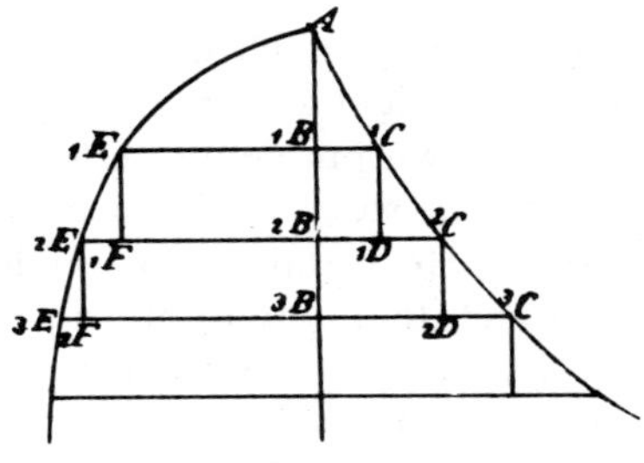

Figure 1.18

and with the help of (1.31) he is able to write

$$t_{C_1C_2} = \frac{C_1C_2}{C_1D_1} t_{C_1D_1}, \qquad t_{C_2C_3} = \frac{C_2C_3}{C_2D_2} t_{C_2D_2}.$$

Combining these and noting that $C_1D_1 = C_2D_2$, he finds

$$t_{C_1D_1} \cdot \frac{D_1C_2}{C_1C_2} = t_{C_2D_2} \frac{D_2C_3}{C_2C_3}$$

that is,

$$F_1E_2 \cdot \frac{D_1C_2}{C_1C_2} = F_2E_3 \cdot \frac{D_2C_3}{C_2C_3} \tag{1.32'}$$

since $F_1E_2 = B_2E_2 - B_1E_1$ is the time for the particle to fall from B_1 to B_2, i.e., from C_1 to D_1.

Recall that the distances $B_1B_2 = B_2B_3 = \ldots$ are "small" and hence that $ds = gt\,dt$; thus, e.g., $B_1B_2 = gt_{AB_1} \cdot t_{B_1B_2}$. From this it follows that

$$1 = \frac{B_1B_2}{B_2B_3} = \frac{B_1E_1}{B_2E_2} \cdot \frac{F_1E_2}{F_2E_3} = \frac{\sqrt{AB_1}}{\sqrt{AB_2}} \cdot \frac{F_1E_2}{F_2E_3}$$

because $F_1E_2 = t_{B_1B_2}$, $F_2E_3 = t_{B_2B_3}$, and $BE^2 \propto AB$; hence with the help of (1.32′) Leibniz has the solution

$$\frac{D_1C_2}{D_2C_3} = \frac{C_1C_2}{C_2C_3} \left(\frac{\sqrt{AB_2}}{\sqrt{AB_1}} \right) \quad \text{or} \quad dy = k'\sqrt{x} \cdot dc$$

in differential form where $dc^2 = dx^2 + dy^2$. But this is Leibniz's relation (1.29).

1.7. John Bernoulli's First Published Solution and Some Related Work

John Bernoulli was apparently much taken with Fermat's principle of least time (see Section 1.1) for light and in an elegant manner saw how to transfer this principle to his own problem—perhaps this is why he was so pleased with the name "brachystochrone" since this notion of least time was basic to his approach. He certainly had at least an inkling of the importance that would attach to a careful pursuit of the analogy between optics and mechanics. His work also must have considerably influenced both Euler and Maupertuis in their enunciation of the principle of least action, as we shall see later.[47] It is also clear that Bernoulli's method bears no resemblance to Newton's for the least resistance problem and is certainly not derived from Newton's.

As we shall see in Section 1.11, by 1697 John had already given a direct analysis of the brachystochrone problem. The method he used, but with a significant modification, provided Carathéodory with a general principle for solving problems of the calculus of variations. This elegant work of John's is never referred to in the modern literature apart from Carathéodory's works. It can be seen from John Bernoulli's two solutions of the brachystochrone problem how remarkable a mathematician he was.

As we saw in Fermat's work, a ray of light moves in a medium of given index of refraction or optical density on a path so determined that it reaches the boundary in the shortest possible time; when it enters a contiguous medium with a different index, it is refracted so that it again moves in the shortest time. First Bernoulli divides up the space between the points A and B into horizontal slices, in each of which he assigns an index of refraction or density inversely proportional to the heavy particle's velocity in that layer. Then at each interface he is able to make use of Snell's law, which tells him exactly how the particle-ray bends as it crosses that interface.

In Figure 1.19 he plots a curve AHE whose ordinate (notice that some of the Continental mathematicians of this era use the term "ordinate" to mean the horizontal distance from the vertical axis to a curve and usually call it y; they generally use x for their vertical distances. James uses exactly the opposite conventions) $CH = t$ measures the velocity or the inverse of

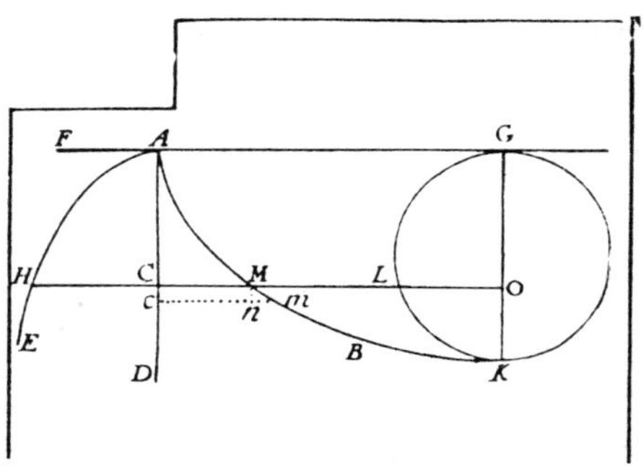

Figure 1.19

[47] It is fortunate that most of the text of both John's and James's papers appears in English translation by D. J. Struik, *Source*, pp. 391–399. (Struik has an omission on p. 395 that makes his footnote 5 meaningless. I give the material correctly below.) In German translation they appear fully in Ostwald, *Klassiker*, No. 46, pp. 1–20. The originals are in John Bernoulli, *CR*, and James Bernoulli, *JB*. There are two enjoyable and excellent essays on the period by Carathéodory [1937], [1945].

the density at the vertical distance $AC = x$ from the starting point A of the particle-ray. He takes his minimizing curve to be AMB and lets the ordinate CM be y. Then he takes c to be a point on the vertical AC near to C and sets $Cc = dx$; the horizontal through c cuts the curve AMB in m, and he takes n to be the point directly below M on cm. Then $Mn = Cc = dx$, and he sets the arc-length $Mm = dz$.

He then notes that "the sines of the angles of inclination with respect to the vertical are proportional to the velocities." It is clear that at the point C the sine of the angle of refraction is $dy/dz = mn/Mm$ and that this must be proportional to the velocity of the light at this point, i.e., to $CH = t$. Hence

$$\frac{dy}{dz} = \frac{t}{a},$$

where a is a proportionality factor. He rewrites this equation in the form

$$\frac{dy}{dx} = \frac{t}{\sqrt{a^2 - t^2}}, \tag{1.33}$$

using the fact that $dz^2 = dx^2 + dy^2$. This is his fundamental differential equation in which it remains for him to relate t and x. He says that[48]:

> I have with one blow solved two fundamental problems, one optical and the other mechanical and have accomplished more than I have asked of others: I have shown that the two problems, which arose from totally different fields of mathematics, nevertheless possess the same nature.
>
> Now we consider a special case, namely the usual hypothesis, that was first introduced and proved by Galileo, according to which the velocity of heavy falling bodies are to each other as the square roots of traversed heights; this then is precisely the case. Under this hypothesis the curve AHE is the parabola $tt = ax$ and $t = (ax)^{1/2}$; if this is substituted into the general equation, there results
>
> $$dy = dx\sqrt{\frac{x}{a-x}} \tag{1.34}$$
>
> from which I conclude that the *Brachystochrone* curve is the *common* Cycloid. If indeed the circle GLK of diameter $= a$ rolls along AG and starts at A, then the point K describes a cycloid which has for its differential equation
>
> $$dy = dx\sqrt{\frac{x}{a-x}}$$
>
> when AC is set to x and CM to y.

Bernoulli now goes to show that this differential equation actually leads

[48] John Bernoulli, *OO*, Vol. I, p. 191. This is reference *CR*.

to the cycloid. To do this, he writes

$$dx\sqrt{\frac{x}{a-x}} = \frac{x\,dx}{\sqrt{ax-x^2}} = \frac{1}{2}\,\frac{a\,dx}{\sqrt{ax-x^2}} - \frac{1}{2}\,\frac{a\,dx-2x\,dx}{\sqrt{ax-x^2}}\,.$$

He then goes on to say[49]:

> However $(a\,dx - 2x\,dx) : 2(ax - xx)^{1/2}$ is the differential quantity whose integral is $(ax - xx)^{1/2}$ or LO; and $a\,dx : 2(ax - xx)^{1/2}$ is the differential of the arc GL. For this reason by integrating the equation $dy = dx(x/(a-x))^{1/2}$ one has $y = CM = [\text{arc}]GL - LO$, and therefore $MO = CO - [\text{arc}]GL + LO$. Since however $CO - [\text{arc}]GL = [\text{arc}]LK$ (because CO = semicircle GLK), it follows that $MO = [\text{arc}]LK + LO$ and by deducting the common value LO, that $ML = [\text{arc}]LK$, which implies that the curve KMA is a cycloid.

Bernoulli next goes on to show, as did Newton, that given two points A and B there always is a cycloid through those points. Bernoulli exclaimed that "Nature tends always to proceed in the simplest way" when he noted that his brachystochrone and Huygens's tautochrone are the same curve.[50] He continues and notes that when the velocity of the body is not proportional to the square root of the distance fallen but to the cube root, then the brachystochrone is algebraic and the tautochrone transcendental; "nevertheless if the velocity is directly proportional to the distance, both will be algebraic: the one a circle, the other a straight line."[51]

In the first case Bernoulli has chosen $t = ax^{1/3}$, and his differential relation (1.33) for the brachystochrone becomes

$$\frac{dy}{dx} = \frac{x^{1/3}}{\sqrt{1-x^{2/3}}}\,.$$

From this we find by integrating that

$$y - c = -(2 + x^{2/3})(1 - x^{2/3})^{1/2}$$

and a little manipulation shows this is expressible as an algebraic equation in x and y of degree 6.

---

[49] John Bernoulli, *OO*, Vol. I, p. 191. To see that this last property defines a cycloid, let $b = a/2$, arc $LK = b\varphi$, and arc $GL = b\pi - b\varphi = bt$. Then $\sin t = \sin\varphi = LO/b$ and $CO = AG = b\pi = y + ML + LO = y + \text{arc } LK + b\sin t$; hence $y = b\pi - \text{arc } LK - b\sin t = b\pi - b\varphi - b\sin t = b(t - \sin t)$. To find x, note that $\cos t = -\cos\varphi = -(b - OK)$ and that $x = AC = GK - OK = 2b - b - b\cos t = b(1 - \cos t)$. (Note that b is the radius of the generating circle.)

[50] Stäckel, in a remark appended to his translation (p. 137, note 8), mentions that this idea of Bernoulli was a pet of eighteenth century thinkers, and he refers the reader to a discussion in Mach [1933].

[51] John Bernoulli, *CR*, p. 192.

In the second case Bernoulli has chosen $t = ax$, and (1.33) becomes for the brachystochrone

$$\frac{dy}{dx} = \frac{x}{\sqrt{1 - x^2}}$$

so that the curve is $x^2 + (y - c)^2 = 1$, a circle as Bernoulli stated.

In the last two paragraphs of his paper Bernoulli reintroduced a concept he had examined in connection with Huygens's theory of light in 1693. In effect, what he did was to find a curve orthogonal to each member of a sheaf or family of curves emanating from a given point. (This presages Kneser's concept of transversality.) Thus in effect he found wave-fronts. In his Figure 1.20 the plane of the paper is vertical, and the curves AB represent the one-parameter family of cycloids passing through the point A. The points B represent the places where heavy particles starting simultaneously from A under the action of gravity would arrive in a given fixed time T. Bernoulli called the curve PB, which is the locus of these points B, a *synchrone*. He shows there is such a curve and says it cuts each cycloid of the family in a right angle at B.

He first poses the problem in this way: "To find the curve which cuts all common Cycloids through a common point normally [at right angles]." He says the problem so phrased is very hard for geometers. However, if it is viewed as the locus of heavy particles descending as we described above, then it is very easy. His construction is ingenious. At the time T the particle falling vertically from A will have reached a point P and another, some point B on the cycloid ABK generated by the circle GLK as shown in Figure 1.20. Bernoulli says that the arc GL is the mean proportional between AP and $GK = 2a$. But the arc GL is the quantity $a\varphi$, where φ is given by the parametric relations (1.28″), (1.28‴); and hence Bernoulli's relation is that

$$a\varphi = \sqrt{2a} \cdot \sqrt{\frac{1}{2} g} \cdot T,$$

since $AP = gT^2/2$; i.e.,

$$\sqrt{g}\, T = \sqrt{a}\, \varphi$$

which, as we shall see shortly, is precisely the correct condition. He lays off the arc GL on the circle and draws a line through the point L parallel to

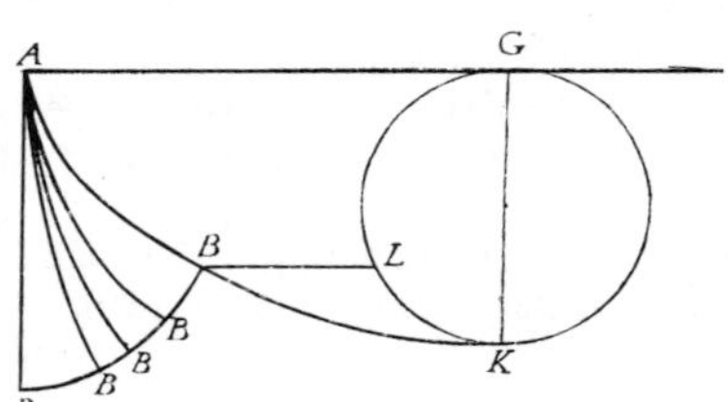

Figure 1.20

AG; the point B where it cuts ABK is the desired point on the locus. Stäckel in a note on pp. 137ff of his translation gives what he terms "The proof in the manner of Johann Bernoulli." In a given fixed time-interval dT it is clear that a particle falling down AP will move so that

$$\frac{dX}{\sqrt{X}} = \sqrt{2g}\cdot dT,$$

where $X = AP$, by Galileo's hypothesis. Moreover, in the same time-interval we know by (1.24) that the particle at $B:(x, y)$ will move along any one of the cycloids (1.28″), (1.28‴) with $x_1 - k = 0$ and $2r = a$ so that

$$dT = \frac{1}{\sqrt{2g}}\,\frac{ds}{\sqrt{x}}\,.$$

Then equating these two values of dT, Stäckel finds that

$$\frac{dX}{\sqrt{X}} = \frac{ds}{\sqrt{x}} = \sqrt{\frac{a}{ax - x^2}}\;dx,$$

or, equivalently, if $u = aX$,

$$\frac{1}{2}\,\frac{du}{\sqrt{u}} = \sqrt{\frac{x}{a - x}}\;dx + \frac{\frac{1}{2}a - x}{\sqrt{ax - x^2}}\;dx.$$

He points out that this means the differential of $u^{1/2}$ is the sum of the differentials of CM and LO in Figure 1.19. (In fact, $CM = y$ and his statement about dy follows at once from the differential equation (1.30) and the value of LO we calculated earlier (on p. 41). Thus it follows that $(AP\cdot GK)^{1/2} = CM + LO = \text{arc } GL$, as Bernoulli asserted.[52]

To determine the analytical condition that a curve $C:\xi = \xi(a)$, $\eta = \eta(a)$ cuts the family of cycloids through the point $A:(x_1, y_1)$ at the points for which the time of fall is a constant T, let us suppose that the points of intersection are given by $\varphi = \varphi(a)$. Then with the help of relations (1.28″) and (1.28‴) with $r = 2a$, we can find the slope $d\eta/d\xi = (\partial\eta/\partial a)/(\partial\xi/\partial a)$. It is easy to see that

$$\frac{d\eta}{d\xi} = \frac{(\varphi - \sin\varphi) + a(1 - \cos\varphi)\varphi_a}{(1 - \cos\varphi) + a\sin\varphi\cdot\varphi_a}\,.$$

This must be the negative reciprocal of dy/dx, the slope of the cycloid. This implies directly that

$$\frac{d\varphi}{da} = -\frac{1}{2a}\,\varphi$$

or $\varphi = c/a^{1/2}$.

[52] Curiously, Struik says that Bernoulli noted that his synchrone is a cycloid. I cannot find this assertion in the original text, nor is it true.

Now if we evaluate our integral (1.24) along the family of cycloids, we find

$$\sqrt{g}\,T = \sqrt{2a}\,\varphi,$$

and hence we see that the orthogonality of the curve C to the family of cycloids is equivalent to the fact that the points of intersection correspond to a constant time of descent.

It is easy to see that Bernoulli's synchrones are transversal to his cycloids. In this same connection it is important to recall that John Bernoulli discovered geodesics on surfaces and discussed this subject in a 1698 letter to Leibniz. He remarked there that "geodesics have always the property of possessing osculating planes which cut the surface at right angles, and that this property leads to the invention of a differential equation for these curves" (see Carathéodory [1937], p. 96). Bernoulli proposed to Euler in 1728 that he study this problem, whose solution Euler presented to the Academy of St. Petersburg in 1732—it is dated November 1728, but Eneström showed it must have been done after April 1729.

1.8. James Bernoulli's Solution

The solution as given by James Bernoulli to his brother's problem is quite different and is a good deal more like Newton's solution to the least resistance problem; moreover, it probably influenced Euler in developing his earlier techniques—until, in fact, Euler saw Lagrange's superior method of variations at which point Euler shifted over and renamed the subject the calculus of variations.[53]

In the preface to his paper James Bernoulli says that he was persuaded to solve his brother's problem by a letter on 13 September 1696 from Leibniz urging him to work on it. Leibniz went on to say that he, Leibniz, had already solved the problem. James Bernoulli says he attained his solution on 6 October 1696.

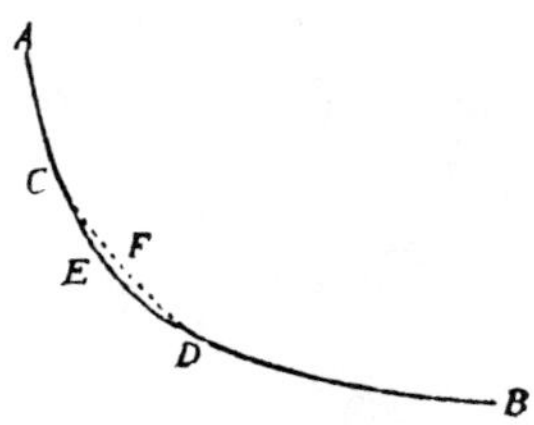

Figure 1.21

[53] James Bernoulli, *JB*, pp. 768–778 or Stäckel, *KLASSIKER*, No. 46, pp. 14–20.

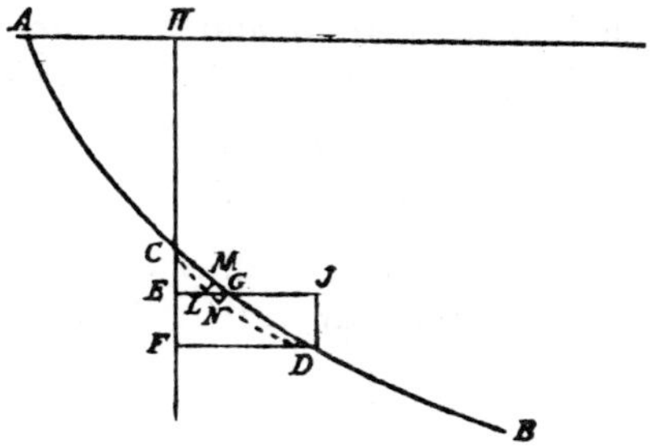

Figure 1.22

First, Bernoulli shows that if an arc AB joining A to B is a minimizing arc, then so also is any subarc CED joining C to D in Figure 1.21. For if another subarc CFD joining the same points gave a lesser value, then the arc $ACFDB = AC + CFD + DB$ would be an arc along which the time of descent was less than on $ACEDB$, which is a contradiction.

Second, Bernoulli proceeds much as did Newton, as we shall see with the help of Figure 1.22. At the arbitrary point C on the minimizing arc he has drawn HF through C perpendicular to AH, the ordinate of C. The point D is "near to" C and $CE = EF$; EJ is parallel to AH and FD; L on EJ is such that LG is the differential of EG; and $EFDJ$ is a parallelogram. Since the arc CGD is the minimizing arc through C and D, then as we saw earlier (on p. 21),

$$tCG + tGD = tCL + tLD,$$

where, for instance, tCG means the time to descend from C to G along the arc CG (this equality is good through terms that are to be retained in the limit), and so

$$tCG - tCL = tLD - tGD. \tag{1.35}$$

Now to evaluate these times, Bernoulli states that

$$\frac{CE}{CG} = \frac{tCE}{tCG}, \qquad \frac{CE}{CL} = \frac{tCE}{tCL}.$$

To see why this is so, notice that $dt = ds/v$, where s is distance fallen, v is velocity, and $v^2 = v_1^2 + 2gs$. Thus

$$tCE = \frac{CE}{\sqrt{v_1^2 + 2g \cdot HC}}, \qquad tCG = \frac{CG}{\sqrt{v_1^2 + 2g \cdot HC}}, \tag{1.36}$$

with a similar expression for tCL.

As a consequence he now can assert that

$$\frac{CE}{CG - CL} = \frac{tCE}{tCG - tCL}.$$

He draws LM normal to CG and has $CL = CM$ approximately since

$$CL \sim CM + \frac{LM^2}{2CG};$$

he also notes that the triangles MLG and CEG are similar. This enables him to conclude that $MG/GL = EG/CG$ ($s''/y' = y''/s'$) and hence that

$$\frac{CE}{GL} = \frac{EG \cdot tCE}{CG \cdot (tCG - tCL)}\ .$$

He then turns to the interval EF, draws GN normal to DL so that $DG = DN$ and, by reasoning similar to that above, concludes that

$$\frac{EF}{GL} = \frac{GJ \cdot tEF}{GD \cdot (tLD - tGD)}\ .$$

Combining these ratios ($EF = CE$), he finds with the help of (1.35) that

$$\frac{EG \cdot tCE}{GJ \cdot tEF} = \frac{CG(tCG - tCL)}{GD(tLD - tGD)} = \frac{CG}{GD}\ .$$

He then asserts that by the law of fall for heavy bodies,

$$\frac{EG \cdot tCE}{GJ \cdot tEF} = \frac{EG/\sqrt{HC}}{GJ/\sqrt{HE}}\ .$$

[To see this, recall relations (1.36), set $v_1 = 0$, and calculate by the same means $tEF = EF/(2gHE)^{1/2}$.] Bernoulli now concludes that

$$\frac{EG/\sqrt{HC}}{GJ/\sqrt{HE}} = \frac{CG}{GD}\ . \tag{1.37}$$

If we write $CG = ds = (dx^2 + dy^2)^{1/2}$, $HC = x$, and $EG = dy$, then this relation of Bernoulli's says that

$$ds = \frac{k}{\sqrt{x}}\, dy,$$

which is exactly Leibniz's defining relation (1.29). Squaring both members of this relation, we find readily that

$$\frac{dy}{dx} = \sqrt{\frac{x}{k^2 - x}}\ ,$$

which is our differential equation (1.27) for the cycloid with $k^2 = 2a$, the diameter of the generating circle.

Bernoulli now says "one seeks to determine the curve whose elements of arc-length are directly proportional to the elements of abscissa and inversely as the square root of the ordinates. I determined that this property of Huygens's Isochrone . . . is the same one that belongs to the cycloid, well known to geometers. I go to show that this is so." He does this geometrically with the help of Figure 1.23. The proof is not complex, nor does it deepen our understanding of the calculus of variations, so I shall omit it. He also gives a proof that there always is a cycloid passing through the points A and B.

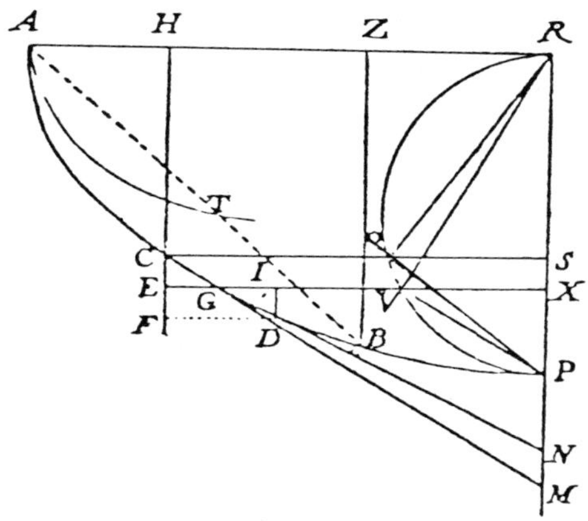

Figure 1.23

1.9. James Bernoulli's Challenge to His Brother

After setting out his proof in detail, James Bernoulli turned to three other problems in the paper that we have been considering. Among these are two very significant ones, both of which served to open still further the field since they involved new ideas. More specifically, what he proposed was to find, among all cycloids through the point A with base AH and intersecting a given vertical ZB, the one down which a heavy particle will fall in the shortest possible time.

His second problem, which is in two parts, is also remarkable for reviving the ancient notion of an isoperimetric condition. (Thus, for example, he points out that among all plane figures of a given perimeter the circle contains the greatest area.) He then proposes the problem, in terms of Figure 1.24, "to find, among all isoperimetrical figures on the base BN, that one BFN which, to be sure, will not enclose the greatest area, but is such that a related curve BZN has this property; its ordinate PZ should be proportional to a power or root of the segment PF or of the arc BF."

As a goad, James said in his May 1697 paper that an unnamed gentleman would offer a prize of 50 ducats for his brother's solutions if John acknowledged within 3 months that he had accepted the challenge and that by the end of the year he had exhibited his solutions by means of quadratures. However, if no one had solved them by year's end he, James, would publish his solutions.

We may infer from this challenge that James Bernoulli had understood how his method of attack on the brachystochrone problem was capable of extension to a considerable class of problems. We will see later that this was indeed so.

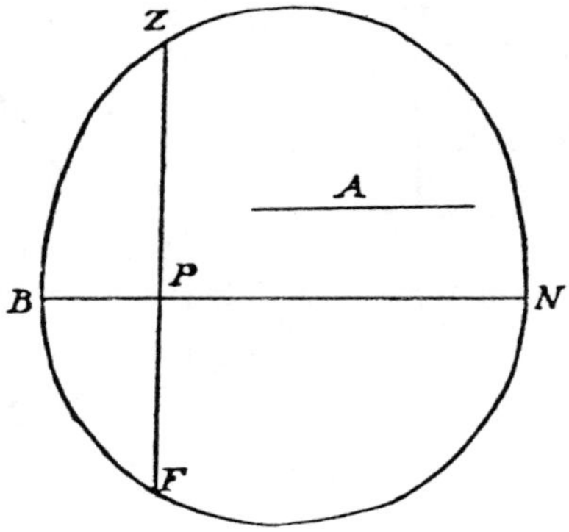

Figure 1.24

In June 1697 John published as a paper the letter to M. Basnage mentioned earlier.[54] In this letter he quite boastfully says that "in place of the three months given me to 'sound the depth of the river' and in place of the balance of the year to achieve the solution, I did not spend more than three minutes in probing the mystery and even going well beyond." He then proposes the more general problem in which PZ is an arbitrary function of PF.[55]

He goes on to say that "As regards the other problem in which one asks among all cycloids which start from the same point and the same horizontal base, that one down which a heavy body arrives in the least time at a given vertical line; it is true that this is properly the problem for whose solution the generous 'no name' promised me the prize of fifty silver ducats (ecus blancs) . . . . Since I also found which cycloid one should take, if in place of the vertical line, one takes any given oblique curve." He proceeds to say that he has sent his solutions to Leibniz and asked him to serve as their judge.

In August 1697 he then himself issued a list of six problems to be solved which widened yet further the class of special cases that had been considered.[56] On 15 October of the same year, probably to fulfill the conditions of the challenge by his brother James, he sent off a letter to a M. Varignon which appeared in print in December.[57] In there he restates his brother's first problem and then proceeds to solve only the first part; that is, he takes up only the case in Figure 1.25 when an arbitrary ordinate PZ is a power n of the corresponding ordinate PF, but he does not consider when it depends on the corresponding arc BF, which is really the key point.

[54] John Bernoulli, *LE*, pp. 194–204.

[55] John Bernoulli, *LE*, p. 202. Notice that John Bernoulli does not discuss the really difficult problem—in fact, it was to be his nemesis—the case when PZ depends on the arc BF.

[56] John Bernoulli, *PA*, pp. 204–205.

[57] John Bernoulli, *SI*.

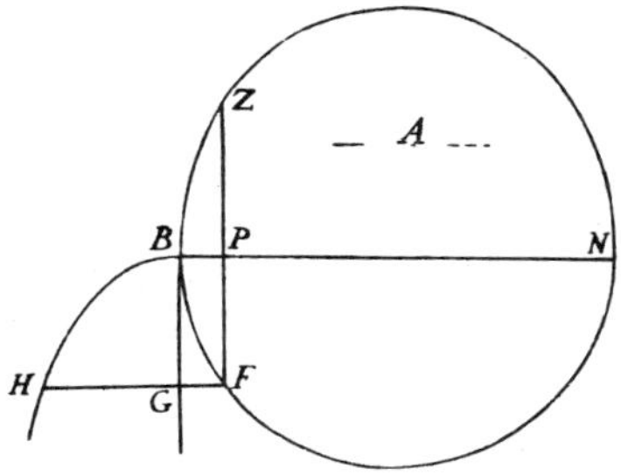

Figure 1.25

He states his result in terms of $PF = BG = x$ and $BP = GF = y$ and says that

$$GF = y = \int \left(x^n \, dx : \sqrt{a^{2n} - x^{2n}} \right)$$

is the solution curve. He goes on to discuss the cases $n = 1/3, 1/2, 1, 2$ and says that when n is a fraction with numerator 1, then for odd denominators the solution is algebraic and for even ones the solution "can always be constructed by the quadrature of a circle." He then generalizes to the case where $PZ = GH$ is an arbitrary function of PF. To do this, he sets

$$b = \int \frac{GH \, dx}{x}$$

and asserts that the solution curve is now expressible as[58]

$$y = \int \frac{b \, dx}{\sqrt{a^2 - b^2}} .$$

From this he attempts largely by rhetoric to cover the case where PZ is a function of the corresponding arc BF. He gives no proof or hint of a proof but merely asserts that this case is covered.

He also gives his solution to his brother's problem of finding among all the cycloids through a given point and with the same horizontal base the one down which a heavy particle will fall most quickly to a given vertical line. He repeats his assertion that he should receive the prize for the solution of this problem. He remarks that the solution follows as a simple corollary to the brachystochrone problem he solved in May 1697. The solution curve, he says, is the cycloid generated by the circle whose circumference is equal to twice the distance between the origin and the given vertical line. But he indicates that he can go further. "If in place of the vertical one supposes any line whatever or any curve, the problem does not become more difficult; since it is clear by the nature of my Synchrone

[58] A few months later in a subsequent paper John Bernoulli corrected this to read "*I call b the ordinate GH*" instead of the definition above. See John Bernoulli, *SI*, p. 210 and *PD*, p. 217.

curve that the desired cycloid will always be that one which interesects the given curve at right angles."[59]

James Bernoulli replied to this paper in a notice in the same journal, *Journal des Savans*, where he wrote that he would undertake to examine his brother's solution, indicate its contradictions, and give the correct solution.[60] There was then a considerable correspondence on this problem which finally culminated in the publication of a note by James Bernoulli in the *Acta Eruditorum* of June 1700 entitled "The solution of isoperimetrical problems". This note was followed in May 1701 by a longer and fuller paper, "Analysis of the large isoperimetrical problem," in the same journal.[61] Perhaps a remark by James about his brother's reasoning might be worth repeating. He comments that his brother argues in this way: "Every man is a stone; every pebble is a man; hence every pebble is a stone."[62] It does not seem to be relevant to discuss this series of papers in detail. Suffice it to say that John's counter to James's papers of 1700 and 1701, just mentioned, appeared in 1706. However, it had been sent sealed to the French Academy on 1 February 1701 by Varignon with the request that it not be opened until after James Bernoulli's solution should appear. For whatever reasons, it did not appear in print until after James's death on 16 August 1705. There is also a long paper by John Bernoulli on the subject which appeared in 1718.[63] This latter paper by John is in part clearly dependent on James's work in 1701. It is discussed in part in Section 1.11.

1.10. James Bernoulli's Method

The March 1701 paper by James Bernoulli contains the basic ideas which his brother was to take up in his 1718 opus and which Euler was to transform into his standard systematic procedure even later. (Incidentally,

[59] John Bernoulli, *SI*, p. 211. In a letter to Basnage (John Bernoulli, *LE*, p. 194), John remarks on the deep relation between the brachystochrone problem and some problems discussed by Huygens in his *Traité de la Lumière*. This is extraordinary and was not again noted until Hamilton's elegant work.

[60] John Bernoulli, *PD*, p. 214 XV.

[61] See John Bernoulli, *SD* and *OO*, Vol. II, pp. 214–234 as well as James Bernoulli, *SO*, pp. 874–887 and *MP*, pp. 897–920. In these papers the brothers also solve the isoperimetrical problem of the shape of the chain hanging under the action of gravity so that its center of gravity will be the lowest possible and show that a cloth filled with liquid will assume the same shape: a catenary. In James Bernoulli's 1701 paper (James Bernoulli, *MD*, p. 914; this is his Problem III) appears the problem of finding among all heavy curves of given length joining two fixed points and hanging freely to find the one with the lowest center of gravity. Bernoulli shows it to be the catenary.

[62] "*Tout homme est pierre; Tout caillou est homme; Donc tout caillou est pierre.*" See John Bernoulli, *OO*, Vol. I, p. 227 and James Bernoulli, *O*, Vol. II, p. 836.

[63] John Bernoulli, *SD* and *RE*.

it is interesting to note that James's paper is dedicated to his four mathematical heroes: L'Hôpital, Leibniz, Newton, and de Duillier!) In this paper we can see James consciously working toward a general theory. The key point he realized, which his brother had not, is that the presence of an isoperimetric condition required another degree of freedom. Thus he permitted and, indeed, required two points on the minimizing arc to vary slightly in ways that we shall note. It is worth remarking that Euler's method was much more systematic and deeper than was the Bernoullis', as we shall see in Chapter 2.

The 1701 paper is entitled "Analysis of the large isoperimetrical problem," and in the preface Bernoulli says that he calls a problem large, not because of its inherent difficulty, but because it extends the limits of science. In any case in this paper he sets up the general apparatus he needs to solve the two isoperimetrical problems he posed his brother (see Section 1.9) and also the isoperimetrical problem of finding the curve of lowest center of gravity. In addition, he attempts unsuccessfully to argue sufficient conditions; that is, he tries in both his 1700 and 1701 papers to reason that his arcs necessarily provide maxima or minima. However, in 1718 John gives an elegant proof of sufficiency (see Section 1.11 below).

James's 1701 paper is somewhat pedantic and tedious, but he was perhaps trying to lay the groundwork for the subject of the calculus of variations in a systematic way. In any case John's paper in 1718 is a more readable account of James's work.

In accord with the isoperimetric condition (Fig. 1.26), James assumes that the arc $BFGC$ is of constant length while he allows the two points F

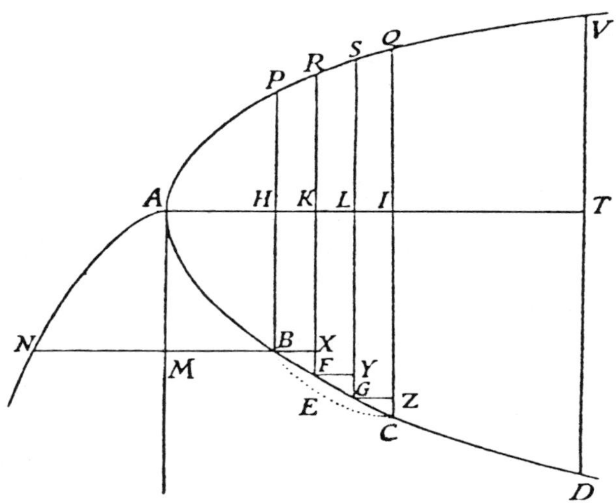

Figure 1.26

and G to vary up or down along the vertical lines KF and LG. He sets

$$\begin{array}{lll} BX = l & FX = p & BF = s \\ FY = m & GY = q & FG = t \\ GZ = n & CZ = r & GC = u \\ \text{and also} & & \text{and thus} \\ HB = b & & \\ KF = f = b + p & & df = dp \\ LG = g = b + p + q & & dg = dp + dq. \end{array}$$

From this he concludes that

$$-\frac{df}{dg} = \frac{rst - qsu}{qsu - ptu}$$

To derive this result, James Bernoulli notes that his hypotheses require that

$$\begin{array}{ll} BX^2 + FX^2 = BF^2, \quad \text{i.e.,} & ll + pp = ss \\ FY^2 + GY^2 = FG^2 \ldots & mm + qq = tt \\ GZ^2 + CZ^2 = GC^2 \ldots & nn + rr = uu \\ BX + FY + GZ = \text{const.} & l + m + n = \text{const.} \\ FX + GY + CZ = \text{const.} & p + q + r = \text{const.} \\ BF + FG + GC = \text{const.} & s + t + u = \text{const.} \end{array}$$

and further that $dl = dm = dn = 0$. From these it follows easily by a little calculation that

$$\begin{array}{llll} \text{I.} & p\,dp = s\,ds & & \\ \text{II.} & q\,dq = t\,dt & \text{V.} & dp + dq + dr = 0 \\ \text{III.} & r\,dr = u\,du & \text{VI.} & ds + dt + du = 0, \end{array} \tag{1.38}$$

and by eliminating dr, ds, dt and du that

$$\frac{dp}{dp + dq} = \frac{rst - qsu}{ptu - qsu}. \tag{1.38'}$$

But $dp = df$ and $dp + dq = dg$, and hence his conclusion follows. This is James's Theorem II.

This gives James his first condition. His brother John in his 1718 paper calls this his (John's) *fundamental equation*.[64] John remarks in his paper that his brother had "demonstrated [this condition] by a very tedious calculation," whereas he reached the conclusion without any. (We will see John's analysis in Section 1.11.)

James then proceeds in his Theorem IV to evaluate relation (1.38′) in terms of differentials of $x = HB$, $y = AH$, and $z = AB$, where these are, respectively, the ordinate, the abscissa, and arc-length at the point B on the

[64] John Bernoulli, *RE*, p. 239.

curve. He finds the ratio to be

$$\frac{dz^2d^2x + dz^2d^3x - dx(d^2x)^2}{dz^2d^2x + 2dx(d^2x)^2}.$$

This calculation is reasonably straightforward and requires mainly the facts that

$$\begin{array}{lll} BX = \quad l = dy & FY = m = dy & GZ = n = dy \\ FX = \quad p = dx & GY = \ q = dx + d^2x & GZ = r = dx + 2d^2x + d^3x \\ BF = \quad s = dz & FB = \ t = dz + d^2z & GC = u = dz + 2d^2z + d^3z \\ dx^2 + dy^2 = dz^2. & & \end{array}$$

In that calculation he also needs to eliminate certain terms as being "infinitesimal" with respect to the others. He finds that

$$rst - qsu = dz^2\,d^2x + dz^2\,d^3x - dx\,dz\,d^2z - dx\,dz\,d^3z,$$

$$qsu - ptu = dz^2\,d^2x + 2dz\,d^2x\,d^2z - dx\,dz\,d^2z - 2dx\,(d^2z)^2$$

and uses the fact that $dx^2 + dy^2 = dz^2$, $dx\,d^2x = dz\,d^2z$, and further that $dx\,d^3x + (d^2x)^2 = dz\,d^3z + (d^2z)^2$, since he considers y as the independent variable and hence $d^2y = 0$. From this he finds that

$$rst - qsu = dy^2\left[d^2x + d^3x - \frac{dx(d^2x)^2}{dz^2}\right]$$

and similarly

$$qsu - ptu = dy^2\left[d^2x + \frac{2dx(d^2x)^2}{dz^2}\right].$$

This allows him to conclude with the help of Theorem II that

$$-\frac{df}{dg} = \frac{dz^2d^2x + dz^2d^3x - dx(d^2x)^2}{dz^2d^2x + 2dx(d^2x)^2}.$$

This is his Theorem IV.

Now James's Problem I is as follows (see Figure 1.26). Given that the lines AT and AM are normal to each other and the curve AN is "arbitrary," seek among all curves on the base AT of a given length and joining the fixed points A and D the one ABD such that at each point B the ordinate $MN = HP$ and that the area ATV under the curve APV is the largest possible. Recall that this was James's first challenge to his brother. To solve the problem, he uses the notations of before and sets $IC = c$. He uses capital letters for lines related to the curve APV; thus $HP = B$, $KR = F$, $LS = G$, and $IQ = C$. He shows in his Theorem VII that if an arc is an extremal, then so is every subarc, and so he has

$$HK \times HP + KL \times KR + LI \times LS = lB + lF + lG$$

a maximum or minimum; i.e., since B is regarded as fixed,

$$dF + dG = 0. \tag{1.39}$$

But in Bernoulli's notation, which is somewhat confused at this point, $F = KR$ is a function of $x = f = HB$ and $G = LS$ of $g = LG = f + df$; and $df = FX = p$, $dg = CZ = r$. In fact, G and F are the same function evaluated at two different points; the latter at f and the former at $f + df$ in the sense that

$$G(g) = F(f) + F'(f)\,df = F(f) + dF. \tag{1.40}$$

Thus Bernoulli's relation (1.39) really states that

$$F'(f)\,df + G'(g)\,dg = 0 \tag{1.41}$$

and

$$-\frac{df}{dg} = \frac{F'(f) + F''(f)\,df}{F'(f)},$$

as we see from (1.40) and (1.41). By his Theorem IV, this gives him the relation

$$-\frac{df}{dg} = \frac{dz^2\,d^2x + dz^2\,d^3x - dx(d^2x)^2}{dz^2\,d^2x + 2dx(d^2x)^2} = \frac{h + dh}{h}$$

where he has set (a is an arbitrary constant he has introduced) $h \equiv aF'(f) = aF'(x)$. After a little calculation this reduces to $h\,dz^2\,d^3x - 3h\,dx(d^2x)^2 = dh\,dz^2\,d^2x$ since $dh\,dx(d^2x)^2$ vanishes in respect to the other quantities. In this he replaces $dx\,d^2x$ by $dz\,d^2z$ since he has fixed dy and finds the equation

$$h\,dz^2\,d^3x - 3h\,dz\,d^2x\,d^2z = dh\,dz^2\,d^2x.$$

To solve this equation, he considers an equation involving parameters m, n, r: $h^m\,dz^n(d^2x)^r = \text{const.}$ He differentiates both members and compares the result to the equation above. He finds $m = -1$, $n = -3$, and $r = 1$ and thus has as a first integral the relation

$$\frac{d^2x}{h\,dz^3} = \text{const.} = \frac{\pm 1}{a^2\,dy}.$$

(He remarks that dy is constant and that the result follows by "the law of homogeneity," i.e., by dimensional analysis.) If the functions h, x, and z are viewed as functions of y, then his relation above states that

$$\frac{d^2x/dy^2}{h(dz/dy)^3} = \frac{\pm 1}{a^2}.$$

With Bernoulli, let $t = a\,dx/dy$ and note that $a^2(dz/dy)^2 = (a^2 + t^2)$ and

$$dz = \frac{\sqrt{a^2 + t^2}}{a}\,dy;$$

Since $h = a\,dF/dx$, $dt/dy = a\,d^2x/dy^2$, and $t\,dy = a\,dx$, he has

$$\pm\frac{1}{a^2} = \frac{a^2(dt/dy)}{h(a^2 + t^2)^{3/2}} = \frac{at\,dt/dx}{h(a^2 + t^2)^{3/2}} = \frac{t\,dt/dx}{(a^2 + t^2)^{3/2}\,dF/dx}.$$

From this he can infer that

$$dF = \pm\frac{a^2t\,dt}{(a^2 + t^2)^{3/2}}$$

or

$$p \equiv F = b \pm \frac{a^2}{(a^2 + t^2)^{1/2}}$$

and thus

$$\frac{dx}{dy} = \frac{\pm 1}{b - p}\sqrt{a^2 - (b - p)^2}\,.$$

(Bernoulli considers in some detail the case $b = a$.) This is his defining differential equation for x as a function of y. Recall that $p = F$ is a given function of x.

James Bernoulli's Problem II is to find among all curves of the same length on the base AT passing through the fixed points A and D that one ABD such that at the arbitrary point B the ordinate HP of the corresponding curve APV is a given function of the arc AB. To solve this problem, he introduces some definitions for various arcs: $AB = \beta$, $AF = AB + BF = \beta + s = \varphi$, and $AG = AF + FG = \beta + s + t = \gamma$. He notes that $d\varphi = ds$ and $d\gamma = ds + dt$. He also sets $HB = B$, $KR = \Phi$, and $LS = \Gamma$. To maximize the area ATV, he notes it is necessary that

$$HK \times HP + KL \times KR + LI \times LS = lB + l\Phi + l\Gamma$$

be a maximum. Thus $d\Phi + d\Gamma = 0$, and he sets $d\Phi = h\,d\varphi/a$ and $d\Gamma = i\,d\gamma/a$. He then has $h\,d\varphi + i\,d\gamma = 0$ or $h\,ds + i\,ds + i\,dt = 0$. But $p\,dp = s\,ds$ and $q\,dq = t\,dt$ from equations (1.38). Hence Bernoulli has the relation $hpt\,dp + ipt\,dp + iqs\,dq = 0$, and so

$$\frac{dp}{dq} = -\frac{iqs}{hpt + ipt}$$

or

$$\frac{df}{dg} = \frac{dp}{dp + dq} = \frac{iqs}{hpt + ipt - iqs}.$$

After some considerable calculation, similar to what he did before, he concludes that $h\,dx\,dz^2\,d^3x = 2h\,dz^2(d^2x)^2 + h\,dx^2(d^2x)^2 + dh\,dx\,dz^2\,d^2x$; and, as before, he obtains the results

$$dy = \frac{q\,dz}{\sqrt{a^2+q^2}} \quad \text{or} \quad dy = \frac{(a-q)\,dz}{\sqrt{b^2-2aq+q^2}},$$

where q plays a role analogous to p above.[65]

In concluding this paper, James considers his third problem of finding the shape of the heavy curve of given length whose center of gravity hangs the lowest. To analyze this problem, he needs to consider two theorems which he calls Theorems III and V. In Figure 1.26 he now allows his variable points F and G to move so that the distances BF, FG, and GC remain fixed in length. Thus F moves on a circle with center at B and radius BF and G on one with center at C and radius GC.

In his previous analysis he fixed $BX = l$, $FY = m$, and $GZ = n$ while $BF = s$, $FG = t$, and $GC = u$ were variable. Now he has the opposite situation, and instead of the equations (1.38), he has

$$\begin{array}{llll} \text{I.} & l\,dl + p\,dp = 0 & \text{IV.} & dl + dm + dn = 0 \\ \text{II.} & m\,dm + q\,dq = 0 & \text{V.} & dp + dq + dr = 0. \\ \text{III.} & n\,dn + r\,dr = 0 & & \end{array}$$

From these he calculates straightforwardly in his Theorem III that

$$-\frac{df}{dg} = \frac{lmr - lnq}{lnq - mnp}, \tag{1.42}$$

where as before df is the variation in KF and dg that in LG.

After a calculation which is tedious but not difficult, James Bernoulli finds as before that

$$lmr - lnq = dz^2\,d^2x + dz^2\,d^3x + \frac{dx\,dz^2(d^2x)^2}{dy^2},$$

$$lnq - mnp = dz^2\,d^2x - \frac{2dx\,dz^2(d^2x)^2}{dy^2},$$

or, equivalently,

$$lmr - lnq = \frac{dz^2\left[dy^2\,d^2x + dy^2\,d^3x + dx(d^2x)^2\right]}{dy^2}$$

$$lnq - mnp = \frac{dz^2\left[dy^2\,d^2x - 2\,dx(d^2x)^2\right]}{dy^2}. \tag{1.42$'$}$$

---

[65] James Bernoulli, *MP*, pp. 915–916.

(John Bernoulli also derives this result, as we shall see later, and calls this his other fundamental equation.)

James now returns to his Figure 1.26 and, as before, lets $HB = b$, $KF = f$, $LG = g$, and the arc $AB = z$; he further supposes that the weight of z is q—recall that the chain is heavy—and notes that the weight of BF is then dq, of FG is $dq + d^2q$, and of GC is $dq + 2d^2q$ (he says that he omits d^3q "since it is superfluous"). Then the moment of this mass with respect to the AT axis is

$$b\,dq + f(dq + d^2q) + g(dq + 2d^2q). \tag{1.43}$$

He now allows F and G to vary on arcs of circles, as above, and observes that under this assumption quantity (1.43) is a minimum, i.e.,

$$df(dq + d^2q) + dg(dq + 2d^2q) = 0;$$

or by relations (1.42) and (1.42′),

$$-\frac{df}{dq} = \frac{dq + 2\,d^2q}{dq + d^2q} = \frac{dy^2\,d^2x + dy^2\,d^3x + dx(d^2x)^2}{dy^2\,d^2x - 2\,dx(d^2x)^2}.$$

This reduces to

$$\frac{d^2q}{dq} = \frac{dy^2\,d^3x + 3\,dx(d^2x)^2}{dy^2\,d^2x}$$

on neglecting "superfluous quantities." Bernoulli succeeds in integrating this equation by considering $dq^m\,dy^n\,d^2x = \text{const.}$ and finds

$$\frac{d^2x}{dq\,dy^3} = \pm\frac{1}{a\,dz^2}.$$

(Recall in the present case that dz is constant and so $dy\,d^2y = -dx\,d^2x$.)

The integration of this equation is carried out by setting $a\,dy = t\,dz$ and noting that $a\,d^2y = dt\cdot dz$ since $d^2z = 0$ in this case. It follows readily that

$$a\,dy\,d^2y = -a\,dx\,d^2x = \frac{t\,dt\,dz^2}{a}$$

so that

$$d^2x = -\frac{t\,dt\,dz^2}{a^2\,dx} = \pm\frac{dq\;y^3}{a\,dz^2} = \pm\frac{t^3\,dq\,dz}{a^4}.$$

These relations then imply that

$$\frac{dq}{dt} = \mp\frac{a^2}{t^2}\cdot\frac{dz}{dx} = \mp\frac{a^3}{t^2\sqrt{a^2 - t^2}}.$$

(This follows since $t^2(dx^2 + dy^2) = a^2\,dy^2$ implies that

$$\frac{dy}{dx} = \frac{\pm t}{\sqrt{a^2 - t^2}}, \qquad \frac{dy}{dz} = \frac{t}{a}$$

and hence that

$$\frac{a^2\,dz}{dx} = \mp \frac{a^3}{\sqrt{a^2 - t^2}} .)$$

It now is clear that

$$(b - q) = \mp \frac{a\sqrt{a^2 - t^2}}{t}$$

or, equivalently, in Bernoulli's terms[66]

$$\frac{dy}{dz} = \frac{t}{a} = \frac{a}{\sqrt{a^2 + (b - q)^2}} .$$

1.11. John Bernoulli's 1718 Paper

The important point in James Bernoulli's paper is his realization that when an isoperimetric condition is appended to a simple problem in the calculus of variations, it is, in general, necessary to allow two ordinates to vary. John Bernoulli's early unsuccessful solutions are characterized by his attempts to use just one variable ordinate. John clearly understood his brother's ideas and in part made them the basis for his 1718 paper.[67] His arguments are more geometrical than James's but nonetheless are fundamentally the same even though he claims they are better.

Let us recreate John's arguments far enough to see how they compare to his brother's. He starts in Figure 1.27 with his Lemma 1 in which a, b, c and e are four points on the given curve and g and i are two others "infinitely near" to b and c on the vertical lines Pb and Rc, respectively. Further the isoperimetric condition implies that

$$ab + bc + ce = ag + gi + ie.$$

Then he concludes that

$$\left(\frac{fb}{ab} - \frac{kc}{bc} \right) \times bg = \left(\frac{kc}{bc} - \frac{le}{ce} \right) \times ci. \tag{1.44}$$

He uses this relation in his discussion of his brother's Problems I and II.

Bernoulli's proof is very simple. He draws the "little lines" bm, gn, io, and ch from the points b, g, i, and c perpendicular to the lines ag, bc, ie,

---

[66] Bernoulli considers in particular the case $b = 0$.

[67] John Bernoulli, *RE*. His argument on pp. 267–269, however, breaks new ground and in part gave Carathéodory ([1904], pp. 71–79) the insight for his elegant approach to the calculus of variations.

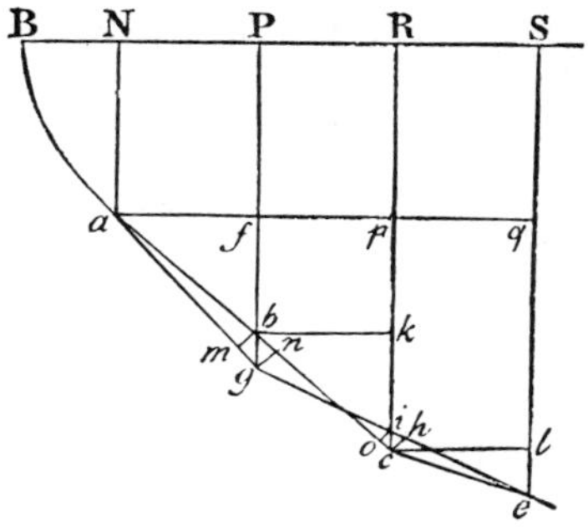

Figure 1.27

and ge and bk and cl parallel to aq. Then there are four pairs of similar triangles: gmb and bfa, bng and ckb, coi and ckb, ihc and elc; hence $gm = (fb \times bg)/ab$, $bn = (kc \times bg)/bc$, $co = (kc \times ci)/bc$, and $ih = (lc \times ci)/ce$. But by hypothesis, $ag - ab + gi - bc + ie - ce = 0$ and therefore $gm - bn - co + ih = 0$. (Notice that this last equality is correct up to terms of the order of $bm^2, \ldots$.) It then follows easily that Bernoulli's conclusion (1.44) is correct. Bernoulli says that "without any calculation" he has the same preliminary theorem "as my brother has demonstrated by a very long calculation." He calls (1.44) a *fundamental* equation, as we noted in Section 1.10, and he carefully notes that this relation is *uniform* in the sense that its two members contain corresponding data displaced by one ordinate. Thus the relation holds for any pair of ordinates in the right-hand member; hence the left-hand member is a constant. (Recall how Newton used this same device on p. 24, above.)

John Bernoulli next establishes his Lemma II. In terms of Figure 1.28, he draws arcs of circles about the centers a and e and chooses g and i on these arcs so that $bc = gi$. He then draws "little lines" gn and io parallel to

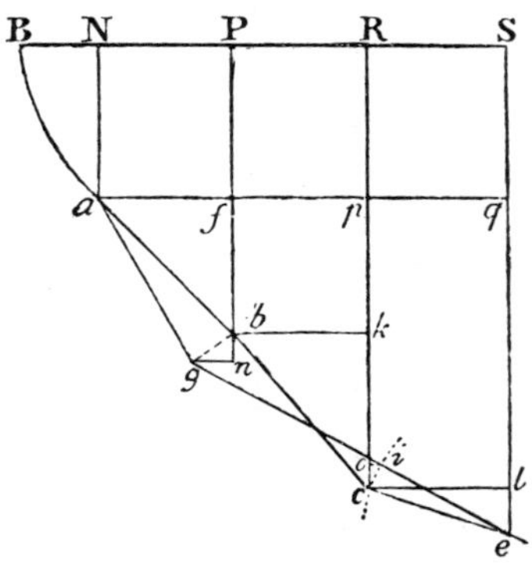

Figure 1.28

aq from *g* to *fb* produced and from *i* to *pc*. He concludes that

$$\left(\frac{fb}{af}-\frac{kc}{bk}\right)\times bn=\left(\frac{kc}{bk}-\frac{le}{cl}\right)\times co. \tag{1.45}$$

Bernoulli's proof is again easy. He notes that

$$bk^2+kc^2=bc^2=gi^2=(bk+gn+oi)^2+(kc-co-bn)^2$$

and so

$$bk\times(gn+oi)=kc\times(co+bn).$$

Since the triangles *afb* and *bng* are similar as are *cle* and *coi*, he finds

$$\frac{af}{fb}=\frac{bn}{gn},\qquad \frac{cl}{le}=\frac{co}{oi}$$

and, as a consequence of the previous relation,

$$\frac{fb\times bn}{af}+\frac{le\times co}{cl}=\frac{kc\times co}{bk}+\frac{kc\times bn}{bk}.$$

From this, his second fundamental relation (1.45) follows directly. He remarks it has the same uniformity property that his relation (1.44) does. He also deduces by similar reasoning his Corollary II to this lemma, which states that

$$\left(-\frac{af}{fb}+\frac{bk}{kc}\right)\times gn=\left(-\frac{bk}{kc}+\frac{cl}{le}\right)\times oi. \tag{1.45'}$$

John Bernoulli now proceeds using his Lemma I to solve his brother's Problem I. To solve this problem, he refers to Figure 1.27 and uses relation (1.44) because of the isoperimetric condition. To find the condition that arises from the fact that the area *BMLET* in Figure 1.29 is the greatest possible, he asserts that

$$f(Pb)+f(Rc)=f(Pg)+f(Ri), \tag{1.46}$$

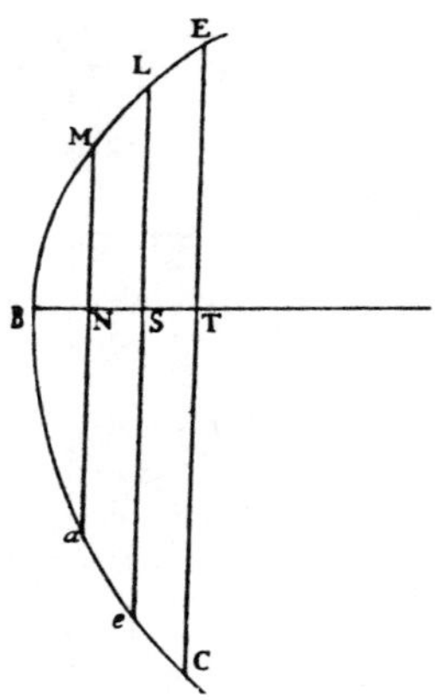

Figure 1.29

where f is the function which relates the ordinates of $BMLE$ to those of $BaeC$; i.e., $f(Na) = MN$, etc., in Figure 1.29. [His reasoning to establish (1.46) is notationally confusing because he is using simultaneously two figures, Figures 1.27 and 1.29, plus a rather obscure notation.] To establish this relation, let $NP = PR = RS$ in Figure 1.27 and then the relation

$$f(Pb) \times NP + f(Rc) \times PR = f(Pg) \times NP + f(Ri) \times PR \qquad (1.47)$$

expresses the fact that the area under the segment of the maximizing arc from N to R is an extremum and hence is not changed appreciably by a small change of the ordinates. Therefore, he can write condition (1.47) in the form $f(Pg) - f(Pb) = f(Rc) - f(Ri)$, which establishes (1.46); and he notes that through terms of the first order, this implies that

$$f'(Pb) \times bg = f'(Pb) \times (Pg - Pb) = f'(Rc) \times (Rc - Ri) = f'(Rc) \times ci. \qquad (1.47')$$

This gives Bernoulli a second condition relating bg and ci. On substituting the value of bg/ci from (1.47′) into (1.44), he finds what he calls the *specific* equation

$$\left(\frac{fb}{ab} - \frac{kc}{bc} \right) \times \frac{1}{f'(Pb)} = \left(\frac{kc}{bc} - \frac{le}{ce} \right) \times \frac{1}{f'(Rc)} . \qquad (1.48)$$

To complete his analysis, Bernoulli now relates the quantities in (1.48) to $y = BN$, $x = Na$, and arc-length $z = Ba$ (Figure 1.29). He now observes that relation (1.48) has the property of uniformity mentioned in connection with relation (1.44), and it can be expressed as

$$-d\left(\frac{dx}{dz} \right) \times \frac{1}{f'(x)} = \text{const.} = \frac{dy}{a} ,$$

by dimensional analysis. This becomes

$$\frac{-dz\, d^2x + dx\, d^2z}{dz^2} \times \frac{1}{f'(x)} = \frac{dy}{a} ;$$

but $dx^2 + dy^2 = dz^2$, $dx\, d^2x = dz\, d^2z$, since dy is a constant and hence $-dy\, d\, dz : dz^2 = \Delta x \times dx : a$, where Δx is Bernoulli's notation for $f'(x)$. He now writes $X = \int f'(x)\, dx$ and notes that since dy is constant, this equation becomes

$$\frac{dy}{dz} = \frac{X \pm C}{a}$$

on integration and, after an obvious manipulation, in his notation[68]

$$y = \int \frac{(X \pm C)\, dx}{\sqrt{aa - (X \pm C)^2}} .$$

[68] John Bernoulli, *RE*, p. 244. Problems I and II are on pp. 102 and 106.

In his Problem II he takes up his brother, James's, second problem: the ordinates in Figure 1.29 such as NM are now functions f of the corresponding arcs such as Ba. Again he has relation (1.44) relating bg and ci, arising from the isoperimetric condition, and he needs to find another such relation, arising from the maximizing condition. In terms of Figure 1.27, he has arcs $abce = agie$ and also relation (1.44). He concludes that the extremal property for the area requires that the relation

$$f(Bab)\cdot NP + f(Babc)\cdot PR = f(Bag)\cdot NP + f(Bagi)\cdot PR,$$

as may be readily seen. Since $NP = PR$,

$$f(Bag) - f(Bab) = f'(Bab)\times mg = f'(Babc)\times ih. \tag{1.47''}$$

Now in the proof of Lemma I, Bernoulli showed that

$$mg = \frac{fb}{ab}\times bg \qquad ih = \frac{le}{ce}\times ci.$$

Using the relations (1.44), (1.47″), and these equations, Bernoulli has his *specific* equation

$$\left(\frac{fb}{ab} - \frac{kc}{bc}\right)\times\frac{ab}{fb}\times\frac{1}{f'(Bab)} = \left(\frac{kc}{bc} - \frac{le}{ce}\right)\times\frac{ce}{le}\times\frac{1}{f'(Babc)}. \tag{1.49}$$

He notes that the uniformity of this equation is not what is needed to proceed from the arc ab to the "following" arc bc and so on, since what appears in the right-hand member outside the parentheses is not bc but ce. He needs to alter this relation. To do this, he multiplies both members of (1.49) by bc/kc and finds

$$\left(\frac{fb}{ab} - \frac{kc}{bc}\right)\times\frac{ab\times bc}{fb\times kc\times f'(Bab)} = \left(\frac{kc}{bc} - \frac{le}{ce}\right)\times\frac{ce\times bc}{le\times kc\times f'(Babc)}. \tag{1.49'}$$

Now it is not hard to see that to terms that are "infinitesimal," $ab = bc = ce$ and $fb = le = kc$. With the help of these, equation (1.49′) becomes

$$\left(\frac{fb}{ab} - \frac{kc}{bc}\right)\times\frac{ab^2}{fb^2}\times\frac{1}{f'(Bab)} = \left(\frac{kc}{bc} - \frac{le}{ce}\right)\times\frac{bc^2}{kc^2}\times\frac{1}{f'(Babc)}, \tag{1.50}$$

which is now uniform, so he can infer that the left-hand member is a constant. He sets $y = BN$, $x = Na$, and $z = BA$ and has

$$-d\left(\frac{dx}{dz}\right)\times\left(\frac{dz}{dx}\right)^2\times\frac{1}{f'(z)} = \frac{dy}{a}.$$

Then, just as before, he reasons that

$$d\left(\frac{dy}{dx}\right) = -\frac{dy\,d^2x}{dx^2} = \frac{f'(z)\,dz}{a},$$

since $d^2y = 0$. He integrates this and finds

$$\frac{dy}{dx} = \frac{(Z \pm c)}{a},$$

where $Z = \int f'(z)\,dz$.

To illustrate how to use his Lemma II, Bernoulli takes up the same problem again using Figure 1.28. He now makes use of his second *fundamental* equation (1.45′) and notes that at a maximum

$$\begin{aligned} &\Phi Bab \times af + \Phi Babc \times bk + \Phi Babce \times cl \\ &\quad = \Phi Bag \times (af - gn) + \Phi Bagi \times (bk + gn + oi) + \Phi Bagie \times (cl - oi), \end{aligned} \tag{1.51}$$

where Φ is the function that determines the ordinates of the desired curve; that is, ΦBab is the ordinate in Figure 1.28 at the point P standing above the axis BS. Thus equation (1.51) expresses the fact that the area under the desired curve between B and S, not shown in Figure 1.29, is a maximum.

Bernoulli points out that $ab = ag$, $bc = gi$, and $ce = ei$ and arcs $Bab = Bag$, $Babc = Bagi$, and $Babce = Bagie$ up to first order terms. He can then write (1.51) in the form

$$(-\Phi Bab + \Phi Babc) \times gn = (-\Phi Babc + \Phi Babce) \times oi, \tag{1.52}$$

which gives him a second relation involving gn and oi. He therefore has, by (1.45′),

$$\frac{-af/fb + bk/kc}{-\Phi Bab + \Phi Babc} = \frac{-bk/kc + cl/le}{-\Phi Babc + \Phi Babce},$$

which has the uniformity he desires. He points out that $bk/kc - af/fb = d(dy/dx)$ and $\Phi Babc - \Phi Bab = d\Phi z$ and hence that

$$d\left(\frac{dy}{dx}\right) \times \frac{1}{d\Phi z} = \frac{1}{a}.$$

This yields the relation

$$a\frac{dy}{dx} = \Phi z \pm c.$$

Bernoulli proceeds in his Problem III to reconsider his brother James's problem of finding among all curves of the same length the one whose center of gravity is the lowest. Since there are no new principles involved, we do not consider his discussion but go on to study of the accomplishments of his great student Euler after examining an extremely important addendum to Bernoulli's paper in *RE* on pp. 267–269, which forms the basis for Carathéodory's method of quickest or geodesic descent (see Section 7.11 for a discussion of that method). In this addendum Bernoulli gives a simple proof that the cycloid is in fact the curve of swiftest descent. He asserts that this work was already completed in 1697 and communicated to Leibniz and some other friends. Bernoulli wrote as follows (pp. 266–267):

> To bring this memoir to a conclusion I proceed to add my direct method for solving the famous problem of quickest descent, not having yet published this method although I had communicated it to several of my friends as early as 1697 when I published my other indirect one. The incomparable Mr. Leibniz to whom I had communicated both, as he has himself testified in the *Leipzig Acts* for the same year 1697, p. 204, found this direct method of such elegance that he counselled me not to publish it for the reasons which then obtained but which do not any longer. I hope it will also please the reader as much that, although the analysis concerns only the radius of curvature or of the osculating circle of the desired curve, what one finds however is the common cycloid having at whatever point such a radius of curvature or of the osculating circle; this method also provides me meanwhile with a synthetic demonstration which with extraordinary and agreeable ease shows that this cycloid is effectively the desired curve of quickest descent.

Let us look at this to see how he indeed proceeded. He examines an arbitrary point M on the curve of quickest descent and draws the radius of curvature MK to the curve at that point; he then finds that the cycloid is characterized by having the horizontal line AL in Figure 1.30 from which the particle starts falling, bisect, MK. In Figure 1.30 he draws through any point M on the curve the line INC, where the angle INL is arbitrary, cutting the horizontal AL in N. He then draws another line Knc intersecting the other one in a very small angle CKc so that the small circular arcs Ce and Mm described about the center K can be considered as small

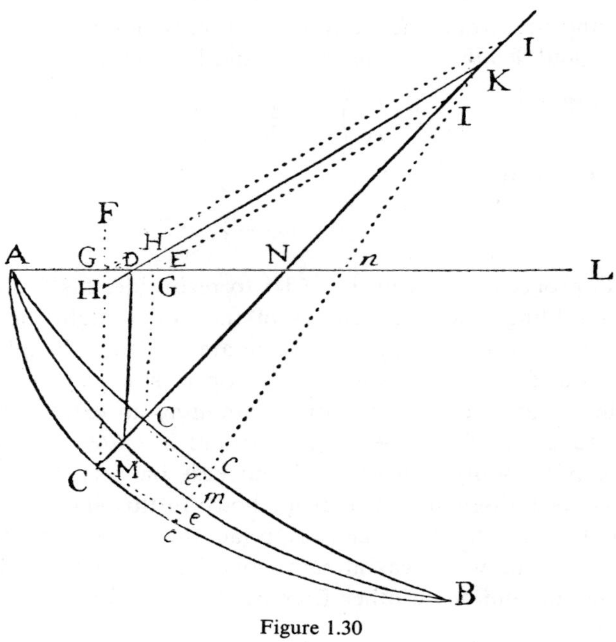

Figure 1.30

straight lines. He now seeks "among an infinitude of these little concentric arcs" the one down which a heavy particle starting from A will fall most quickly. To do this, he sets $NK = a$ and $MN = x$ and draws the vertical MD. Then he defines m by the ratio

$$1 : m = MN : MD$$

so that $MD = mx$ and n by the ratio

$$1 : n = cK : Ce = MK : Mm$$

so that $Mm = nx + na$. He remarks that m is finite and n infinitesimal and asks x to be so chosen that (ds/v)

$$\frac{n(x+a)}{\sqrt{m}\cdot\sqrt{x}}$$

is a minimum. He concludes that $x = a$, that is, that at each point of the curve of quickest descent the radius of curvature is bisected by the horizontal AL. This, he points out, is characteristic of the cycloid.

In the next paragraph (p. 268) Bernoulli says his method can be generalized from the case where "their velocities are not in the ratio of the square roots of their heights . . . but in the ratio of any function . . . of their heights."

Bernoulli then turns to his synthetic demonstration; i.e., he proceeds to show that the cycloid actually furnishes the least time of fall. To do this, he considers MK and mK, two normals to the cycloid AMB through M and a neighboring point m; they pass through the center of curvature K so that $MN = NK$. He also considers a comparison curve ACB between A and B, extends MK and mK to cut this curve in C and c, and draws the small circular arc Ce with center K as in Figure 1.30. He now erects the perpendiculars CG and MD to the horizontal AL, draws GI parallel to DK, and extends DK to a point H where it meets CG and CG to a point F so that $MD : CH = CH : CF$. It follows from the similarity of triangles that $CN = NI$ since $MN = NK$ (recall that this is a characterizing property of the cycloid). Bernoulli then argues that $CN^2 + NK^2 > 2CN \cdot NK$ and as a consequence $CN^2 + NK^2 + 2CN \cdot NK > 4CN \cdot NK = CI \cdot MK$; hence he has shown that $CK^2 > CI \cdot MK$ or $MK : CK < CK : CI$. He next proceeds to note that $MK : CK = MD : CH = CH : CF$ by the way F was chosen and also that $CK : CI = CH : CG$. From these facts he concludes that $CH : CF < CH : CG$ or $CG < CF$.

To complete his proof Bernoulli first compares the time of fall—starting from A—along Mm to that along Ce. He has for this ratio $Mm/(MD)^{1/2} : Ce/(CG)^{1/2}$ since the time of fall varies directly as the distance and inversely as the square root of the height. But it is also true that (see the definition of F above)

$$\frac{\sqrt{MD}}{\sqrt{CF}} = \frac{MD}{CH} = \frac{MK}{CK} = \frac{Mm}{Ce}\ ;$$

and thus the ratio of the times of fall along Mm and Ce becomes

$$\frac{Mm}{Ce} \cdot \frac{\sqrt{CG}}{\sqrt{MD}} = \frac{\sqrt{MD}}{\sqrt{CF}} \cdot \frac{\sqrt{CG}}{\sqrt{MD}} = \frac{\sqrt{CG}}{\sqrt{CF}} < 1.$$

The time of fall along the arc Mm of the cycloid is then less than that along the circular arc Ce. Bernoulli next remarks that the time of fall along Ce is greater than the time along Cc. To demonstrate this, he remarks that Cc is the hypotenuse of the right triangle Cce. (It follows from this that $Cc > Ce$, and hence with the same velocity a particle will descend Ce more rapidly than it will Cc.) The time of descent along Mm is then less than the time along Cc, and consequently the total time of descent along the cycloid is less than the time on "any other curve ACB passing through the same points A, B" (p. 269). This very elegant sufficiency proof is probably the first in history.

2. Euler

2.1. Introduction

It is not quite certain when Euler first became seriously interested in the calculus of variations. Carathéodory, who edited Euler's magnificent 1744 opus, *The Method of Finding Plane Curves that Show Some Property of Maximum or Minimum* . . . , believed it unlikely that it occurred during his period in Basel with John Bernoulli.[1] However, we should note that Euler considered in 1732 and 1736 problems more or less arising out of James Bernoulli's isoperimetric problems; and even as early as the end of 1728 or early in 1729, as Eneström showed, he wrote "On finding the equation of geodesic curves." In effect, Euler in 1744, following John Bernoulli, examined the question of end-curves that cut a family of geodesics so that they have equal length. He showed that the end-curves must be orthogonal to the geodesics. This is the precursor of the so-called envelope theorem of Chapter 7 below and plays a key part in Carathéodory's work.

The introduction to Euler's great work of 1744 (Euler, I, XXIV, pp. lvi–lxii) contains a listing of the 100 special problems taken up by Euler to illustrate his methods; these were arranged in 11 convenient categories by Carathéodory, the editor of Volumes XXIV and XXV in Series I. No interested reader should overlook this "Complete listing of Euler's examples in the calculus of variations."

Perhaps the greatest things that Euler did in his 1744 book were to set up a general apparatus or procedure for writing down the so-called Euler differential equation or first necessary condition and to enunciate and discuss the *principle of least action*, which he first discovered prior to 15 April 1743. In a letter of that date Daniel Bernoulli congratulated him on this work, and Euler wrote it down in the latter part of 1743. (The book itself, *Methodus Inveniendi*, was sent to the publisher Bousquet in May 1743 and its two appendices, in December of that same year.) Carathéodory pointed out that Maupertuis's paper on least action was presented to the

[1] Euler, I, XXIV, p. viii and Euler I, XXV, p. vii and Carathéodory [1937], [1945].

Academy of Paris on 25 April 1744 and "had the greatest similarity to the above mentioned ideas of Euler's." Carathéodory was of the opinion that fundamentally Euler made no effort to establish his claim to this great discovery because "the true reason for this notion of his lay above all in his philosophical attitude."[2] (We shall discuss the principle in Section 2.7.) Euler's little (pp. 298–308 of Euler, I, XXIV) paper is the first presentation by anyone of this great discovery and is a truly remarkable historical document. (It appears as Appendix II to I, XXIV.) Here we see in nascent form the first deep insight—apart from Fermat's principle of least time—into the role of the calculus of variations in physics. This role has since been of great importance to both subjects in a number of ways, and it has become of paramount significance in quantum mechanics.

Another important contribution Euler made to the calculus of variations was to change the subject from a discussion of essentially special cases to a discussion of very general classes of problems, even by modern standards. In fact, he took the fairly special methods of James and John Bernoulli and transformed these into a whole new brand of mathematics. It is interesting also that he had not yet completely understood the significance of allowing the end-points of his admissible arcs to vary, and he repeatedly is vague on this point. This is not too surprising since he was investigating how to find the family of extremals for each general type of problem, irrespective of the end-conditions. He clearly did not look for so-called transversality conditions arising from variability of the end-points.

Euler's methods were themselves displaced, as we shall see later, in 1755 by Lagrange's elegant technique of "variations." In Lagrange's honor Euler then renamed the subject the *calculus of variations* and dropped his methods in favor of Lagrange's.

2.2. The Simplest Problems

In Chapters I and II Euler considers the basic problem of finding among all plane curves $y = y(x)$, with $0 \leqslant x \leqslant a$, the one which furnishes the definite integral $\int Z\,dx$ a maximum or minimum, where Z is a "determinate" function of $x, y, p = dy/dx, q = dp/dx, r = dq/dx$, etc.[3] To carry out this program, Euler in Figure 2.1 first considers the abscissa AZ

[2] Euler, I, XXIV, pp. lv, x. Carathéodory has done a great deal of investigation of the Euler-Maupertuis correspondence on this general subject, and the interested reader would do well to read Carathéodory's discussion of this and may also wish to consult Brunet, *ACT*, p. 8, who attributes priority to Maupertuis. I believe that Carathéodory's attribution to Euler is sound, for reasons that we shall examine later (pp. 101ff).

[3] Euler I, XXIV, p. 18. This is "Proposition II. Theorem." (Throughout the entire text Euler is lax in specifying end-conditions.) The integral here is understood to be evaluated between the limits of $x = 0$ and $x = a$.

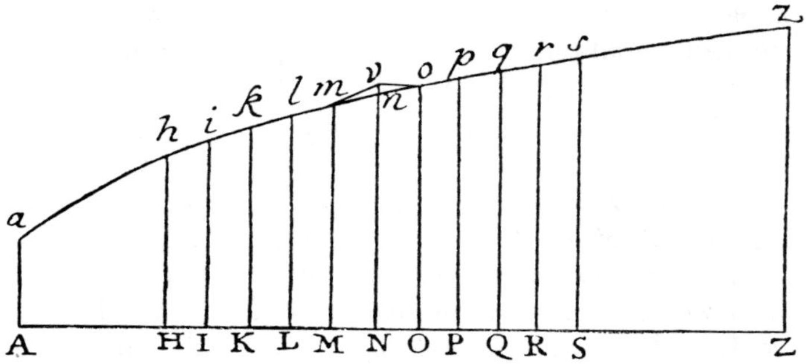

Figure 2.1

to be divided into "infinitely" many small and equal subintervals $dx = HI = IK = KL = LM$, etc. by the points $\ldots x_{\prime\prime}, x_{\prime}, x = AM, x', x'', \ldots$ (we will actually choose finitely many) with the corresponding ordinates

$$\begin{array}{ll} Mm = y & Mm = y \\ Nn = y' & Ll = y_{\prime} \\ Oo = y'' & Kk = y_{\prime\prime} \\ Pp = y''' & Ii = y_{\prime\prime\prime} \\ Qq = y^{\text{iv}} & Hh = y_{\text{iv}} \\ \text{etc.} & \text{etc.} \end{array}$$

He next approximates to his derivative with the help of finite differences and writes

$$\begin{array}{ll} p = \dfrac{y' - y}{dx} & p = \dfrac{y' - y}{dx} \\[2ex] p' = \dfrac{y'' - y'}{dx} & p_{\prime} = \dfrac{y - y_{\prime}}{dx} \\[2ex] p'' = \dfrac{y''' - y''}{dx} & p_{\prime\prime} = \dfrac{y_{\prime} - y_{\prime\prime}}{dx} \\[2ex] p''' = \dfrac{y^{\text{iv}} - y'''}{dx} & p_{\prime\prime\prime} = \dfrac{y_{\prime\prime} - y_{\prime\prime\prime}}{dx} \\[2ex] \text{etc.} & \text{etc.} \\[2ex] q = \dfrac{y'' - 2y' + y}{dx^2} & q = \dfrac{y'' - 2y' + y}{dx^2} \\[2ex] q' = \dfrac{y''' - 2y'' + y'}{dx^2} & q_{\prime} = \dfrac{y' - 2y + y_{\prime}}{dx^2} \\[2ex] q'' = \dfrac{y^{\text{iv}} - 2y''' + y''}{dx^2} & q_{\prime\prime} = \dfrac{y - 2y_{\prime} + y_{\prime\prime}}{dx^2} \\[2ex] \text{etc.} & \text{etc.} \end{array}$$

with comparable formulas for r, s, t, etc. (We know that by a proper passage to the limit all Euler's results are completely valid.)

Finally on p. 25 he approximates to his integral $\int_0^a Z\,dx$ by the sum

$$\int_0^{x_{,}} Z\,dx + Z\,dx + Z'\,dx + Z''\,dx + Z'''\,dx + \cdots, \tag{2.1}$$

where he has set $Z = Z(x, y, p, q, r, \ldots)$, $Z' = Z(x', y', p', q', r', \ldots)$, $Z'' = Z(x'', y'', p'', q'', r'', \ldots)$, etc., and supposes that anz is an extremal. Then the derivative of expression (2.1) with respect to the ordinate y', viewed as a variable, must vanish at the value $y' = Nn$. To evaluate this derivative, Euler first begins Chapter II with a calculation of the effects on y, p, q, r, s, t, etc., of altering y'. He shows these straightforward consequences in the table below, where he has set $dy' = n\nu$ in Figure 2.1.[4]

Quant.	*Increm.*	*Quant.*	*Increm.*
y'	$+n\nu$	$s_{,,,}$	$+\frac{n\nu}{dx^4}$
p	$+\frac{n\nu}{dx}$	$s_{,,}$	$-\frac{4n\nu}{dx^4}$
p'	$-\frac{n\nu}{dx}$	$s_{,}$	$+\frac{6n\nu}{dx^4}$
$q_{,}$	$+\frac{n\nu}{dx^2}$	s	$-\frac{4n\nu}{dx^4}$
q	$-\frac{2n\nu}{dx^2}$	s'	$+\frac{n\nu}{dx^4}$
q'	$+\frac{n\nu}{dx^2}$	t_{IV}	$+\frac{n\nu}{dx^5}$
$r_{,,}$	$+\frac{n\nu}{dx^3}$	$t_{,,,}$	$-\frac{5n\nu}{dx^5}$
$r_{,}$	$-\frac{3n\nu}{dx^3}$	$t_{,,}$	$+\frac{10n\nu}{dx^5}$
r	$+\frac{3n\nu}{dx^3}$	$t_{,}$	$-\frac{10n\nu}{dx^5}$
r'	$-\frac{n\nu}{dx^3}$	t	$+\frac{5n\nu}{dx^5}$
		t'	$-\frac{n\nu}{dx^5}$

[4] Euler, I, XXIV, pp. 30ff. In the succeeding pages he works through a number of cases for his integrand Z. One of these is $dZ = M\,dx + N\,dy$.

Euler then uses this in "Proposition III, Problem" (p. 39) to find what happens in the case when Z is of the form

$$dZ = M\,dx + N\,dy + P\,dp,$$

and the integral $\int Z\,dx$ is an extremum. (Here Z is a function of x, y, and p.) If we consult his table, we see that the effect of altering y' is to vary both p and p'.[5] This implies that the portion of $\int_0^a Z\,dx$, or rather of the expression (2.1), that is affected is

$$Z\,dx + Z'\,dx = Z(x, y, p)\,dx + Z(x', y', p')\,dx. \tag{2.2}$$

Euler now views expression (2.2) as a function of the variable y' and asks that this expression be an extremum. He then knows that its derivative with respect to y' vanishes. To calculate this derivative, he notes that the change in $Z\,dx$ caused by varying y' is $P \cdot nv(nv = y' - y$ in Figure 2.1) and in $Z'\,dx$ is $N' \cdot nv\,dx - P' \cdot nv$. This implies that for an extremum $P + N'\,dx - P' = 0$, i.e., $N' - (P' - P)/dx = 0$ or in the limit[6]

$$N - \frac{dP}{dx} = 0. \tag{2.3}$$

Let us examine how Euler treats the case where Z depends on x, y, p, q, and r. In this case Euler has $Z = M\,dx + N\,dy + P\,dp + Q\,dq + R\,dr$. Once again the integral J is to be viewed as a function of y', and hence $dJ/dy' = 0$ is necessary for an extremum. To calculate this derivative, note first that $J = (\cdots + Z_{\prime\prime} + Z_{\prime} + Z + Z' + Z'' + \cdots)\,dx$.

$$\frac{\partial p}{\partial y'} = \frac{1}{dx}, \qquad \frac{\partial p'}{\partial y'} = -\frac{1}{dx}, \qquad \frac{\partial q'}{\partial y'} = \frac{1}{dx^2},$$

$$\frac{\partial q}{\partial y'} = -\frac{2}{dx^2}, \qquad \frac{\partial q_{\prime}}{\partial y'} = \frac{1}{dx^2},$$

$$\frac{\partial r_{\prime\prime}}{\partial y'} = \frac{1}{dx^3},$$

$$\frac{\partial r_{\prime}}{\partial y'} = -\frac{3}{dx^3} = -\frac{\partial r}{\partial y'}, \qquad \frac{\partial r'}{\partial y'} = -\frac{1}{dx^3}.$$

[5] It is worth noting that Taylor in his *MET* introduced this method of representing the dependence of Z on its variables. He also discussed the isoperimetric problem "on principles not different from those of the Bernoullis, but with some alteration of symbolic notation." See Woodhouse, *COV*, p. 29. This book is quite archaic and unsatisfactory in many respects.

[6] Euler I, XXIV, p. 39. Had Euler varied some ordinate other than y', he clearly would have found again his differential equation (2.3).

It is then clear that

$$\frac{\partial Z_{,,}}{\partial y'} = R_{,,} \cdot \frac{1}{dx^3}, \qquad \frac{\partial Z_{,}}{\partial y'} = Q_{,} \cdot \frac{1}{dx^2} - 3R_{,} \cdot \frac{1}{dx^3},$$

$$\frac{\partial Z}{\partial y'} = P \cdot \frac{1}{dx} - 2Q \cdot \frac{1}{dx^2} + 3R \cdot \frac{1}{dx^3},$$

$$\frac{\partial Z'}{\partial y'} = N' - P' \cdot \frac{1}{dx} + Q' \frac{1}{dx^2} - R' \frac{1}{dx^3}.$$

Putting all this material together, we have

$$0 = \frac{dJ}{dy'} = \frac{d}{dy'}\left(Z_{,,} + Z_{,} + Z + Z'\right) dx = dx\left(N' - \frac{\Delta P}{dx} + \frac{\Delta^2 Q}{dx^2} - \frac{\Delta^3 R}{dx^3}\right),$$

where ΔP, e.g., stands for $P' - P$, etc. (Euler views these differences as differentials.) Then by a passage to the limit, the Euler differential equation

$$N - \frac{dP}{dx} + \frac{d^2 Q}{dx^2} - \frac{d^3 R}{dx^3} = 0$$

results.

To show how to use his differential equation, Euler considers a number of examples. Let us look at his Example III, which is, in his notation, to minimize the integral

$$\int \frac{dx\sqrt{1 + pp}}{\sqrt{x}}.$$

He points out that in this case

$$dZ = -\frac{dx\sqrt{1 + pp}}{2x\sqrt{x}} + \frac{p\,dp}{\sqrt{x(1 + pp)}},$$

and hence that

$$M = -\frac{\sqrt{1 + pp}}{2x\sqrt{x}}, \qquad N = 0, \qquad P = \frac{p}{\sqrt{x(1 + pp)}}.$$

His differential equation (2.3) can be easily integrated, and he finds that

$$P = \frac{p}{\sqrt{x(1 + pp)}} = \text{const.} = \frac{1}{\sqrt{a}}:$$

which reduces to $ap^2 = x + p^2 x$. This gives Euler the differential equation

$$p = \frac{dy}{dx} = \sqrt{\frac{x}{a - x}},$$

whose integration leads directly to the cycloid, as we have seen earlier.[7]

[7] Euler, I, XXIV, pp. 44–45.

Later in this chapter Euler sets out a mnemonic rule for writing down his differential equation, which was to have a profound effect on Lagrange, as we shall see later (p. 111). His rule is that for the case when $dZ = M\,dx + N\,dy + P\,dp$, in the differential set $M\,dx$ equal to zero, leave $N\,dy$ unchanged, and write $-p\,dP$ instead of $P\,dp$. The resulting expression should be set equal to zero. Clearly this rule gives for Euler's differential equation the result

$$N\,dy - p\,dP = 0; \tag{2.4}$$

but $dy = p\,dx$, and hence it reduces to (2.3) directly.[8]

Next Euler considers the case where Z depends on x, y, p, and q. He decomposes $\int Z\,dx$ into two sums $Z\,dx + Z'\,dx + Z''\,dx + Z'''\,dx + \cdots$ and $Z_{,}\,dx + Z_{,,}\,dx + Z_{,,,}\,dx + \cdots$ so that all appropriate terms are included. By differentiation and passage to the limit, he finds his equation in this case is

$$N - \frac{dP}{dx} + \frac{d^2Q}{dx^2} = 0. \tag{2.5}$$

Again he gives a number of examples of how to use this Euler equation, which we do not have the space to discuss.

It is interesting that Euler did not completely understand the fact that his condition is a necessary but not a sufficient one. In his discussion (e.g., on pp. 36–38) it is clear that he felt his condition was sufficient to ensure an extremum, and that by evaluating the integral along an extremal he could decide whether it was a maximum or a minimum. (His calculations in connection with Example II are not clear to me.)

2.3. More General Problems

Euler considers in Chapter III a more complex set of problems than he has previously. Typically he considers a definite integral—from 0 to a—

$$\int Z\,dx, \tag{2.6}$$

where Z now depends not only on x, y, p, etc., as in his two earlier chapters, but also on a quantity Π defined as

$$\Pi = \int [Z]\,dx, \tag{2.6'}$$

where this integral is evaluated between 0 and x and the integrands are given by the differential equations

$$dZ = L\,d\Pi + M\,dx + N\,dy + P\,dp + \cdots,$$
$$d[Z] = [M]\,dx + [N]\,dy + [P]\,dp + \cdots.$$

[8] Euler, I, XXIV, pp. 51–52.

This is the first instance of a Lagrange problem: to minimize

$$\int Z(x, y, y', \Pi)\, dx$$

subject to the side-condition $\Pi' - [Z](x, y, y') = 0$; this leads Euler to generalize his necessary condition. (In this, $y' = dy/dx$ and $\Pi' = d\Pi/dx$.) What he finds is a simple—but the first—instance of the *multiplier* rule. Perhaps Lagrange multipliers ought to be called *Euler–Lagrange multipliers* just as the first necessary condition for general problems is often referred to as the *Euler–Lagrange equation*. It is interesting that the procedure he used in Chapter II is valid here with trivial modification. The only essential difference is that the procedure is somewhat more involved and complicated notationally.

A typical problem Euler wished to handle by the method of this new chapter is that in which $Z = A(x, y)(1 + p^2)^{1/2}\, dx = A(x, y)\, ds$, where s is arc-length along the curve $y = y(x)$, or more generally, that in which $Z = A(x, y, s)\, ds$. In both cases he sets

$$\Pi = s = \int_0^x \sqrt{1 + p^2}\, dx.$$

Let us look at how Euler handled the general case described in (2.6), (2.6′) and was thereby able to "solve" quite complex problems of the calculus of variations in the sense of finding their families of extremals. To this end, again partition the interval from A to Z into finitely—Euler says infinitely—many equal subintervals by the points $\ldots x_{\mathrm{iv}}, x_{,,,}, x_{,,}, x_{,}, x, x', x'', x''', x^{\mathrm{iv}} \ldots$, and call the corresponding ordinates $\ldots y_{\mathrm{iv}}, y_{,,,}, y_{,,}, y_{,}, y, y', y'', y''', y^{\mathrm{iv}} \ldots$. Also for definiteness, write $[Z] = f(x, y, p, q)$ and note that

$$\begin{aligned}
[Z_{,,}] &= f(x_{,,}, y_{,,}, p_{,,}, q_{,,}), \qquad [Z_{,}] = f(x_{,}, y_{,}, p_{,}, q_{,}),\\
[Z] &= f(x, y, p, q), \qquad [Z'] = f(x', y', p', q'),\\
[Z''] &= f(x'', y'', p'', q''), \ldots; \qquad \Pi' = \Pi + [Z]\, dx,\\
\Pi'' &= \Pi' + [Z']\, dx = \Pi + [Z]\, dx + [Z']\, dx,\\
\Pi''' &= \Pi'' + [Z'']\, dx\\
&= \Pi + [Z]\, dx + [Z']\, dx + [Z'']\, dx, \ldots,\\
J &= \int_0^{x_{,}} Z\, dx + Z\, dx + Z'\, dx + \cdots + Z^{\mathrm{v}}\, dx + \cdots.
\end{aligned}$$

In these expressions we should note that Π itself is a constant with respect to y'', the ordinate Euler chooses to vary this time. As a result the condition for an extremum is $dJ/dy'' = 0$ when we view J as a function of y'' alone.[9]

[9] Euler, I, XXIV, pp. xviii–xx, 85–90. What I give below is not precisely Euler's procedure but is modeled closely by Carathéodory on it and is hopefully a little clearer. It is, in essence, Carathéodory's.

We first calculate that

$$\frac{\partial \Pi}{\partial y''} = 0, \quad \frac{\partial \Pi'}{\partial y''} = [Q] \cdot \frac{1}{dx}, \quad \frac{\partial \Pi''}{\partial y''} = [P'] - [Q] \cdot \frac{1}{dx} - 2\Delta[Q] \cdot \frac{1}{dx},$$

$$\frac{\partial \Pi'''}{\partial y''} = \frac{\partial \Pi^{\text{iv}}}{\partial y''} = \cdots = [N''] \, dx - \Delta[P'] + \Delta^2[Q] \cdot \frac{1}{dx}$$

and

$$\frac{\partial Z^{(\nu)}}{\partial y''} = L^{(\nu)} \frac{\partial \Pi^{(\nu)}}{\partial y''} + N^{(\nu)} \frac{\partial y^{(\nu)}}{\partial y''} + P^{(\nu)} \frac{\partial p^{(\nu)}}{\partial y''} + Q^{(\nu)} \frac{\partial q^{(\nu)}}{\partial y''}$$
$$(\nu = 0, 1, 2, \ldots).$$

It remains now only to evaluate these latter expressions with the help of our prior ones. Note that

$$\frac{\partial Z}{\partial y''} = Q \frac{1}{dx^2}, \qquad \frac{\partial Z'}{\partial y''} = L'[Q] \frac{1}{dx} + P' \frac{1}{dx} - 2Q' \frac{1}{dx^2},$$

$$\frac{\partial Z''}{\partial y''} = L''\left\{ [P'] - [Q] \frac{1}{dx} - 2\Delta[Q] \frac{1}{dx^2} \right\} + N'' - P'' \frac{1}{dx} + Q'' \frac{1}{dx^2},$$

$$\frac{\partial Z^{(\nu)}}{\partial y''} = L^{(\nu)} \, dx \left\{ [N''] - \Delta[P'] \frac{1}{dx} + \Delta^2[Q] \frac{1}{dx^2} \right\} \quad (\nu = 3, 4, 5, \ldots).$$

Since the sum of all these derivatives must be zero along a minimizing arc, Euler deduces the condition

$$O = N - \frac{dP}{dx} + \frac{d^2Q}{dx^2} + L[P] - \frac{dL}{dx}[Q] - 2L \frac{d[Q]}{dx}$$
$$+ \left([N] - \frac{d[P]}{dx} + \frac{d^2[Q]}{dx^2} \right) \int_x^a L \, dx. \tag{2.7}$$

This is, in effect, the conclusion Euler reaches on p. 95 to his Problem, Proposition III, on p. 94. He prefers to write

$$\int_x^a L \, dx = \int_0^a L \, dx - \int_0^x L \, dx = H - \int_0^x L \, dx,$$

and thus expresses relation (2.7) in the neat form

$$0 = [N]\left(H - \int L \, dx\right) - \frac{d \cdot [P]\left(H - \int L \, dx\right)}{dx} + \frac{dd \cdot [Q]\left(H - \int L \, dx\right)}{dx^2}$$
$$+ \text{etc.} + N - \frac{dP}{dx} + \frac{ddQ}{dx^2} - \text{etc.}, \tag{2.7'}$$

where $\int L \, dx = \int_0^x L \, dx$.

Let us examine his Example I of Section 15 on p. 90. He lets $dZ = L \, d\Pi$, and $d[Z] = dy$ and seeks among all curves through given end-points the one which renders the integral $\int Z \, dx$ an extremum. [Notice that Z is a function of Π alone and that he tacitly assumes that $Z(0) = 0$.] In the expression $d[Z] = [M]\,dx + [N]\,dy + [P]\,dp + \cdots$, he has $[M] = 0$, $[N] = 1$, $[P] = 0$, $[Q] = 0$, etc. From this and the fundamental relation (2.7) above it is clear that

$$\int_x^a L\,dx = 0$$

i.e., that $L \equiv 0$; but this means that $Z \equiv 0$, $\Pi = \text{const.}$, and that $y \equiv 0$.

His Example I of Section 26 (p. 97) is also of interest. Euler wishes now to render the integral

$$\int yx\,dx \int y\,dx = \int_0^a yx\,dx \int_0^x y\,dx$$

an extremum among all admissible curves $y = y(x)$ $(0 \leqslant x \leqslant a)$. Here he has

$$\Pi = \int_0^x y\,dx, \qquad [Z] = y, \qquad [N] = 1, \qquad [P] = [Q] = 0.$$

Moreover, he has $Z = yx\Pi$ and $dZ = yx\,d\Pi + y\Pi\,dx + x\Pi\,dy$ so that $L = yx$, $M = y\Pi$, and $N = x\Pi$, $P = Q = 0$. It follows therefore that Euler's equation (2.7) is now, in his notation, $0 = (H - \int yx\,dx) + x\Pi$ or

$$\int_0^x yx\,dx = H + x\int_0^x y\,dx,$$

where H is a constant, namely, zero. Euler notes that this equation has as its only solution the curve $y = 0$, the x-axis.

Notice that Euler's Proposition III above was to find in the class of arcs defined for $0 \leqslant x \leqslant a$ and satisfying the differential equations

$$\frac{d\Pi}{dx} = [Z](x, y, p, q),$$

$$\frac{dy}{dx} = p, \qquad \frac{dp}{dx} = q$$

one which minimizes or maximizes the integral

$$\int_0^a Z(x, y, p, q, \Pi)\,dx.$$

If we write

$$\lambda(x) = \int_x^a L(\xi)\,d\xi$$

in Euler's equation (2.7) or (2.7′), it is not difficult to see that it assumes the equivalent form

$$\begin{gathered}(N + \lambda[N]) - \frac{d}{dx}(P + \lambda[P]) + \frac{d^2}{dx^2}(Q + \lambda[Q]) = 0,\\ \frac{d\lambda}{dx} = -L, \qquad \lambda(a) = 0.\end{gathered} \tag{2.8}$$

Here λ is clearly a "Lagrange multiplier."

Euler's Proposition IV, Problem on p. 102 is concerned with a yet more general problem. Euler's apparatus is quite adequate to handle this new problem, and even more complex ones. He has an integrand Z as before, expressed as $dZ = L\,d\Pi + M\,dx + N\,dy + P\,dp + Q\,dq$ with $\Pi = \int[Z]\,dx$. In this case $[Z]$ is expressed in the form

$$d[Z] = [L]d\pi + [M]\,dx + [N]\,dy + [P]\,dp + [Q]\,dq$$

with $\pi = \int[z]\,dx$ and

$$d[z] = [m]\,dx + [n]\,dy + [p]\,dp + [q]\,dq.$$

He again wishes to find his condition that $\int Z\,dx$ an extremum. He arrives without undue effort at his equation (p. 107). The details are very similar to what we have seen above, only they are somewhat more complicated:

$$\begin{aligned} 0 = \left(N - \frac{dP}{dx} + \frac{d\,dQ}{dx^2} - \text{etc.}\right) \\ + \left[[N]\left(H - \int L\,dx\right) - \frac{d\cdot[P]\left(H - \int L\,dx\right)}{dx} \right. \\ \left. + \frac{dd\cdot[Q]\left(H - \int L\,dx\right)}{dx^2} - \text{etc.}\right] \\ + \left[[n]\left(G - \int[L]\,dx\left(H - \int L\,dx\right)\right) \right. \\ - \frac{d\cdot[p]\left(G - \int[L]\,dx\left(H - \int L\,dx\right)\right)}{dx} \\ \left. + \frac{dd\cdot[q]\left(G - \int[L]\,dx\right)\left(H - \int L\,dx\right)}{dx^2} + \text{etc.}\right]. \end{aligned} \tag{2.9}$$

(The interval of integration is from 0 to x.) In his Corollary 2 on p. 108 Euler simplifies this equation notationally by introducing the new variables —the interval of integration is again from 0 and x—

$$T = H - \int L\,dx, \qquad V = G - \int[L]\,dx\left(H - \int L\,dx\right).$$

In terms of these the equation (2.9) becomes

$$\begin{aligned} 0 = N + [N]T + [n]V - \frac{d(P + [P]T + [p]V)}{dx} \\ + \frac{dd(Q + [Q]T + [q]V)}{dx^2} - \text{etc.} \end{aligned}$$

Incidentally, there is no reason why Euler could not have allowed Z to have depended not only on Π, x, y, p, q, but also on π. In fact, Euler remarks in his Corollary 4 in Section 35 (pp. 108–109) that Z need not depend upon just one indefinite integral Π but can depend upon many. He thus had obtained the multiplier rule for a quite general problem.

Euler next discusses a modification of the Brachystochrone problem first suggested by the Bernoullis. It is to find the curve, joining two points in a vertical plane, in a resisting medium down which a heavy particle will descend in the least time. Euler carries out his analysis by generalizing the problem still further. He wishes, in fact, to permit Π to be defined not by the explicit differential equation

$$\frac{d\Pi}{dx} = [Z], \qquad d[Z] = [M]\,dx + [N]\,dy + [P]\,dp + [Q]\,dq$$

but rather by the implicit one

$$\frac{d\Pi}{dx} = [Z], \qquad d[Z] = [L]\,d\Pi + [M]\,dx + [N]\,dy + [P]\,dp + [Q]\,dq.$$[10]

In this situation (p. 114) he finds that his equation becomes

$$0 = N + [N]V - \frac{d(P + [P]V)}{dx} + \frac{dd(Q + [Q]V)}{dx^2} - \text{etc.}, \tag{2.10}$$

where

$$V = e^{-\int [L]\,dx}\left(H - \int e^{\int [L]\,dx} L\,dx\right). \tag{2.10$'$}$$

(The limits of integration are, as usual, 0 and x. The analysis is somewhat complicated in this case and occupies pp. 110–114 but is not really different from his earlier ones.) Note that V satisfies the differential equation $dV/dx = -[L]V - L$. Carathéodory points out (p. xxii) in his Introduction to Euler, I, XXIV that Euler, in effect, has worked out the first necessary condition for the so-called Lagrange problem.[11]

In Section 45, Examples I and II (pp. 117–123) Euler then takes up the elegant problem of finding the shape of the brachystochrone curve in case the medium through which the heavy particle falls resists the motion depending only on the velocity v. Here he assumes that the resistance function R is proportional to v^{2n}. He lets $\Pi = v^2/2$ and notes that

$$\frac{d\Pi}{dx} = g - R(v)\sqrt{1+p^2} = g - \alpha\Pi^n\sqrt{1+p^2} \tag{2.11}$$

where g is the acceleration due to gravity which acts along the AZ axis in Figure 2.2.[12] Equation (2.11) is equivalent to a statement of the conservation of energy with the mass = 1.

---

[10] Euler, I, XXIV, p. xxi, pp. 110ff. This is Proposition V. Problem.

[11] For a formulation of the problem see, e.g., Bolza, *VOR*, pp. 542ff, where it is discussed in detail.

[12] Actually, Euler takes Π to be v^2, but this is inconsistent with his definition of the centrifugal force below. It appears to be a misprint but occurs in several places throughout the text. It may have to do with Leibniz's ideas on kinetic energy. See Hiebert, *HR*, p. 81.

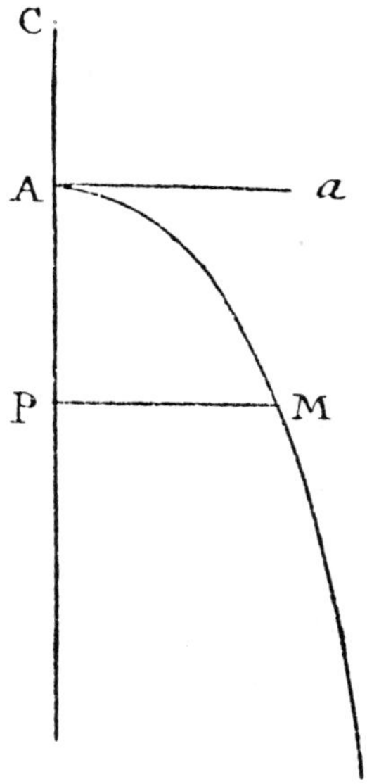

Figure 2.2

In Example I Euler considers the interesting problem of seeking among all curves in a vertical plane, which pass through two given points, the one down which a heavy particle, starting with a given initial velocity, will fall so that it has the greatest possible terminal velocity or terminal kinetic energy Π (see Figure 2.2). To solve this problem, Euler notes that $Z = [Z] = g - \alpha\Pi^n(1 + p^2)^{1/2}$ and hence that

$$[L] = -\alpha n\Pi^{n-1}\sqrt{1 + pp}\,, \qquad [M] = 0 = [N],$$
$$[P] = -\frac{\alpha\Pi^n p}{\sqrt{1 + pp}}\,, \qquad [Q] = 0,$$

etc. His differential equation (2.10) becomes in this case

$$0 = -d\cdot[P]e^{-\int [L]\,dx} \quad \text{or} \quad C = [P]e^{-\int [L]\,dx},$$

and thus $[L]\,dx = d[P]/[P]$ since $L = [L]$ in (2.10′). He now replaces $[L]$ and $[P]$ by their values as given in the relation $-\int[L]\,dx = lC - l[P]$, where he uses l for *logarithm*. He finds

$$\int \alpha n\Pi^{n-1}\,dx\sqrt{1 + pp} = +lC - l(-\alpha) - l\Pi^n - lp + l\sqrt{1 + pp}\,.$$

This gives, on differentiation and simplification, the differential equation

$$0 = n\,d\Pi + \alpha n\Pi^n\,dx\sqrt{1 + pp} + \frac{\Pi\,dp}{p(1 + pp)}\,.$$

Euler notes with the help of (2.11) that $d\Pi + \alpha\Pi^n\,dx(1 + pp)^{1/2} = g\,dx$, and so he finds

$$0 = ng\,dx + \frac{\Pi\,dp}{p(1+pp)} \quad \text{or} \quad \Pi = -\frac{ngp\,dx(1+pp)}{dp}. \tag{2.12}$$

To simplify notations, he now sets $dx = -t\,dp/ng$ and has

$$\Pi = pt(1+pp).$$

He then calculates $d\Pi$, which after some manipulation yields the equation

$$\frac{np\,dt(1+pp) + t\,dp(n+1+3npp)}{nt^{n+1}p^{n+2}(1+pp)^{n+1/2}} = \frac{\alpha\,dp}{ngp^2}.$$

On integration, this becomes

$$g = (\alpha + \beta p)t^n p^n (1+pp)^{n-1/2},$$

where β is a constant of integration. From this, the definition of t, and the fact that $dy = p\,dx$, he finds the differential equations

$$dx = \frac{-dp}{np(1+pp)^{1-1/2n}\sqrt[n]{g^{n-1}(\alpha+\beta p)}},$$

$$dy = \frac{-dp}{n(1+pp)^{1-1/2n}\sqrt[n]{g^{n-1}(\alpha+\beta p)}}, \qquad \Pi = \sqrt[n]{\frac{g\sqrt{1+pp}}{\alpha+\beta p}}.$$

He concludes that

$$x = -\frac{1}{ng}\int \frac{dp}{p(1+pp)}\sqrt[n]{\frac{g\sqrt{1+pp}}{\alpha+\beta p}},$$

$$y = -\frac{1}{ng}\int \frac{dp}{1+pp}\sqrt[n]{\frac{g\sqrt{1+pp}}{\alpha+\beta p}}.$$

Euler notes that $\Pi = v^2/2$ can never vanish and hence, by (2.12), that dp is always negative so that the curve is concave toward the axis as in the figure. The vertical axis is AP and the horizontal one is Aa. At the start of the motion $p = \infty$ and, by definition, $\Pi = b$. Then $\beta = g/b^n$, and he sets $\alpha = 1/k^n$. This gives

$$x = -\frac{bk}{ng}\int \frac{dp}{p(1+pp)}\sqrt[n]{\frac{g\sqrt{1+pp}}{b^n + gk^n p}},$$

$$y = -\frac{bk}{ng}\int \frac{dp}{1+pp}\sqrt[n]{\frac{g\sqrt{1+pp}}{b^n + gk^n p}}, \qquad \Pi = bk\sqrt[n]{\frac{g\sqrt{1+pp}}{b^n + gk^n p}}.$$

Euler defines the centrifugal force to be the component of the total force on the particle acting along the normal. This gives

$$F = \frac{v^2}{\rho} = \frac{2\Pi}{\rho},$$

where ρ is the radius of curvature; i.e.,

$$\frac{1}{\rho} = \frac{-dp}{dx(1+pp)^{3/2}}.$$

The normal force on the particle due to the action of gravity alone—ignoring the reaction of the curve—is clearly

$$\frac{g\,dy}{ds} = \frac{gp}{\sqrt{1+p^2}} = G.$$

It is perhaps relevant to point out that in the general case the reaction of the curve to the heavy particle is given by[13]

$$N = \frac{2n-1}{2n} \cdot \frac{v^2}{\rho};$$

and thus Euler's centrifugal force F—the total normal force—is then exactly equal to the normal force due to gravity G plus the reaction N of the curve, as it should be. Euler then shows by direct calculation that $F = 2nG$ and takes up in a little more detail the cases $n = 1$ and $n = \frac{1}{2}$. Note that in the latter case $F = G$.

In the former case, $n = 1$, he has

$$x = -bk\int \frac{dp}{p(b+gkp)\sqrt{1+pp}}, \qquad y = -bk\int \frac{dp}{(b+gkp)\sqrt{1+pp}};$$

and the arc-length is given by

$$s = AM = -bk\int \frac{dp}{p(b+gkp)} = C + kl\frac{b+gkp}{p}.$$

He determines the constant C to be $-klgk$ by using the fact that at the starting point A of the motion in Figure 2.2, $p = \infty$ and $s = 0$. Thus

$$s = kl\frac{b+gkp}{gkp}.$$

He finds in this case that

$$p = \frac{b}{gk(e^{s/k}-1)} = \frac{dy}{dx}$$

---

[13] Bolza, *VOR*, p. 580. (Note that the n in Bolza's and Euler's accounts differ by a factor 2.) The solutions to Examples I and II below ($n = 1$, $n = 1/2$) are very similar.

and that

$$y=\frac{bk}{\sqrt{b^2+g^2k^2}}\log\frac{(b+gkp)\left(b+\sqrt{b^2+g^2k^2}\right)}{gk\left(bp-gk+\sqrt{(b^2+g^2k^2)(1+p^2)}\right)}.$$

In the latter case, $n=\frac{1}{2}$, he calculates without difficulty that

$$x=-2gbk\int\frac{dp}{p\left(\sqrt{b}+gp\sqrt{k}\right)^2},\qquad y=-2gbk\int\frac{dp}{\left(\sqrt{b}+gp\sqrt{k}\right)^2}$$

and hence that

$$x=-gy\sqrt{\frac{k}{b}}+2gk\log\frac{2b\sqrt{k}}{2b\sqrt{k}-y\sqrt{b}},\qquad y=\frac{2b\sqrt{k}}{\sqrt{b}+gp\sqrt{k}}.$$

Euler remarks correctly that in this case the minimizing curve is the same as that for a free but heavy particle falling under the action of gravity through the resisting medium.

In his Example II on p. 121 Euler takes up the problem of finding that plane curve down which a heavy particle will fall in the shortest possible time through a resisting medium; he assumes, as before, that the law of resistance is given by the $2n$th power of the velocity. The integral to be minimized is now

$$\int_0^a\frac{dx\sqrt{1+p^2}}{\sqrt{\Pi}},$$

where the quantity $\Pi=v^2/2$. Since $Z=(1+p^2)^{1/2}/\Pi^{1/2}$ and $[Z]=g-\alpha\Pi^n(1+p^2)^{1/2}$, he has

$$L=-\frac{\sqrt{1+p^2}}{2\Pi\sqrt{\Pi}},\qquad M=N=0,\qquad P=\frac{p}{\sqrt{\Pi(1+p^2)}},$$

$$[L]=-\alpha n\Pi^{n-1}\sqrt{1+p^2},\qquad [M]=[N]=0,\qquad [P]=-\frac{\alpha\Pi^n p}{\sqrt{1+p^2}}.$$

From these, Euler finds that the multiplier V in the Euler–Lagrange equation (2.10) is of the form

$$V=\exp\left(\alpha n\int_0^x\Pi^{n-1}dx\sqrt{1+p^2}\right)$$
$$\cdot\left[\int_0^x\exp\left(-\alpha n\int_0^x\Pi^{n-1}dx\sqrt{1+p^2}\right)\frac{dx\sqrt{1+p^2}}{2\Pi\sqrt{\Pi}}-H\right]\tag{2.13}$$

where H is the value of the integral in the brackets evaluated from 0 to a. Hence V vanishes for $x=a$.

This enables him to show that

$$dV = \alpha n V \Pi^{n-1} dx\sqrt{1+p^2} + \frac{dx\sqrt{1+p^2}}{2\Pi\sqrt{\Pi}}, \tag{2.13'}$$

also his equation (2.10) above in this case has the form

$$d(P+[P]V)=0 \quad \text{or} \quad V=\frac{C-P}{[P]}.$$

(Thus for $x = a$, $C = P$.) This implies that

$$V = \frac{C\sqrt{1+p^2} - p/\sqrt{\Pi}}{-\alpha\Pi^n p} = \frac{1}{\alpha\Pi^n\sqrt{\Pi}} - \frac{C\sqrt{1+p^2}}{\alpha\Pi^n p}, \tag{2.14}$$

or

$$\begin{aligned} dV &= \frac{-(n+\frac{1}{2})\,d\Pi}{\alpha\Pi^{n+1}\sqrt{\Pi}} + \frac{nC\,d\Pi\sqrt{1+p^2}}{\alpha\Pi^{n+1}p} + \frac{C\,dp}{\alpha\Pi^n p^2\sqrt{1+p^2}} \\ &= \frac{dx\sqrt{1+p^2}}{2\Pi\sqrt{\Pi}} + \frac{n\,dx\sqrt{1+p^2}}{\Pi\sqrt{\Pi}} - \frac{nC(1+p^2)\,dx}{\Pi p}. \end{aligned} \tag{2.15}$$

The latter equation follows easily from (2.13′) and (2.14). But, by hypothesis, $d\Pi = g\,dx - \alpha\Pi^n\,dx(1+p^2)^{1/2}$, and so Euler finds with the help of relations (2.15) that

$$\frac{C\,dp}{p^2\sqrt{1+p^2}} = \frac{(n+\frac{1}{2})g\,dx}{\Pi\sqrt{\Pi}} - \frac{nCg\,dx\sqrt{1+p^2}}{\Pi p}$$

or equivalently that

$$\frac{C\,dp}{gp^2} = \frac{(n+\frac{1}{2})\,dx\sqrt{1+p^2}}{\Pi\sqrt{\Pi}} - \frac{nC\,dx(1+p^2)}{\Pi p}.$$

Notice that the right-hand member of this last equation is precisely dV in (2.15). This leads Euler to conclude immediately that

$$dV = \frac{C\,dp}{gp^2}, \qquad V = D - \frac{C}{gp} = \frac{1}{\alpha\Pi^n\sqrt{\Pi}} - \frac{C\sqrt{1+p^2}}{\alpha\Pi^n p}.$$

He now has the two relations between p, x, and Π

$$\frac{C\,dp}{gp^2} = \frac{(n+\frac{1}{2})\,dx\sqrt{1+p^2}}{\Pi\sqrt{\Pi}} - \frac{nC\,dx(1+p^2)}{\Pi p},$$

$$\alpha D - \frac{\alpha C}{gp} = \frac{1}{\Pi^n\sqrt{\Pi}} - \frac{C\sqrt{1+p^2}}{\Pi^2 p}.$$

If he were to eliminate Π between these equations, he would have a relation between p and x, which would define the minimizing curve. Euler prefers, however, to treat Π as a parameter and writes

$$x = \int \frac{d\Pi}{g - \alpha\Pi^n\sqrt{1 + pp}}, \qquad y = \int \frac{p\,d\Pi}{g - \alpha\Pi^n\sqrt{1 + pp}}.$$

This discussion essentially completes Euler's Chapter III. There is a scholium attached, but it mainly summarizes the first three chapters.

2.4. Invariance Questions

It is remarkable that as early as 1744 Euler was already concerned with the problem of the invariance of his fundamental equation or necessary condition. In the first part of Chapter IV he indicates that his fundamental condition remains invariant under "general" transformations of the coordinate axes.[14] According to Carathéodory, "The first question which is discussed in Chapter IV deals with the geometrical significance of the variables x, y that enter into a variational problem." This is the essence of Euler's Proposition I. Problem together with its corollaries and scholium. Following this, he considers a number of examples where x, y are not related by being cartesian, rectangular coordinates, and shows the utility of his ideas on covariance. It is truly in keeping with Euler's genius that he should have worked at ideas that were only to be satisfactorily and completely discussed in modern times. His first problem (p. 125) is to find the form of his necessary condition when y is given as a function of x $(0 \leqslant x \leqslant a)$ in a coordinate system which is not necessarily cartesian; thus, e.g., it might be polar. Euler lists his results on pp. 127 and 128 in a scholium and observes that his equation is invariant in all preceding situations.

To illustrate this point, he starts with the case of polar coordinates in Figure 2.3 with a pole at C. The lines CA and CM are radii, and he seeks the shortest curve joining two fixed points in this coordinate system. In Figure 2.3 the angle $ACM = x$ and the radius $CM = y$; he then draws an arc BSs of a circle of radius $CB = 1$ and notes that when m is a point on the curve neighboring to M and $Cn = CM$, then $dx = Ss$ and $dy = mn$. Then the similar triangles CSs and CMn are such that $1/dx = CM/Mn = y/y\,dx$. Euler now reasons that

$$Mm = \left(dy^2 + y^2\,dx^2\right)^{1/2} = dx\left(y^2 + p^2\right)^{1/2}$$

[14] I am indebted to Professor A. Weil for the observation that Euler's interest in this topic probably stemmed from Leibniz's inquiries into the behavior of $\int y\,dx$ under coordinate transformations.

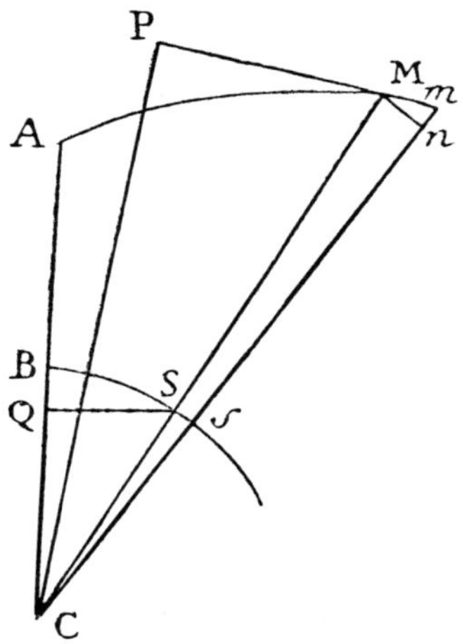

Figure 2.3

since $p = dy/dx$. (He now has shown the well-known formula for arc-length in polar coordinates, namely: $ds/d\phi = [r^2 + (dr/d\phi)^2]^{1/2}$.) Thus the length of the curve AM, which is to be minimum, is given as

$$\int_0^a \sqrt{y^2 + p^2}\, dx$$

and $Z = (y^2 + p^2)^{1/2}$. It is then clear that

$$M = 0, \qquad N = \frac{y}{\sqrt{y^2 + p^2}}, \qquad P = \frac{p}{\sqrt{y^2 + p^2}}, \qquad 0 = Q = R = \cdots.$$

Euler knows by his invariance property that $N - dP/dx = 0$ or $N\,dy = p\,dP$; but $dZ = N\,dy + P\,dp$ and thus $dZ = p\,dP + P\,dp = d(pP)$, i.e., $Z + C = Pp$. In the present case this means that

$$C + \sqrt{y^2 + p^2} = \frac{p^2}{\sqrt{y^2 + p^2}};$$

i.e.,

$$\frac{y^2}{\sqrt{y^2 + p^2}} = \text{const.} = b. \tag{2.16}$$

From this last relation Euler infers that the minimizing curve is a straight line. To do this, he notes in effect that his relation (2.16) states that the length of the perpendicular CP dropped from the pole C upon the tangent MP to the curve AM at the point M is a constant and thus that the curve must be the straight line joining the end-points.[15]

---

[15] Williamson, *DC*, pp. 223–224.

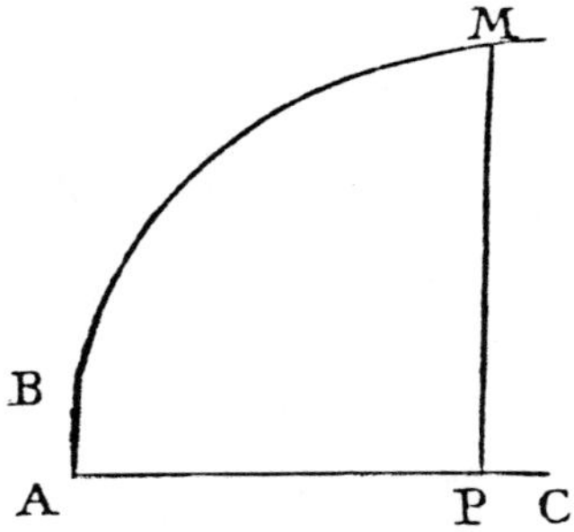

Figure 2.4

His next example is derived from the famous one of the Greeks: among all admissible curves ABM in Figure 2.4, enclosing a given area $ABMP$, to find the one of least length. This time he sets the area $ABMP = x$, the ordinate $PM = y$, and the abscissa $AP = t$. Then

$$x = \int y\, dt$$

and the arc-length BM will be $\int(dy^2 + dx^2/y^2)^{1/2}$, since $dt = dx/y$ and the lines AP and PM form a cartesian coordinate system, $ds^2 = dt^2 + dy^2$, in that system.

Now if $p = dy/dx$, Euler has

$$Z = \int \sqrt{\frac{1}{y^2} + p^2}\; dx$$

or

$$M = 0, \qquad N = \frac{-1}{y^2\sqrt{1 + y^2p^2}}, \qquad P = \frac{yp}{\sqrt{1 + y^2p^2}}, \qquad Q = 0,$$

etc. From these relations he has immediately his condition that

$$b = y\sqrt{1 + y^2p^2}^{\,1/2};$$

i.e.,

$$p = \frac{\sqrt{b^2 - y^2}}{y^2} = \frac{dy}{dx} = \frac{dy}{y\,dt}.$$

Euler integrates this equation and writes $t = c \pm (b^2 - y^2)^{1/2}$; he has shown the solution curve to be an arc of a circle with center somewhere on the line AP, for example, at C in Figure 2.4.

His Example IV (p. 133) is interesting. He seeks among all curves on a given concave or convex surface the geodesic joining two fixed points thereon. In his Figure 2.5 Euler shows a plane section APQ of his surface with the line AP extended as one coordinate axis. At an arbitrary point Q

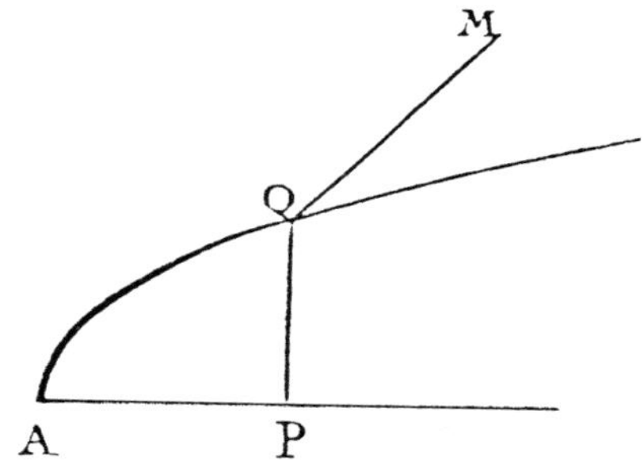

Figure 2.5

he draws QM perpendicular to the plane of the paper, and he assumes M is the point where it cuts the surface. Then he sets $AP = x$, $PQ = y$, and $QM = z$, and he comments that z is a function of x and y—it defines, in fact, the equation of the surface. He writes it as

$$dz = T\,dx + V\,dy,$$

where T and V are themselves functions of x and y, which he exhibits by writing $dT = E\,dx + F\,dy$, $dV = F\,dx + G\,dy$. (Note that the function F in each of these relations is $\partial^2 z/\partial x\,dy = \partial^2 z/\partial y\,dx$, under reasonable continuity considerations.) Given these facts, Euler now sees that arc-length along a curve on the surface is given by the relation

$$ds = \sqrt{dx^2 + dy^2 + dz^2} = \sqrt{dx^2 + dy^2 + (T\,dx + V\,dy)^2}\ ;$$

and thus that the integral to be minimized is

$$\int_0^a dx\sqrt{1 + p^2 + T^2 + 2TVp + V^2p^2}\ .$$

He then calculates M, N, P and writes out his condition in the form

$$\frac{TF\,dx + VFp\,dx + TGp\,dx + VGp^2\,dx}{\sqrt{1 + p^2 + T^2 + 2TVp + V^2p^2}} = d\cdot\frac{p + TV + V^2p}{\sqrt{1 + p^2 + T^2 + 2TVp + V^2p^2}}\ .$$

With the help of the relations $dV = F\,dx + G\,dy$ and $dy = p\,dx$, this equation may be reduced to the form

$$\frac{T\,dV + Vp\,dV}{\sqrt{1 + p^2 + (T + Vp)^2}} = d\cdot\frac{p + TV + V^2p}{\sqrt{1 + p^2 + (T + Vp)^2}}\ ;$$

when the differentiation is carried out and the results simplified, Euler finds that

$$dp = \frac{(Tp - V)(dT + p\,dV)}{1 + T^2 + V^2}.$$

Consequently, since $dp/dx = d^2y/dx^2$, he has

$$dx\,d^2y = \frac{(T\,dy - V\,dx)(dx\,dT + dy\,dV)}{1 + T^2 + V^2}. \tag{2.17}$$

To integrate this equation, he differentiates the relation $dz = T\,dx + V\,dy$ and substitutes in (2.17) above to find, after a little calculation, that

$$dx\,d^2y + T\,dz\,d^2y = T\,dy\,d^2z - V\,dx\,d^2z$$

and hence that

$$\frac{d^2y}{d^2z} = \frac{T\,dy - V\,dx}{dx + T\,dz}. \tag{2.18}$$

Next he multiplies both members of this result by $dz = T\,dx + V\,dy$ and finds

$$T\,dx^2\,d^2y + V\,dx\,dy\,d^2y + T\,dz^2\,d^2y = T\,dy\,dz\,d^2z - V\,dx\,dz\,d^2z.$$

It follows then that

$$Td^2y(dx^2 + dy^2 + dz^2) = (dz\,d^2z + dy\,d^2y)(T\,dy - V\,dx).$$

Euler multiplies both members by dx, replaces $T\,dx$ by $dz - V\,dy$, and adds $-V\,dz^2\,d^2z$ to both members of the remaining equation, finding

$$\begin{aligned} d^2y(dx^2 + dy^2 + dz^2) &- V\,dz(dy\,d^2y + dz\,d^2z) \\ &= dy(dy\,d^2y + dz\,d^2z) - V\,d^2z(dx^2 + dy^2 + dz^2). \end{aligned}$$

From this he infers with the help of (2.18) that

$$\frac{dy\,d^2y + dz\,d^2z}{dx^2 + dy^2 + dz^2} = \frac{T\,d^2y}{T\,dy - V\,dx} = \frac{Td^2z}{dx + T\,dz} = \frac{d^2y + V\,d^2z}{dy + V\,dz}. \tag{2.19}$$

At this point Euler makes the simplifying assumption that the surface is a figure of revolution given in the form $y^2 + z^2 = X^2(x)$ so that $dz = X\,dX/z - y\,dy/z$. Since he has $T = X\,dX/z\,dx$, $V = -y/z$, the first and last members of (2.19) can be expressed as

$$\frac{dy\,d^2y + dz\,d^2z}{dx^2 + dy^2 + dz^2} = \frac{z\,d^2y - y\,d^2z}{z\,dy - y\,dz},$$

whose integral is expressible as $b(dx^2 + dy^2 + dz^2)^{1/2} = z\,dy - y\,dz$. (Since Euler has chosen x as his independent variable, $d^2x = 0$.) He sets $dX = v\,dx$ and finds directly that

$$dz = \frac{Xv\,dx - y\,dy}{\sqrt{X^2 - y^2}}, \qquad z\,dy - y\,dz = \frac{X^2dy - Xyv\,dx}{\sqrt{X^2 - y^2}};$$

he then has an expression for $(dx^2 + dy^2 + dz^2)^{1/2}$ in terms of dx, dy, which can be solved for dy. This yields after considerable manipulation the relation

$$dy = \frac{yv\,dx}{X} \pm \frac{b\,dx\sqrt{(1+v^2)(y^2-X^2)}}{X\sqrt{b^2-X^2}}.$$

Finally he sets $y = Xt$ and has—in his notation—

$$\frac{dt}{\sqrt{tt-1}} = \frac{b\,dx\sqrt{1+vv}}{X\sqrt{bb-XX}}.$$

In closing his discussion he remarks that both X and v are functions of x alone and hence that the variables are separated.[16]

In the second part of this chapter Euler takes up another topic. He considers now three indefinite integrals

$$\int Z\,dx, \quad \int X\,dx, \quad \int Y\,dx, \tag{2.20}$$

in which Z, X, and Y are given functions of x, y, z, p, q, etc. as well as of a number of indefinite integrals as in his Chapter III. When the integrals are evaluated at $x = 0$ and a, Euler calls them A, B, and C and considers a given function W of these quantities. He then seeks to find the arc $y = y(x)$ $(0 \leqslant x \leqslant a)$ such that $W(A, B, C)$ has an extreme value.[17]

Prior to taking up this general result he works through his simpler Propositions II and III. Problems (pp. 140, 146) where he seeks to minimize or maximize $A \cdot B$ and A/B. To handle these, he supposes some particular ordinate y is increased by an amount nv and remarks that A and B will change by amounts dA and dB. He concludes that the fraction A/B will go into $(A + dA)/(B + dB)$ and that along the extremal he has

$$\frac{A + dA}{B + dB} = \frac{A}{B}$$

and hence that $B\,dA = A\,dB$ is a necessary condition. In Proposition II he remarks, by similar reasoning, that $A \cdot B$ will go into

$$(A + dA) \cdot (B + dB) = A \cdot B$$

and hence that $A\,dB + B\,dA = 0$ is now the necessary condition.

In his third example (p. 151) he takes up the heavy, hanging chain problem. Here he seeks to minimize the height of the center of gravity which can be expressed as

$$\frac{\int_0^a x\,dx\sqrt{1+p^2}}{\int_0^a dx\sqrt{1+p^2}},$$

---

[16] Euler, I, XXIV, p. 136.

[17] Euler, I, XXIV, pp. 140–163. His general result is stated in Proposition IV. Problem on p. 155.

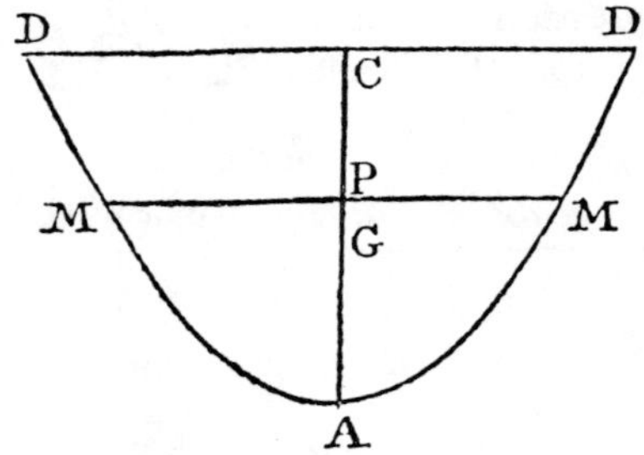

Figure 2.6

where $AP = x$ and $PM = y$ in Figure 2.6. Here $A = \int x\,dx(1+p^2)^{1/2}$, $B = \int dx(1+p^2)^{1/2}$. To calculate dA, for example, he writes the integral as the sum

$$A = \cdots + x_{,} dx\sqrt{1+p_{,}^2} + x\,dx\sqrt{1+p^2} + \cdots$$

and notes that the ordinate y enters precisely into $p_{,}$ and p and no others. It then follows that

$$dA = \left[\frac{x_{,}p_{,}}{\sqrt{1+p_{,}^2}} - \frac{xp}{\sqrt{1+p^2}}\right] \cdot nv = -nv \cdot d\frac{xp}{\sqrt{1+p^2}}.$$

Similarly, he finds that

$$dB = -nv \cdot d\frac{p}{\sqrt{1+p^2}}.$$

He therefore finds that

$$d\frac{xp}{\sqrt{1+p^2}} = cd\frac{p}{\sqrt{1+p^2}},$$

where $c = A/B$ is a constant. It then follows that $b(1+p^2)^{1/2} = (c-x)p$ and hence that

$$y = \int \frac{b\,dx}{\sqrt{(c-x)^2 - b^2}},$$

which is the equation of a catenary.

In his general result, Proposition IV. Problem (p. 155) concerning a function $W(A, B, C)$ he proceeds much as just indicated. He varies an ordinate y by nv and calculates dA, dB, and dC. He substitutes $A + dA$, $B + dB$, and $C + dC$ into W and states that now $W + dW = W$; i.e., $dW = 0$ is his necessary condition. In his Corollary 4 he assumes that W is a function of the variables A, B, C, etc. and remarks that his necessary condition then becomes

$$dW = 0 = F\,dA + G\,dB + H\,dC + \cdots,$$

where F, G, H, etc. are the relevant partial derivatives of W.

The chapter closes with four examples of this technique. The first of these is to find the extremals for the expression

$$\sqrt{1 + p^2(a)}\int_0^a y\,dx + y(a)\int_0^a dx\sqrt{1 + p^2}\,.$$

Euler sets $f = y(a)$, $g = (1 + p^2(a))^{1/2}$, and

$$A = \int_0^a y\,dx, \qquad B = \int_0^a dx\sqrt{1 + p^2}\,.$$

He now proceeds to calculate that, in his notation,

$$dA = nv \cdot dx, \qquad dB = -nv \cdot d \cdot \frac{p}{\sqrt{1 + pp}}\,.$$

His condition is then $g\,dA + f\,dB = 0$, which implies that

$$g\,dx = fd\,\frac{p}{\sqrt{1 + pp}}\,;$$

or if $c = f/g$,

$$x + b = \frac{cp}{\sqrt{1 + pp}}\,;$$

then by solving for p and integrating, Euler has the result

$$y = h \pm \sqrt{c^2 - (x + b)^2}\,.$$

To find the value of the constant c, the radius of this circle, he notes by definition that

$$\frac{f}{g} = \frac{y}{\sqrt{1 + p^2}}\bigg|^{x=a} = c$$

and that

$$y(a) = h \pm \sqrt{c^2 - (a + b)^2}\,, \qquad \sqrt{1 + p^2}\,\bigg|^{a} = \frac{c}{\sqrt{c^2 - (a + b)^2}}\,.$$

It follows that

$$c^2 = h\sqrt{c^2 - (a + b)^2} \pm \left(c^2 - (a + b)^2\right),$$

which determines the relation between a and c. To fix the constants of integration, Euler sets $h = 0$ and $b = -c$. The circle is then

$$y = \sqrt{2cx - x^2}\,.$$

It clearly has the axis for a diameter and passes through the origin. In the equation above relating a and c Euler takes the upper sign and concludes that $c = a$. He then considers yet another case, but this is not really relevant here.

2.5. Isoperimetric Problems

In Euler's Chapter V we find him explaining how to handle problems in which the class of admissible arcs must satisfy an accessory condition, e.g., an isoperimetric condition. To handle such situations, Euler, just as the Bernoulli brothers did, now allows two "consecutive" ordinates to vary independently. In Figure 2.7 these ordinates are $Nn = y^{\text{iv}}$ and $Oo = y^{\text{v}}$ and they are varied by $n\nu$ and $o\omega$, respectively. He points out that after he calculates the variations in the quantities y, p, q, r, s, etc., he finds each one is of the form $n\nu \cdot I + o\omega \cdot K$.[18] He has, e.g.,

$$
\begin{aligned}
&d \cdot y^{\text{iv}} = n\nu, \qquad d \cdot y^{\text{v}} = o\omega,\\
&d \cdot p''' = \frac{n\nu}{dx}, \quad d \cdot p^{\text{iv}} = -\frac{n\nu}{dx} + \frac{o\omega}{dx}, \quad d \cdot p^{\text{v}} = -\frac{o\omega}{dx},\\
&d \cdot q'' = \frac{n\nu}{dx^2}, \quad d \cdot q''' = -\frac{2n\nu}{dx^2} + \frac{o\omega}{dx^2}, \quad d \cdot q^{\text{iv}} = \frac{n\nu}{dx^2} - \frac{2o\omega}{dx^2},\\
&d \cdot q^{\text{v}} = \frac{o\omega}{dx^2}.
\end{aligned}
$$

He observes that I is the expression that would be present if only Nn had been varied, and it is therefore the same value that has been already

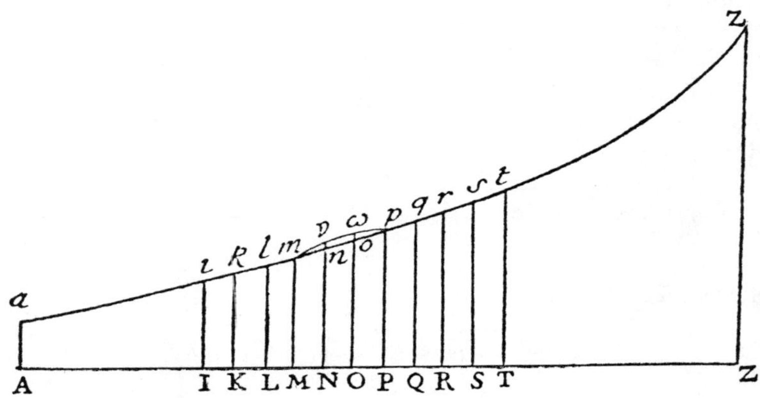

Figure 2.7

[18] Euler, I, XXIV, p. 172. Here he tabulates these variations in detail.

calculated. He further observes that K is the expression that would be present if only Oo had been varied and is therefore the value of I at the next ordinate; this value he calls I' and writes $K = I' = I + dI$.

We saw earlier in the case where only one ordinate $Nn = y'$ was varied that the nonzero part of the variation of the integral $\int Z\,dx$, or rather finite sum which replaced the integral, was an expression of the form $dA \cdot dy' = I \cdot nv$. Now the relevant part is of the form

$$dA \cdot dy^{\text{iv}} + dA' \cdot dy^{\text{v}} = nv \cdot dA + o\omega \cdot dA'. \tag{2.21}$$

If the accessory condition is given by another integral being a constant, then its relevant part will clearly be of the form[19]

$$dB \cdot dy^{\text{iv}} + dB' \cdot dy^{\text{v}} = nv \cdot dB + o\omega \cdot dB'. \tag{2.21'}$$

Euler further remarks in Proposition IV. Problem (p. 174) that along an extremal both (2.21) and (2.21′) above must vanish and hence for arbitrary quantities α, β

$$nv \cdot \alpha\, dA + o\omega \cdot \alpha\, dA' = 0, \qquad nv \cdot \beta\, dB + o\omega \cdot \beta\, dB' = 0.$$

He then has for proper choice of α, β

$$\alpha\, dA + \beta\, dB = 0 \tag{2.22}$$

and equally

$$\alpha\, dA' + \beta\, dB' = 0. \tag{2.22'}$$

This gives Euler the condition

$$\frac{nv}{o\omega} = -\frac{dA'}{dA} = -\frac{dB'}{dB}\,.$$

from which he infers that

$$\frac{d^2A}{dA} = \frac{d^2B}{dB}\,, \tag{2.23}$$

since $dA' = dA + d^2A$ and $dB' = dB + d^2B$. On integrating (2.23), he finds that $\log dA - \log dB = \log C$ and hence that $dA = C\,dB$. But by the relation (2.22) he notes that $C = -\beta/\alpha$. He observes that there are also quantities α', β' such that $\alpha'\, dA' + \beta'\, dB' = 0$—this is, in essence, a consequence of John Bernoulli's law of uniformity. It then follows directly that

$$\frac{\alpha'}{\alpha} = \frac{\beta'}{\beta}$$

is a constant.[20]

In the Scholium 1 that Euler appends to this proposition, he notes that there are two closely associated problems which are essentially equivalent: the first is to maximize or minimize a function V in the class of arcs for

---

[19] Euler, I, XXIV, pp. 174–175.

[20] Euler, I, XXIV, p. xxiii. Euler essentially points out the result on p. 175. We shall see in the next section how Euler tried to generalize this result.

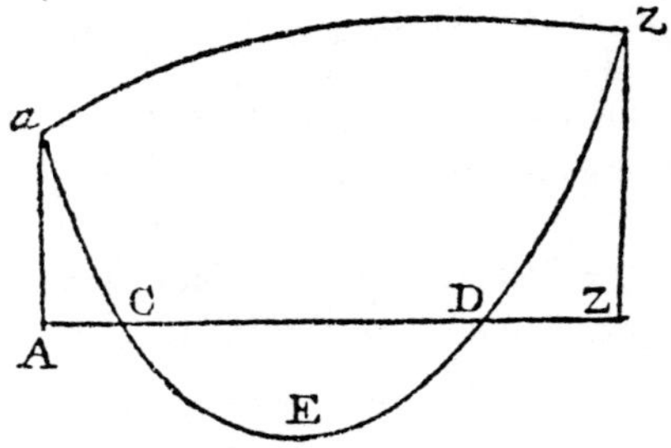

Figure 2.8

which another functional W has a given constant value; the second is to maximize or minimize W in the class of arcs for which V has a given constant value. It is easy to see that both problems have the same set of extremals. Bolza observed that A. Mayer showed "that the two problems are equivalent in respect to the usual necessary conditions for an extremal."[21] Mayer called this result *the reciprocity theorem for isoperimetric problems*. (See pp. 295ff below.)

To illustrate his procedure, Euler, in his usual manner, works through a considerable number of examples. We will look at his Example V (pp. 184–185): to find among all curves az containing an equal area $aAZz$ (Figure 2.8), the one which when rotated about the axis AZ generates the solid of least superficial area. Analytically, the problem is to find an arc $y(x)$ $(0 \leqslant x \leqslant a)$ such that

$$\int y\,dx = \text{const.}, \qquad \int y\,dx\sqrt{1+p^2} = \min.$$

He then finds at once the Euler condition

$$n\,dx = dx\sqrt{1+p^2} - d\cdot\frac{yp}{\sqrt{1+p^2}},$$

in which $\alpha/\beta = -n$. This yields the condition

$$\sqrt{1+p^2} = \frac{y}{ny+b}.$$

This is equivalent to the differential equation

$$p = \frac{dy}{dx} = \frac{\sqrt{y^2-(ny+b)^2}}{ny+b}.$$

For $b = 0$, the solution of this equation is the straight line joining the points A and Z. For the case when the multiplier n is zero, the differential

[21] Bolza, *VOR*, p. 488, Mayer [1877], p. 60; Kneser, *LV*, pp. 131, 136.

equation becomes

$$dx = \frac{b\,dy}{\sqrt{y^2 - b^2}}\ ;$$

this is a catenary since its solution is of the form

$$\cosh\left(\frac{x+k}{b}\right) = \frac{y}{b}\ .$$

For $n = -1$, the differential equation becomes

$$dx = \frac{(b-y)\,dy}{\sqrt{2by - y^2}}\ ,$$

and the curve is expressible as

$$x = c + \frac{2b-y}{3b}\sqrt{2by - b^2}$$

or equivalently as

$$9b(x-c)^2 = (2b-y)^2(2y-b).$$

Euler remarks that this is a curve of the third order and of Newton's type 68.

After working through 11 examples, Euler moves to the more general case, Proposition V. Problem, where Z contains not only the variables x, y, p, q, etc. but also an indefinite integral; thus

$$dZ = L\,d\Pi + M\,dx + N\,dy + P\,dp + Q\,dq + \cdots,$$

$$\Pi = \int [Z]\,dx, \qquad d[Z] = [M]\,dx + [N]\,dy + [P]\,dp + [Q]\,dq + \cdots.$$

The problem Euler poses is to make $\int_0^a Z\,dx$ a maximum or a minimum subject to the condition that $\int_0^a [Z]\,dx$ is a given constant. He now shows, by analogy with the method described in his Chapter IV (see Section 2.4 above) and what he just showed, that his necessary condition becomes, in his notation,

$$0 = N + \left(\alpha + H - \int L\,dx\right)[N] - \frac{d\left(P + \left(\alpha + H - \int L\,dx\right)[P]\right)}{dx}$$
$$+ \frac{dd\left(Q + \left(\alpha + H - \int L\,dx\right)[Q]\right)}{dx^2} - \text{etc.}, \tag{2.24}$$

where $H = \int_0^a L\,dx$. His proof is not difficult. He notes that the differential of Π is

$$n\nu \cdot dx\left([N] - \frac{d[P]}{dx} + \frac{d^2[Q]}{dx^2} + \text{etc.}\right)$$

and of Z is

$$nv \cdot dx\left(N + [N]V - \frac{d(P + [P]V)}{dx} + \frac{d^2(Q + [Q]V)}{dx^2} - \text{etc.}\right),$$

where $V = H - \int L\,dx$. From these he concludes there is a constant α such that relation (2.24) obtains.

He now combines α and H into a new constant C and writes[22]

$$0 = N + \left(C - \int L\,dx\right)[N] - \frac{d\left(P + \left(C - \int L dx\right)[P]\right)}{dx} + \frac{dd\left(Q + \left(C - \int L\,dx\right)[Q]\right)}{dx^2} - \text{etc.} \tag{2.24$'$}$$

In his Scholium (pp. 196ff) Euler considers the case where $\Pi = s$ is arc-length; i.e., $[Z] = (1 + \mathrm{p}^2)^{1/2}$ and

$$dZ = L\,ds + M\,dx + N\,dy + P\,dp + \cdots.$$

In this case Euler finds, in his notation, that

$$\frac{1}{dx} d \cdot \frac{\left(C - \int L\,dx\right)p}{\sqrt{1 + pp}} = N - \frac{dP}{dx} + \frac{d\,dQ}{dx^2} - \text{etc.} \tag{2.25}$$

since $[P] = p/(1 + p^2)^{1/2}$. The result follows then immediately from (2.24′). In the case when $N = 0$, Euler sees that the differential equation (2.25) can be integrated and finds

$$A + \frac{\left(C - \int L\,dx\right)p}{\sqrt{1 + p^2}} = -P + \frac{dQ}{dx} - \cdots. \tag{2.26}$$

In the case when $M = 0$ instead, he multiplies both members of (2.25) by $p\,dx = dy$ and has

$$pd \cdot \frac{\left(C - \int L\,dx\right)p}{\sqrt{1 + p^2}} = N\,dy - p\,dP + p\,\frac{d\,dQ}{dx} - \text{etc.}$$

to which he adds $L\,ds = dZ - N\,dy - P\,dp - Q\,dq + \cdots$ and integrates,

---

[22] Carathéodory in his Introduction on p. xxiv, notes that if one sets $d\lambda/dx = -L$, then Euler's relation (2.24) above may be written in the form

$$0 = (N + [N]\lambda) - \frac{d}{dx}(P + [P]\lambda) + \frac{d^2}{dx^2}(Q + [Q]\lambda) - \cdots,$$

$$\frac{d\lambda}{dx} = -L,$$

which is equivalent to (2.8) above (see p. 76) except that $\lambda(a) = 0$ is no longer valid.

finding

$$\int\left[L\,dx\sqrt{1+p^2}+pd\cdot\frac{\left(C-\int L\,dx\right)p}{\sqrt{1+p^2}}\right]$$
$$=-A+Z-Pp-Qq+p\frac{dQ}{dx}+\cdots.$$

But the left-hand member of this equation is integrable in the form

$$-\frac{C-\int L\,dx}{\sqrt{1+p^2}},$$

as can be readily verified. He thus has in the case $M=0$ the relation

$$\frac{C-\int L\,dx}{\sqrt{1+pp}}=A-Z+Pp+Qq-\frac{p\,dQ}{dx}.$$

On the other hand, if both $M=0$ and $N=0$, he multiplies both members of equation (2.26) by $dp=q\,dx$ and finds

$$A\,dp+\frac{\left(C-\int L\,dx\right)p\,dp}{\sqrt{1+pp}}=-P\,dp+q\,dQ. \tag{2.27}$$

Recall that $dZ=L\,dx(1+pp)^{1/2}+P\,dp+Q\,dq$. From these two relations Euler finds the new one

$$dZ+A\,dp-L\,dx\sqrt{1+pp}+\frac{\left(C-\int L\,dx\right)p\,dp}{\sqrt{1+pp}}=q\,dQ+Q\,dq$$

and hence upon integration,

$$Z+B+Ap+\left(C-\int L\,dx\right)\sqrt{1+pp}=Qq.$$

Combining this with (2.26), he concludes that

$$A\,dx-B\,dy=Z\,dy-P\,dx-Pp\,dy+dQ+pp\,dQ-Qp\,dp; \tag{2.28}$$

notice that relation (2.28) is free of the integral $\int L\,dx$.

This is very useful to him in the five examples that follow the scholium. Thus, e.g., in Example I (p. 198) he has $Z=s^n$ and thus $dZ=ns^{n-1}\,ds$. Here he has $L=ns^{n-1}$, $M=0$, $N=0$, $P=0$, etc., and relation (2.28) becomes in the present situation

$$A\,dx-B\,dy=Z\,dy=s^n\,dy; \tag{2.29}$$

moreover, $A^2ds^2 = A^2(dx^2 + dy^2) = dy^2(A^2 + (B + s^n)^2)$ so that

$$dy = \frac{A\,ds}{\sqrt{A^2 + (B + s^n)^2}}, \qquad dx = \frac{(B + s^n)\,ds}{\sqrt{A^2 + (B + s^n)^2}}. \tag{2.30}$$

These relations can be reexpressed with the help of $dy = p\,dx$ by noting in (2.29) that $s^n p = A - Bp$ and hence that

$$ds = dx\sqrt{1 + pp} = -\frac{A\,dp(A - Bp)^{(1-n)/n}}{np^{(1+n)/n}}.$$

Euler concludes that

$$x = -\frac{A}{n}\int \frac{dp(A - Bp)^{(1-n)/n}}{p^{(1+n)/n}\sqrt{1 + pp}}, \qquad y = -\frac{A}{n}\int \frac{dp(A - Bp)^{(1-n)/n}}{p^{1/n}\sqrt{1 + pp}}. \tag{2.31}$$

In the case $n = 1$ he finds from the second of the equations (2.30) that

$$x = \sqrt{A^2 + B^2 + 2Bs + ss} - \sqrt{A^2 + B^2},$$

or by setting $B = b$ and $(A^2 + B^2)^{1/2} = c$,

$$x + c = \sqrt{c^2 + 2bs + ss}.$$

From the first of equations (2.31) he has

$$x = \frac{A\sqrt{1 + pp}}{p} + b$$

or (the values b, c are constants of integration)

$$p = \frac{c}{\sqrt{(x - b)^2 - c^2}} = \frac{dy}{dx}.$$

This leads directly to the final relation

$$y = \int \frac{c\,dx}{\sqrt{(x - b)^2 - c^2}},$$

which is a catenary.

In Proposition VI. Problem on p. 207 Euler turns to a further generalization. He considers the variational problem that arises when A and B are two given definite integrals such that A is to have a given value and a function $\Phi(A, B)$ is to be an extremum. To handle this situation, Euler first observes that $d\Phi$ can be expressed as $d\Phi = \alpha\,dA + \beta\,dB$; and that since Φ is to be an extremum subject to the condition $A = \text{const.}$, there must exist a constant γ such that

$$(\alpha + \gamma)\,dA + \beta\,dB = 0$$

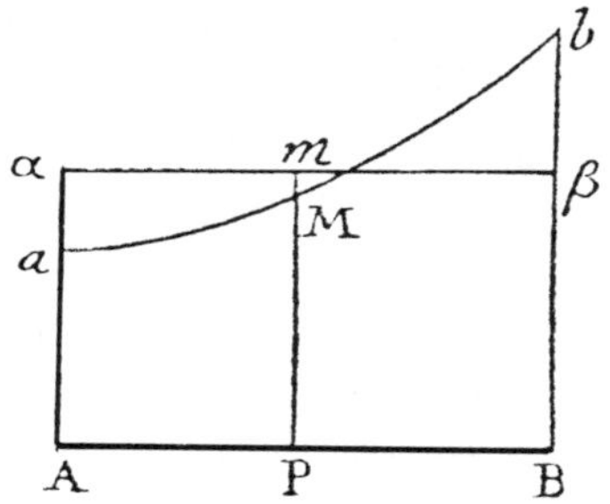

Figure 2.9

for the desired curve. He notes that α, β have fixed values, and he writes $\xi = (\alpha + \gamma)\delta$, $\eta = \beta\delta$; from this he concludes that there are values ξ, η such that along the extremal arc

$$\xi \, dA + \eta \, dB = 0. \tag{2.32}$$

To illustrate this principle, he works through four problems. The first of these is, as stated by Euler, to find among all curves aMb in Figure 2.9 with the axis AB and with fixed area under the curve the one which minimizes the expression

$$\frac{\int yy \, dx}{\int y \, dx}.$$

Thus he seeks the curve aMb having the lowest center of gravity for the figure $AabB$. He lets $AP = x$ and $PM = y$. His denominator is now the integral A and his numerator B. It then follows, in his terms, that their differentials are $n\nu \cdot dx \cdot 1$ and $n\nu \cdot dx \cdot 2y$, and his relation (2.32) above becomes $\xi + 2\eta y = 0$ or $y = \text{const}$. He concludes that the line $\alpha\beta$ in the figure is the solution curve.

2.6. Isoperimetric Problems, Continuation

In his Chapter VI Euler seeks to examine the variational problem when there are any number of accessory conditions such as a finite set of functionals each of which much have a given fixed value. To carry out his analysis, he starts by establishing a lemma in Proposition I. Theorem (p. 216). It states, in effect, that if α and β are constants and A and B definite integrals or functionals, then the minimizing arc for the problem of minimizing $\alpha A + \beta B$ is the same as that for the problem of minimizing B subject to the condition that A has a fixed value.

To prove this result, Euler considers the arc Q that maximizes $\alpha A + \beta B$ and any other curve R on the same base from 0 to a which gives A the same value as Q does. Then $\alpha A + \beta B$ cannot have a greater value on R than on Q, i.e.,

$$\alpha A(Q) + \beta B(Q) = (\alpha A + \beta B)(Q) \geqslant (\alpha A + \beta B)(R) = \alpha A(R) + \beta B(R).$$

But $A(Q) = A(R)$ and hence $\beta B(Q) \geqslant \beta B(R)$. Thus Q is the desired arc for the problem of maximizing B subject to the condition that A is a constant. The converse is clearly trivial to establish. [Euler has tacitly assumed that $\beta B(Q) \geqslant \beta B(R)$ implies $B(Q) \geqslant B(R)$; this clearly depends on the sign of β and is not discussed by Euler. The converse is noted in his Corollary 1.]

In Proposition II. Theorem he generalizes this result to the case of three integrals A, B, and C. He now shows that the arc which maximizes $\alpha A + \beta B + \gamma C$ is the same as that for the problem of maximizing C subject to the conditions that both A and B have fixed values. The proof is similar to the one above: let Q be the maximizing arc for $\alpha A + \beta B + \gamma C$ and R any other arc such that $A(Q) = A(R)$ and $B(Q) = B(R)$; then since $\alpha A(Q) + \beta B(Q) + \gamma C(Q) \geqslant \alpha A(R) + \beta B(R) + \gamma C(R)$, it follows directly that $\gamma C(Q) \geqslant \gamma C(R)$. He also notes the converse.

He unfortunately gets into difficulties when he attempts to take up the major result of the chapter, Proposition III. Problem (p. 221). He seeks to find a necessary condition that a functional C be an extremum among all arcs on a common base that give the functionals A and B fixed and preassigned values. To do this, he attempts to reason by analogy with his results on multipliers in his Chapter V (see Section 2.5 above). To carry the analysis through, he chooses (Figure 2.7) to vary the ordinates Nn, Oo, Pp by "small" amounts $n\nu$, $o\omega$, and $p\pi$, respectively. This enables him to calculate the variations of the functionals A, B, and C. He finds directly that

$$\begin{aligned} 0 &= n\nu \cdot P + o\omega \cdot P' + p\pi \cdot P'', \\ 0 &= n\nu \cdot Q + o\omega \cdot Q' + p\pi \cdot Q'', \\ 0 &= n\nu \cdot R + o\omega \cdot R' + p\pi \cdot R'', \end{aligned}$$

where, P, Q, and R are his designations for dA, dB, and dC, the variations caused by $n\nu$, etc. He then finds quantities α, β, and γ such that

$$\begin{aligned} 0 &= \alpha P + \beta Q + \gamma R, \\ 0 &= \alpha P' + \beta Q' + \gamma R', \\ 0 &= \alpha P'' + \beta Q'' + \gamma R''. \end{aligned} \tag{2.33}$$

At this point Euler fails to establish the fact that α, β, and γ are constants. The analogy with the argument in Section 2.5 above fails, and the proof is flawed.[23] As Carathéodory so aptly comments (p. xxvi), "it is a

[23] After writing down system (2.33), Euler curiously leaps to the demonstration that if α, β, γ are constants, then the second and third equation in (2.33) are consequences of the first.

pity . . . the work, which contains so many novel ideas, should end in this fashion on a discordant note."

2.7. The Principle of Least Action

In his first appendix, "On elastic curves," Euler discusses curves satisfying differential equations of the form $dy = x^2\,dx/(a^4 - x^4)^{1/2}$. These curves were apparently first studied by James Bernoulli in 1703.[24] It is not germane to our ends to pursue this topic here. Instead, let us look at his second appendix, "On the motion of bodies in a non-resisting medium, determined by the method of maxima and minima." (He uses the term *projectiles* or *projected bodies* instead of bodies.) The paper is of the greatest importance because it contains the first publication of the principle of least action.[25] Unfortunately, as mentioned earlier, this important mechanical principle now is usually attributed to Maupertuis for curious reasons. I intend to postpone to the next section a discussion of the relation, if any, between the works of Euler and Maupertuis and instead proceed directly to an examination of Euler's paper. As formulated by him, the principle of least action is this: let the mass of the projected body be M—it is in reality a point mass which moves in a plane—, let half the square of its velocity be v; [He has a somewhat bothersome inconsistency here again in respect to how he defines velocity. Actually, the text says "v is the square of the velocity," but this creates problems later when he calculates centrifugal force as $2v/\rho$, where ρ is the radius of curvature (see p. 78)], and let the element of arc-length along the prescribed path be ds. Among all curves passing through the same end-points the desired one makes the integral $\int M\,ds\,v^{1/2}$ a minimum; or for constant M, $\int ds\,v^{1/2}$ a minimum. Euler remarks that if the distance ds is traversed in the time dt, then $ds = dt\,(2v)^{1/2}$, and his integral may be expressed as[26]

$$\int ds\sqrt{v} = \sqrt{2}\int v\,dt.$$

[24] James Bernoulli, *CL*. Example VI in Euler I, XXIV, p. 185 takes up the problem of finding among all curves az of the same length the one which, when rotated about the axis AZ (in Figure 2.8), contains the greatest volume. Euler shows that the solution is expressible as

$$x = \int \frac{(yy + bc)\,dy}{\sqrt{b^4 - (yy + bc)^2}}\,.$$

He remarks that the radius of curvature of this curve "is inversely proportional to the ordinate y; whence it is clear that the curve in question is an Elastic one." (p. 186).

[25] Euler, I, XXIV, pp. 298–308. The interested reader should read the very good account by Carathéodory on p. li.

[26] The interested reader may also wish to read Brunet, *ACT*. This is a small historical study of the principle of least action and in particular the works of Euler and Maupertuis on the subject.

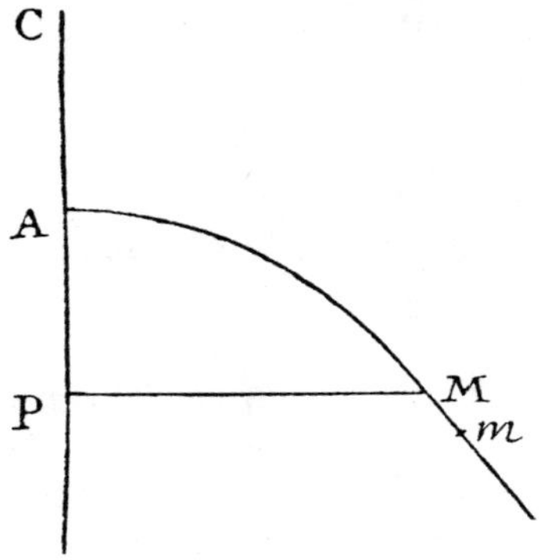

Figure 2.10

He first observes that if there are no forces acting on the particle and its initial velocity is a constant b, then his principle implies that s is a minimum, and so the motion must be uniform and along the straight line joining the end-points; this is, of course, one of Newton's laws of motion.

In case the only force acting is gravity (Figure 2.10) Euler sets $AP = x$, $PM = y$, and $Mm = ds$. He therefore has $dv = g\,dx$—remember that v is half the *square* of the velocity—and so $v = a + gx$. The path of the particle is then determined by minimizing the integral

$$\int ds\sqrt{a + gx} = \int dx\sqrt{(a + gx)(1 + pp)}.$$

Euler remarks that

$$N = 0, \qquad P = \frac{p\sqrt{a + gx}}{\sqrt{1 + pp}}$$

and as a consequence $P = \text{const.} = C^{1/2}$. It then follows easily that $dy = dx\, C^{1/2}/(a - C + gx)^{1/2}$ or

$$y = \frac{2}{g}\sqrt{C(a - C + gx)}\,,$$

a parabola. In the Figure 2.10 it is evident that $y = 0$ at $x = 0$ since the origin is at the point A; it is then so that $a = C$ and $y = 2(ax/g)^{1/2}$. In this expression $(2a)^{1/2}$ is the initial value of the velocity in the x direction. Hence Euler's parabola becomes

$$y = \sqrt{\frac{2v_0^2}{g}\,x},$$

which is precisely the familiar formula.

Euler next assumes that the downward force is no longer a constant but a function X of x and works through the analysis. He remarks that

$dv = X\,dx$ and so $v = A + \int X\,dx$. The action integral becomes

$$\int dx\sqrt{\left(A + \int X\,dx\right)(1 + p^2)}\ ,$$

and a little calculation shows that

$$\frac{dy}{dx} = p = \frac{\sqrt{C}}{\sqrt{A - C + \int X\,dx}}$$

or

$$y = \int \frac{dx\sqrt{C}}{\sqrt{A - C + \int X\,dx}}\ .$$

Euler notes that the curve is a horizontal line when $\int X\,dx = C - A$.

He then turns to the case where a force $Y(y)$ acts horizontally and a force $X(x)$ vertically. Then it is clear that $dv = -X\,dx - Y\,dy$, where he has changed his convention on the signs of the forces. He finds that $v = A - \int X\,dx - \int Y\,dy$, and the action integral becomes

$$\int dx\sqrt{(1 + pp)\left(A - \int X\,dx - \int Y\,dy\right)}\ ,$$

from which he concludes that

$$N = -\frac{Y\sqrt{1 + p^2}}{2\sqrt{A - \int X\,dx - \int Y\,dy}}\ , \qquad P = \frac{p\sqrt{A - \int X\,dx - \int Y\,dy}}{\sqrt{1 + p^2}}\ .$$

His necessary condition becomes, after a little calculation,

$$\frac{dp\sqrt{A - \int X\,dx - \int Y\,dy}}{\sqrt{(1 + p^2)^3}} = \frac{X\,dy - Y\,dx}{2\sqrt{(1 + p^2)\left(A - \int X\,dx - \int Y\,dy\right)}}$$

or

$$\frac{2\,dp}{1 + p^2} = \frac{X\,dy - Y\,dx}{A - \int X\,dx - \int Y\,dy} = \frac{X\,dy - Y\,dx}{v}\ . \tag{2.34}$$

He rewrites this as

$$\frac{2v\,dp}{\sqrt{(1 + p^2)^3}} = \frac{X\,dy - Y\,dx}{\sqrt{1 + p^2}}\ ;$$

and he remarks that the radius of curvature is

$$r = -\frac{\sqrt{(1+p^2)^3}\,dx}{dp}.$$

This transforms the differential equation into the form

$$\frac{2v}{r} = \frac{Y\,dx - X\,dy}{ds},$$

where $2v/r$ is what Euler calls the *centrifugal force* and the expression $(Y\,dx - X\,dy)/ds$ is an expression for the normal force. Thus the equation above represents the balancing of forces on the particle, exactly as it should, showing in this case the relation of the principle to Newton's laws of motion, i.e., showing that the principle produces as its extremals the dynamically correct curves.

To integrate equation (2.34), Euler finds the integrating factor

$$\frac{p\left(A - \int X\,dx - \int Y\,dy\right)}{(1+p^2)}.$$

This gives him the new relation

$$\frac{-p^2\int X\,dx + \int Y\,dy - A}{1+p^2} = C$$

or equivalently

$$p = \frac{\sqrt{B + \int Y\,dy}}{\sqrt{C + \int X\,dx}},$$

where $B = -A - C$. Since $p = dy/dx$, he writes

$$\int \frac{dy}{\sqrt{B + \int Y\,dy}} = \int \frac{dx}{\sqrt{C + \int X\,dx}}$$

and notes that the variables are separated.

He then turns to the elegant case of central forces to show that his principle yields the usual results in this case as well. In Figure 2.11 he sets $CP = QM = x$, $PM = y$, and $CM = (x^2 + y^2)^{1/2} = t$ and assumes the central force T is a function of t alone. The components of force along MQ and MP are, respectively, Tx/t and Ty/t so that

$$dv = -\frac{Tx\,dx}{t} - \frac{Ty\,dy}{t} = -T\,dt,$$

$$v = A - \int T\,dt.$$

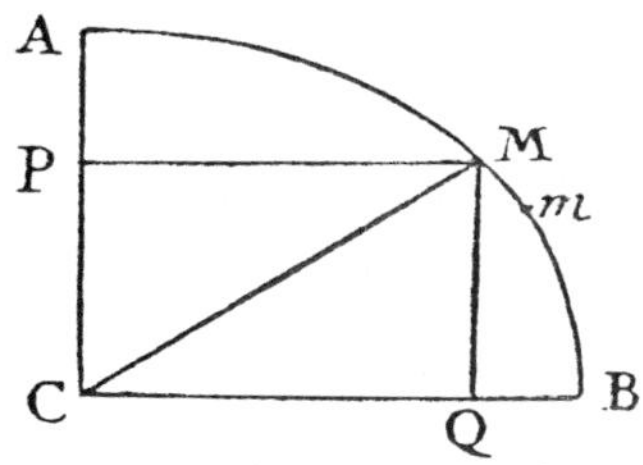

Figure 2.11

The action integral then becomes, in his notation

$$\int dx\,\sqrt{(1+pp)(A-\int T\,dx)}\ .$$

A little calculation shows that Euler's condition becomes

$$-\frac{Ty\,dx\sqrt{1+p^2}}{2t\sqrt{A-\int T\,dx}}=\frac{dp\sqrt{A-\int T\,dt}}{\sqrt{(1+p^2)^3}}-\frac{pT\,dt}{2\sqrt{(1+p^2)\left(A-\int T\,dx\right)}},$$

which reduces to the simpler form

$$\frac{T(x\,dy-y\,dx)}{2t\left(A-\int T\,dt\right)}=\frac{dp}{1+p^2}\ .$$

To eliminate variables, Euler now expresses dx and dy in terms of dt and has

$$\frac{T\,dt}{2\left(A-\int T\,dt\right)}=\frac{dp(py+x)}{(1+p^2)(px-y)}=\frac{x\,dp}{px-y}-\frac{p\,dp}{1+p^2}\ .$$

Upon integration, he finds

$$\frac{C}{\sqrt{A-\int T\,dt}}=\frac{px-y}{\sqrt{1+p^2}}=\frac{x\,dy-y\,dx}{ds},$$

and he observes that the right-hand member is the perpendicular distance D from the point C in Figure 2.11 to the tangent to the curve at M. Thus the velocity is inversely proportional to this distance D. He correctly says that this is a most important property of the motion.

In fact, we can at once see that in polar coordinates with C as pole and CB as axis this proportion says that

$$\frac{ds}{d\tau}\cdot r^2\,\frac{d\theta}{ds}=\text{const.}$$

or[27]

$$r^2 \frac{d\theta}{d\tau} = \text{const.}$$

Let us now denote by A the area swept out by the radius vector starting from the axis CB. Then the last relation above states that

$$2\frac{dA}{d\tau} = r^2 \frac{d\theta}{d\tau} = \text{const.}$$

The quantity $dA/d\tau$ is called the *areal velocity*, and our relation above says the areal velocity is constant when the force acting is a central one and conversely. In essence this is Kepler's law that equal areas are swept out by the radius vector in equal intervals of time.

Euler considers two other problems which need not concern us. He also observes that his principle is applicable to the case of many bodies or particles. Instead of discussing any of this further, let us quote Euler's statements at the beginning of his paper (p. 298):

> Since all the effects of Nature follow a certain law of maxima or minima, there is no doubt that, on the curved paths, which the bodies describe under the action of certain forces, some maximum or minimum property ought to obtain. What this property is, nevertheless, does not appear easy to define *a priori* by proceeding from the principles of metaphysics; but since it may be possible to determine these same curved paths by means of a direct method, that very thing which is a maximum or minimum along these curves can be obtained with due attention being exhibited. But above all the effect arising from the disturbing forces ought especially to be regarded; since this [effect] consists of the motion produced in the body, it is consonant with the truth that this same motion or rather the aggregate of all motions, which are present in the body ought to be a minimum. Although this conclusion does not seem sufficiently confirmed, nevertheless if I show that it agrees with a truth known *a priori* so much weight will result that all doubts which could originate on this subject will completely vanish. Even better when its truth will have been shown, it will be very easy to undertake studies in the profound laws of Nature and their final causes, and to corroborate this with the firmest arguments.

Again at the end of the appendix (p. 308) Euler muses on the generality of his principle of least action. He remarks there that the principle seems to run into a difficulty when one considers motion in a resisting medium. He has some difficulty in trying to explicate the reason for this; but Lagrange in his *Mécanique Analytique* shows in 1788 that the principle is, in general, valid for conservative forces provided that all constraints are independent of time. In the contrary case the action integral may, in fact, not be an extremum.[28]

---

[27] W. D. MacMillan, *SAD*, pp. 268–269. Note in the relations above that τ denotes time.
[28] Lagrange, *MA*, Part 1, Section 3, Paragraph 13.

There was also a consideration of the principle by Poisson in his *Traité de Mécanique*, but it remained for Jacobi in 1842/43 to set the matter on a completely sound basis.[29] In this connection, as we shall see below (p. 182), Hamilton took a different path than Jacobi and developed a more general principle known usually as *Hamilton's principle*. When a potential function U exists, this principle is concerned with the extremals of the integral

$$\int_{t_1}^{t_2}(T+U)\,dt=\int_{t_1}^{t_2}(T-V)\,dt$$

where T is the kinetic energy and $V=-U$ is the potential energy (Lagrange introduced the function U into the subject in 1773 in his *Oeuvres*, Vol. VI, p. 335. It was Green who called it by the name *potential*); the function $L=T-V$ is often called the *Lagrangian function*, and $-L$ was called the *kinetic potential* by Helmholtz ($H=T+V$ is called the *Hamiltonian function* and plays a major role both in the calculus of variations and in physics, as we shall see later). Notice that the principle of least action assumes an interesting form for conservative forces.[30] When a potential function exists, it is easy to show that

$$T+V=T-U=\text{const.}=h.$$

In this case the least action integral assumes the form

$$\int\sqrt{h-V}\cdot ds.$$

Finally we note in passing that Gauss developed a principle of least constraints:[31]

> Its purpose is to compare the position of a constrained particle with the position which the particle would have had if it had been free during the interval of time dt immediately preceding the instant of comparison. The difference between the two positions evidently is due to the constraints, and Gauss' principle asserts that of all possible geometrical displacements which are compatible with the constraints the smallest one is the one which actually occurs dynamically.

---

[29] See Poisson, *TM* and Jacobi, *WERKE*, Vol. IV, pp. 289–294.

[30] The definition of a conservative force is this: "Let W be the work required to move a system of bodies from a certain configuration A to a certain other configuration B *against* the system of forces F. . . . If the work done *by* the system of forces . . . when it is returned from the configuration B to . . . A also is equal to W, then the system of forces is said to be *conservative*." See MacMillan, *SAD*, p. 49. Examples of nonconservative or dissipative forces are friction and viscosity. In these cases energy is dissipated in the form of heat and radiates away.

[31] MacMillian, *SAD*, p. 419.

2.8. Maupertuis on Least Action

On 20 February 1740 Maupertuis presented a paper, "The law of equilibrium for bodies," where he set forth several "laws of Nature," some already known but the last due to him.[32] The first he states is this: "*In any assemblage of bodies their common center of gravity is as low as possible.*" Another, he says, is "that of the conservation of *living forces*."[33] This principle is a brief enumeration of John Bernoulli's *conservatio virium vivarum*. It says, in effect, that kinetic energy cannot be destroyed without causing a change, that is, without being converted into another form. This statement was by 1750 expressed as the law of conservation of energy and represents a gradual deepening of understanding, starting with Galileo's knowledge of particles sliding down an inclined plane, through Huygens, Newton, John Bernoulli and then Daniel Bernoulli, to Lagrange.

Maupertuis goes on to state this law in this way[34]:

> In every system of particles which mutually interact, the sum of the products of each mass by the square of its velocity, which is called the living force, must remain invariant.

From this statement he proceeds to his own law of equilibrium:

> Consider a system of heavy particles which are attracted to centers by forces, each of which acts towards its center, as the nth power of its distance to the center. In order that all the particles remain in equilibrium, it is necessary that the sum of the products of each mass by the intensity of its force and by the $(n + 1)$st power of its distance to the center of its force (which is called the sum of the equilibrium forces) must be a maximum or minimum.

He observes that he can reduce all other laws of statics to this principle, but he does not here discuss any relation it may have to dynamics. In fact, he does not take up the principle of least action until his 1744 paper, entitled "Harmony between different laws of nature which have, up to now, appeared incompatible."[35] What he says is that "Nature, in the production of its effects, acts always by the simplest means." He amplifies this by remarking that "the path . . . is that along which the quantity of action is the least." (He means by "quantity of action" the action integral

---

[32] Maupertuis [1740], pp. 170–176.

[33] In the *Acta Eruditorum* of 1695 Leibniz introduced the concept of *vis viva*, or living force. It is, by definition, twice the kinetic energy, i.e., mv^2, where m is the mass and v the velocity of a particle. There is an allied concept of *vis mortua* or dead force which is essentially potential energy.

[34] Maupertuis [1740].

[35] Maupertuis [1744], pp. 417–426.

or rather the sum of the products of the velocities by the distances traveled.) He uses this principle, in this paper, to show that Fermat's principle of least time is a consequence of his result by establishing Snell's law.

This is not the place to discuss the fascinating interactions between Euler and Maupertuis in subsequent years, but it is unbelievable to me that Maupertuis understood his 1740 paper to subsume his 1744 one. I therefore believe strongly that Euler has the priority even though he was able to show in a 1751 paper that they are equivalent. It was not until 1743 that d'Alembert enunciated his ideas on virtual work, displacements, and velocities, which served to unify the laws of statics and dynamics. Without such concepts I do not understand how Maupertuis could have seen the connection between his 1740 and 1744 papers. The former deals with the problem of minimizing or maximizing the potential energy and the latter, with the action which is formulated in terms of the kinetic energy. As we mentioned above, it was not until Lagrange showed kinetic plus potential energy to be constant for conservative systems that the relation between these concepts becomes completely clear.

3. Lagrange and Legendre

3.1. Lagrange's First Letter to Euler

On 12 August 1755 a 19-year-old, one Ludovico de la Grange Tournier of Turin, wrote Euler a brief letter to which was attached an appendix containing mathematical details of a very beautiful and revolutionary idea (see Lagrange [1755] and Euler [1755]). He saw how to eliminate from Euler's methods of 1744 the tedium and need for geometrical insight and to reduce the entire process to a quite analytic machine or apparatus, which could turn out the necessary condition of Euler and more, almost automatically. This basic idea of Lagrange ushered in a new epoch in the calculus of variations. Indeed after seeing Lagrange's work, Euler dropped his own method, espoused that of Lagrange, and renamed the subject the *calculus of variations*.[1]

In the summary to his first paper using variations, Euler says "Even though the author of this [Euler] had meditated a long time and had revealed to friends his desire yet the glory of first discovery was reserved to the very penetrating geometer of Turin La Grange, who having used analysis alone, has clearly attained the very same solution which the author had deduced by geometrical considerations."[2]

From 1755 until 1760 Lagrange worked at his ideas until he brought them out in the 1760/61 issue of the *Miscellanea Taurinensia*.[3] He also published a number of other papers on the subject, which reached their ultimate culmination at his hands in 1788 in his great *Mecânique Analytique*, where he developed *inter alia* the so-called Lagrange multiplier rule.

---

[1] See Euler [1764], p. 145. This volume (I, XXV) contains a number of papers written by Euler both before and after his *Methodus* . . . of 1744. After Lagrange's publication Euler wrote several papers on our subject, the first of which appeared in 1764. It, and two others as well, are largely expositions of Lagrange's elegant method.

[2] Euler [1764], p. 142.

[3] Lagrange [1760]. It is also of interest to see [1760′], pp. 365–468, where he disucsses the principle of least action and its relation to the calculus of variations.

In his August 1755 letter Lagrange says to Euler that his departure point was the remark in Euler I, XXIV, p. 52, where Euler said "A method is therefore desired, free of geometric and linear solutions by which it is evident that, in such investigations of maxima and minima, $-p\,dP$ ought to be written in place of $P\,dp$." (I do not know what Euler meant by the word "linear"; perhaps it referred to his use of polygons for comparison purposes.) To do this, Lagrange produced what he thought of as a new form of differential which came to be known as a variation. He envisaged it as acting on functions of a real variable such as $F(y)$ as well as on operators such as $\int Z\,dx$. To distinguish his new process, which may be viewed as an operator, from the usual differential he wrote it as δ instead of d. In his letter he thought of it in connection with a function of x, y, y', etc. in this way: the variable x is kept fixed and the ordinary differential is calculated, but with the differential arguments δy, $\delta y'$, etc. (Later Euler, Lagrange, and Legendre extended the process to the case where x may vary as well; see p. 115 below.)

He then asserts that the commutation rules

$$\delta dF(y) = d\delta F(y), \qquad \delta d^m F(y) = d^m \delta F(y) \qquad (m = 1, 2, \ldots) \tag{3.1}$$

and

$$\delta \int Z = \int \delta Z \tag{3.1'}$$

hold, where he writes the differential of the integrand Z as did Euler, i.e., in the form

$$dZ = M\,dx + N\,dy + P\,d^2y + Q\,d^3y + R\,d^4y + \cdots$$

In this letter he writes the corresponding variation of Z as

$$\delta Z = N\,\delta y + P\,\delta dy + Q\,\delta d^2y + R\,\delta d^3y + \cdots.$$

To make (3.1′) clear, it is evident that we may regard $\int Z$ as being evaluated along a family of arcs $y = y(x, \alpha)$; the δ process consists of differentiation with respect to the parameter α; thus if the given arc corresponds to $\alpha = 0$, then $\delta y(x) = \alpha \cdot (\partial y(x, \alpha)/\partial\alpha|^{\alpha=0})$. (It is, of course, true that $\delta \int Z$ is the first differential of the operator $\int Z$ and need not be formulated in terms of families of arcs. Notice also that Lagrange has not yet thought of varying the limits of integration; he does this in later work.) In the right-hand member Lagrange views Z as a function of $x, y, dy, d^2y, \ldots$ and, in effect, differentiates with respect to α again at $\alpha = 0$ and then integrates the resulting expression with respect to x. In this way he has freed himself from the variations such as $n\nu$, $o\omega$, etc. used by Euler to change selected ordinates. Here there is no need for a decision about how many ordinates must be altered or for a replacement of derivatives by differences. In other words, instead of Euler's highly specialized comparison curves, Lagrange has recourse to very general ones.

To what extent Lagrange viewed his variation operator as a means for effecting a comparison of curves or as a purely formalistic construct I can form no precise reasoned opinion; but I incline toward the latter since he seems to treat his variation in that way. It is clear, however, in a 1771 paper by Euler that he had, by then, the idea; Euler imagined a family of comparison curves $y = y(x, t)$ and took for his variation δy the quantity $dt(dy/dt)$, evaluated at the value of t corresponding to the given curve.[4] It is also the case that Lagrange followed the same course in his later work.

In any case having his commutation relations (3.1) and (3.1′), he sets down some results, which follow from the usual rules for integration by parts; in his notation they are these:

$$\int z\,du = zu - \int u\,dz,$$

$$\int z\,d^2u = z\,du - u\,dz + \int u\,d^2z,$$

$$\int z\,d^3u = z\,d^2u - dz\,du + u\,d^2z - \int u\,d^3z, \tag{3.2}$$

$$\int u \int z = \int u \times \int z - \int z \int u.$$

(His last result is to be understood to mean that

$$\int_0^a u(\xi)\,d\xi \int_0^\xi z(x)\,dx = \int_0^a u(x)\,dx \cdot \int_0^a z(x)\,dx - \int_0^a z(\xi)\,d\xi \int_0^\xi u(x)\,dx.)$$

In the last result he writes $H = \int_0^a u\,dx$ and $V = H - \int_0^x u\,dx$. This gives

$$\int_0^a u\,d\xi \int_0^\xi z\,dx = \int_0^a Vz\,dx.$$

Given these relations, he states his problem: *to find a relation between x and y for which the integral* $\int_0^a Z$ *is a maximum or minimum.* His solution is carried out elegantly by means of his formalism. Having

$$dZ = M\,dx + N\,dy + P\,d^2y + Q\,d^3y + R\,d^4y + \cdots,$$

he writes

$$\delta Z = N\,\delta y + P\delta\,dy + Q\delta\,d^2y + R\delta\,d^3y + \cdots$$

in accordance with his rules. It then follows from $\delta \int Z = \int \delta Z$, with the help of relations (3.2) that

$$\begin{aligned}\delta \int Z &= \int N\,\delta y + \int P\,\delta dy + \int Q\,\delta d^2y + \cdots \\ &= \int N\,\delta y + \int P\,d\delta y + \int Q\,d^2\delta y + \cdots \\ &= \int N\,\delta y + P\,\delta y - \int dP\,\delta y + Q\,d\delta y - dQ\,\delta y + \int d^2Q\,\delta y + \cdots \end{aligned} \tag{3.3}$$

[4] Euler [1771], p. 208.

and hence that

$$\delta \int Z = \int (N - dP + d^2Q - \cdots)\,\delta y + (P - dQ + \cdots)\,\delta y$$
$$+ (Q - \cdots)\,d\delta y + \cdots ; \tag{3.4}$$

in these relations the integrals are evaluated between the limits $x = 0$ and $x = a$, as are the other terms. Then if the comparison curves are such that $0 = \delta y = d\delta y = \cdots$ at $x = 0$ and $x = a$, Lagrange has

$$\delta \int Z = \int (N - dP + d^2Q - d^3R \ldots)\,\delta y. \tag{3.5}$$

Since δy is arbitrary on the interval from 0 to a, it follows that

$$N - dP + d^2Q - d^3R \ldots = 0. \tag{3.6}$$

It is interesting that Lagrange tacitly assumes the condition $\delta \int Z = 0$ to be necessary for an extremum. He does not give at this time any discussion of the point. In the event that the conditions $0 = \delta y = \delta dy = \cdots$ do not obtain at the end-points, he also has the further necessary condition that the expression

$$(P - dQ + \cdots)\,\delta y + (Q - \cdots)\,d\delta y + \cdots = 0$$

must hold when evaluated between the limits $x = 0$ and $x = a$. (This leads to the transversality condition.) He does not, as yet, quite know what to do with this. Later, as we shall see, he begins to understand the situation somewhat better and, so makes a substantial step toward establishing another fundamental necessary condition for an extremum.

Lagrange then turns in [1755] to the results in Chapter III of Euler's *Methodus Inveniendi* . . . to show how simply he can produce them. He has

$$dZ = L\,d\pi + M\,dx + N\,dy + P\,d^2y + Q\,d^3y + \cdots,$$

$$\pi = \int_0^x (Z), \tag{3.7}$$

$$d(Z) = (M)\,dx + (N)\,dy + (P)\,d^2y + (Q)\,d^3y + \cdots \tag{3.8}$$

and so he can write at once

$$\delta Z = L\,\delta\pi + N\,\delta y + P\delta\,dy + Q\,\delta d^2y + \cdots,$$

$$\delta(Z) = (N)\,\delta y + (P)\,\delta dy + (Q)\,\delta d^2y + \cdots,$$

$$\delta\pi = \int (N)\,\delta y + \int (P)\,\delta dy + \cdots;$$

hence

$$\delta \int Z = \int N\,\delta y + \int P\,d\delta y + \int Q\,d^2y + \cdots + \int L \int (N)\,\delta y$$
$$+ \int L \int (P)\,d\delta y + \cdots.$$

He sets $H = \int_0^a L$ and $V = H - \int_0^x L$ and finds easily that

$$\delta\int_0^a Z = \int_0^a [N + (N)V]\,\delta y + \int_0^a [P + (P)V]\,d\delta y + \int_0^a [Q + (Q)V]\,d^2\delta y + \cdots.$$

By his previous argument in connection with equations (3.3)–(3.6), i.e., by integrations by parts, he concludes that

$$N + (N)V - d[P + (P)V] + d^2[Q + (Q)V] - \cdots = 0.$$

He observes that a very similar argument would suffice to handle Euler's Proposition IV in Chapter III (I, XXIV, p. 102). Instead, he goes to the latter's Proposition V (p. 110), which is "also wonderfully easy to solve." This is the case where not only Z but also (Z) involves π. He now has relations (3.7) as before, but here

$$\delta(Z) = (L)\,\delta\pi + (N)\,\delta y + (P)\,\delta^2 y + \cdots$$

and so

$$\delta\pi = \delta\int (Z) = \int \delta(Z) = \int [(L)\,\delta\pi + (N)\,\delta y + \cdots].$$

By means of these two relations he can write

$$d\delta\pi = (L)\,\delta\pi\,dx + (N)\,\delta y\,dx + (P)\,d\delta y\,dx + \cdots.$$

(Actually, he omitted the factors dx in the right-hand member.) For notational purposes, he sets $(N)\delta y + (P)\,d\delta y + \cdots = V$ and has the differential equation

$$d\delta\pi - (L)\,\delta\pi\,dx = V\,dx$$

to solve. He finds for his solution (in his notation)

$$\delta\pi = e^{\int(L)}\int V e^{-\int(L)},$$

and hence he has

$$\delta\int Z = \int N\,\delta y + \int P\,d\delta y + \cdots + \int e^{\int(L)} L \int e^{-\int(L)}(N)\,\delta y + \int e^{\int(L)} L \int e^{-\int(L)}(P)\,\delta y + \cdots.$$

As before, he sets $H = \int_0^a e^{\int(L)} L$ and $V = H - \int_0^x e^{\int(L)} L$ and concludes that

$$N + (N)e^{-\int(L)}V - d\left[P + (P)e^{-\int(L)}V\right] + \cdots = 0;$$

or if $e^{-\int(L)}V = S$,

$$N + (N)S - d[P + (P)S] + d^2[Q + (Q)S] \cdots = 0.$$

(Note that this V is not the same quantity he called V just above.)

3.2. Lagrange's First Paper

Lagrange corresponded with Euler a good deal before he first published his fundamental results in Volume II of the *Miscellanea Taurinensia*, the Memoirs of the Turin Academy, for the years 1760/61, which appeared in 1762. From then on he continued for some years to publish on the subject in that journal.[5] In the meantime Euler read two papers on his ideas to the Berlin Academy on 9 and 16 September 1756 and presented them to the St. Petersburg Academy in 1760, and they appeared in 1762.[6]

In his early work Lagrange states, in effect, correctly that his variation δZ of a function is the same as its differential except that it differs in the differential variables. Thus if Z is a function of x and $dZ = m\,dx$, then $\delta Z = m\,\delta x$ "and the same will be true for other equations." His Problem I concerns an indefinite integral $\int Z$, where Z is a given function of $x, y, z, dx, dy, dz, d^2x, d^2y, d^2z, \ldots$. He then seeks the curve which maximizes or minimizes the (definite) integral $\int Z$.

His proof is curious. He says, "according to the well-known methods of maxima and minima it will be necessary to differentiate, regarding the quantities $x, y, z, dx, dy, dz, d^2x, d^2y, d^2z, \ldots$ as variables, and to set the differential which results from this equal to zero. Designating this variation by δ one will have, at first, the equation for a maximum or a minimum $\delta \int Z = 0$, or what is equivalent, $\int \delta Z = 0$." To calculate the value of this last integral, Lagrange writes

$$\begin{aligned}\delta Z &= n\,\delta x + p\,\delta dx + q\,\delta d^2x + r\,\delta d^3x + \cdots \\ &\quad + N\,\delta y + P\,\delta dy + Q\,\delta d^2y + R\,\delta d^3y + \cdots \\ &\quad + \nu\,\delta z + \omega\,\delta dz + \chi\,\delta d^2z + \rho\,\delta d^3z + \cdots.\end{aligned}$$

[Notice that he is considering his arcs to be in three-space and parametric form: $x = x(t)$, $y = y(t)$, $z = z(t)$.] He notes that $\delta dx = d\delta x$, $\delta d^2x = d^2\delta x$, etc., that

$$\begin{aligned}&\int p\,d\delta x = p\,\delta x - \int dp\,\delta x, \\ &\int q\,d^2\delta x = q\,d\delta x - dq\,\delta x + \int d^2q\,\delta x, \\ &\int r\,d^3\delta x = r\,d^2\delta x - dr\,d\delta x + d^2r\,\delta x - \int d^3r\,\delta x,\end{aligned}$$

[5] Lagrange [1760]. In the *Memoirs* of the Turin Academy he published not only the 1760/61 paper but a second one that appeared in Vol. IV, 1766/69. The reader may also wish to consult his paper on pp. 4–20 of Vol. I, which he wrote in 1759 (Lagrange [1759]) on maxima and minima of real functions, especially his discussion of the second differential.

[6] Euler [1764], [1764′], and [1771]. There are also some other interesting papers by him on the subject in the same volume.

and hence that

$$\begin{aligned}\int \delta Z = 0 = &\int (n - dp + d^2q - d^3r + \cdots)\,\delta x \\ &+ \int (N - dP + d^2Q - d^3R + \cdots)\,\delta y \\ &+ \int (\nu - d\omega + d^2\chi - d^3\rho + \cdots)\,\delta z \\ &+ (p - dq + d^2r - \cdots)\,\delta x + (q - dr + \cdots)\,d\delta x \\ &+ (r - \cdots)\,d^2\delta x + \cdots \\ &+ (P - dQ + d^2R - \cdots)\,\delta y \\ &+ (Q - dR + \cdots)\,d\delta y + (R - \cdots)\,d^2\delta y + \cdots \\ &+ (\omega - d\chi + d^2\rho - \cdots)\,\delta z + (\chi - d\rho + \cdots)\,d\delta z \\ &+ (\rho - \cdots)\,d^2\delta z + \cdots .\end{aligned} \tag{3.9}$$

From this relation he concludes first that

$$\begin{aligned}(n - dp + d^2q - d^3r + \cdots)\,\delta x + (N - dP + d^2Q - d^3R + \cdots)\,\delta y \\ + (\nu - d\omega + d^2\chi - d^3\rho + \cdots)\,\delta z = 0\end{aligned} \tag{3.10}$$

and second that

$$\begin{aligned}(p - dq + d^2r - \cdots)\,\delta x + (q - dr + \cdots)\,d\delta x \\ + (r - \cdots)\,d^2\delta x + \cdots + (P - dQ + d^2R - \cdots)\,\delta y \\ + (Q - dR + \cdots)\,d\delta y + (R - \cdots)\,d^2\delta y \\ + (\omega - d\chi + d^2\rho - \cdots)\,\delta z + (\chi - d\rho - \cdots)\,d\delta z \\ + (\rho - \cdots)\,d^2\delta z + \cdots = 0,\end{aligned} \tag{3.10$'$}$$

where these last relations hold when the expression is understood to mean its value at the terminal point less its value at the initial point. He calls the value at the latter point $'M$ and at the former M'. Then the condition is that $M' - 'M = 0$. When there are no constraints on the variation δx, δy, δz at the end-points, Lagrange observes that relation (3.10) becomes the three equations

$$\begin{aligned} n - dp + d^2q - d^3r + \cdots &= 0, \\ N - dP + d^2Q - d^3R + \cdots &= 0, \\ \nu - d\omega + d^2\chi - d^3\rho + \cdots &= 0. \end{aligned} \tag{3.11}$$

[As we recall in the parametric case, the integrand $f(x, y, x', y')$ is such that for $\kappa > 0$, $f(x, y, \kappa x', \kappa y', \kappa z') = \kappa f(x, y, x', y', z')$; hence $f_{x'}x' + f_{y'}y' + f_{z'}z' = f$, where $x' = dx/dt$, $y' = dy/dt$, $z' = dz/dt$, and t is the parameter. We have, therefore, $f_{x'}x' + f_{y'}y' + f_{z'}z' = f$.]

Suppose now that the arc in question is assumed at one end-point to lie upon a given surface. In that case Lagrange's condition (3.10′)—I take the case where f contains only x, y, z, x', y', z' since this does not restrict the generality of the discussion—becomes $0 = f_{x'}\,\delta x + f_{y'}\,\delta y + f_{z'}\,\delta z$ where δx, δy, δz are the direction cosines of the normal to the bounding surface at the point where the extremal intersects it. But we can write this equation as

$$0 = f\,dt + f_{x'}(\delta x - x'\,dt) + f_{y'}(\delta y - y'\,dt) + f_{z'}(\delta z - z'\,dt)$$

since $f_{x'}x' + f_{y'}y' + f_{z'}z' = f$. The result is the familiar transversality condition.

Lagrange does not enlighten us on why he felt that it was necessary to go to curves expressed in parametric form to find the correct transversality conditions. Certainly, later writers seemed to comprehend the matter less than he, and the situation was really not completely clarified until much later, as we shall see.

Having set out his relations, Lagrange now considers the brachystochrone problem to illustrate the power of his process ([1760], p. 339). He takes x as the vertical axis and y, z as horizontal. Then the integral to be minimized is

$$\int \frac{\sqrt{dx^2 + dy^2 + dz^2}}{\sqrt{x}} .$$

From this he calculates directly that

$$n = -\frac{ds}{2x\sqrt{x}}, \qquad p = \frac{dx}{\sqrt{x}\,ds}, \qquad P = \frac{dy}{\sqrt{x}\,ds}, \qquad \omega = \frac{dz}{\sqrt{x}\,ds},$$

and all the other quantities $q, r, N, Q, \ldots$ are zero. He now recognizes several problems. The first is to find among all admissible curves the one providing the time of least descent. He has with the help of the equations (3.11) the relations

$$-\frac{ds}{2x\sqrt{x}} - d\frac{dx}{\sqrt{x}\,ds} = 0, \qquad -d\frac{dy}{\sqrt{x}\,ds} = 0, \qquad -d\frac{dz}{\sqrt{x}\,ds} = 0. \quad (3.12)$$

If the second of these is multiplied by $2\,dy/x^{1/2}\,ds$ and the third by $2\,dz/x^{1/2}\,ds$, and if the results are added, there results the equation

$$d\left(\frac{1}{x} - \frac{dx^2}{x\,ds^2}\right) = 0,$$

since $dx^2 = ds^2 - dy^2 - dz^2$. It is not difficult to show that the first of equations (3.12) follows from this. (It suffices to carry out the indicated differentiations.) If, moreover, the second and third of the equations (3.12) are integrated and the results divided one by the other, it follows that

$$\frac{dy}{dz} = \frac{\sqrt{b}}{\sqrt{a}},$$

which implies the motion is in a vertical plane. Lagrange consequently replaces y, z by a single variable t such that $t = (y^2 + z^2)^{1/2}$. With the help of this he writes

$$z = t\frac{\sqrt{a}}{\sqrt{a+b}}, \qquad y = t\frac{\sqrt{b}}{\sqrt{a+b}}, \qquad dy = dt\frac{\sqrt{b}}{\sqrt{a+b}}, \qquad ds = \sqrt{dx^2 + dt^2},$$

and after a little calculation, he remarks that

$$dt = \frac{\sqrt{x}\,dx}{\sqrt{c-x}}, \qquad c = \frac{ab}{a+b},$$

which is the equation of a cycloid generated by a circle of diameter c rolling on a horizontal line.

Lagrange now turns to the end-conditions (3.10′). He next takes up the case of the brachystochrone when the first end-point is fixed, and the problem is to find among all curves through that point the one which goes down to a fixed horizontal plane in the shortest time. Then $'M = 0$, and $M' = 0$ is equivalent to the condition that

$$\frac{dx}{\sqrt{x}\,ds}\,\delta x + \frac{dy}{\sqrt{x}\,ds}\,\delta y + \frac{dz}{\sqrt{x}\,ds}\,\delta z = 0 \tag{3.13}$$

when the quantities are evaluated on the plane, i.e., $\delta x = 0, \delta y, \delta z$ arbitrary. But this gives (in his words)

$$\frac{dy}{\sqrt{x}\,ds} = 0, \qquad \frac{dz}{\sqrt{x}\,ds} = 0, \qquad a = \infty, \qquad b = \infty,$$

"which will transform the cycloid into a vertical line." However, if the plane is not horizontal but vertical and orthogonal to the y- or z-axis, then $\delta y = 0$ or $\delta z = 0$: in the former case

$$\frac{dx}{\sqrt{x}\,ds} = 0, \qquad \frac{dy}{\sqrt{x}\,ds} = 0$$

and in the latter,

$$\frac{dx}{\sqrt{x}\,ds} = 0, \qquad \frac{dy}{\sqrt{x}\,ds} = 0.$$

In both cases a, b can be found and the cycloid cuts the plane at right angles.

He next supposes that the second end-point is constrained to lie on an arbitrary surface. Let the surface be given by the relation $dz = T\,dx + U\,dy$ so that $\delta z = T\,\delta x + U\,\delta y$. Now the condition $M' = 0$ becomes

$$\frac{dx}{\sqrt{x}\,ds}\,\delta x + \frac{dy}{\sqrt{x}\,ds}\,\delta y + \frac{dz}{\sqrt{x}\,ds}\,(T\,\delta x + U\,\delta y) = 0;$$

this says that on the surface

$$\left(\frac{dx}{\sqrt{x}\,ds}+T\frac{dz}{\sqrt{x}\,ds}\right)\delta x+\left(\frac{dy}{\sqrt{x}\,ds}+U\frac{dz}{\sqrt{x}\,ds}\right)dy=0,$$

which implies that

$$dx + T\,dz = 0, \qquad dy + U\,dz = 0.$$

But this means that the minimizing curve cuts the surface in a right angle.

If now the brachystochrone is to pass from one surface to another, then Lagrange remarks that both $'M = 0$ and $M' = 0$ for the first and last points on the curve. He says it follows simply that the minimizing curve cuts each surface orthogonally. His analysis of the case when the first end-point lies on a curve and the second one is fixed, was correctly criticized by Borda and Lagrange later amended his reasoning to conform to Borda's.[7]

We see here how Lagrange has begun to master the transversality condition and also has no fear of proceeding in three-space. We can also observe how formalistically he proceeds, but also how adroitly and carefully. In what follows he continues to exploit his ability to handle problems with variable end-points. Inasmuch as his procedure is so formalistic it is quite remarkable that it leads to the correct result. The whole procedure shows, however, how immature was the real understanding of some aspects of analysis in the eighteenth century and yet how bold and how inspired and inspiring were its best practitioners.

Let us see how Lagrange copes with the problem of finding the solution to the brachystochrone problem when all the admissible curves are constrained to be on a given surface $dz = p\,dx + q\,dy$. This gives him the constraint $\delta z = p\,\delta x + q\,\delta y$ on the variations $\delta x, \delta y, \delta z$, which must hold along each curve. He goes back to relation (3.10) and substitutes for δz its value above; he now has δx and δy arbitrary so he can obtain the results

$$-pd\frac{dz}{\sqrt{x}\,ds}-\frac{ds}{\sqrt{x}\,2x}-d\frac{dx}{\sqrt{x}\,ds}=0,$$

$$-qd\frac{dz}{\sqrt{x}\,ds}-d\frac{dy}{\sqrt{x}\,ds}=0,$$

for the brachystochrone problem. Lagrange notes with the help of the relation $dz = p\,dx + q\,dy$ that these two equations are equivalent; thus the

[7] Borda [1767], p. 558 and Lagrange [1766/69], p. 63. The criticism was based on Lagrange's failure to notice that when a particle descends from a fixed curve, the line $x = (x_1 - v_1^2)/2g$ on which the cycloid-producing circle rolls is not any longer fixed. The difficulty arose largely because Lagrange failed to specify the initial velocity v_1. We discuss the point later in more detail. The reader may also wish to read Bliss, *COV*, pp. 78–79. The essential point is that the initial curve at its intersection point with the minimizing curve must have its direction parallel to that of the final curve at the second point which the minimizing curve cuts orthogonally.

solution curve can be found by choosing either of these equations above and the equation of the surface $dz = p\,dx + q\,dy$.

If now both end-points vanish, then matters are quite simple. However, if one of them is arbitrary, then relation (3.10′) needs to be coupled with $\delta z = p\,\delta x + q\,\delta y$. This gives in the present case the relations

$$p\frac{dz}{\sqrt{x}\,ds} + \frac{dx}{\sqrt{x}\,ds} = 0, \qquad q\frac{dz}{\sqrt{x}\,ds} + \frac{dy}{\sqrt{x}\,ds} = 0,$$

which must be fulfilled at one or both of the end-points.

If the second end-point is allowed to vary along a curve on the surface and if that curve is determined by $dy = m\,dx$, then relation (3.10′) is changed since now $\delta y = m\,\delta x$ is yet another constraint. In this case it is easy to see that (3.10′) becomes

$$p\frac{dz}{\sqrt{x}\,ds} + \frac{dx}{\sqrt{x}\,ds} + \left(q\frac{dz}{\sqrt{x}\,ds} + \frac{dy}{\sqrt{y}\,ds}\right)m = 0,$$

or $(p + qm)\,dz + dx + m\,dy = 0$. As Lagrange says, "an equation which contains the necessary condition that the brachystochrone cuts the given curve in a right angle."

At this point Lagrange ([1760], p. 345), remarks that " Euler is the first who has given the general formula for finding the curve along which a given integral expression has its greatest or least value . . . ; but the formulas of this author are less general than ours: 1. because he only permits the single variable y in the integrand Z to vary; 2. because he assumes that the first and last points of the curve are fixed."

In Section IX of his paper [1760] Lagrange takes up his Problem II: to find conditions that the expression $\int Z$ be a maximum or a minimum provided that Z is an "algebraic" function of the variables $x, y, z, dx, dy, dz, d^2x, d^2y, \ldots$ and of $\Pi = \int Z'$, where Z' is another such function of $x, y, z, dx, dy, dz, d^2x, d^2y, \ldots$. He first notes that

$$\begin{aligned}\delta Z &= L\,d\Pi + n\,\delta x + p\,\delta dx + p\,\delta d^2x + \cdots\\ &\quad + N\,\delta y + P\,\delta dy + Q\,\delta d^2y + \cdots\\ &\quad + \nu\,dz + \omega\,\delta dz + \chi\,\delta d^2z + \cdots,\\ \delta Z' &= n'\,\delta x + p'\,\delta dx + q'\,\delta d^2x + \cdots\\ &\quad + N'\,\delta y + P'\,\delta dy + Q'\,\delta d^2y + \cdots\\ &\quad + \nu'\,\delta z + \omega'\,\delta dz + \chi'\,\delta d^2z + \cdots.\end{aligned}$$

He therefore has

$$\begin{aligned}\delta\int Z = \int\delta Z = &\int(n\,\delta x + p\,\delta dx + q\,\delta d^2x + \cdots)\\ &+ \int L\int(n'\,\delta x + p'\,\delta dx + q'\,\delta d^2x + \cdots)\end{aligned}$$

and remarks that the first integral is, just as before, transformable into the form

$$\int (n - dp + d^2q - \cdots)\delta x + (p - dq + \cdots)\delta x + (q - \cdots)\, d\delta x + \cdots .$$

The second he expresses, with the help of an integration by parts, in the form

$$\int L \cdot \int (n'\,\delta x + p'\,\delta dx + q'\,\delta d^2x + \cdots)$$
$$- \int \Big[\int L \cdot (n'\,\delta x + p'\,\delta dx + q'\,\delta d^2x + \cdots) \Big];$$

just as Euler did, Lagrange lets H be the value of $\int L$ over the whole interval. Then the expression above becomes

$$\begin{aligned} &\int \Big[\Big(H - \int L\Big)(n'\,\delta x + p'\,\delta dx + q'\,\delta d^2x + \cdots) \Big] \\ &\quad = \int \Big[n'\Big(H - \int L\Big) - dp'\Big(H - \int L\Big) + d^2q'\Big(H - \int L\Big) - \cdots \Big] \delta x \\ &\qquad + \Big[p'\Big(H - \int L\Big) - dq'\Big(H - \int L\Big) + \cdots \Big] \delta x \\ &\qquad + \Big[q'\Big(H - \int L\Big) - \cdots \Big] d\delta x + \cdots . \end{aligned}$$

To simplify notations, set

$$n + n'\Big(H - \int L\Big) = (n),$$
$$p + p'\Big(H - \int L\Big) = (p),$$
$$q + q'\Big(H - \int L\Big) = (q),$$

with corresponding definitions for $(N), (P), (Q), (\nu), (\omega), (\chi)$. Then

$$\begin{aligned} \delta \int Z = &\int \big[(n) - d(p) + d^2(q) - \cdots \big] \delta x \\ &+ \int \big[(N) - d(P) + d^2(Q) - \cdots \big] \delta y \\ &+ \int \big[(\nu) - d(\omega) + d^2(\chi) - \cdots \big] \delta z \\ &+ \big[(p) - d(q) + \cdots \big] \delta x + \big[(q) - \cdots \big] \delta dx + \cdots \\ &+ \big[(P) - d(Q) + \cdots \big] \delta y + \big[(Q) - \cdots \big] \delta dy + \cdots \\ &+ \big[(\omega) - d(\chi) + \cdots \big] \delta z \\ &+ \big[(\chi) - \cdots \big] \delta dz + \cdots = 0. \end{aligned} \tag{3.14}$$

This form for $\delta \int Z$ is quite similar to that for Problem I above (p. 116).

In a corollary Lagrange observes that if Z' depends on another integral $\Pi' = \int Z''$ with

$$\delta Z' = L'\,d\Pi' + n'\,\delta x + p'\,\delta dx + \cdots, \delta Z'' = n''\,\delta x + p''\,\delta dx + \cdots,$$

then equation (3.14) is still valid provided that the quantity (n) there is augmented by $n''[H' - \int(H - \int L)L']$, (p) by $p''[H' - \int(H - \int L)L']$, etc. Clearly, this process can be generalized.

In Section XI Lagrange poses his Problem III: "*To find the equation for a maximum or minimum of the formula* $\int Z$, *when* Z *is simply given by means of a differential equation which contains no differentials of* Z *beyond the first*."

To solve the problem, Lagrange imagines Z is defined by an equation whose variation is

$$\begin{aligned} \delta\,dZ + T\,\delta Z = n\,\delta x + p\,\delta dx + \cdots + N\,\delta y + P\,\delta dy + \cdots \\ + \nu\,\delta z + \omega\,\delta dz + \cdots. \end{aligned}$$

Since $\delta\,dZ = d\,\delta Z$, he views this as a linear differential equation of the first order, and he has

$$\delta Z = e^{-\int T}\int e^{\int T}(n\,\delta x + p\,\delta dx + \cdots);$$

it then follows that

$$\delta\int Z = \int e^{-\int T}\int e^{\int T}(n\,\delta x + p\,\delta dx + \cdots).$$

He now throws the problem back onto the preceding one with the help of the definitions

$$ne^{\int T}\left(G - \int e^{-\int T}\right) = (n),$$

$$pe^{\int T}\left(G - \int e^{-\int T}\right) = (p),$$

$$qe^{\int T}\left(G - \int e^{-\int T}\right) = (q), \ldots .$$

His solution is thence given by relation (3.14) and its consequences.

Lagrange remarks ([1760], p. 351) that "the formulas, which are the object of the two preceding problems, are analogous to those that M. Euler has treated in Chapter III of his work on this subject." In a corollary he proceeds to the case where the differential equation is of order 2 or higher.

In Appendix I he makes a modest start on considering multiple integral problems of the calculus of variations. Specifically, he says "By the methods which have been explained one can also seek the maxima and minima for curved surfaces in a most general manner that has not been done till now."[8] This then is still another of his great innovations.

[8] Lagrange [1760], p. 353.

He starts with the example of finding the surface of least area among all those bounded by a given curve. (This is the so-called problem of Plateau.) To analyze the situation, Lagrange assumes the surface is represented by the equation (in his notation)

$$dz = p\,dx + q\,dy = \frac{dz}{dx}\,dx + \frac{dz}{dy}\,dy$$

and writes the integral to be minimized in the nonparametric form

$$\int\int dx\,dy\sqrt{1+p^2+q^2}\ ,$$

"where the two integral signs indicate two successive integrations, the one with respect to x and the other with respect to y, or inversely." (Although Lagrange does not make clear the character of the region of x, y-space over which the integration takes place, let us assume that $x_1 \leqslant x \leqslant x_2$; $y_1 \leqslant y \leqslant y_2$.) He then writes

$$\begin{aligned} 0 = \delta\int\int dx\,dy\sqrt{1+p^2+q^2} &= \int\int \delta\left(dx\,dy\sqrt{1+p^2+q^2}\right) \\ = \int\int dx\,dy\,\frac{p\,\delta p + q\,\delta q}{\sqrt{1+p^2+q^2}} &= \int\int dx\,dy\,\frac{p}{\sqrt{1+p^2+q^2}}\,\frac{\delta dz}{dx} \\ &\quad + \int\int dx\,dy\,\frac{q}{\sqrt{1+p^2+q^2}}\,\frac{\delta dz}{dy}\ . \end{aligned}$$

Lagrange observes that $\delta p = \delta dz/dx = d\delta z/dx$, and hence he can write

$$\begin{aligned} &\int_{x_1}^{x_2} dx\,\frac{p}{\sqrt{1+p^2+q^2}}\,\frac{d\delta z}{dx} \\ &\qquad = \frac{p}{\sqrt{1+p^2+q^2}}\,\delta z\Big|_{x_1}^{x_2} - \int_{x_1}^{x_2} d\,\frac{p}{\sqrt{1+p^2+q^2}}\,\delta z. \end{aligned}$$

(I have inserted limits of integration into Lagrange's notation.) He assumes that the surface is such that $\delta z = 0$ at $x = x_1$ and $x = x_2$; i.e., the surface passes through a fixed boundary curve. Then he writes

$$\frac{p}{\sqrt{1+p^2+q^2}} = P$$

and the double integral being considered (in his notation) becomes

$$\int\int dx\,dy\,\frac{p}{\sqrt{1+p^2+q^2}}\,\frac{d\delta z}{dx} = -\int dy\int dx\,\frac{dP}{dx}\,\delta z = -\int\int dx\,dy\,\frac{dP}{dx}\,\delta z.$$

(Notice that he does not have a notation for partial derivatives.) Similarly, he lets

$$\frac{q}{\sqrt{1+p^2+q^2}} = Q$$

and finds

$$\int\int dx\,dy \frac{q}{\sqrt{1+p^2+q^2}} \frac{d\delta z}{dy} = -\int\int dx\,dy \frac{dQ}{dy}\,\delta z;$$

thus

$$-\int\int dx\,dy\left[\frac{dP}{dx} + \frac{dQ}{dy}\right]\delta z = 0,$$

"independently of δz." This enables him to conclude that

$$\frac{\partial P}{\partial x} + \frac{\partial Q}{\partial y} = 0$$

which says that $P\,dy - Q\,dx$ is an exact differential; hence he says "the problem is therefore reduced to seeking p and q by the conditions that

$$p\,dx + q\,dy \quad \text{and} \quad \frac{p\,dy - q\,dx}{\sqrt{1+p^2+q^2}}$$

are each exact differentials." He notes that the choice of p and q each constant satisfies these conditions but that this is only a very special case "since the general solution must be such that the boundary curve of the surface can be determined at will."

Now he turns to the case where the desired surface is to be a minimum among all those of given volume. In this case

$$\int\int dx\,dy\sqrt{1+p^2+q^2} = \text{min.,} \qquad \int\int z\,dx\,dy = \text{const.,}$$

and

$$\delta\left(\int\int z\,dx\,dy\right) = 0, \qquad \delta\left(\int\int dx\,dy\sqrt{1+p^2+q^2}\right) = 0.$$

This is equivalent to the conditions

$$\int\int dx\,dy\,\delta z = 0, \qquad \int\int dx\,dy\left[\frac{dP}{dx} + \frac{dQ}{dy}\right]\delta z = 0.$$

He multiplies the former integral by an arbitrary coefficient k and adds it to the latter, finding the relation

$$k + \frac{dP}{dx} + \frac{dQ}{dy} = 0.$$

This is expressible by saying that $(P + kx)\,dy - Q\,dx$ must be an exact differential—note that k is a constant. His condition is now that

$$\frac{p\,dy - q\,dx}{\sqrt{1 + p^2 + q^2}} + kx\,dy, \qquad p\,dx + q\,dy \tag{3.15}$$

are both exact differentials.

He next takes the sphere $(z - a)^2 + (y - b)^2 + (x - c)^2 = r^2$ and observes that

$$dz = \frac{(y - b)\,dy + (x - c)\,dx}{\sqrt{r^2 - (y - b)^2 - (x - c)^2}},$$

so that

$$p = \frac{x - c}{\sqrt{r^2 - (y - b)^2 - (x - c)^2}}, \qquad q = \frac{y - b}{\sqrt{r^2 - (y - b)^2 - (x - c)^2}}.$$

In this situation his conditions (3.15) above become

$$\frac{p\,dy - q\,dx}{\sqrt{1 + p^2 + q^2}} + kx\,dy = \left(\frac{1}{r} + k\right)x\,dy - \frac{1}{r}\,y\,dx + \frac{b\,dx - c\,dy}{r},$$

and this is an exact differential provided that

$$\frac{1}{r} + k = -\frac{1}{r}$$

In his Appendix II (pp. 357ff) Lagrange gives a quite different but also elegant application of his methodology. Specifically, he seeks "to find that polygon, whose area is greatest, among all those which have a fixed number of given sides." He points out that "the method of this memoir is also applicable to these sorts of questions . . . ." Since his analysis is somewhat sketchy and depends on finite difference techniques, perhaps we should give a little discussion to expand his remarks.

Let the coordinates of the vertices of the given polygon be $(x_0, y_0), (x_1, y_1), \ldots, (x_{I-1}, y_{I-1}), (x_I, y_I) = (x_0, y_0)$, and let (x_0, y_0) and (x_{I-1}, y_{I-1}) lie on the x-axis. To evaluate the area, Lagrange decomposes it into trapezoids with altitudes $x_{i+1} - x_i$ and bases $y_{i+1} + y_i$. (Note that some of these may have negative areas.) The area of such a trapezoid is

$$\frac{y_{i+1} + y_i}{2}(x_{i+1} - x_i) = \left[y_i + \frac{y_{i+1} - y_i}{2}\right](x_{i+1} - x_i),$$

and the total area is

$$\int\left(y + \frac{1}{2}\,dy\right)dx = \sum_{i=0}^{I-1}\left[y_i + \frac{y_{i+1} - y_i}{2}\right](x_{i+1} - x_i), \tag{3.16}$$

where $dy = y_{i+1} - y_i$ and $dx = x_{i+1} - x_i$. (Lagrange uses an integral sign here to denote a sum.) The condition for a minimum is then that the variation of this sum must vanish. This implies that

$$0=\int\left[(\delta y)\,dx+\tfrac{1}{2}(\delta dy)\,dx+\left(y+\tfrac{1}{2}dy\right)\delta dx\right];\tag{3.17}$$

but the length of each side, $(dx^2+dy^2)^{1/2}$, is fixed and consequently

$$\delta\sqrt{dx^2+dy^2}=\frac{dx\,\delta dx+dy\,\delta dy}{\sqrt{dx^2+dy^2}}=0.$$

In other words,

$$\delta\,dx=-\frac{dy\,\delta dy}{dx}.\tag{3.18}$$

When this value is substituted into equation (3.17), the result is

$$\int\left[dx\,\delta y+z\,\delta dy\right]=0,\tag{3.19}$$

where Lagrange has defined z with the help of the relation

$$z=\frac{1}{2}dx-\frac{y\,dy}{dx}-\frac{1}{2}\frac{dy^2}{dx}.$$

To simplify result (3.18), we need to develop two formulas for the difference of a product. It is clear that for differences, $df=f_{i+1}-f_i$,

$$d(uv)=v\,du+u\,dv+du\cdot dv=v\,du+(u+du)dv=v\,du+u'\,dv,$$

where $u'=u+du$, since

$$\begin{aligned}u_{i+1}v_{i+1}-u_iv_i&=v_i(u_{i+1}-u_i)+u_i(v_{i+1}-v_i)+(u_{i+1}-u_i)(v_{i+1}-v_i)\\&=v_i(u_{i+1}-u_i)+u_{i+1}(v_{i+1}-v_i).\end{aligned}$$

It follows from this that

$$\begin{aligned}\sum_{i=0}^{I-1}d(u_iv_i)=u_Iv_I-u_0v_0&=\sum_{i=0}^{I-1}v_i\,du_i+\sum_{i=0}^{I-1}u_{i+1}\,dv_i\\&=\sum_{i=0}^{I-1}v_i\,du_i+\sum_{i=0}^{I-1}u_i'\,dv_i,\end{aligned}\tag{3.20}$$

where $u_i'=u_i+du_i=u_{i+1}$. Furthermore, by interchanging u and v, we find that

$$\begin{aligned}u_Iv_I-u_0v_0&=\sum_{i=0}^{I-1}u_i\,dv_i+\sum_{i=0}^{I-1}v_{i+1}\,du_i=\sum_{i=0}^{I-1}u_i\,dv_i+\sum_{j=1}^{I}v_j\,du_{j-1}\\&=\sum_{j=0}^{I-1}u_i\,dv_i+\sum_{j=0}^{I-1}v_j\,d'u_j+v_I(u_I-u_{I-1})-v_0(u_0-u_{-1}),\end{aligned}$$

where, by definition, $u_I=u_0$, $v_I=v_0$, $u_{I-1}=u_{-1}, d'u_j=du_{j-1}, du_j'=du_{j+1}$. In this case, then,

$$\sum_{i=0}^{I-1}d(u_iv_i)=\sum_{i=0}^{I-1}u_i\,dv_i+\sum_{i=0}^{I-1}v_i\,d'u_i.\tag{3.20$'$}$$

In (3.20) let us set $u = \delta y$ and $v = z$; then Lagrange writes the resulting relation symbolically as $\int z\, d\delta y = z\,\delta y - \int dz\,\delta y'$; next in (3.20′) set $u = z$ and $v = \delta y$ and note that the result is

$$\int z\, d\delta y = z\,\delta y - \int d'z\,\delta y \tag{3.21}$$

Lagrange returns to his condition (3.19) above and writes it as

$$z\,\delta y + \int (dx - d'z)\,\delta y = 0.$$

Recall that the expression $z\,\delta y$ above in Lagrange's relation is in reality $z\,\delta y|_0^{I-1}$, i.e., $z\,\delta y$ evaluated at the last end-point minus $z\,\delta y$ evaluated at the first one. Since he keeps those points fixed on the x-axis, he has $\delta y|^0 = \delta y|^{I-1} = 0$ and so $\int (dx - d'z)\delta y = 0$ holds for δy arbitrary. He concludes from this that $dx - d'z = 0$ and consequently that

$$a = x - {}'z = x' - z = x + dx - z = x + dx - \frac{1}{2}\,dx + \frac{y\,dy}{dx} + \frac{1}{2}\,\frac{dy^2}{dx},$$

i.e., that

$$a\,dx = (x + \tfrac{1}{2}dx)\,dx + (y + \tfrac{1}{2}dy)\,dy = \tfrac{1}{2}d(x^2) + \tfrac{1}{2}d(y^2).$$

He says then that by integration he has

$$2ax + r^2 = x^2 + y^2,$$

which is the equation of a circle with center on the x-axis. He concludes that the desired polygon must be inscribed in the semicircle bounded by that axis.

To proceed, Lagrange remarks that the last δx must be zero when the base of the polygon is fixed; but it is given by the equation

$$\delta x = -\int \frac{dy\,d\delta y}{dx}, \tag{3.22}$$

as we see from relation (3.18), and so the sum above must vanish as well as

$$\int \left[dx\,\delta y + \left(\frac{1}{2}\,dx - \frac{y\,dy}{dx} - \frac{1}{2}\,\frac{dy^2}{dx} \right) d\delta y \right]. \tag{3.23}$$

To find the consequences of these conditions, he multiplies (3.22) by an "undetermined coefficient" k and adds the result to (3.23), finding

$$\int \left[dx\,\delta y + \left(k\,\frac{dy}{dx} + \frac{1}{2}\,dx - \frac{1}{2}\,\frac{dy^2}{dx} \right) d\delta y \right] = 0.$$

To simplify, call the summand in parentheses z. It results, as before, that $a = x + dx - z$ or

$$a\,dx = k\,dy + \tfrac{1}{2}d(x^2) + \tfrac{1}{2}d(y^2),$$

which gives the result

$$ax + b^2 = ky + \tfrac{1}{2}x^2 + \tfrac{1}{2}y^2.$$

He sums up in the theorem (p. 360) that "*the largest polygon that can be formed with given sides is that which can be inscribed in a circle.*" He goes on to say that Cramer demonstrated this result synthetically in the 1752 *Memoirs* of the Berlin Academy.[9]

He goes on to consider the case when the individual sides are not given but only their sum; i.e., he fixes the perimeter. Then the variation of $\int(dx^2 + dy^2)^{1/2}$ must vanish and

$$\int \frac{dx\,\delta dx + dy\,\delta dy}{\sqrt{dx^2 + dy^2}} = 0, \qquad \int\left[dx\,\delta y + \frac{1}{2}\,dx\,\delta dy + \left(y + \frac{1}{2}\,dy\right)\delta dx\right] = 0,$$

as we see with the help of (3.17). Again he introduces a multiplier k and says that

$$\int\left[dx\,\delta y + \left(\frac{1}{2}\,dx + \frac{k\,dy}{\sqrt{dx^2 + dy^2}}\right)\delta dy + \left(y + \frac{1}{2}\,dy + \frac{k\,dx}{\sqrt{dx^2 + dy^2}}\right)\delta dx\right] = 0.$$

To simplify notations, he calls the coefficient of δdy, z and of δdx, u. Then $\int[dx\,\delta y + z\,\delta dy + u\,\delta dx] = 0$, which can be transformed, as before, into

$$z\,\delta y + u\,\delta x - \int\left[(dx - d'z)\,\delta y - d'u\,\delta x\right] = 0.$$

From this he concludes that $dx - d'z = 0$ and $d'u = 0$, which gives the relations $x - {}'z = a$ or $x + dx - z = a$, ${}'u = b$ or $u = b$. These mean that

$$x + \frac{1}{2}\,dx - \frac{k\,dy}{\sqrt{dx^2 + dy^2}} = a, \qquad y + \frac{1}{2}\,dy + \frac{k\,dx}{\sqrt{dx^2 + dy^2}} = b. \tag{3.24}$$

Lagrange multiplies the first by dx and the second by dy and adds, finding

$$(x + \tfrac{1}{2}dx)\,dx + (y + \tfrac{1}{2}dy)\,dy = a\,dx + b\,dy$$

or by integrating, i.e., summing,

$$\tfrac{1}{2}x^2 + \tfrac{1}{2}\,y^2 = ax + by + r^2,$$

which is the equation of a circle.

To simplify, he returns to equations (3.24) and writes

$$\frac{k^2\,dy^2}{dx^2 + dy^2} = \left(a - x - \frac{1}{2}\,dx\right)^2, \qquad \frac{k^2\,dx^2}{dx^2 + dy^2} = \left(b - y - \frac{1}{2}\,dy\right)^2.$$

[9] Cramer [1752].

This gives the relation

$$\begin{aligned} k^2 &= (a - x - \tfrac{1}{2}dx)^2 + (b - y - \tfrac{1}{2}dy)^2 \\ &= a^2 + b^2 - 2ax - 2by + x^2 + y^2 \\ &\quad - (a - x)\,dx - (b - y)\,dy + \tfrac{1}{4}dx^2 + \tfrac{1}{4}dy^2 \\ &= a^2 + b^2 + r^2 - \tfrac{1}{4}dx^2 - \tfrac{1}{4}dy^2, \end{aligned}$$

since

$$x^2 + y^2 - 2ax - 2by = r^2, \qquad (a - x)\,dx + (b - y)\,dy = \tfrac{1}{2}dx^2 + \tfrac{1}{2}dy^2.$$

He now concludes (p. 362) that

$$\sqrt{dx^2 + dy^2} = 2\sqrt{a^2 + b^2 + r^2 - k^2}\ ,$$

"Which shows that all sides of the polygon must be equal to each other and consequently the polygon must be regular." He goes on in his last paragraph to clean up a few details. If the first and last x and y are fixed, then $z\,\delta y$ and $u\,\delta x$ vanish. But, he continues, if the base is fixed and of length c, then the corresponding ordinate y will not be fixed and it is necessary to make $u = 0$ and $z = 0$ when $x = c$; this gives $b = 0$ and $c = a$, and the base becomes the diameter of the circumscribing circle.

In the same issue of the *Memoirs* of the Turin Academy Lagrange proceeded to apply his variational method to the solution of a variety of dynamical problems. In this paper he generalized Euler's principle of least action and used his own techniques to solve the resulting variational problems.[10] Since this work would carry us too far afield, I leave the topic and turn to more relevant work of Lagrange.

3.3. Lagrange's Second Paper

In Volume IV (1766/69) of the *Memoirs* of the Turin Academy, *Miscellanea Taurinensia*, Lagrange carries his ideas forward to more complex situations.[11] In the introduction to his paper Lagrange takes the opportunity to attack a hostile critic, one Fontaine, and to discuss two others who failed to give him credit; they are Franciscan monks named Le-Seur and Jacquier who published a two-volume treatise on integral calculus in Parma containing a chapter on Lagrange's method. They not only did not make any mention of his memoir of 1762, but actually transcribed many pages from it "without naming the person as they have

[10] Lagrange [1760′].

[11] Lagrange [1766/69].

done in other places in the same volume . . . but as, by the citation of the memoirs of M. Euler . . . they appear to wish to attribute to him this method, I believe I must remark that I am the first author and that I do not share the credit with any person."

Lagrange goes on to correct another misstatement by these men (Vol. II, p. 531) regarding Euler's principle of least action. They said Euler had shown that on the trajectories described by any number of bodies whatsoever, the action integral is always a maximum or a minimum. Lagrange says that "he [Euler] has seen that the trajectory of a body being acted upon by any central forces is the same as the curve which one would find by supposing the action integral to be a maximum or minimum . . . . The application of this beautiful theorem to any system of bodies and above all the way it serves to solve with the greatest ease and generality all the problems of dynamics is completely due to me . . . "[12]

Lagrange then turns to a calculation he will use later: he supposes that ϕ is a function given by "a differential equation of arbitrary degree between ϕ and $z, y, z, \ldots$ and the differentials of these quantities." He calls this differential equation $\Phi = 0$ and proceeds to find $\delta\Phi$. (His procedure is quite formalistic.) In what follows the reader should realize that $x, y, z, \ldots$ are functions defining a curve

$$x = x(t), \qquad y = y(t), \qquad z = z(t), \ldots$$

whose end-points are $(a, b, c, \ldots)$ and $(l, m, n, \ldots)$.

He writes the variation of Φ in the form

$$\begin{aligned}\delta\Phi = p\,\delta\phi + p'\,\delta d\phi + p''\,\delta d^2\phi + p'''\,\delta d^3\phi + \cdots \\ + q\,\delta x + q'\,\delta dx + q''\,\delta d^2x + q'''\,\delta d^3x + \cdots \\ + r\,\delta y + r'\,\delta dy + r''\,\delta d^2y + r'''\,\delta d^3y + \cdots \\ + s\,dz + s'\,\delta dz + s''\,\delta d^2z + s'''\,\delta d^3z + \cdots = 0,\end{aligned}$$

where $p, p', p'', \ldots, q, q', q'', \ldots$ are given functions of $\phi, x, y, \ldots, d\phi, dx, dy, \ldots$. Since $\delta\, d\phi = d\,\delta\phi$, etc., he writes this equation as

$$\begin{aligned}p\,\delta\phi + p'\,d\delta\phi + p''\,d^2\delta\phi + p'''\,d^3\delta\phi + \cdots \\ + q\,\delta x + q'\,d\delta x + q''\,d^2\delta x + q'''\,d^3\delta x + \cdots \\ + r\,\delta y + r'\,d\delta y + r''\,d^2\delta y + r'''\,d^3\delta y + \cdots \\ + s\,\delta z + s'\,d\delta z + s''\,d^2\delta z + s'''\,d^3\delta z + \cdots = 0,\end{aligned} \tag{3.25}$$

multiplies both members by an as yet undetermined function ξ and integrates the result. He then integrates by parts in his usual fashion and

[12] Lagrange [1766/69], p. 39.

introduces the following notations to simplify his results:

$$P = p\xi - d(p'\xi) + d^2(p''\xi) - d^3(p'''\xi) + \cdots,$$
$$P' = p'\xi - d(p''\xi) + d^2(p'''\xi) - \cdots, \ldots$$
$$Q = q\xi - d(q'\xi) + d^2(q''\xi) - d^3(q'''\xi) + \cdots,$$
$$Q' = q'\xi - d(q''\xi) + d^2(q'''\xi) - \cdots,$$

with similar definitions for $R, R', \ldots, S, S', \ldots$. This gives the relation

$$\begin{aligned}\int (P\,\delta\phi + Q\,\delta x + R\,\delta y + S\,\delta z + \cdots)\\ + P'\,\delta\phi + P''\,d\delta\phi + P'''\,d^2\delta\phi + \cdots\\ + Q'\,\delta x + Q''\,d\delta x + Q'''\,d^2\delta x + \cdots\\ + R'\,\delta y + R''\,d\delta y + R'''\,d^2\delta y + \cdots\\ + S'\,\delta z + S''\,d\delta z + S'''\,d^2\delta z + \cdots = \text{const.},\end{aligned} \tag{3.26}$$

where the expression outside of the integral sign is the value at the upper limit less its value at the lower limit. Lagrange next calls the integrand Ψ and the part outside the integral, Π. This gives the relation

$$\Pi + \int \Psi = \text{const.};$$

next he lets Γ be the value of Π at the lower limit of integration and Δ its value at the upper limit. Then clearly $\Delta = \Gamma - \int \Psi$; and he proceeds to choose ξ so that

$$0 = P = p\xi - d(p'\xi) + d^2(p''\xi) - d^3(p'''\xi) + \cdots.$$

Lagrange views this as a linear differential equation, and he selects the arbitrary constants that enter into the general solution in such a way that at the second end-point of the interval of integration the quantities $P'', P''', \ldots$ in Δ are zero. It is not hard to see that there is one more constant than there are equations $P'' = 0$, $P''' = 0, \ldots$ (note that P' is not set equal to zero). Then equation (3.26) becomes in Lagrange's notation

$$\begin{aligned}(P'\,\delta\phi) = \big[\, & P'\,\delta\phi + P''\,d\delta\phi + P'''\,d^2\delta\phi + \cdots + Q'\,\delta x\\ & + Q''\,d\delta x + Q'''\,d^2\delta x + \cdots + R'\,\delta y + R''\,d\delta y + R'''\,d^2\delta y + \cdots\\ & + S'\,\delta z + S''\,d\delta z + S'''\,d^2\delta z + \cdots \big]\\ & - (Q'\,\delta x + Q''\,d\delta x + Q'''\,d^2\delta x + \cdots + R'\,\delta y + R''\,d\delta y\\ & + R'''\,d^2\delta y + \cdots + S'\,\delta z + S''\,d\delta z + S'''\,d^2\delta z + \cdots)\\ & - \int (Q\,\delta x + R\,\delta y + S\,\delta z + \cdots),\end{aligned} \tag{3.27}$$

where Lagrange has used square brackets outside the integral sign to indicate that the quantities are evaluated at the first end-point of the interval and round brackets at the second end-point. Relation (3.27) is valid provided that ξ has been so chosen that

$$0 = p\xi - d(p'\xi) + d^2(p''\xi) - d^3(p'''\xi) + \cdots,$$

$$0 = (P''), \qquad 0 = (P'''), \cdots.$$

Lagrange now turns in Section III of this paper to the question of maximizing or minimizing the function ϕ in $\Phi = 0$, which has just been discussed. To make things more precise, he notes that ϕ may even contain indefinite integrals "and the circumstances surrounding [each] question will determine the initial point at which these integrals will be supposed to start." He assumes initial conditions to be $x = a, y = b, z = c, \ldots$ and says that ϕ and its differentials are then functions of $a, b, c, \ldots, da, db, dc, \ldots$ as well as of the constants of integration that enter into the determination of ϕ. (This is the case when, e.g., the integrand depends on the limits of integration.) He goes on to say that if there are μ such constants of integration, i.e., if the highest differential of ϕ in Φ is $d^\mu\phi$, then the quantities $\phi, d\phi, d^2\phi, \ldots, d^{\mu-1}\phi$ at the point $x = a$, $y = b$, $z = c, \ldots$ are "arbitrary and can be supposed to be given."

To fix things, Lagrange now assumes that ϕ is to be a maximum or minimum when the other limit of integration is fixed at $x = l$, $y = m, z = n, \ldots$. Then at this point $(\delta\phi) = 0$ in relation (3.27), where the integral in that formula is evaluated over the interval from $(a, b, c \ldots)$ to $(l, m, n, \ldots)$. He then concludes that the relations

$$0 = \int (Q\,\delta x + R\,\delta y + S\,\delta z + \cdots), \tag{3.28}$$

$$\begin{aligned} 0 = \big[& P'\delta\phi + P''\,d\delta\phi + P'''\,d^2\delta\phi + \cdots + Q'\,\delta x + Q''\,d\delta x \\ & + Q'''\,d^2\delta x + \cdots + R'\,\delta y + R''\,d\delta y + R'''\,d^2\delta y + \cdots \\ & + S'\,\delta z + S''\,d\delta z + S'''\,d^2\delta z + \cdots \big] \\ & - (Q'\,\delta x + Q''\,d\delta x + Q'''\,d^2\delta x + \cdots + R'\,\delta y + R''\,d\delta y \\ & + R'''\,d^2\delta y + \cdots + S'\,\delta z + S''\,d\delta z + S'''\,d^2\delta z + \cdots) \end{aligned} \tag{3.29}$$

must hold separately in the variations, $\delta x, \delta y, \delta z, \ldots$; and hence that for $a \leqslant x \leqslant l$, $b \leqslant y \leqslant m$, $c \leqslant z \leqslant n, \ldots,$

$$Q\,\delta x + R\,\delta y + S\,\delta z + \cdots = 0. \tag{3.28$'$}$$

He now proceeds to discuss this result. If there is no relation between the variables $x, y, z, \ldots,$ then $\delta x, \delta y, \delta z, \ldots$ are independent and $Q = 0$, $R = 0$, $S = 0, \ldots$. However, if there is a relation between $x, y, z, \ldots$ of the form

$$X\,dx + Y\,dy + Z\,dz + \cdots = 0,$$

then he has (note that $X\,\delta x + Y\,\delta y + Z\,\delta z + \cdots = 0$)

$$RX - QY = 0,\ SX - QZ = 0, \cdots .$$

He says "In general it will be necessary to reduce the differentials $\delta x, \delta y, \delta z, \ldots$ to the smallest possible number and to equate to zero the coefficients of each of those that remain, and these equations appended to the given equations . . . will serve to find the necessary relation between the variables $x, y, z, \ldots$ in order that the function ϕ must be the largest or smallest."

(In what follows, recall that $d\delta\phi, d^2\delta\phi, \ldots, d\delta x, d^2\delta x, \ldots$ are the same as $\delta d\phi, \delta d^2\phi, \ldots, \delta dx, \delta d^2x, \ldots$.) Lagrange now proceeds to revise (3.29); he sets $f = \phi(a, b, c, \ldots)$; $F^{(i)} = P^{(i)}$, $A^{(i)} = Q^{(i)}$, $B^{(i)} = R^{(i)}$, and $C^{(i)} = S^{(i)}, \ldots$ $(i = {}', {}'', {}''', \ldots)$ at the point $(a, b, c, \ldots)$ and $L^{(i)} = Q^{(i)}, M^{(i)} = R^{(i)}, N^{(i)} = S^{(i)}$ at $(l, m, n \ldots)$. In these terms his relation (3.29) above is expressible as

$$\begin{aligned} F'\,\delta f &+ F''\,\delta df + F'''\,\delta d^2 f + \cdots + A'\,\delta a + A''\,\delta da \\ &+ A'''\,\delta d^2 a + \cdots + B'\,\delta b + B''\,\delta db + B'''\,\delta d^2 b + \cdots \\ &- L'\,\delta l - L''\,\delta dl - L'''\,\delta d^2 l - \cdots - M'\,\delta m \\ &- M''\,\delta dm - M'''\,\delta d^2 m - \cdots - N'\,\delta n - N''\,\delta dn - N'''\,\delta d^2 n \\ &- \cdots = 0. \end{aligned} \tag{3.29'}$$

In closing the section he remarks that in order to use this relation, one must examine the problem at hand and find out what conditions, if any, exist between $f, a, b, c, \ldots, l, m, n, \ldots$ and their differentials.

In Section IV ([1766/1769], p. 47) Lagrange notes that

$$\delta f = \pi\,\delta a + \rho\,\delta b + \sigma\,\delta c + \cdots, \qquad \delta df = \pi'\,\delta a + \rho'\,\delta b + \cdots$$

and that these should be substituted into relation (3.29′) and then terms rearranged so that a linear equation in $\delta a, \delta b, \delta c, \ldots, \delta l, \delta m, \delta n, \ldots$ results.

If, moreover, the function ϕ depends not only on $x, y, z, \ldots$ and their differentials, but also on the parameters $a, b, c, \ldots, l, m, n, \ldots$ and their differentials, then the value of $\delta\Phi$ itself will have a more complex form. In that case expression (3.25) for $\delta\Phi$ must be altered by the addition of an expression of the form

$$\begin{aligned} \alpha\,\delta a &+ \alpha'\,\delta da + \alpha''\,\delta d^2 a + \cdots \\ &+ \beta\,\delta b + \beta'\,\delta db + \beta''\,\delta d^2 b + \cdots \\ &+ \gamma\,\delta c + \gamma'\,\delta dc + \gamma''\,\delta d^2 c + \cdots \\ &+ \lambda\,\delta l + \lambda'\,\delta dl + \lambda''\,\delta d^2 l + \cdots \\ &+ \mu\,\delta m + \mu'\,\delta dm + \mu''\,\delta d^2 m + \cdots \\ &+ \nu\,\delta n + \nu'\,\delta dn + \nu''\,\delta d^2 n + \cdots . \end{aligned}$$

This means that the expression

$$\begin{aligned}\delta a\int\alpha\xi+\delta da\int\alpha'\xi+\delta d^2a\int\alpha''\xi+\cdots\\+\delta b\int\beta\xi+\delta db\int\beta'\xi+\delta d^2b\int\beta''\xi+\cdots\\+\delta c\int\gamma\xi+\delta dc\int\gamma'\xi+\delta d^2c\int\gamma''\xi+\cdots\\+\delta l\int\lambda\xi+\delta dl\int\lambda'\xi+\delta d^2l\int\lambda''\xi+\cdots\\+\delta m\int\mu\xi+\delta dm\int\mu'\xi+\delta d^2m\int\mu''\xi+\cdots\\+\delta n\int\nu\xi+\delta dn\int\nu'\xi+\delta d^2n\int\nu''\xi+\cdots.\end{aligned}$$

must be appended to relation (3.26). This causes a consequent alteration in (3.29′); e.g., instead of $A'\,\delta a$, the result is now $(A'-\int\alpha\xi)\,\delta a$. (It should be remarked that the case when ϕ contains the end-points of the arc as parameters is an important one.)

Lagrange considers next the case when the quantities $f, df, \ldots$ are given functions of the parameters $a, b, c, \ldots$ and derives the resulting condition. However, let us skip to his Section VI (p. 52) where he takes up the simple example $\phi=\int Z$, Z being a function of $x, y, z, \ldots$ and their differentials. This gives

$$\Phi = Z - d\phi, \qquad \delta\Phi = \delta Z - \delta d\phi.$$

(The integral is understood to vanish at the point $(a, b, c, \ldots)$.) He then expresses δZ as

$$\begin{aligned}\delta Z = q\,\delta x + q'\,\delta dx + q''\,\delta d^2x + \cdots\\+ r\,\delta y + r'\,\delta dy + r''\,\delta d^2y + \cdots\\+ s\,\delta z + s'\,\delta dz + s''\,\delta d^2z + \cdots,\end{aligned}\tag{3.30}$$

so that in expression (3.25) for $d\Phi$ he has

$$p=0, \qquad p'=-1, \qquad p''=0, \qquad p'''=0, \ldots.$$

This gives him values for $P, P', P'', \ldots$ in terms of ξ

$$P=-d\xi, \qquad P'=\xi, \qquad P''=0, \qquad P'''=0, \ldots.$$

When he sets $P=0$, he finds, of course, that $\xi=\text{const.}$, which he takes to be 1—notice that automatically $(P'')=0, (P''')=0, \ldots$. For an extremum he consequently has relations (3.28′) and (3.29′). In the latter he has $F'=1, F''=0, F'''=0, \ldots$ since $P'=\xi=1, P''=0, P'''=0, \ldots$. He further notes that at $x=a, y=b, z=c, \ldots, \phi=0$ so $f\equiv 0$ and $\delta f\equiv 0$; this implies that there are no terms in (3.29′) containing $\delta f, \delta df, \ldots$. He notes that the resulting necessary conditions are precisely those of his previous memoir (see pp. 132ff above).

Next in Section VII (p. 54) he takes up the case in which $\phi=\int Z$ again, but Z now contains, in addition to its usual variables, an additional one in the form of an indefinite integral $(\phi) = \int(Z)$, where (Z) depends on $x, y, z, \ldots$ and their differentials. He writes δZ as $\delta Z = \delta V + \pi\,\delta(\phi)$, where δV is of the form (3.30). It follows that

$$\delta V + \pi\,\delta(\phi) - \delta d\phi = 0, \; \delta d(\phi) = d\delta(\phi) = \delta(Z),$$

and consequently that

$$0 = d\delta(\phi) + d\frac{\delta V}{\pi} - d\frac{\delta d\phi}{\pi} = \delta(Z) + d\frac{\delta V}{\pi} - d\frac{\delta d\phi}{\pi}, \tag{3.31}$$

where $\delta(Z)$ is of the form

$$\begin{aligned}\delta(Z) = (q)\delta x + (q')\delta dx + (q'')\delta d^2x + \cdots \\ + (r)\delta y + (r')\delta dy + (r'')\delta d^2y + \cdots \\ + (s)\delta z + (s')\delta dz + (s'')\delta d^2z + \cdots.\end{aligned} \tag{3.32}$$

He multiplies both members of (3.31) by the undetermined function ξ and integrates, finding

$$\xi\frac{\delta V}{\pi} - \xi\frac{\delta d\phi}{\pi} + \int\left[\xi\delta(Z) - \frac{d\xi}{\pi}\delta V + \frac{d\xi}{\pi}\delta d\phi\right] = \text{const.}$$

After substituting the values of δV and $\delta(Z)$ into this equation, he finds, in the notation he used earlier, that

$$P = -d\frac{d\xi}{\pi}, \qquad P' = \frac{d\xi}{\pi}, \qquad P'' = -\frac{\xi}{\pi}, \qquad P''' = 0, \ldots$$

and, as before, he sets $P = 0$ to determine ξ, finding

$$\xi = h + g\int\pi,$$

where g and h are arbitrary constants. As he did before, he now proceeds to set $(P'') = 0$ and notes that this requires ξ to be zero at $x = l, y = m, \ldots$. To do this, he lets Π be the value of $\int\pi$ at the point $(l, m, \ldots)$ and observes that $h + g\Pi = 0$ is the condition in question. He sets $g = -1$ and has

$$\xi = \Pi - \int\pi,$$

$$P' = -1, \qquad P'' = -\frac{\Pi - \int\pi}{\pi}.$$

These conclusions agree also with those of Section IX of Lagrange's first memoir (see p. 120ff above).

Since ϕ again vanishes at the first end-point, $f = \phi(a, b, c, \ldots) = 0$, and hence $\delta f = 0$. This gives the result that $\delta df = \delta d\phi = \delta Z$ at the initial point; it is also clear that at that point $(\phi) = \int(Z) = 0$, so $\delta(\phi) = 0$. Thus the value of δdf is expressible as

$$\begin{aligned} q\,\delta x &+ q'\,\delta dx + q''\,\delta d^2x + \cdots \\ &+ r\,\delta y + r'\,\delta dy + r''\,\delta d^2y + \cdots \\ &+ s\delta z + s'\,\delta dz + s''\,\delta d^2z + \cdots, \end{aligned}$$

$x, y, z, \ldots$ being evaluated at $a, b, c \ldots .$

In Section VIII (p. 57) Lagrange summarizes the advantages "of my method of variations for the solution of problems of *maxima* and *minima*," as follows:

> 1. The simplicity and generality of the calculation as can be seen by comparison of this method to that which M. Euler has given in his excellent work entitled *Methodus inveniendi* . . . .
> 2. The fact that the procedure leads to precise conditions whose solutions serve to resolve the problem at hand.

As a case in point, Lagrange recalculates the problem of the brachystochrone whose end-points are free to move along two arbitrary curves in the same plane. He chooses the coordinates of his first curve to be a, b; of his second one, l, m; and of his minimizing arc, x, y. He further lets u be the velocity of the heavy particle so that $u = (2(y - k))^{1/2}$, where k is an arbitrary constant and the gravity constant $g = 1$. At the beginning of the motion $y = b$ and $u = (2h)^{1/2}$, where $k = b - h$. Then $\int ds/u$ is the integral to be minimized. He therefore has in his previous notations

$$\phi = \int \frac{ds}{u}, \qquad Z = \frac{ds}{u}$$

and since $ds = (dx^2 + dy^2)^{1/2}$, $u = (2(y - k))^{1/2}$,

$$\delta Z = -\frac{ds\,\delta y}{u^3} + \frac{dx\,\delta dx}{u\,ds} + \frac{dy\,\delta dy}{u\,ds},$$

$$q = 0, \qquad q' = \frac{dx}{u\,ds}, \qquad q'' = 0, \ldots$$

$$r = -\frac{ds}{u^3}, \qquad r' = \frac{dy}{u\,ds}, \qquad r'' = 0, \ldots .$$

Now, by what has preceded, he has ($\xi = 1$)

$$Q = -d\frac{dx}{u\,ds}, \qquad Q' = \frac{dx}{u\,ds}, \qquad Q'' = 0, \ldots,$$

$$R = -\frac{ds}{u^3} - d\frac{dy}{u\,ds}, \qquad R' = \frac{dy}{u\,ds}, \qquad R'' = 0, \ldots .$$

He concludes with the help of (3.28′) from these relations that for $\delta x, \delta y$ arbitrary,

$$Q\,\delta x + R\,\delta y = 0$$

so that $Q = 0$ and $R = 0$, either of which will serve to find the extremals.

He takes $Q = 0$ and has

$$\frac{dx}{ds} = fu = f\sqrt{2(y-k)}$$

for f a constant of integration; this gives him

$$dx = \frac{f\,dy\sqrt{2(y-k)}}{\sqrt{1-2f^2(y-k)}}$$

as the differential equation for the brachystochrone. He also has, as the condition when there are no parameters except (a, b), (l, m),

$$A'\,\delta a + B'\,\delta b - L'\,\delta l - M'\,\delta m = 0, \tag{3.33}$$

where A', B' are the values of Q', R' at the first point (a, b) of the curve and L', M' at the last point (l, m). But in the case with k viewed as a parameter, Lagrange returns to his analysis on p. 133 above, where he showed how to alter the condition (3.29′).

In the present case he has (recall $k = b - h$)

$$\Phi = Z - d\phi = \frac{ds}{\sqrt{2(y-k)}} - d\phi$$

and "it will be necessary to adjoin to the value of $\delta\Phi$ the term

$$\frac{ds\,\delta k}{2\sqrt{2(y-k)^3}} = \frac{ds}{u^3}\,\delta k;$$

then because $\xi = 1$ one has to adjoin the quantity $-\delta k \int ds/u^3$ to ... [(3.33)], which becomes consequently

$$A'\,\delta a + B'\,\delta b - L'\,\delta l - M'\,\delta m - \delta k \int \frac{ds}{u^3} = 0,$$

the integral $\int ds/u^3$ being supposed evaluated from the first to the last point of the curve." Since $dx/u\,ds = f$, $Q' = f$ and $A' = L' = f$; furthermore, since $R = 0$,

$$-\frac{ds}{u^3} - d\frac{dy}{u\,ds} = 0$$

or

$$\int \frac{ds}{u^3} + \frac{dy}{u\,ds} = \text{const.},$$

which implies that

$$\int \frac{ds}{u^3} + R' = \text{const.}$$

By evaluating this integral at the first point on the curve, he finds $R' = B'$ and at the last point, $R' = M'$. Thus $\int ds/u^3$ calculated over this interval is

$B' - M'$. He also has $k = b - h$, and if h is a function of a, b with $dh = G\,da + H\,db$, then $\delta k = \delta b - \delta h = \delta b - G\,\delta a - H\,\delta b$.

Given these things the valid version of (3.33) is, in this instance,

$$\left[f - (M' - B')G\right]\delta a + \left[M' - (M' - B')H\right]\delta b - f\,\delta l - M'\,\delta m = 0;$$

and since the end-curves are independent, Lagrange has the relations

$$\left[f - (M' - B')G\right]\delta a + \left[M' - (M' - B')H\right]\delta b = 0,$$
$$f\,\delta l + M'\,\delta m = 0.$$

He then defines the two end-curves with the help of the differential equations $da = \varepsilon\,db$, $dl = \eta\,dm$ and as a result has $\delta a = \varepsilon\,\delta b$, $\delta l = \eta\,\delta m$. It follows from these that his conditions are now expressible as

$$\left[f - (M' - B')G\right]da + \left[M' - (M' - B')H\right]db = 0, \qquad f\,dl + M'\,dm = 0.$$

(To see this, Lagrange notes that $\varepsilon = \delta a/\delta b = da/db$, $\eta = \delta l/\delta m = dl/dm$.)

Lagrange now supposes the height h, which corresponds to the initial velocity is equal to b, so that the bead will start to move on the brachystochrone with the same velocity it would have acquired in falling freely down the axis. Then $G = 0$, $H = 1$, and these relations above simplify to

$$f\,da + B'\,db = 0, \qquad f\,dl + M'\,dm = 0;$$

but $f = dx/u\,ds$ and $B' = dy/u\,ds$ at the initial point, and $M' = dy/u\,ds$ at the terminal point. Then he has

$$dx \cdot da + dy \cdot db = 0; \qquad \text{i.e.,} \quad \frac{dx}{dy} = -\frac{db}{da}$$

at the first point on the minimizing arc and

$$dx \cdot dl + dy \cdot dm = 0, \qquad \text{i.e.,} \quad \frac{dx}{dy} = -\frac{dm}{dl}$$

at the final point. These imply that the minimizing arc, in this case, cuts each bounding curve orthogonally.

In the case that the initial velocity is zero, then $h = 0$ and therefore $G = H = 0$ so that

$$f\,da + M'\,db = 0, \qquad f\,dl + M'\,dm = 0.$$

Since the latter relation is exactly as it was above, it follows that the minimizing curve must cut the second curve orthogonally. The first one, however, implies that

$$\frac{da}{db} = -\frac{M'}{f} = \frac{dl}{dm};$$

this means that the slope of the first curve at the point the minimizing arc cuts it is equal to the slope of the second one at the point where the minimizing curve cuts it. This is the conclusion that Borda pointed out in correcting Lagrange's first paper.[13]

[13] Borda [1767].

3.4. Legendre's Analysis of the Second Variation

The next really significant investigation was made by Legendre in 1786 in a memoir presented to the Paris Academy.[14] Legendre became interested in the problem of deciding whether a given extremal is a minimizing or a maximizing arc. To do this, he considered the second variation and found his well-known transformation of the second variation as well as his condition.

To do this, he considers an integrand function v of x, y and $p = dy/dx$ and writes

$$\delta\int v\,dx = \int dx\,\delta v = \left(\frac{\partial v}{\partial p}\,\delta y\right)^{I} - \left(\frac{\partial v}{\partial p}\,\delta y\right)^{O} + \int dx\,\delta y\left[\frac{\partial v}{\partial y} - \frac{1}{dx}\,d\left(\frac{\partial v}{\partial p}\right)\right],$$

where the superscripts O and I are used to signify that the quantity indicated is evaluated at the beginning and end of the interval integration. Along an extremal he concludes it is necessary that

$$\frac{\partial v}{\partial y} - d\left(\frac{\partial v}{\partial p}\right) = 0, \qquad \left(\frac{\partial v}{\partial p}\,\delta y\right)^{I} - \left(\frac{\partial v}{\partial p}\,\delta y\right)^{O} = 0. \tag{3.34}$$

(Notice that Legendre seems to have invented the symbolism for a partial derivative. He says "In order to avoid ambiguity I designate the coefficient of dx in the differential of v by $\partial v/\partial x$ and the coefficient of dx in the exact differential by dx/dx.") Next he expands $v(x, y + \delta y, p + \delta p)$ using a Taylor expansion. The total variation of v is

$$\Delta v = v(x, y + \delta y, p + \delta p) - v = \frac{\partial v}{\partial y}\,\delta y + \frac{\partial v}{\partial p}\,\delta p$$
$$+ \frac{1}{2}\,\frac{\partial^2 v}{\partial y^2}\,\delta y^2 + \frac{1}{2}\,\frac{\partial^2 v}{\partial y\,\partial p}\cdot 2\delta y\,\delta p + \frac{1}{2}\,\frac{\partial^2 v}{\partial p^2}\,\partial p^2 + \cdots.$$

For functions v satisfying equations (3.34) he has, through terms of the second order,

$$\Delta\int v\,dx = \int dx\left(\frac{1}{2}\,\frac{\partial^2 v}{\partial y^2}\cdot\delta y^2 + \frac{1}{2}\,\frac{\partial^2 v}{\partial y\,\partial p}\cdot 2\delta y\,\delta p + \frac{1}{2}\,\frac{\partial^2 v}{\partial p^2}\cdot\delta p^2\right),$$

which he writes in the form $\int dx(P\,\delta y^2 + 2Q\,\delta y\,\delta p + R\,\delta p^2)$. (He does not use Δ but δ, which makes for some confusion.)

Legendre now employs a very neat trick to find his necessary condition. He arbitrarily writes the last expression as

$$\Delta\int v\,dx = \int dx(P\,\delta y^2 + 2Q\,\delta y\,\delta p + R\,\delta p^2)$$
$$+ \int dx\left(\frac{d\alpha}{dx}\,\delta y^2 + 2\alpha\,\delta y\,\delta p\right) - (\alpha\,\delta y^2)^{I} + (\alpha\,\delta y^2)^{O},$$

[14] Legendre [1786].

where α is an as yet undetermined function of x. (Notice that what he has added to the variation has the value zero.) This gives him

$$\Delta \int v\,dx = (\alpha\,\delta y^2)^O - (\alpha\,\delta y^2)^I$$
$$+ \int dx\left[\left(P + \frac{d\alpha}{dx}\right)\delta y^2 + 2(Q+\alpha)\,\delta y\,\delta p + R\,\delta p^2\right],$$

and he asks that the integrand in this expression be a perfect square, i.e., that α be chosen so that

$$\left(P + \frac{d\alpha}{dx}\right)R = (Q+\alpha)^2. \tag{3.35}$$

If he can so determine α, he can write

$$\Delta \int v\,dx = (\alpha\,\delta y^2)^O - (\alpha\,\delta y^2)^I + \int R\,dx\left(\delta p + \frac{Q+\alpha}{R}\,\delta y\right)^2.$$

He now chooses δy at each end-point so that the expression outside the integral sign vanishes or has the same sign as R; and he concludes that for a minimum, it is necessary the coefficient

$$\frac{\partial^2 v}{\partial p^2} = R$$

must be positive (presumably he means nonnegative) and for a maximum negative (nonpositive). He also asserts, incorrectly, that this condition is sufficient. There is a theorem that says this is true for *regular* problems, i.e., for problems for which $v_{pp} > 0$ for all admissible sets (x, y, p) and for which every (x, y, p) with $p_1 < p < p_2$ is admissible whenever the sets (x, y, p_1) and (x, y, p_2) have this property. Many applications are indeed regular. (He disregards the case where $R \equiv 0$ since it is so special.) He concludes that the Euler condition plus $R > 0$ are sufficient for a minimum.

[As we shall see below, Legendre's argument is not flawless, but to him goes the credit for focusing attention on the second variation. First Lagrange showed, in general, that $R > 0$ is not sufficient, as asserted by Legendre; and then Jacobi made a beautiful analysis of the problem and uncovered yet another necessary condition in so doing (see Sections 3.5 and 4.2 below).] Legendre proceeds next to the case where v depends not only on x, y and $p = dy/dx$, but also on $q = dp/dx$. He gives the same sort of argument as before and concludes that now the relevant quantity is $\partial^2 v/\partial q^2$.

Having shown this, he now proceeds in a formalistic way to vary not only the dependent variable y and its derivative p, but also the independent variable x. This forces him to some understanding of parametric problems. In effect, what he does is to introduce for an integrand the function $V(x, y, \dot{x}, \dot{y}) = v(x, y, \dot{y}/\dot{x})\dot{x}$, where curves are now defined parametrically as $x = x(t)$, $y = y(t)$ and $\dot{x} = dx/dt$, $\dot{y} = dy/dt$. This integrand is positively homogeneous of degree 1 in $\dot{x}$, $\dot{y}$ in the sense that

$$V(x, y, k\dot{x}, k\dot{y}) = kV(x, y, \dot{x}, \dot{y})$$

for $k > 0$. Legendre then calculates the second-order terms that enter into $\Delta\int V\,dt$ and finds

$$\int dx\left\{\frac{1}{2}\frac{\partial^2 v}{\partial x^2}\delta x^2 + \frac{1}{2}\frac{\partial^2 y}{\partial x\,\partial y}\cdot 2\delta x\,\delta y + \frac{1}{2}\frac{\partial^2 v}{\partial y^2}\cdot\delta y^2\right.$$
$$\left. + \frac{1}{2}\frac{\partial^2 v}{\partial x\,\partial p}\cdot 2\delta x\,\delta p + \frac{1}{2}\frac{\partial^2 v}{\partial y\,\partial p}\cdot 2\delta y\cdot\delta p + \frac{1}{2}\frac{\partial^2 v}{\partial p^2}\delta p^2\right\}$$
$$+ \int\left(\frac{\partial v}{\partial x}\delta x\,d\delta x + \frac{\partial v}{\partial y}\delta y\,d\delta x\right) \tag{3.36}$$

as the value. From his point of view the important thing about this expression is that it does not contain a term of the form $(\partial v/\partial p)\,\delta p\,d\delta x$ since his method would not then be applicable.

To simplify notations, Legendre writes the integrand of the first integral as

$$F\,\delta x^2 + 2G\,\delta x\,\delta y + J\,\delta y^2 + 2H\,\delta x\,\delta p + 2K\,\delta y\,\delta p + L\,\delta p^2.$$

He says that since $d\delta y = \delta p\,dx + p\,d\delta x$, the differential of the expression $\alpha\,\delta x^2 + 2\beta\delta x\,\delta y + \gamma\,\delta y^2$, for any functions α, β, γ of x, is

$$\begin{aligned} &d\alpha\,\delta x^2 + 2d\beta\,\delta x\,\delta y + d\gamma\,\delta y^2 \\ &\quad + 2\beta\,dx\,\delta x\,\delta p + 2\gamma\,dx\,\delta y\,\delta p \\ &\quad + (2\alpha + 2\beta p)\,\delta x\,d\delta x + (2\beta + 2\gamma p)\,\delta y\,d\delta x. \end{aligned}$$

In accord with what he did in the previous case, in effect, he notices that $\int d(\alpha\,\delta x^2 + 2\beta\,\delta x\,\delta y + \gamma\,\delta y^2) - (\alpha\,\delta x^2 + 2\beta\,\delta x\,\delta y + \gamma\,\delta y^2)|_0^I = 0$. Then he adds this expression to (3.36). He now wishes to choose α, β, γ so that the modified quadratic functional (3.36) becomes

$$(\alpha\,\delta x^2 + 2\beta\,\delta x\,\delta y + \gamma\,\delta y^2)^O - (\alpha\,\delta x^2 + 2\beta\,\delta x\,\delta y + \gamma\,\delta y^2)^I$$
$$+ \int L\,dx(\delta p + \mu\delta y + \lambda\,\delta x)^2.$$

where λ, μ are to be determined along with α, β, γ. This will be the case provided that

$$L\mu = K + \gamma, \qquad L\lambda = H + \beta,$$
$$L\mu^2 = J + \frac{d\gamma}{dx}, \qquad L\mu\lambda = G + \frac{d\beta}{dx}, \qquad L\lambda^2 = F + \frac{d\alpha}{dx} \tag{3.37}$$

together with

$$\frac{\partial v}{\partial x} + 2\beta p + 2\alpha = 0, \qquad \frac{\partial v}{\partial y} + 2\gamma p + 2\beta = 0. \tag{3.37'}$$

These relations he views as equations to determine the α, β, γ, μ, λ. To find them, he proceeds by first expressing the differentials of $\partial v/\partial x$, $\partial v/\partial y$, and $\partial v/\partial p$ with the help of the functions F, G, H, J, K, L in the form

$$d\left(\frac{\partial v}{\partial x}\right) = 2F\,dx + 2G\,dy + 2H\,dp,$$

$$d\left(\frac{\partial v}{\partial y}\right) = 2G\,dx + 2J\,dy + 2K\,dp,$$

$$d\left(\frac{\partial v}{\partial p}\right) = 2H\,dx + 2K\,dy + 2L\,dp.$$

Next he differentiates relations (3.37′) and finds that

$$F + Gp + H\frac{dp}{dx} + \frac{d\alpha}{dx} + p\frac{d\beta}{dx} + \beta\frac{dp}{dx} = 0,$$

$$G + Jp + K\frac{dp}{dx} + \frac{d\beta}{dx} + p\frac{d\gamma}{dx} + \gamma\frac{dp}{dx} = 0.$$

With the help of equations (3.37), these become

$$L\lambda^2 + L\mu\lambda p + L\lambda\frac{dp}{dx} = 0, \qquad L\mu\lambda + L\mu^2 p + L\mu\frac{dp}{dx} = 0,$$

i.e., if $L \neq 0$ and $\lambda \neq 0$ or $\mu \neq 0$, then

$$\frac{dp}{dx} + \mu p + \lambda = 0.$$

Last he notes with the help of (3.37) that the Euler differential equation is expressible as

$$\begin{aligned} 0 = \frac{d}{dx}\left(\frac{\partial v}{\partial p}\right) - \frac{\partial v}{\partial y} &= \frac{\partial^2 v}{\partial x\,\partial p} + p\frac{\partial^2 v}{\partial y\,\partial p} + \frac{\partial^2 v}{\partial p^2}\frac{dp}{dx} - \frac{\partial v}{\partial y} \\ &= 2H + 2Kp + L\frac{dp}{dx} - \frac{\partial v}{\partial y} \\ &= 2H + 2Kp + L\frac{dp}{dx} + 2\gamma p + 2\beta, \end{aligned}$$

which implies that

$$0 = L\lambda + L\mu p + L\frac{dp}{dx}.$$

In other words, instead of seven independent equations (3.37), (3.37′) to determine the quantities $\alpha, \beta, \gamma, \lambda, \mu$, there are only five, which may be written as

$$\left(J + \frac{d\gamma}{dx}\right)L = (K + \gamma)^2,$$

$$\beta = -\gamma p - \frac{1}{2}\frac{\partial v}{\partial y}, \qquad \alpha = -\beta p - \frac{1}{2}\frac{\partial v}{\partial x},$$

$$\mu = \frac{K + \gamma}{L}, \qquad \lambda = \frac{H + \beta}{L}.$$

Legendre remarks that γ contains an arbitrary constant by means of which the expression

$$(\alpha\,\delta x^2 + 2\beta\,\delta x\,\delta y + \gamma\,\delta y)^{O} - (\alpha\,\delta x^2 + 2\beta\,\delta x\,\delta y + \gamma\,\delta y^2)^{I}$$

can be made to vanish.

We skip over some of his analysis, which contains an error, and go to Section VI of his paper, where he gives some examples. Instead of looking at all these, we shall concern ourselves only with his first one: the surface of least resistance of Newton. Legendre considers the integral to be minimized to be of the form ($p = dy/dx$)

$$\int \frac{y\,dy^3}{dx^2 + dy^2} = \int \frac{p^3 y}{1 + p^2}\,dx, \tag{3.38}$$

subject to the condition that the admissible arcs pass through two given points. As we know, this leads to the Euler equation

$$\frac{p^3 y}{(1+p^2)^2} = a,$$

and if $lp = \log p$, to the equations

$$x = a\left(\frac{3}{4p^4} + \frac{1}{p^2} + lp\right) + b, \qquad y = \frac{a(1+p^2)^2}{p^3}.$$

Legendre points out that this two-parameter family of extremals consists of curves each of which has a cusp as in Figure 3.1. At the cuspidal point the tangent makes an angle of 60° with the GD axis. The portion FB has for an asymptote the curve $y^4 = 64ax^3/27$ and the portion FN, the curve $x - b = a\log(y/a)$. He calculates $\partial^2 v/\partial \mathrm{p}^2$ and finds

$$\frac{\partial^2 v}{\partial p^2} = \frac{2py(3-p^2)}{(1+p^2)^3}.$$

Now on FB we recall that p varies from 0 to $3^{1/2}$ so that $\partial^2 v/\partial p^2 \geqslant 0$, and on FN it varies from $3^{1/2}$ to ∞ so that $\partial^2 v/\partial p^2 \leqslant 0$. He concludes that on the former the integral is a minimum and on the latter a maximum.

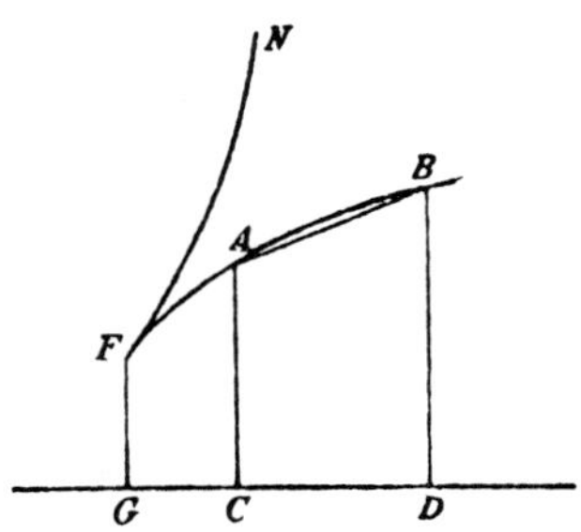

Figure 3.1

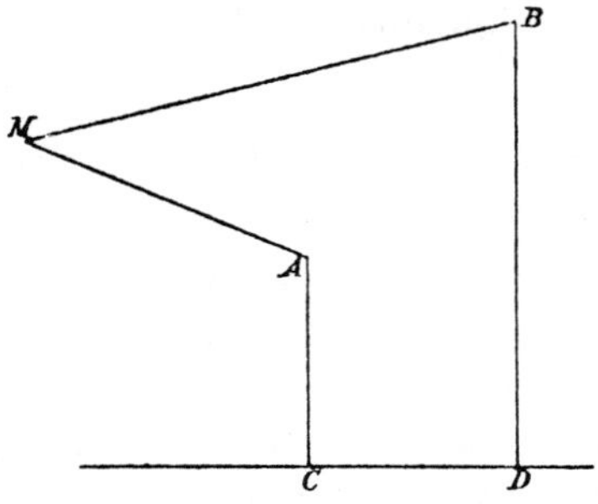

Figure 3.2

In Figure 3.1 Legendre now considers whether a portion of *FB* or *FN* can be drawn through two given points *A* and *B*. (The lines *AC* and *BD* are normal to *GD*.) If the angle *ABD* is greater than 30°, it is possible to pass through *A* and *B* a portion of the form *FN* and when the angle is less than 30°, a portion of the form *FB*. However, when the angle is equal to 30°, it is not possible to pass a portion of the solution curve through the given points, and hence there is no solution. (We have seen above on pp. 27ff how these results follow.)

Legendre now goes on to construct his famous "zigzag" solution to Newton's problem. In Figure 3.2 he draws *AM* and *MB* so that they have numerically equal slopes. He then reasons that Newton's integral (3.38) has the value

$$\frac{BD^2 - AC^2}{2} \sin^2 J, \tag{3.39}$$

where *J* is the angle *MB* makes with the *CD*-axis. From this he points out that by moving *M* farther and farther to the left in Figure 3.2, the value of integral (3.38) can be made as small as desired. He goes beyond this and says if one wants a solution for which the abscissa does not go outside the interval from *C* to *D*, then it suffices to draw a zigzag *AB*, as in Figure 3.3, whose sides all make numerically equal angles *J* with *CD*; then integral (3.38) is again given by (3.39) and can be made as small as desired. He gives an example to show how to achieve an arbitrarily large value for

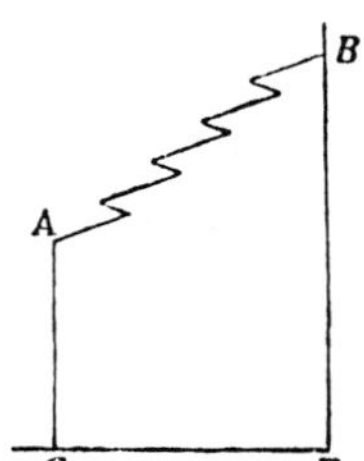

Figure 3.3

(3.39) and how an absolute minimum can be achieved by restricting the length of the curve. (Recall how Newton restricted the slopes x' and y' on p. 16 above to ensure a minimum.)

3.5. Excursus

After Lagrange wrote his two papers in the *Memoirs* of the Turin Academy, he brought out two more works on our subject, which should be mentioned. The first of these appears in his *Theory of Analytic Functions*, the first edition of which came out in 1797 and the second in 1813; it is the latter edition which appears in his collected works. The second paper appears in his *Lectures on the Calculus of Functions*, specifically in lectures 21 and 22 of 1806. This volume of lectures were intended "to serve as a commentary and supplement to the first Part of the *Theory of Analytic Functions*."[15]

In his *Theory* Lagrange criticizes Legendre's method (*TAF*, p. 305):

> In this way one would achieve the same result as is given by the method proposed in the Memoirs of the Academy of Sciences of 1786 for distinguishing maxima from minima in the Calculus of Variations. But after what we have said above it would be necessary for the correctness of this result that one be able to make sure that the value of v [this is the negative of Legendre's function α in equation (3.34)] could not become infinite for a value of x lying between the given values a and b, which will most often be impossible to do because of the difficulty of finding the function $v(x)$. Without this requirement, although the quantity
>
> $$\omega^2 M + \omega\omega' N + \omega'^2 P$$
>
> [this is the second variation, and $\delta y = \omega$], will have become
>
> $$P\left(\omega' + \frac{\omega N}{2P}\right)^2,$$
>
> and that it may, consequently, always be positive or negative according as to what the value of P shall be, one will not ever be certain of the positive or negative state of its primitive function [its integral].

To see the force of Lagrange's objection, notice equation (4.11) below, where v is defined by Jacobi. In that equation the factor $1/u$ enters in such a way that precisely for $u = 0$, v is not well defined. These values of x define the conjugate points as we shall see.

To illustrate his point, Lagrange considers the integrand function

$$f(x, y, y') = ny^2 + 2myy' + y'^2 \qquad (n = -k^2),$$

[15] Lagrange, *TAF*, *LCF*.

where m, n are constants, and finds the extremals to be the family $y = g\sin(kx + h)$ $(a \leqslant x \leqslant b)$. It is clear that the second variation is

$$\int_a^b (n\omega^2 + 2m\omega\omega' + \omega'^2)\,dx, \tag{3.40}$$

and Legendre's differential equation (3.35) becomes $\alpha' = k^2 + (m + \alpha)^2$. To solve this equation, Lagrange sets $k\rho = m + \alpha$ and finds $\rho' = k(1 + \rho^2)$, from which he infers that $\rho = \tan(kx + d)$. He points out (TAF, p. 307), that ρ and hence α become infinite for $kx + d = (2p + 1)\pi/2$ and further states that "one will not be assured of the existence of a minimum if the quantity $(b - a)k$ is larger than the value of two right angles." This is, in an interesting way, a presage of Jacobi's condition on conjugate points.

Lagrange proceeds by supposing that in (3.40) the variation ω is $i\sin x$, where i is some constant. Then the integrand becomes

$$i^2\left(\frac{1+n}{2} + \frac{1-n}{2}\cos 2x + m\sin 2x\right),$$

and its indefinite integral or primitive is

$$i^2\left(\frac{1+n}{2}x + \frac{1-n}{4}\sin 2x - \frac{m}{2}\cos 2x\right) + c, \tag{3.41}$$

where c is to be determined so that this expression vanishes for $x = a$. He now wishes to examine (3.41) for $a = 0$, $b = \pi$; he has $\omega(a) = \omega(b) = 0$, which is required since he assumes that the admissible curves have fixed end-points. He sees that (3.41) evaluated at $x = \pi$ is $i^2(1 + n)D$, "D representing the right angle." Now if $n = -k^2$ and $k > 1$, he concludes that the second variation will be negative even though $\partial^2 f/\partial y'^2 > 0$. (See Jacobi's analysis in Section 4.2.)

(It is perhaps worth noting how ungracious Lagrange was in not mentioning Legendre's name anywhere in his paper.) Legendre's attempt to assert that his condition is both necessary and sufficient is not correct, nor is his proof that it is necessary; nevertheless, small alteration or addition to that proof suffices to make it rigorous, as Weierstrass showed in 1879.

Let us examine Weierstrass's proof in his *Lectures*.[16] The problem at hand is this: given that the integral

$$\int_a^b f(x, y, y')\,dx$$

is a minimum along an arc $y(x)$ $(a \leqslant x \leqslant b)$ it is necessary that $R = \partial^2 f/\partial y'^2$, evaluated along the minimizing arc, be nonnegative.

Suppose there were a point c with $a < c < b$ at which $R = \partial^2 f/\partial y'^2 < 0$.

[16] Bolza, *VOR*, pp. 55–58.

Then by continuity considerations there would be some closed interval $[c-\delta, c+\delta]$ inside of $[a, b]$ in which $R(x)<0$. Now consider Legendre's differential equation (3.35) above, which we shall write as

$$\frac{d\alpha}{dx} = -P + \frac{(Q+\alpha)^2}{R}, \qquad \alpha(c) = \alpha_0,$$

where $P = \partial^2 f/\partial y^2$, $Q = \partial^2 f/\partial y\,\partial y'$, $R = \partial^2 f/\partial y'^2$, and α_0 is any initial value of α. We know that in a neighborhood of (c, α_0) the right-hand member of this equation would be continuous and have a continuous derivative with respect to α, provided we assume that in a region of (x, y, p)-space the function f is of class C'''. (We assume that the minimizing arc lies inside this region.) Then there would be an interval $[c-\varepsilon, c+\varepsilon]$ inside of $[c-\delta, c+\delta]$ and a solution $\alpha = A(x)$ of Legendre's differential equation above, of class C' on the interval with $A(c) = \alpha_0$.

Recall that Legendre has expressed the second variation as

$$\int_a^b R\left(\eta' + \frac{Q+A}{R}\eta\right)^2 dx,$$

where A is the solution of Legendre's equation above. Let us now choose a variation η so that

$$\eta = \begin{cases} (x-c+\varepsilon)^2(x-c-\varepsilon)^2 & \text{on } [c-\varepsilon, c+\varepsilon] \\ 0 & \text{elsewhere.} \end{cases}$$

Then η is of class C', vanishes at $x = a$ and b, and gives to the second variation the value

$$\int_{c-\varepsilon}^{c+\varepsilon} R\left(\eta' + \frac{Q+A}{R}\eta\right)^2 dx.$$

But $R(x) < 0$ on the interval $[c-\varepsilon, c+\varepsilon]$, and thus this integral $\leqslant 0$, i.e., the second variation, would be nonpositive. The equality sign above, moreover, could hold only if

$$\eta' + \frac{Q+A}{R}\eta \equiv 0$$

identically on the interval, as can be seen by continuity considerations. But we have

$$\lim_{x=c-\varepsilon} \frac{\eta' + \dfrac{Q+A}{R}\eta}{(x-c+\varepsilon)(x-c-\varepsilon)} = -4\varepsilon \neq 0;$$

thus the integrand is not zero near $x = c-\varepsilon$. It then follows at once that $R(c) < 0$ is impossible, and hence it is necessary that $R(x) \geqslant 0$ on $[a, b]$.

3.6. The Euler–Lagrange Multiplier Rule

As we saw above, first Euler and then Lagrange, in imitation of him, discovered the form of the multiplier rule when there is a side-condition of the form $y' - \phi(x, y) = 0$. However, their initial papers on the subject contain no mention of multipliers. It was not until 1788 that Lagrange in his *Mécanique Analytique* apparently saw the multiplier rule in its generality and understood, at least operationally, how to determine the multipliers. Here is how he gave expression to his rule in the first volume of the *Méchanique Analytique* portion on Statics, Part 1, Section IV, Article 2 (*MA*, Vol. XI, pp. 77ff.):

> Let $L = 0$, $M = 0$, $N = 0$, &c. be the different equations of condition which are given by the nature of the system, the quantities L, M, N, &c. being finite functions of the variables x, y, z, x', y', z', &c; by differentiating these equations one will have these, $dL = 0$, $dM = 0$, $dN = 0$, &c., which will give the relation that must exist between the differentials of the same variables . . . .
>
> Now as these equations [$dL = 0$, $dM = 0$, $dN = 0$, &c.] must serve only to eliminate an equal number of differentials from the equation of virtual velocity, after which the coefficients of the remaining differentials must each be equal to zero, it is not difficult to prove by the theory of elimination for linear equations that one would have the same result if one simply adjoins to the equation of virtual velocity the different equations of condition $dL = 0$, $dM = 0$, $dN = 0$, &c., each multiplied by an undetermined coefficient; after this, one equates to zero the sum of all the terms that one finds multiplied by the same differential.

Later in Part 2, Section IV, Article 2 (Vol. XI, p. 336) which is on dynamics, Lagrange continues on the same "tack." He remarks that "One should adjoin to the first member of the general formula . . . the quantity $\lambda\, dL + \mu\, dM + \nu\, dN + \cdots$ in which λ, μ ν, &c, are undetermined coefficients; then the variations $\delta\xi$, $\delta\psi$, $\delta\phi$, . . . can be regarded as independent and arbitrary."

In his *Theory* Lagrange again states the Lagrange method of undetermined multipliers for problems of the following sort: A function f of n variables $x, y, z, \ldots$ is to be minimized subject to the condition that the variables are not independent but must satisfy the equation of condition[17]

$$\phi(x, y, z, \ldots) = 0.$$

His argument is quite straightforward. He replaces $x, y, z \ldots$ by $x + p$, $y + q$, $z + r, \ldots$ both in f and ϕ and notices that the condition $\phi = 0$ must hold at this point also. He then expands both f and ϕ by Taylor's theorem and

[17] Lagrange, *TAF*, IX, pp. 290–292.

has for ϕ the expansion

$$p\phi'(x) + q\phi'(y) + r\phi'(z) + \cdots + \tfrac{1}{2}p^2\phi''(x) + pq\phi''(x, y) + \tfrac{1}{2}q^2\phi''(y) + \cdots = 0, \tag{3.42}$$

where his notation is somewhat peculiar. By $\phi'(x), \phi'(y), \phi'(z), \ldots$ he means $\partial\phi/\partial x, \partial\phi/\partial y, \partial\phi/\partial z, \ldots,$ and by $\phi''(x), \phi''(x, y), \phi''(y), \ldots$ he means $\partial^2\phi/\partial x^2, \partial^2\phi/\partial x\partial y, \partial^2\phi/\partial y^2, \ldots$. He now says that one could solve (3.42) for p and substitute that value in the expansion for $f(x + p, y + q, z + r, \ldots)$ or, better yet, multiply (3.42) by an undetermined quantity a and find as a necessary condition that, in his notation,

$$f'(x) + a\phi'(x) = 0, \qquad f'(y) + a\phi'(y) = 0, \qquad f'(z) + a\phi'(z) = 0, \ldots;$$

this system can be reduced by one equation by eliminating a.

Lagrange now states (Vol. IX, p. 292) a general principle: "When a function of many variables must be a maximum or minimum and there are one or more equations between the variables, it will suffice to adjoin to the proposed function the functions which must be null, each multiplied by an undetermined quantity, and to seek afterwards the maximum or minimum as if the variables were independent; the equations which one will find, combined with the given equations, will serve to determine all the unknowns."

Then later in Volume IX (p. 312) and in Volume X (pp. 416ff) he takes up the Euler–Lagrange equations for the Lagrange multiplier rule. In the former place he considers the problem of minimizing the integral

$$\int_a^b f(x, y, y', \ldots, z, z', \ldots)\, dx, \tag{3.43}$$

subject to the side-condition

$$\phi(x, y, y', \ldots, z, z', \ldots) = 0. \tag{3.44}$$

He says that from the discussion above, one simply adjoins the ϕ, multiplied by a variable Δ, to f and regards the variables as being independent. He then finds, in his notation, the two equations

$$f'(y) - [f'(y')]' + [f'(y')]'' - \cdots + \Delta\phi'(y) - [\Delta\phi'(y')]' + [\Delta\phi'(y'')]'' - \cdots = 0,$$

$$f'(z) - [f'(z')]' + [f'(z'')]'' - \cdots + \Delta\phi'(z) - [\Delta\phi'(z')]' + [\Delta\phi'(z'')]'' - \cdots = 0.$$

He comments that by eliminating Δ between these equations, one has an equation which combined with $\phi = 0$ serves to determine the values of y and z as functions of x. Alternately, the equations above serve to determine Δ as a function of x and the extremal arc y as a function of x.

He also takes up the isoperimetric case and considers the integral (3.43) being a maximum or minimum subject now to the condition that

$$\int_a^b \phi(x, y, y', \ldots, z, z', \ldots)\, dx = \text{const.} \tag{3.44$'$}$$

Here he says that $f + \Delta\phi$ is again considered, but that now Δ is a constant, which can be determined from the value of the constant in (3.44′).

In his Lecture 22 Lagrange again discusses the so-called Lagrange problem.[18] He takes as before an integral (3.43) and the side-condition (3.44). He says "it is considerably simpler to employ the multipliers in the way that they are used in the *Mécanique Analytique*, which is totally based upon the calculus of variations." He then goes on to another most important case. He considers now that his integral (3.43) has adjoined to it one or more end-conditions which he writes in the form

$$\Phi(x_0, y_0, y_0', \ldots, z_0, z_0', \ldots, x_1, y_1, y_1', \ldots, z_1, z_1', \ldots) = 0,$$

where $x_0, y_0, y_0', \ldots, z_0, z_0', \ldots$ and $x_1, y_1, y_1', \ldots, z_1, z', \ldots$ are the values of $x, y, y', \ldots, z, z', \ldots$ at the ends of the arc being examined. He remarks that the variation of $\Phi = 0$ is

$$\begin{gathered} \dot{x}_0\Phi'(x_0) + \dot{y}_0\Phi'(y_0) + \dot{z}_0\Phi'(z_0) + \cdots \\ + \dot{x}_1\Phi'(x_1) + \dot{y}_1\Phi'(y_1) + \dot{z}_1\Phi'(z_1) + \cdots = 0. \end{gathered}$$

(Symbols such as $\dot{x}_0$ and $\dot{x}_1$ are here used by Lagrange to mean the variations of x_0, x_1, etc.)

Lagrange's procedure is to multiply this equation and any others of the same sort by undetermined coefficients $\alpha, \beta, \ldots$ and to adjoin them to the terms outside the integral sign in the first variation of (3.43). Then he says all the variables $\dot{x}_0, \dot{y}_0, \dot{y}_0', \dot{y}_0'', \ldots, \dot{y}_1, \dot{y}_1', \dot{y}_1'', \ldots$ may be viewed as independent. He then can equate to zero each of their coefficients.

In Chapter 6 below there is a discussion of Mayer's work on the multiplier rule (see Section 6.5 below). The result due to Mayer is the first nearly complete—it contains one important gap—proof of the rule. Lagrange's is flawed by his unsupported statement that "the variations . . . can be regarded as independent and arbitrary."

---

[18] Lagrange, *TCF*, Vol. X, pp. 414–421; Bolza, *VOR*, pp. 543–546, and 566–569.

4 Jacobi and His School

4.1. Excursus

Before proceeding to a detailed examination of Jacobi's 1836 paper, it is probably desirable that we review the situation surrounding the second variation from the point of view of hindsight. This is particularly true in the case of Jacobi and his commentators since Jacobi stated his results with little or no indication as to why they are so. I do not know why he chose this means of announcing his results, unless he was anxious to publish before someone preceded him. In any case let us look at some results in connection with the second variation that should help to illuminate Jacobi's paper.

To do this, consider the simplest problem of the calculus of variations: to maximize or minimize the integral

$$\int_{x_1}^{x_2} f(x, y, y')\,dx \tag{4.1}$$

subject to the condition that

$$y(x_1) = y_1, \qquad y(x_2) = y_2. \tag{4.2}$$

To make matters a little more precise, it will be supposed that f is defined on a region R of real values (x, y, y') and has there continuous derivatives up to and including those of order 4. An arc $y = y(x)$, $x_1 \leqslant x \leqslant x_2$ is admissible if it is made up of a finite number of pieces, each of which has $(x, y(x), y'(x))$ in R. Furthermore, it will be supposed hereafter that the minimizing arc $y = y(x)$ is such that along it $R \equiv \partial^2 f/\partial y'^2 \neq 0$.

Suppose that the Euler equation

$$f_y = \frac{d}{dx} f_{y'} \tag{4.3}$$

has the two-parameter family of solutions $y = y(x, a, b)$, where, as before, we write $f_y = \partial f/\partial y$, $f_{y'} = \partial f/\partial y'$. It is not difficult to show that there is such a family containing the minimizing arc $y = y(x)$ $(x_1 \leqslant x \leqslant x_2)$ for values a_0, b_0 such that the determinant

$$\begin{vmatrix} y_a(x,a_0,b_0) & y_b(x,a_0,b_0) \\ y'_a(x,a_0,b_0) & y'_b(x,a_0,b_0) \end{vmatrix}$$

is different from zero, where y' means dy/dx and y_a, y_b mean $\partial y/\partial a$, $\partial y/\partial b$, respectively.

As we know, the first variation of (4.1) can be expressed in the form

$$\int_{x_1}^{x_2} (f_y \cdot \eta + f_{y'} \cdot \eta')\,dx = f_{y'} \cdot \eta \Big|_{x_1}^{x_2} + \int_{x_1}^{x_2} \Big(f_y - \frac{d}{dx} f_{y'} \Big) \eta\,dx, \tag{4.4}$$

and the second in the form

$$\int_{x_1}^{x_2} \big(f_{yy}\eta^2 + 2f_{yy'}\eta\eta' + f_{y'y'}\eta'^2 \big)\,dx = \int_{x_1}^{x_2} 2\Omega(x,\eta,\eta')\,dx$$

$$= \int_{x_1}^{x_2} (\Omega_\eta \cdot \eta + \Omega_{\eta'} \cdot \eta')\,dx = \Omega_{\eta'} \cdot \eta \Big|_{x_1}^{x_2} + \int_{x_1}^{x_2} \Big(\Omega_\eta - \frac{d}{dx} \Omega_{\eta'} \Big) \eta\,dx,$$

$$= \Omega_{\eta'} \cdot \eta \Big|_{x_1}^{x_2} + \int_{x_1}^{x_2} \eta \Psi(\eta)\,dx, \tag{4.5}$$

where

$$2\Omega(x,\eta,\eta') = f_{yy}\eta^2 + 2f_{yy'}\eta\eta' + f_{y'y'}\eta'^2 = P\eta^2 + 2Q\eta\eta' + R\eta'^2, \tag{4.6}$$

$$\Psi(\eta) = \Omega_\eta - \frac{d}{dx}\Omega_{\eta'} = (P - Q')\eta - \frac{d}{dx}(R\eta')$$

$$= (P - Q')\eta - R'\eta' - R\eta'' = 0. \tag{4.7}$$

Consider now Euler's differential equation (4.3) above, whose coefficients are evaluated along the family of extremals $y = y(x, a, b)$. Then (4.3) is an identity not only in x but also a and b. If this relation is differentiated with respect to a or b and if $\eta \equiv y_a$ or y_b, then clearly

$$f_{yy}\eta + f_{yy'}\eta' = \frac{d}{dx}(f_{yy'}\eta + f_{y'y'}\eta')$$

or equivalently

$$(P - Q')\eta - \frac{d}{dx}(R\eta') = 0;$$

i.e., relation (4.7), which is called the *Jacobi differential equation*, is valid for $\eta = \partial y/\partial a$ or $\partial y/\partial b$ and hence for a linear combination

$$u = \alpha y_a + \beta y_b$$

with α, β arbitrary constants. Notice that the coefficient of η'' in (4.7) is $R = f_{y'y'}$ and that this quantity is, by Legendre's condition, always of one sign along a maximizing or minimizing arc. (Since $y = y(x) = y(x, a_0, b_0)$ is a minimizing arc, it follows that $R \geqslant 0$ on $[x_1, x_2]$ and by the assumption that $R \neq 0$ on this arc that $R > 0$ on $[x_1, x_2]$.)

An interesting and often important choice for α, β above is

$$\alpha = ky_b(x_1, a_0, b_0), \qquad \beta = -ky_a(x_1, a_0, b_0).$$

In this case u is expressible as the determinant.

$$u(x) = \Delta(x, x_1) = \begin{vmatrix} y_a(x, a_0, b_0) & y_b(x, a_0, b_0) \\ y_a(x_1, a_0, b_0) & y_b(x_1, a_0, b_0) \end{vmatrix}, \tag{4.8}$$

provided we set the constant $k = 1$. The importance of this choice lies in the result

Theorem. *If there is a point x_3 between x_1 and x_2 or at x_2 for which $\Delta = 0$, then the second variation* (4.5) *vanishes for a suitably chosen $\eta \not\equiv 0$.*

The proof of this result is easy if η is defined as

$$\eta(x) = \begin{cases} \Delta(x, x_1) & [x_1, x_3], \\ 0 & [x_3, x_2]. \end{cases} \tag{4.9}$$

We see at once that $\eta(x_1) = 0 = \eta(x_2)$, and except at the possible corner $x = x_3$, η satisfies Jacobi's differential equation (4.7) since both y_a and y_b do. Then with the help of (4.7) and (4.5) we have $\Psi(\eta) = 0$ on $[x_1, x_2]$, and the second variation vanishes for this choice of η. Moreover, this η is not identically 0 as can be seen directly from the existence theorem for second-order differential equations. In fact, for Euler's equation, it is well known that the family of extremals can be so chosen that at $x = x_1$ the determinant

$$D(x, a, b) = \begin{vmatrix} y_a'(x, a, b) & y_b'(x, a, b) \\ y_a(x, a, b) & y_b(x, a, b) \end{vmatrix} \neq 0.$$

It is clear, moreover, that $d\Delta(x, x_1)/dx|^{x_1} = \Delta'(x, x_1)|^{x_1} = D(x_1, a_0, b_0)$, and so Δ' is different from zero at x_1. It is then evident that Δ is not identically zero. Any points $x_3 \neq x_1$ at which the determinant Δ of (4.8) vanish are called *conjugate* points to the point x_1. Thus the theorem above guarantees that if there is a point x_3 conjugate to x_1 on $[x_1, x_2]$, then the second variation cannot be positive for all $\eta \neq 0$ vanishing at x_1 and x_2.[1] In fact, we can show that there are such variations η which give it negative values, as we shall now see with the help of the Weierstrass–Erdmann corner condition applied to the accessory minimum problem. To do this, express the Jacobi differential equation in the integral form

$$\Omega_{\eta'} = \int_{x_1}^{x} \Omega_\eta \, dx + d,$$

where d is a constant.

[1] Since for a minimum the second variation must be nonnegative for all variations vanishing at x_1 and x_2, the problem of finding an η which makes the second variation a minimum is often called the *accessory minimum problem*.

It is clear that if η is a minimizing arc for the accessory problem, then $\Omega_{\eta'}(x, \eta, \eta')$ must be a continuous function of x at each point on $[x_1, x_2]$. Thus

$$R \cdot \eta'(x-0) = \Omega_{\eta'}(x, \eta, \eta'(x-0)) = \Omega_{\eta'}(x, \eta, \eta'(x+0)) = R \cdot \eta'(x+0),$$

and consequently η' is itself continuous since $R \neq 0$ by hypothesis. This means that η can have no corners on the interval (x_1, x_2).

Suppose now that there is an x_3 in (x_1, x_2) for which $\Delta = 0$. Then the function η defined by relations (4.9) cannot be a minimizing arc for the accessory problem unless it has no corner at $x = x_3$. If it were, then not only would $\Delta(x_3, x_1) = 0$, but also $\Delta'(x_3, x_1) = 0$. Since $u = \Delta(x, x_1)$ satisfies the Jacobi differential equation, this would mean that both u and u' vanish at $x = x_3$. As a consequence, $u \equiv 0$; i.e., $\Delta \equiv 0$, which contradicts the fact that $\Delta'(x_1, x_1) \neq 0$. We know then that η is not a minimizing arc but that it does make the second variation vanish. There must therefore exist another variation which gives the second variation negative values.

However, if $y = y(x, a_0, b_0)$ is a minimizing arc for the original problem, the second variation must be nonnegative for all variations η such that

$$\eta(x_1) = 0 = \eta(x_2).$$

This implies the result

Theorem. *If $y = y(x)$ is a minimizing arc for the problem described, there can be no point x_3 between x_1 and x_2, which is conjugate to x_1.*

It is also not complicated to establish a weak form of the converse of this theorem with the help of a result of Jacobi's. (In fact, as shown below, a stronger form is also true.) Notice that for u, v arbitrary functions of x

$$u\Psi(v) - v\Psi(u) = -\frac{d}{dx}\left[R(uv' - u'v)\right],$$

and if $\Psi(u) = 0$, where u is a solution of Jacobi's equation (4.7) above, it follows that

$$u\Psi(v) = -\frac{d}{dx}\left[R(uv' - u'v)\right].$$

Now choose $v = \eta = pu$, where p is an arbitrary function of class C'' vanishing at x_1 and x_2; then

$$pu\Psi(pu) = -p\frac{d}{dx}(Rp'u^2) = -\frac{d}{dx}(Rpp'u^2) + R(p'u)^2.$$

Consequently, for η arbitrary but vanishing at x_1 and x_2, the second variation can be expressed as

$$\int_{x_1}^{x_2} \eta\Psi(\eta)\,dx = \int_{x_1}^{x_2} R\left(\eta' - \frac{u'}{u}\eta\right)^2 dx = \int_{x_1}^{x_2} R\,\frac{(u\eta' - u'\eta)^2}{u^2}\,dx, \quad (4.10)$$

as we see with the help of relations (4.5). (This expression, as will be seen, is related to Legendre's form for the second variation. It is often called *Jacobi's form*. If there is no point x_3 between x_1 and x_2 conjugate to x_1, then $u(x) = \Delta(x, x_1)$ does not vanish on (x_1, x_2), and hence (4.10) implies that the second variation is positive—recall the assumption that $f_{y'y'} = R \neq 0$ on $[x_1, x_2]$—for all variations η vanishing at x_1 and x_2.

Theorem. *Let $y = y(x)$ define an extremal arc on the interval $[x_1, x_2]$ embedded in the two-parameter family $y = y(x, a, b)$ described earlier and suppose that there is no point x_3 between x_1 and x_2 conjugate to x_1. Then the second variation defined along $y(x)$ is positive for all variations η vanishing at x_1 and x_2.*

Suppose now that a one-parameter family of curves $y = y(x, \alpha)$, all passing through the point (x_1, y_1), is chosen so that

$$y(x, \alpha_0) = y(x) = y(x, a_0, b_0)$$

and having $y_\alpha(x_1, \alpha_0) \neq 0$. Let us see how this family is then related to the two-parameter family $y = y(x, a, b)$ when $y_b(x_1, a_0, b_0) \neq 0$. Consider the equation $y_1 - y(x_1, a, b) = 0$. It can be solved for b as a function of $a, b = B(a)$, so that $y = y(x, a, B(a))$ with $b_0 = B(a_0)$ is a one-parameter family of extremals all passing through (x_1, y_1). It can, moreover, be shown that near $\alpha = \alpha_0$ the two families $y = y(x, a, B(a))$ and $y = y(x, \alpha)$ can be related; in fact, there is a function $A(a)$ such that

$$y(x, A(a)) = y[x, a, B(a)], \qquad A(a_0) = \alpha_0,$$

for a near a_0.

It is easy to see that if $k = A_a / B_a$ and $\kappa = -k y_a(x_1)$, then

$$\Delta(x, x_1) = \begin{vmatrix} y_a(x, a_0, b_0) & y_b(x, a_0, b_0) \\ y_a(x_1, a_0, b_0) & y_b(x_1, a_0, b_0) \end{vmatrix}$$

$$= k \begin{vmatrix} y_a(x, a_0, b_0) & y_\alpha(x, a_0) \\ y_a(x_1, a_0, b_0) & 0 \end{vmatrix} = \kappa y_\alpha(x, \alpha_0),$$

since $y(x_1, \alpha) \equiv y_1$. From this it follows that the points x_3 conjugate to x_1 are equally given by the zeros $\neq x_1$ of Δ or of y_α. Then consider the curve determined by the parametric equations $y = y(x, \alpha)$, $y_\alpha(x, \alpha) = 0$. This curve is well known to be the envelope of the family $y = y(x, \alpha)$. It is patent that along this envelope y_α vanishes, and so the point x where the minimizing arc meets the envelope is conjugate to x_1 since $\Delta(x, x_1) = \kappa y_\alpha(x, \alpha_0) = 0$. If the envelope degenerates into a single point (x_3, y_3), then the curves of the family all pass through the points (x_1, y_1) and

(x_3, y_3), and no minimum can exist over any interval containing $[x_1, x_3]$.[2] This result, which we discuss below, has great elegance in that it gives a remarkable geometrical significance to what perhaps might have otherwise seemed like an analytical shortcoming.

It is also possible to establish a converse of this result. Suppose that x_3 is conjugate to x_1 in the sense that $y_\alpha(x_3, \alpha_0) = 0$, and that $y_{\alpha x}(x_3, \alpha_0) \neq 0$. Then there is a unique solution $x = \xi(\alpha)$ of $y_\alpha(x, \alpha) = 0$ through (x_3, α_0) and $\xi'(\alpha) = -y_{\alpha\alpha}/y_{\alpha x}$. We now can see that the curve $x = \xi(\alpha)$, $y = \eta(\alpha) = y[\xi(\alpha), \alpha]$ is the envelope of the family of extremals $y = y(x, \alpha)$ near (x_3, α_0). To do this, notice that along this curve $y_\alpha \equiv 0$, and the conclusion then follows directly.

4.2. Jacobi's Paper of 1836

It is curious that it was not for 50 years after Legendre's discovery of his necessary condition that the next major step in the calculus of variations occurred. Jacobi made this great step forward in a short paper containing his results essentially without proofs or only bare suggestions as to them.[3] It may be that he was anxious to publish his results to ensure his intellectual priority; but, in any case, his paper gave rise to a whole school of commentators who worked to fill in the gaps. This group includes Bertrand, Clebsch, Delaunay, Hesse, V.-A. Lebesgue, Mainardi, and Spitzer. Among these perhaps Hesse gave the most complete account.[4]

Jacobi begins by considering the simplest case: the integral in question is

$$\int f\left(x, y, \frac{dy}{dx}\right) dx.$$

He writes the second variation as

$$\int\left(\frac{\partial^2 f}{\partial y^2} ww + 2\frac{\partial^2 f}{\partial y\, \partial y'} ww' + \frac{\partial^2 f}{\partial y'^2} w'w'\right) dx$$

[2] These results on envelopes were apparently first considered for geodesics on a surface by Darboux, (*TDS*, Vol. II, p. 88), by Zermelo ([1894], p. 96), and later by Kneser ([1898], p. 27, *LV*, p. 93, or *LV'*, p. 116). Note that great circles on a sphere are instances of extremals for a problem in which the envelope is a fixed point.

[3] Jacobi [1838]. It is noteworthy, perhaps that as early as 1770 Laplace interested himself in the second variation but got nowhere.

[4] Bertrand [1841], Clebsch [1858], [1858′], Delaunay [1841], Hesse [1857], V.-A. Lebesgue [1841], Mainardi [1852], and Spitzer [1854]. There were also a considerable number of others who worked on consequences of these papers, but the next major figure was clearly Weierstrass, and the men in between Jacobi and him only prepared the way.

and says that it is necessary for an extremum that $\partial f/\partial y'^2$ always have the same sign. He then turns attention to the function v of Lagrange—$v = -\alpha$, the function in (3.35) above Legendre used—defined by the differential equation

$$\frac{\partial^2 f}{\partial y'^2}\left(\frac{\partial^2 f}{\partial y^2}+\frac{dv}{dx}\right)=\left(\frac{\partial^2 f}{\partial y \partial y'}+v\right)^2.$$

To find the complete solution of this equation, which enables one to transform the second variation Jacobi considers certain families of extremals and their partial derivatives with respect to the parameters defining these families. In the present case he takes a two-parameter family

$$y = y(x, a, b)$$

of extremals containing the minimizing or maximizing arc for $(a, b) = (a_0, b_0)$ and out of it forms another family

$$u = \alpha y_a(x, a_0, b_0) + \beta y_b(x, a_0, b_0).$$

The u clearly are solutions of the Jacobi differential equation (4.7) above. He then remarks that the function v above can be found now by setting

$$v = -\left(\frac{\partial^2 f}{\partial y \partial y'}+\frac{1}{u}\frac{\partial^2 f}{\partial y'^2}\frac{du}{dx}\right) \tag{4.11}$$

and that this expression for v contains just one arbitrary constant, β/α.

Notice that the function v becomes infinite precisely when u vanishes, since $u' \neq 0$ at such points. [Otherwise, u, a solution of (4.7), would be identically zero.] This is precisely Lagrange's objection to Legendre's result: v becomes infinite just at the values x_3 conjugate to x_1.

To see that the function $\alpha = -v$ in (4.11) satisfies Legendre's equation (3.35) above, let $P = \partial^2 f/\partial y^2$, $Q = \partial^2 f/\partial y\, \partial y'$, $R = \partial^2 f/\partial y'^2$. Then since u satisfies Jacobi's equation (4.7) above and since $u'/u = (v + Q)/R$ by (4.11),

$$\begin{aligned} v' &= -Q' + \frac{u'^2}{u^2}R - \frac{1}{u}(Ru')' = -Q' + \frac{u'^2}{u^2}R + Q' - P \\ &= -P + \frac{1}{R}(v+Q)^2, \end{aligned} \tag{4.12}$$

which is Legendre's equation. It is then easy to see, as we showed in the proof earlier (p. 140 above), that Jacobi's function v transforms the second variation into the form (4.10). As we see below, this is only a special case of a general result Jacobi found.

Jacobi then turns to the case where the integrand is of the form $f(x, y, y', y'')$. Even though we no longer consider such problems in this form, it is important to trace Jacobi's reasoning since it led him to discover self-adjoint differential equations, as we shall see. In this case Euler's differential equation becomes

$$f_y - \frac{d}{dx} f_{y'} + \frac{d^2}{dx^2} f_{y''} = 0 \tag{4.13}$$

and, in general, a complete family of extremals will contain four parameters a, a_1, a_2, a_3. He sets $\delta y = \omega$ and has $\delta y' = \omega'$ and $\delta y'' = \omega''$. He then expresses the second variation in the usual form

$$\int\left(\frac{\partial^2 f}{\partial y^2} \omega\omega + 2 \frac{\partial^2 f}{\partial y\, \partial y'} \omega\omega' + 2 \frac{\partial^2 f}{\partial y\, \partial y''} \omega\omega'' + \frac{\partial^2 f}{\partial y'^2} \omega'\omega' \right.$$

$$\left. + 2 \frac{\partial^2 f}{\partial y'\, \partial y''} \omega'\omega'' + \frac{\partial^2 f}{\partial y''^2} \omega''\omega'' \right) dx.$$

Now for a maximum or a minimum, Jacobi notes that $f_{y''y''}$ must have the same sign on the interval $[x_1, x_2]$. To simplify notations, let us write his integrand as

$$2\Omega(\omega, \omega', \omega'') = P\omega^2 + 2Q\omega\omega' + R\omega'^2 + 2S\omega\omega'' + 2T\omega'\omega'' + U\omega''^2.$$

He then introduces Legendre's differential equations by means of three functions v, v_1, v_2 with the help of the relations

$$\left(P + \frac{dv}{dx}\right)\left(R + \frac{dv_2}{dx} + 2v_1\right) = \left(Q + v + \frac{dv_1}{dx}\right)^2,$$

$$U\left(P + \frac{dv}{dx}\right) = (S + v_1)^2, \qquad U\left(R + \frac{dv_2}{dx} + 2v_1\right) = (T + v_2)^2.$$

[Actually, these equations appear in Lagrange's *Theory* (*TAF*); Jacobi never refers to Legendre.] He then proceeds to define two functions

$$u = \alpha \frac{\partial y}{\partial a} + \alpha_1 \frac{\partial y}{\partial a_1} + \alpha_2 \frac{\partial y}{\partial a_2} + \alpha_3 \frac{\partial y}{\partial a_3},$$

$$u_1 = \beta \frac{\partial y}{\partial a} + \beta_1 \frac{\partial y}{\partial a_1} + \beta_2 \frac{\partial y}{\partial a_2} + \beta_3 \frac{\partial y}{\partial a_3}, \tag{4.14}$$

which are clearly solutions of Jacobi's differential equation in the present case.

In fact, if we set $\eta = \partial y / \partial c$, where $c = a$, a_1, a_2 or a_3, and differentiate both members of (4.13) with respect to c, we find

$$\Omega(x, \eta, \eta', \eta'') \equiv f_{yy}\eta + f_{yy'}\eta' + f_{yy''}\eta'' - \frac{d}{dx}(f_{y'y}\eta + f_{y'y'}\eta' + f_{y'y''}\eta'')$$

$$+ \frac{d^2}{dx^2}(f_{y''y}\eta + 2f_{y''y'}\eta' + f_{y''y''}\eta'') = 0,$$

or

$$\Omega_\eta - \frac{d}{dx}\Omega_{\eta'} + \frac{d^2}{dx^2}\Omega_{\eta''} = 0.$$

We can then write

$$\Psi(\eta) \equiv P\eta + Q\eta' + S\eta'' - \frac{d}{dx}(Q\eta + R\eta' + T\eta'') + \frac{d^2}{dx^2}(S\eta + T\eta' + U\eta'')$$
$$= (P - Q' + S'')\eta - \left[(R - 2S - T')\eta'\right]' + (U\eta'')'' = 0, \tag{4.15}$$

the generalized form of Jacobi's equation (4.7) above. The second variation is then representable as

$$\int_{x_1}^{x_2} \eta\Psi(\eta)\,dx. \tag{4.16}$$

Since (4.15) is a linear differential equation of the fourth order having both u and u_1 as solutions, it is evident that there must be relationships between the parameters $\alpha, \alpha_1, \alpha_2, \alpha_3, \beta, \beta_1, \beta_2, \beta_3$. In fact, Jacobi remarks that even the six quantities $\alpha\beta_1 - \alpha_1\beta$, $\alpha\beta_2 - \alpha_2\beta$, $\alpha\beta_3 - \alpha_3\beta$, $\alpha_2\beta_3 - \alpha_3\beta_2$, $\alpha_3\beta_1 - \alpha_1\beta_3$, $\alpha_1\beta_2 - \alpha_2\beta_1$ are not independent but that he is not going to set out the relationship.

By analogy with what he did before, Jacobi relates Legendre's functions v, v_1, v_2 to u and u_1 with the help of the relations

$$v_2 = -T - U\cdot\frac{uu_1'' - u_1u''}{uu_1' - u_1u'}, \qquad v_1 = -S + U\cdot\frac{u'u_1'' - u_1'u''}{uu_1' - u_1u'},$$
$$v = -v_1' - Q - U\cdot\frac{(uu_1'' - u_1u'')(u'u_1'' - u_1'u'')}{(uu_1' - u_1u')^2}.$$

Let us see how he transforms the second variation with the help of these functions. To do this, he first states a very elegant result on self-adjoint differential operators. The way he establishes this theorem is not indicated. What he says is, however, quite interesting and revealing. He considers the differential operator

$$Ay + \frac{d(A_1y')}{dx} + \frac{d^2(A_2y'')}{dx^2} + \frac{d^3(A_3y''')}{dx^3} + \cdots + \frac{d^n(A_ny^{(n)})}{dx^n} = Y,$$

where $y^{(m)} = d^my/dx^m$ and A, A_1, etc. are given functions of x. He then makes the following profound observations on the equation we now know as Jacobi's differential equation[5]:

> If y is any integral of the equation $Y = 0$, and if one sets $u = ty$, then the expression, in which $u^{(m)} = d^mu/dx^m$,
>
> $$y\left[Au + \frac{d(A_1u')}{dx} + \frac{d^2(A_2u'')}{dx^2} + \cdots + \frac{d^n(A_nu^{(n)})}{dx^n}\right] = yU$$

[5] Jacobi [1838], p. 44. As we mentioned earlier, numerous papers were written giving proofs of his results. In addition, a very elaborate literature grew up about the second variation, whose purpose was to show how, by suitable transformations, it could be changed into a form that was surely positive. Most of this literature is now quite obsolete and can be substantially simplified by newer understandings. In fact most of this modern approach grew out of Weierstrass's concepts and of a better way to formulate general problems in the calculus of variations, as we shall see below.

will be integrable, i.e., one can specify its integral without knowing t, and this integral has again the form of Y, except that n becomes 1 smaller; one has, namely:

$$\int yU\,dx = Bt' + \frac{d(B_1 t'')}{dx} + \frac{d^2(B_2 t''')}{dx^2} + \cdots + \frac{d^{n-1}(B_{n-1}t^{(n)})}{dx^{n-1}},$$

where $t^{(m)} = d^m t/dx^m$ and the functions B themselves can, in general, be specified in terms of the u and the functions A and their derivatives. The proof of this theorem is not without difficulty. I have found the general expression for the functions B; nevertheless it suffices for the previous application only to show in particular that $\int yU\,dx$ has the given form without needing to know the functions B themselves.

The metaphysics of the obtained result, if I may make use of a french expression, rests roughly upon the following considerations. As is well-known the first variation assumes the form $\int V\delta y\,dx$, where $V = 0$ is the equation to be integrated [Euler's differential equation]. After this the second variation takes the form: $\int \delta V \delta y\,dx$. If the sign of the second variation is not to change [viewed as a function of δy], it must not vanish so that the equation $\delta V = 0$ [this is Jacobi's differential equation], which is linear in δy, must not have any solution $\delta y[\neq 0]$ satisfying the conditions which the nature of the problem imposes upon δy. It is thus seen that the equation $\delta V = 0$ plays a key role in this investigation and in this case its connection with the differential equation to be integrated is soon perceived to find the criteria for a maximum or minimum. It is also seen immediately that each partial derivative of y, a solution of $V = 0$, with respect to any parameter it contains is a solution δy of $\delta V = 0$. The general expression for the solution δy of the differential equation $\delta V = 0$ is therefore found by forming a linear combination of all these partial derivatives of y. The equation $\delta V = 0$, whose complete solution is found in this way, can be brought into the form of the equation $Y = 0$ above, as can be shown, provided δy is written instead of y in it; with the help of known properties of this type of equation the second variation can be transformed by repeated integrations by parts into another expression, which contains a perfect square under the integral sign. This is the transformation of the second variation that one has, in this way, struggled to achieve.

With the help of these remarks Jacobi now illustrates the use of his theorem in the case where the integrand is $f(x, y, y', y'')$. He begins by representing the second variation in the form $\int \delta V\,\delta y\,dx$, where he has

$$\delta V = A\delta y + \frac{d(A_1\delta y')}{dx} + \frac{d^2(A_2\delta y'')}{dx^2},$$

and $\delta V = 0$ for $\delta y = u$, one of the solutions given in (4.14). This form for δV follows directly from the expression (4.16) for the second variation with $A = P - Q' + S''$, $A_1 = -(R - 2S - T')$, $A_2 = U$. He now uses his result above on self-adjoint expressions to write

$$\delta' y = \frac{\delta y}{u}, \qquad u\,\delta V = d\left[B\delta' y' + \frac{d(B_1\delta' y'')}{dx}\right]$$

and thus to express the second variation as

$$\int \delta V \,\delta y\, dx = \int u\,\delta V\,\delta' y\, dx$$

$$= \left[B\,\delta' y' + \frac{d(B_1\,\delta' y'')}{dx} \right] \delta' y - \int \left[B\,\delta' y' + \frac{d(B_1\,\delta' y'')}{dx} \right] \delta' y'\, dx,$$

where he has used the expressions $\delta' y', \delta' y''$ to stand for the derivatives $d(\delta' y)/dx$, $d(\delta' y')/dx = d^2(\delta' y)/dx^2$. In the right-hand member of this equation the integral, as well as the expression outside the integral sign, are understood to be evaluated between the given limits.

In the equations above he calls the integral $\int V_1\,\delta' y'\, dx$ and says that $V_1 = 0$ if $\delta' y = u_1/u$, i.e., if $\delta' y' = (uu_1' - u_1 u')/u^2$; here u and u_1 are given by equations (4.14). In order to use his theorem again he defines a new variation $\delta'' y = u^2\,\delta' y'/(uu_1' - u' u_1)$. Notice that

$$V_1 = B\,\delta' y' + d(B_1\,\delta' y'')/dx$$

is self-adjoint, and consequently Jacobi can again apply his theorem. He finds this time that

$$\int V_1\,\delta' y'\, dx = \int V_1 \left(\frac{uu_1' - u_1 u'}{u^2} \right) \delta'' y\, dx = C\,\delta'' y' \cdot \delta'' y - \int C (\delta' y')^2\, dx,$$

and comments that now $\delta' y'$ appears only as a square. Moreover, he remarks without any indication why it is so, that

$$B_1 = u^2 A_2, \qquad C = \left(\frac{uu_1' - u_1 u'}{u^2} \right)^2 B_1 = \frac{(uu_1' - u_1 u')^2}{u^2} A_2$$

and that $A_2 = \partial^2 f/\partial y''^2$ (see Section 4.2 below). He is now able to conclude that A_2 must always be positive [nonnegative] for a minimum and negative [nonpositive] for a maximum. He then observes that one must examine whether $\delta'' y'$ remains finite on the interval between the end-points. This, he says, can be done by an examination of u and u_1 as soon as y, the complete solution of $V = 0$, has been found.

Jacobi now proceeds to formulate a geometrical interpretation for his condition. He also states a sufficiency theorem which is only partially correct as he has formulated it. (His result is, in fact, only true for a "weak" minimum, i.e., for the case when the comparison arcs are close to the given arc in both position and direction. Weierstrass gave an example in 1879 to show this; but it was not until the turn of the century that Kneser in his *Lehrbuch*, *LV*, explicitly distinguished between weak and strong minima.) Jacobi's words are these[6]:

---

[6] Jacobi [1838], pp. 46–47. Stäckel in *Ostwald's Klassiker*, No. 47, p. 108 tells us Bertrand tried to argue that even if a conjugate point is passed, the property of a curve to furnish an extremum may still hold. This is false, as Weierstrass proved in his 1879 lectures. See Section 5.5 below.

> Even though an understanding of the indicated analysis requires a fairly profound understanding of the integral calculus, yet the criteria derived by it for deciding whether a solution furnishes a maximum or a minimum are very elementary. I shall consider the case in which we have under the integral sign y and its derivatives through the nth order, the end-values of $y, y', \ldots, y^{(n-1)}$, as well as the end-points themselves being given. The arbitrary constants appearing in the solutions of the differential equation of order $2n$ [Euler's equation for this case] are to be determined by substituting the end-values into those solutions with their $2n$ arbitrary constants; since doing this, however, requires the solution of equations there will, in general, be several ways for their determination, so that a number of curves may be found, which satisfy the same end-conditions and the same differential equation. Let one of these be chosen; consider one end-point as fixed and proceed along the curve to succeeding points. If now one of these succeeding points is taken as the second end-point, then as mentioned earlier it may happen that other curves can be drawn having the same values $y', y'', \ldots, y^{(n-1)}$ at both end-points and satisfying the aforementioned differential equation. As soon then as such a point is reached while progressing along the curve in which other curves [extremals through the given first end-point] coincide with it; or, as can be said, approach indefinitely near to it, the limit is reached at or beyond which the [interval of] integration must not extend if a maximum or minimum is to be found. If, however, the [interval of] integration does not reach this limit, then a maximum or minimum will always be found provided that $\partial^2 f/\partial y^{(n)2}$ does not change sign on the interval.

To understand what he meant, consider the simplest problem of the calculus of variations: the integrand f depends only on x, y, y'. In the first case Jacobi has, in effect, said that there is a family of extremals, say, $y = y(x, \alpha)$, passing through two fixed end-points (x, y_1) and (x_2, y_2) and containing the given curve for $\alpha = \alpha_0$. The variation $\eta(x) = y_\alpha(x, \alpha_0)$ will then vanish at x_1 and x_2 but not be identically in between and will satisfy Jacobi's differential equation (4.7), as we saw. This η must then make the second variation vanish, and hence a conjugate point has been reached at x_2. In the second Jacobi has said that there is a family of extremals $y = y(x, \alpha)$ which pass through the first end-point but are each tangent to a given fixed curve, say, $x = \xi(\alpha)$, $y = \eta(\alpha)$ near to $\alpha = \alpha_0$.

It is not hard to see that the point $x_2 = \xi(\alpha_0)$, $y_2 = \eta(\alpha_0)$ is the limiting position of the intersection point of the curves $y = y(x, \alpha_0 + h)$ with $y = y(x, \alpha_0)$ as h approaches zero. Let their intersection be given as $(x_2(h), y_2(h))$; then clearly

$$y[x_2(h), \alpha_0] = y[x_2(h), \alpha_0 + h]$$

and consequently

$$y_\alpha[x_2(0), \alpha_0] = 0, y_2 - y[x_2(0), \alpha_0] = 0,$$

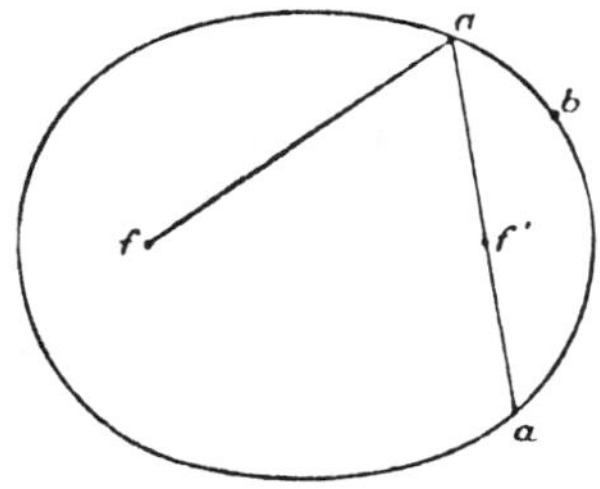

Figure 4.1

from which it follows by the definition of the envelope that $x_2(0) = \xi(\alpha_0)$, $y_2(0) = \eta(\alpha_0)$, i.e., $\lim x_2(h) = x_2$, $\lim y_2(h) = y_2$, as asserted.[7]

To illustrate how his condition operates, Jacobi discusses an example from dynamics involving the principle of least action. The problem involves the motion of a planet about the sun. Jacobi says in the first place that Lagrange was wrong in believing the action integral could ever be a maximum; he further says that it is not even always a minimum. He proceeds to observe that if the end-points are not properly chosen with respect to each other, the integral will be neither a minimum or maximum. In Figure 4.1 he imagines a planet is at a point a somewhere between perihelion, the point nearest to the sun f, and aphelion, the point furthest from f. It starts at a and proceeds to b along the ellipse whose major axis has length $2A$ and whose focus f is at the sun's location. The other focus is then to be determined by the property of the ellipse that $af + af' = 2A = bf + bf'$. Jacobi uses this fact to draw circles with centers at a and b and with radii $2A - af$ and $2A - bf$, respectively. In general, these circles intersect in two points and in one point when they are tangent. This latter case occurs only when the line ab passes through the focus f'. Thus in general there are two ellipses and in the latter case, just one. (In the next section we discuss Jacobi's point in detail.)

He draws the chord aa' through the focus f' and says "in accordance with the general rule the other end-point b must lie between a and a', if the ellipse is to make the integral appearing in the principle of least action a minimum. If b falls on a' the second variation cannot become negative, but zero, so that the variation of the integral is of the third order and so may be positive or negative. [In Section 7.5 we mention Erdmann's analysis of the third variation for such cases.] If b falls outside of a', the second variation itself can [will] become negative."

Jacobi then turns to the case when the initial point a lies between aphelion and perihelion (see Figure 4.2). He now draws the chord aa' through the sun f and rotates the ellipse about afa', finding "infinitely

[7] There is a nice discussion of the geometrical significance of conjugate points in Bliss, *LEC*, pp. 34–36. See also Section 4.1 above.

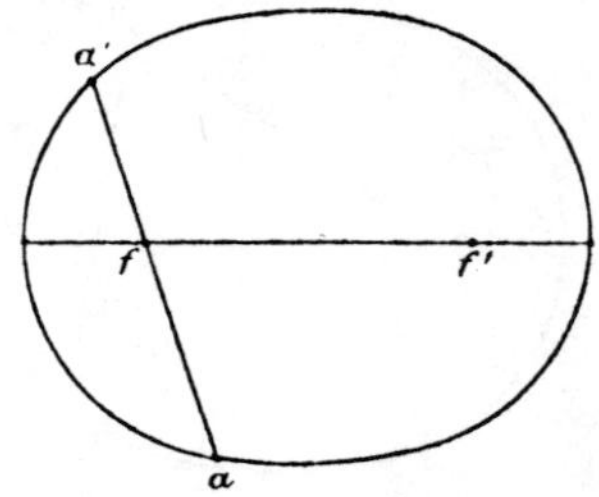

Figure 4.2

many solutions to the problem. If, therefore, the second end-point in the latter case lies beyond a', there will be a space-curve joining the two end-points for which $\int v\,ds$ has a smaller value then for the ellipse."

4.3. Excursus on Planetary Motion

To understand what Jacobi was saying analytically and to see the envelope of the family of ellipses through his first end-point a, let us work through the material in a little detail. To this end we suppose that in Figures 4.3 and 4.4 the sun is the pole in a polar coordinate system whose base line passes through a for $r = r_0$, $\varphi = 0$. In the former figure a lies between perihelion and aphelion and in the latter, between aphelion and perihelion. At the point a we suppose the initial velocity is v_0 and is in the direction α. Finally we have for the force function U—so named by Jacobi—the value k^2m/r. (Recall that $U_r = dU/dr = f$, the force.) But the law of living force of Lagrange—conservation of energy—states that if v is the

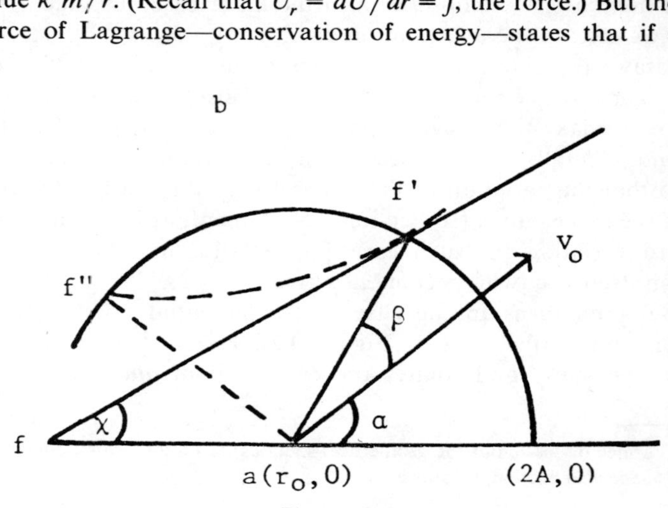

Figure 4.3

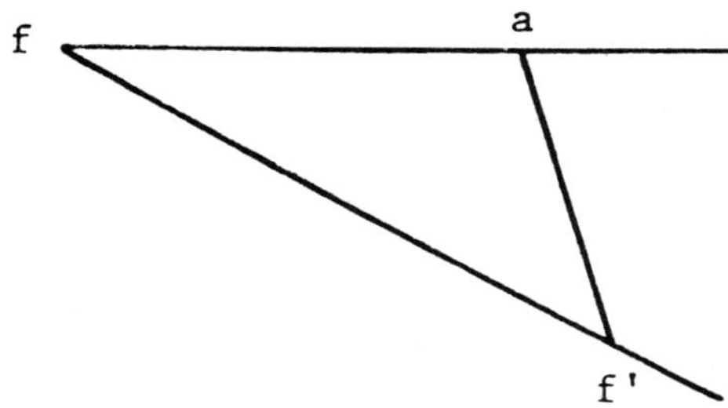

Figure 4.4

velocity and T the kinetic energy

$$\tfrac{1}{2}mv^2 = T = (U + H), \tag{4.17}$$

where H is a constant. Thus the action integral is in terms of arc-length

$$m\int_{s_1}^{s_2} v\,ds = \int_{t_1}^{t_2} [2m(U + H)]^{1/2}(r'^2 + r^2\varphi'^2)^{1/2}\,dt,$$

as can easily be seen by calculating $s' = ds/dt$ in polar coordinates.

It is not difficult to see that the Euler differential equations for the problem are

$$\frac{d}{dt}\left\{\left[2m\left(\frac{k^2m}{r} + H\right)\right]^{1/2}(r'^2 + r^2\varphi'^2)^{-1/2}r^2\varphi'\right\} = 0,$$

$$\frac{d}{dt}\left\{\left[2m\left(\frac{k^2m}{r} + H\right)\right]^{1/2}(r'^2 + r^2\varphi'^2)^{-1/2}\cdot r'\right\}$$

$$= -\frac{k^2m^2}{r^2}\left[2m\left(\frac{k^2m}{r} + H\right)\right]^{-1/2}(r'^2 + r^2\varphi'^2)^{1/2}$$

$$+\left[2m\left(\frac{k^2m}{r} + H\right)\right]^{1/2}(r'^2 + r^2\varphi'^2)^{-1/2}\cdot r\varphi'^2.$$

At this point it is important to make use of the fact that by (4.17)

$$2\left(\frac{k^2m}{r} + H\right) = mv^2 = m(r'^2 + r^2\varphi'^2).$$

This gives the simplified equations

$$mr^2\varphi' = mh, \qquad r'' = -\frac{k^2}{r^2} + r\varphi'^2 = -\frac{k^2}{r^2} + \frac{h^2}{r^3}.$$

(The first relation implies that the areal velocity is constant and is a statement of Kepler's law that the radius vector sweeps out equal areas in equal intervals of time.) It is now convenient to change both the independent and the dependent variables from t, r to $\varphi, u = 1/r$. After a little calculation, the equations above become

$$\varphi' = hu^2, \qquad u_{\varphi\varphi} + u = \frac{k^2}{h^2} = \frac{1}{p},$$

where, e.g., $u_{\varphi\varphi} = \partial^2 u/\partial\varphi^2$. The solution of the second equation is clearly the conic

$$u = \frac{1}{p}\left[1 - e\cos(\varphi - \chi)\right] \tag{4.18}$$

where e, χ are arbitrary constants, which we proceed to calculate. [Note that $r_0(1 - e\cos\chi) = p$ since all the ellipses pass through the point $a : (r_0, 0)$ and that $af + af' = 2A$.]

It is now desirable to evaluate the constant H. To do this, we first notice that $v^2 = r'^2 + r^2\varphi'^2 = h^2(u_\varphi^2 + u^2)$ and that (4.17) becomes

$$\tfrac{1}{2} mh^2\left(u_\varphi^2 + u^2\right) = k^2 mu + H.$$

Now if u approaches zero, then u_φ^2 approaches $2H/mh^2$; thus $H < 0$ implies that r must remain finite. In Jacobi's case of planetary motion the conic (4.18) must be an ellipse and $H < 0$. At perihelion, $\varphi = \chi$, $r_\varphi = 0$ since the tangent of the angle between the radius vector and the tangent to the curve at that point is r/r_φ and since that angle is $\pi/2$ at both perihelion and aphelion. Thus at perihelion

$$H = \frac{m}{2} r^2\varphi'^2 - \frac{k^2 m}{r} = \frac{h^2 m}{2r^2} - \frac{k^2 m}{r} = \frac{k^2}{r}\left(\frac{p}{2r} - 1\right)$$

$$= \frac{k^2}{A(1+e)}\left[\frac{A(1-e^2)}{2A(1+e)} - 1\right] = -\frac{k^2 m}{2A}$$

since there $r = A(1+e)$ and always $p = A(1 - e^2)$ for an ellipse, where e is the eccentricity. From this and (4.17) it follows easily that

$$v^2 = k^2\left(\frac{2}{r} - \frac{1}{A}\right).$$

With these facts at hand we can now proceed. The first thing is to show in Figure 4.3 that $\beta = \alpha$. To do this, note that

$$v_x = r'\cos\varphi - r\sin\varphi\cdot\varphi', \qquad v_y = r'\sin\varphi + r\cos\varphi\cdot\varphi'$$

and that at $\varphi = 0$

$$v_x = v_0\cos\alpha = r' = r_\varphi\cdot\varphi' = \frac{he}{p}\sin\chi, \qquad v_y = v_0\sin\alpha = \frac{h}{r_0},$$

$$\frac{k^2}{Ar_0}(2A - r_0)\sin 2\alpha = v_0^2\sin 2\alpha = \frac{2h^2 e}{pr_0}\sin\chi;$$

moreover, by the so-called law of sines for the triangle aff' in Figure 4.3

$$\frac{(2A - r_0)}{2Ae}\sin(\alpha + \beta) = \sin\chi.$$

Putting these together, we see that $\sin(\alpha + \beta) = \sin 2\alpha$, and therefore we have $\beta = \alpha$. Thus we have the relations

$$p = \frac{h^2}{k^2} = r_0^2 \frac{v_0^2}{k^2} \sin^2\alpha = r_0\left(\frac{2A - r_0}{A}\right)\sin^2\alpha = r_0\kappa\sin^2\alpha,$$

$$\kappa = 2 - \frac{r_0}{A}, \qquad e\sin\chi = \frac{\kappa}{2}\sin 2\alpha,$$

$$e\cos\chi = 1 - \frac{p}{r_0} = 1 - \kappa\sin^2\alpha.$$

With the help of these relations the one-parameter family (4.18) of ellipses passing through the point a may be expressed in terms of the parameter α as

$$f(r,\varphi,\alpha) \equiv r\left[1 - (1 - \kappa\sin^2\alpha)\cos\varphi - \frac{\kappa}{2}\sin 2\alpha\sin\varphi\right] - r_0\kappa\sin^2\alpha = 0. \tag{4.19}$$

This is the family of extremals whose envelope we wish to find; to do this, we calculate

$$\frac{\partial f}{\partial\alpha} = f_\alpha = r\kappa(\sin 2\alpha\cos\varphi - \cos 2\alpha\sin\varphi) - r_0\kappa\sin 2\alpha = 0. \tag{4.20}$$

[It is not difficult to show that these are the straight lines through the points a and f'; the locus of intersections of these lines (4.20) with the ellipses (4.19) is clearly the locus of points conjugate to x_1. This locus is also the envelope.]

To find this envelope, let us substitute the value of r from (4.20) into (4.19). After some manipulation, we find that $2\tan\varphi/2 = \kappa\tan\alpha$. We can use this relation to eliminate α from either (4.19) or (4.20). One way to do this is to set

$$\sin\alpha = 2\lambda\sin\frac{\varphi}{2}, \qquad \cos\alpha = \kappa\lambda\cos\frac{\varphi}{2},$$

in (4.20). This gives us the partial result for $\varphi \neq 0, \pi$ that

$$r = \frac{2\kappa r_0\lambda^2}{(1 - \kappa^2\lambda^2) + \kappa(2 - \kappa)\lambda^2\cos\varphi};$$

but

$$1 = \left(4\sin^2\frac{\varphi}{2} + \kappa^2\cos^2\frac{\varphi}{2}\right)\lambda^2 = \left[\left(2 + \frac{\kappa^2}{2}\right) - \left(2 - \frac{\kappa^2}{2}\right)\cos\varphi\right]\lambda^2$$

and consequently

$$1 - \kappa^2\lambda^2 = \frac{(4 - \kappa^2)}{2}\lambda^2 - \frac{(4 - \kappa^2)}{2}\lambda^2\cos\varphi,$$

which implies directly that the equation of the envelope is the ellipse

$$r = \frac{4\kappa r_0}{(4-\kappa^2)-(2-\kappa)^2\cos\varphi} = \frac{4A(2A-r_0)}{(4A-r_0)-r_0\cos\varphi} \tag{4.21}$$

It is clear that this ellipse passes through the point $r = 2A$, $\varphi = 0$, that the sun is at one focus, and that the point a is at the other since the major axis has length $= (4A - r_0)$ and eccentricity $= r_0/(4A - r_0)$. Consider now Jacobi's point b in Figure 4.3; if it lies inside ellipse (4.21), then the circles about a and b with radii $2A - af = 2A - r_0$ and $2A - bf = 2A - r_1$ intersect in two points f' and f''. Thus there are, in general, two members of the family (4.19) of ellipses through a that also pass through b; if b lies on the envelope, there is only one; and if b lies outside, there are none since the circles of radii r_0, r_1 about a and b do not meet in any point f'.

As Figure 4.3 is drawn, point a lies between perihelion and aphelion. However, if a lies between aphelion and perihelion instead, the situation is quite different. This is the situation as depicted in Figure 4.2. In this case point a' conjugate to a (Figure 4.2) is on the line segment afa' through the sun at f. If b coincides with a', then not only is the elliptical arc through a and a' (Figure 4.2) an extremal, but so are all rotations of this arc about the line afa'. The envelope has here degenerated into the a'.

Jacobi uses these conclusions to state the rule at the end of Section 4.2 above: in the case when a lies between perihelion and aphelion, draw the chord aa' from a through the focus f'; then point b must lie between points a, a' if the action integral $\int v\, ds$ is to be a minimum (see Figure 4.1). If b lies at a', then the second variation of the integral can be made zero, and decision must be made on the basis of the third variation. "If b lies beyond a', then the second variation itself can become negative." He next takes up the case where the first end-point a lies between aphelion and perihelion. Now point a' is determined by the chord drawn from a through the focus f as in Figure 4.2 since there are infinitely many solutions obtainable by rotating the ellipse about the segment afa'. If the second end-point lies beyond a', there will be a space curve through the given end-points for which $\int v\, ds$ is smaller than for the ellipse.

4.4. V.-A. Lebesgue's Proof

In 1839 V.-A. Lebesgue published a paper entitled, "On a formula of Vandermonde, and its applications to the demonstration of a theorem of Jacobi," which we will review in enough detail to understand his proof.[8] It

[8] V.-A. Lebesgue [1841].

is perhaps somewhat less tedious than is the paper of Delaunay on the same subject which appears in the same volume of *Liouville's Journal*.[9]

The basic result due to Vandermonde that is relevant in this paper is derived from the formula

$$[x+y]^n=[x]^n+n[x]^{n-1}[y]+\frac{n,n-1}{1,2}[x]^{n-2}[y]^2+\cdots+[y]^n,$$

where $[x]^n = x(x-1)\dots(x-n+1)$ and

$$(n,n-1)/1,2 = n(n-1)/1\cdot 2.^{10}$$

Lebesgue divides both members of this relation by $1.2.3\dots n$, and writes

$$(x,m)=\frac{x(x-1)\dots(x-m+1)}{m!},\qquad (x,0)=1.$$

In these terms he has the basic combinatoric identity

$$(x-y,n)=\sum_{z=0}^{n}(-1)^z(x,x-n+z)(y-1+z,z);\tag{4.22}$$

as we shall see, he uses this as a tool in establishing his main result:

Theorem. *Let $A_0, A_1, \dots, A_n$, as well as y and t, be any functions of x; one will always have (by employing Lagrange's notation)*

$$y\left\{A_0(ty)+\left[A_1(ty)'\right]'+\left[A_2(ty)''\right]''+\cdots+\left[A_n(ty)^{(n)}\right]^{(n)}\right\}$$

$$=B_0t+(B_1t')'+(B_2t'')''+\cdots+\left(B_nt^{(n)}\right)^{(n)}.\tag{4.23}$$

To prove this result and to determine the form of the B_i, Lebesgue first chooses t to be a constant and has

$$B_0=y\left\{A_0y+(A,y')'+(A_2y'')''+\cdots+\left(A_ny^{(n)}\right)^{(n)}\right\};$$

therefore, if y satisfies the self-adjoint differential equation

$$A_0y+(A_1y')'+(A_2y'')''+\cdots+\left(A_ny^{(n)}\right)^{(n)}=0,$$

then $B_0 = 0$. He remarks that this is Jacobi's case and that inasmuch as Jacobi did not publish the values of the functions B_i, he will do so here. In what follows he writes for simplicity

$$A_i^{(\alpha)}=\frac{d^\alpha A_i}{dx^\alpha},\qquad y^{(\beta)}=\frac{d^\beta y}{dx^\beta}=y^\beta,\qquad y^{(\gamma)}=\frac{d^\gamma y}{dx^\gamma}=y^\gamma.$$

---

[9] Delaunay [1841]. Delaunay later [1843] wrote a prize paper entitled "Mémoire sur le calcul des variations." It is concerned with the problem of finding among all curves with continuously varying tangents and constant curvature, which join two fixed points, the one with greatest or least arc-length. There is a very nice discussion of this problem in Carathéodory [1930].

[10] Vandermonde [1838], p. 47.

He then displays equation (4.23) in the form

$$y\left(C_0t + C_1t' + C_2t'' + \cdots + C_{2n}t^{2n}\right) = B_0t + (B_1t')' + (B_2t'')'' + \cdots + \left(B_nt^{(n)}\right)^{(n)},$$

in which the C_j are functions of the y and its derivatives; and then setting $D_j = C_jy$ ($j = 0, 1, \ldots, 2n$), he has

$$D_0t + D_1t' + D_2t'' + \cdots + D_{2n}t^{(2n)} = B_0t + (B_1t')' + \cdots + \left(B_nt^{(n)}\right)^{(n)}. \tag{4.24}$$

He now goes to calculate the functions C and D by means of equation (4.23). Note that $B_0 = D_0$, and hence he finds

$$D_1t' + D_2t'' + \cdots + D_{2n}t^{(2n)} = (B_1t')' + (B_2t'')'' + \cdots + \left(B_nt^{(n)}\right)^{(n)}. \tag{4.25}$$

Since the right-hand member of this relation is an exact derivative, the usual integrability condition implies that

$$D_1 - D_2' + D_3'' - \cdots - D_{2n}^{(2n-1)} = 0.$$

Integrating both members of (4.25) and making use of repeated integrations by parts, Lebesgue finds

$$\left[D_2 - D_3' + \cdots + D_{2n}^{(2n-2)}\right]t' + \left[D_3 - D_4' + \cdots - D_{2n}^{(2n-3)}\right]t'' + \cdots$$
$$= B_1t' + (B_nt'')' + \cdots,$$

From this he can conclude that $B_1 = D_2 - D_3' + D_4'' + \cdots + D_{2n}^{(2n-2)}$ and that

$$\left[D_3 - D_4' + \cdots - D_{2n}^{(2n-3)}\right]t'' + \left[D_4 - D_5' + \cdots + D_{2n}^{(2n-4)}\right]t''' + \cdots$$
$$= (B_2t'')' + (B_3t''')'' + \cdots,$$

whose right-hand member is again an exact derivative. In this case the integrability condition requires that

$$D_3 - 2D_4' + 3D_5'' - \cdots - (2n-2)D_{2n}^{(2n-3)} = 0$$

and, as before, leads to the new partial result

$$\left[D_4 - 2D_5' + 3D_6'' + \cdots + (2n-3)D_{2n}^{(2n-4)}\right]t'' + \cdots$$
$$= B_2t'' + (B_3t''')' + \cdots,$$

from which it follows that

$$B_2 = D_4 - 2D_5' + 3D_6'' - \cdots + (2n-3)D_{2n}^{(2n-4)}.$$

By continuing in this fashion he concludes that if Jacobi's relation (4.23) holds, then

$$0 = D_{2a-1} - (a,1)D_{2a}' + (a+1,2)D_{2a+1}'' + \cdots - (2n-a, 2n-2a+1)D_{2n}^{(2n-2a+1)} \tag{4.26}$$

and

$$B_a = D_{2a} - (a,1)D'_{2a+1} + (a+1,2)D''_{2a+2} + \cdots + (2n-a-1,2n-2a)D_{2n}^{(2n-2a)}. \tag{4.26'}$$

To simplify notations, he sets

$$(D,b,a) = D_b - (a,1)D'_{b+1} + (a+1,2)D''_{b+2} - \cdots, \tag{4.27}$$

where the series terminates in the term containing D_{2n}.

Lebesgue now notes that "it will suffice to prove that for $b = 2a - 1$, one has $(D,b,a) = 0$, and to calculate (D,b,a) for $b = 2a$, which will give B_a." He continues by saying that the calculation of (D,b,a) will be simplified as follows. Since equations (4.26) must be satisfied for all functions A and since they are built up of terms of the form $MA_i^{(\alpha)}y^{(\beta)}y^{(\gamma)}$ ($i = 1,2,\ldots,$; M a constant), he observes that all terms in A_i must cancel out. He continues,[11] "It will suffice then to verify the equations [(4.26)] for the case in which the equation [(4.23)] reduces to

$$y\left\{A_0ty + \left[A_i(ty)^{(i)}\right]^{(i)}\right\} = B_0t + (B_1t')' + \cdots."$$

What Lebesgue means is this: call for the moment the coefficients B_j, derived from the equation above, $B_{j,i}$. Then the coefficients B_j are sums of the $B_{j,i}$ (i.e., $B_j = \sum B_{j,i}$).

To find the $B_{j,i}$, he now expresses the equation (4.27) in the form

$$\begin{aligned}(D,b,a) = {} & \left[C_b - (a,1)C'_{b+1} + (a+1,2)C''_{b+2} - \cdots\right]y \\ & - (a,1)\left[C_{b+1} - (a+1,1)C'_{b+2} + (a+2,2)C''_{b+3} + \cdots\right]y' \\ & + (a+1,2)\left[C_{b+2} - (a+2,1)C'_{b+3} + (a+3,2)C''_{b+4} + \cdots\right]y'' + \cdots,\end{aligned}$$

since $D_k = C_ky$. Moreover, each quantity in brackets ends with a term containing some derivative of C_{2i}, and the last such quantity reduces to the single term C_{2i}. He reasons from this that the general term of (D,b,a) is given by

$$(-1)^\beta(a+\beta-1,\beta)\left[C_{b+\beta} - (a+\beta,1)C'_{b+\beta+1} + (a+\beta+1,2)C''_{b+\beta+2} - \cdots\right]y^\beta.$$

He next sets $b + \beta = c$, $a + \beta = d$ and notes that the coefficient of $(-1)^\beta(d-1,\beta)y^\beta$ above can be written in the form

$$C_c - (d,1)C'_{c+1} + (d+1,2)C''_{c+2} - \cdots = f_0y + f_1y' + f_2y'' + \cdots + f_\alpha y^\alpha + \cdots. \tag{4.28}$$

[11] V.-A. Lebesgue [1841], p. 27.

He must now calculate the coefficient f_α by finding the appropriate terms in C_c, C'_{c+1}, C''_{c+2}, etc: i.e., the terms in the coefficient of $y^\alpha = y^{(\alpha)}$. To carry out this program, he first observes that the general term of $[A_i(ty)^{(i)}]^{(i)}$ is $(i, z)A_i^{(i-z)}(ty)^{(i+z)}$ and of $(ty)^{(i+z)}$, $(i + z, k)y^{(i+z-k)}t^{(k)}$. Therefore the general term of C_k is $(i, z)(i + z, k)A_i^{(i-z)} \cdot y^{(i+z-k)}$; and the term containing $y^{(\alpha)}$ is found by putting $i + z - k = \alpha$. This gives $(i, \alpha + k - i)(\alpha + k, k)A_i^{2i-k-\alpha}y^\alpha$. Lebesgue then writes out the expansions

$$
\begin{aligned}
C_c &= \cdots + (i, \alpha + c - i)(\alpha + c, c)A_i^{2i-c-\alpha}y^\alpha \cdots,\\
C_{c+1} &= \cdots + (i, \alpha + c - i)(\alpha + c, c + 1)A_i^{2i-c-\alpha}y^{\alpha-1}\\
&\quad + (i, \alpha + c + 1 - i)(\alpha + c + 1, c + 1)A_i^{2i-c-1-\alpha}y^\alpha \cdots,\\
C_{c+2} &= \cdots + (i, \alpha + c - i)(\alpha + c, c + 2)A_i^{2i-c-\alpha}y^{\alpha-2}\\
&\quad + (i, \alpha + c + 1 - i)(\alpha + c + 1, c + 2)A_i^{2i-c-1-\alpha}y^{\alpha-1}\\
&\quad + (i, \alpha + c + 2 - i)(\alpha + c + 2, c + 2)A_i^{2i-c-2\alpha}y^\alpha \cdots,
\end{aligned}
$$

etc. and portrays them in the triangular array

$$
\begin{aligned}
KA_i^\lambda y^\alpha &= C_c,\\
LA_i^{\lambda-1}y^\alpha + L'A_i^\lambda y^{\alpha-1} &= C_{c+1},\\
MA_i^{\lambda-2}y^\alpha + M'A_i^{\lambda-1}y^{\alpha-1} + M''A_i^\lambda y^{\alpha-2} &= C_{c+2},\\
NA_i^{\lambda-3}y^\alpha + N'A_i^{\lambda-2}y^{\alpha-1} + N''A_i^{\lambda-1}y^{\alpha-2} + N'''A_i^\lambda y^{\alpha-3} &= C_{c+3},
\end{aligned}
\tag{4.29}
$$

where the coefficients $K, L, L', \ldots$ are given in the equations above and $\lambda = 2i - c - \alpha$. (Lebesgue's array is different from this but is incorrect. His reasoning about the array is also flawed; however, his results are correct.) The diagonals in this array determined by a fixed value of the superscript on y may be used to evaluate f_α. The terms in $y^{(\alpha)}$ in this array contribute to f_α the value

$$
\begin{aligned}
&(i, \alpha + c - i)\big[(\alpha + c, c) - (d, 1)(\alpha + c, c + 1) + (d + 1, 2)(\alpha + c, c + 2)\\
&\qquad + \cdots\big]A_i^\lambda\\
&= (i, \alpha + c - i)\sum_{z=0}^{\alpha}(-1)^z(\alpha + c, c + z)(d - 1 + z, z)A_i^\lambda\\
&= (i, \alpha + c - i)(\alpha + c - d, \alpha)A_i^{2i-c-\alpha},
\end{aligned}
\tag{4.30}
$$

as can be seen with the help of Vandermonde's relation (4.22) above with $n = \alpha$, $x = \alpha + c$, $y = d$. The next term in f_α comes from the second diagonal of the array in (4.29) given by the exponent $\alpha - 1$ on y, provided that the relevant coefficents are multiplied by the factors $1, (2, 1), (3, 1), \ldots$, which enter when the appropriate quantities have been formed. Thus in the formation of f_α the coefficients $-(d, 1), (d + 1, 2), -(d + 2, 3), \ldots$, which enter into (4.30), are altered for the next term. They are now multiplied by $1, (2, 1), (3, 1), \ldots$ and become $-(d, 1), +(d, 1)(d + 1, 1)$,

$-(d, 1)(d+1, 2), + \cdots$. It is also clear that in (4.30) c needs to be increased to $c+1$. It follows, consequently, that the second diagonal contributes to f_α the term $-(d, 1)(i, \alpha+c+1-i)(\alpha+c-d, \alpha)A_i^{2i-c-\alpha}$; by analogous reasoning, the third diagonal contributes

$$+(d+1, 2)(i, \alpha+c+2-i)(\alpha+c-d, \alpha)A_i^{2i-c-\alpha},$$

and so on. This gives for the total coefficient, f_α, the value

$$\begin{aligned} A_i^{2i-c-\alpha}(\alpha+c-d, \alpha) &\sum_{z=0}^{2i-c-\alpha}(-1)^z(i, \alpha+c-i+z)(d-1+z, z) \\ &= (\alpha+c-d, \alpha)(i-d, 2i-c-\alpha)A_i^{2i-c-\alpha}; \end{aligned}$$

this follows from (4.22) by setting $x=i$, $y=d$, $n=2i-c-\alpha$.

Recall that $c=b+\beta$ and $d=a+\beta$ and that what we just calculated when multiplied by $(-1)^\beta(a+\beta-1, \beta)$ and $y^{(\beta)}$ gives the general term in (D, b, a). This is then

$$\begin{aligned} &(-1)^\beta(a+\beta-1, \beta)(b-a+\alpha, \alpha)(i-a-\beta, 2i-b-\alpha-\beta) \\ &\quad \times A_i^{2i-b-\alpha-\beta}y^{(\alpha)}y^{(\beta)}. \end{aligned}$$

This gives Lebesgue the term in $y^\alpha \cdot y^\beta$. To find in (D, b, a) the term in $y^\beta y^\alpha$, he appeals to symmetry and finds

$$\begin{aligned} &(-1)^\alpha(a+\alpha-1, \alpha)(b-a+\beta, \beta)(i-a-\alpha, 2i-b-\alpha-\beta) \\ &\quad \times A_i^{2i-b-\alpha-\beta}y^\beta y^\alpha \end{aligned}$$

These two terms may be combined into $Q \cdot A_i^{2i-b-\alpha-\beta}y^\alpha y^\beta$ by setting

$$\begin{aligned} Q &= (-1)^\alpha(a+\alpha-1, \alpha)(b-a+\beta, \beta)(i-a-\alpha, 2i-b-\alpha-\beta) \\ &\quad + (-1)^\beta(a+\beta-1, \beta)(b-a+\alpha, \alpha)(i-a-\beta, 2i-b-\alpha-\beta). \end{aligned} \tag{4.31}$$

He now wishes to show for $b=2a-1$ that $Q=0$ and to calculate Q for $b=2a$.

To these ends he sets $b=2a-1$, $h=2i-2a+1-\alpha-\beta$, $p=i-a-\alpha$, $q=i-a-\beta$, and so $h-1=p+q$. Then the expression (4.31) becomes,

$$\left[(-1)^\alpha(p, h)+(-1)^\beta(q, h)\right](a+\alpha-1, \alpha)(a+\beta-1, \beta).$$

Now if both $p>0$ and $q>0$, then $p<h$ and $q<h$ and consequently $(p, h)=(q, h)=0$, by definition. If $p=0$ or $q=0$, then $(p, h)=0$ or $(q, h)=0$. Suppose that $p=0$; it is clear that $q=h-1$, and therefore (q, h) is also 0. Thus the expression above vanishes for $p \geqslant 0$, $q \geqslant 0$. Next suppose that one of the numbers p, q is positive and the other negative, and let $\alpha<\beta$. Then clearly $p>0$ and $q<0$. He sets $q'=-q$ and notes that

$q' = p - h + 1$. It follows then that

$$(p,h) = \frac{p(p-1)\dots q'}{1\cdot 2\dots h} = \frac{q'(q'+1)\dots(q'+h-1)}{1\cdot 2\dots h}$$
$$= (-1)^h(-q',h) = (-1)^h(q,h),$$

and thus $(-1)^\alpha(p,h) + (-1)^\beta(q,h) = (q,h)[(-1)^{\alpha+h} + (-1)^\beta]$. But $2(i - a) + 1 = (\alpha + h) + \beta$ from the definition of h, and therefore the pair $\alpha + h$ and β must consist of an even and an odd number, which means that the expression above is zero. It follows directly that

$$Q = (a+\alpha-1,\alpha)(a+\beta-1,\,\beta)\big[(-1)^\alpha(p,h) + (-1)^\beta(q,h)\big] = 0.$$

Lebesgue observes that the equations of condition (4.26) must be satisfied and that (4.26′) defines B_a, which completes the proof of the theorem.

To exhibit the B_a in an easily calculable form, Lebesgue goes on to the case $b = 2a$ in (D,b,a). The general term in B_a is $QA_i^h y^\alpha y^\beta$, as has been shown, where $h = 2i - 2a - \alpha - \beta$ and

$$Q = (-1)^\alpha(a+\alpha-1,\alpha)(a+\beta,\,\beta)(i-a-\alpha,h)$$
$$+ (-1)^\beta(a+\beta-1,\,\beta)(a+\alpha,\alpha)(i-a-\beta,h) \qquad (\alpha \neq \beta),$$
$$Q = (-1)^\alpha(a+\alpha-1,\alpha)(a+\alpha,\alpha)(\tfrac{1}{2}h,h) \qquad (\alpha = \beta). \tag{4.32}$$

It is evident that for $\alpha = \beta$ and $h > 0$ the expression $Q = 0$ since $(h/2,h) = 0$. But if $h = 0$, $(h/2,h) = 1$ by definition, $i = a + \alpha$, and

$$Q = (-1)^{i-a}(i-1,i-a)(i,i-a).$$

If $\alpha \neq \beta$, Q can be simplified by setting $i - a - \alpha = p$, $i - a - \beta = q$ (note $p + q = h$). Now if $p > 0$, $q > 0$, then $p < h$, $q < h$ and $(p,h) = 0 = (q,h)$, $Q = 0$. But if $p > 0$, $q < 0$, set $q' = -q$ and suppose that $\alpha < \beta$. It is clear that $p - q' = h$ and

$$(q,h) = (-q',h) = (-1)^h\frac{q'(q'+1)\dots(p-1)}{1\cdot 2\dots h} = (-1)^h(p-1,h)$$
$$= (-1)^h\frac{-q}{p}(p,h).$$

Moreover,

$$(a+\alpha,\alpha) = \frac{\alpha+a}{a}(a+\alpha-1,\alpha), \qquad (a+\beta,\,\beta) = \frac{\beta+a}{a}(a+\beta-1,\,\beta);$$

and, as a consequence,

$$Q = (a+\alpha-1,\alpha)(a+\beta-1,\,\beta)(p,h)$$
$$\times\left[(-1)^\alpha\frac{\beta+a}{a} - (-1)^{\beta+h}\frac{\alpha+a}{a}\cdot\frac{q}{p}\right].$$

Much as before, he now reasons that $2(i - a) = \alpha + (\beta + h)$ shows that α and $\beta + h$ have the same parity—both are even or odd—so that $(-1)^\alpha$

$=(-1)^{\beta+h}$. In addition, he has

$$\frac{\beta+a}{a}-\frac{q}{p}\frac{\alpha+a}{a}=\frac{i(\beta-\alpha)}{ap},$$

and hence

$$Q=(-1)^{\alpha}\frac{i(\beta-\alpha)}{ap}(a+\alpha-1,\alpha)(a+\beta-1,\ \beta)(p,h).$$

This relation also holds for $q=0$; then $p=h$ and (4.32) becomes

$$Q=(-1)^{\alpha}(a+\alpha-1,\alpha)(a+\beta,\ \beta)$$

since $(q,h)=0$. Lebesgue next considers $a=0$ and notes that $Q=0$ except when $\alpha=0$ or $\beta=0$. In these cases $Q=(i,h)$, as may be seen directly or by noting that $D_0=C_0y$.

Lebesgue next shows how to use these results to calculate the B_a for the sixth-order case:

$$y\{A_0(ty)+[A_1(ty)']'+[A_2(ty)'']''+[A_3(ty)''']'''\}$$
$$=B_0t+(B_1t')'+(B_2t'')''+(B_3t''')'''.$$

He takes $a=3$ and considers the equation $2(i-a)=h+\alpha+\beta$, with i,a given and $h,\alpha,\ \beta$ unknowns such that $\alpha\leqslant i-a,\ \beta\geqslant i-a$ and $h,\alpha,\ \beta$ are nonnegative integers. It is also clear from a comparison of equation (4.28) with (4.23) that $i\leqslant n$, which is 3 in this case. He then has at once $h=\alpha=\beta=0$ and $Q=1$ since $\alpha=\beta$ and $B_3=A_3y\cdot y$. To find B_2, set $a=2$ in $2(i-a)=h+\alpha+\beta$; thus $i=2$ or 3. The former case yields the term $A_2y\cdot y$. The latter case is more complex since $2=h+\alpha+\beta$ has three solutions $(h,\alpha,\ \beta)$: $(0,0,2)$, $(0,1,1)$, and $(1,0,1)$. In these cases $Q=9,\ -6,3$, respectively, and

$$B_2=A_2y^2+9A_3yy''-6A_3y'y'+3A_3'yy'.$$

To find B_1, he sets $a=1$ and needs to consider the cases when $i=1,2,3$. For $i=1$, he has $(h,\alpha,\ \beta)=(0,0,0)$ and the term A_1y^2; for $i=2$, he has $(h,\alpha,\ \beta)=(0,0,2)$, $(0,1,1)$, and $(1,0,1)$, which correspond to $Q=4,\ -2$, and 2 and the terms $4A_2yy''-2A_2y'y'+2A_2'yy'$. For $i=3$, there are six solutions: $(0,0,4)$, $(0,1,3)$, $(0,2,2)$, $(1,0,3)$, $(1,1,2)$, and $(2,0,2)$, corresponding to the values of Q 6, -6, 3, 9, -3, and 3, respectively; thus he has the terms

$$6A_3yy''''-6A_3y'y'''+3A_3y''y''+9A_3'yy'''-3A_3'y'y''+3A_3''yy''.$$

The complete B_1 is then the sum of all the terms

$$B_1=A_1y^2+4A_2yy''-2A_2y'y'+2A_2'yy'+6A_3yy''''-6A_3y'y'''$$
$$+3A_3y''y''+9A_3'y'y'''-3A_3'y'y''+3A_3yy''.$$

Finally, since

$$B_0=y[A_0y+(A_1y')'+(A_2y'')''+(A_3y''')'''],$$

its general term is $(i, \alpha - i)A_i^{2i-\alpha} yy^{\alpha}$ and

$$B_0 = A_0 y^2 + A_1 yy'' + A_1' yy' + A_2 yy'''' + 2A_2' yy''' + A_2'' yy'' + A_3 yy^{\mathrm{vi}} + 3A_3' yy^{\mathrm{v}} + 3A_3'' yy'''' + A_3''' yy'''.$$

This completes Lebesgue's paper. As we see, his method is tedious, and the various other authors who have attempted the solution have run into more or less the same difficulties in their papers. Since Jacobi's result is no longer central to the calculus of variations, I do not discuss the other methods that were used.

4.5. Hamilton–Jacobi Theory

In the same paper we discussed previously, Jacobi remarks "[William Rowan] Hamilton has shown that the problems of mechanics, for which the principle of living force is valid, can be reduced to the integration of a partial differential equation of the first order."[12] Jacobi proceeds by saying that Hamilton actually requires the integration of two partial differential equations of the first order. He remarks, however, that it can be easily shown that it is sufficient to know a complete integral of one of them. He further says that he has extended without difficulty the result to the case where the force function U (recall that this is the function whose partial derivatives give the components of force in given directions; e.g., $\partial U/\partial x$ is the component of force along the x-axis) contains the time explicitly, in which case the principle of living force is not valid. But before quoting more of Jacobi's letter to Encke, we need to learn the details.

Hamilton's beautiful results appeared first in the *Philosophical Transactions* for 1834 and 1835.[13] In the introduction to his 1834 paper Hamilton says

> In the methods commonly employed, the determination of the motion of a free point in space, under the influence of accelerating forces, depends on the integration of three equations in ordinary differentials of the second order; and the determination of the motions of a system of free points, attracting or repelling one another, depends on the integration of a system of such equations, in number threefold the number of the attracting or repelling points, unless we previously diminish by unity this latter number, by considering only relative motions . . . In the method of

[12] Jacobi [1838], p. 50. The paper is a letter to Encke, who was then secretary of the mathematics–physics section of the Berlin Academy of Sciences. Recall that the principle of living force is what we now call the principle of conservation of energy. (We discuss Hamilton's results below in this section.)

[13] W. Hamilton [1834], [1835].

> the present essay, this problem is reduced to the search and differentiation of a single function, which satisfies two partial differential equations of the first order and of the second degree.

He then proceeds to find these partial differential equations. (As we see below, Jacobi shows that only one of these equations is needed.) He considers a system of n heavy particles, each of mass m_i and coordinates (x_i, y_i, z_i) $(i = 1, 2, \ldots, n)$. To carry out his analysis, he supposes that there exists a force function U and that $T = \frac{1}{2}\sum \cdot m(x'^2 + y'^2 + z'^2)$ is the kinetic energy so that the "celebrated law of living forces"

$$T = U + H, \tag{4.33}$$

holds for the system; i.e., it is conservative. In this relation H is a constant for each given set of initial conditions, but it may vary with those conditions. He takes these conditions to be (a_i, b_i, c_i) $(i = 1, 2, \ldots, n)$, the initial positions of the n particles, and (a_i', b_i', c_i'), $(i = 1, 2, \ldots, n)$, their initial velocities along the coordinate directions. He considers now the variation of T, $\delta T = \delta U + \delta H$, and integrates the results, finding, that since $\delta U = \sum \cdot m(x''\,\delta x + y''\,\delta y + z''\,\delta z)$,

$$\begin{aligned}&\int \sum \cdot m(dx \cdot \delta x' + dy \cdot \delta y' + dz \cdot \delta z')\\ &\quad = \int \sum \cdot m(dx' \cdot \delta x + dy' \cdot \delta y + \delta z' \cdot \delta z) + \int \delta H \cdot dt.\end{aligned} \tag{4.34}$$

When V, the action integral which Hamilton calls the *accumulated living force*, is defined as a function of x, y, z, a, b, c and H, by

$$V = \int \sum \cdot m(x'\,dx + y'\,dy + z'\,dz) = \int_0^t 2T\,dt, \tag{4.35}$$

Hamilton finds that

$$\delta V = \sum \cdot m(x'\delta x + y'\delta y + z'\delta z) - \sum \cdot m(a'\delta a + b'\delta b + c'\delta c) + t\delta H.$$

From this he infers that

$$\frac{\delta V}{\delta x_i} = m_i x_i', \qquad \frac{\delta V}{\partial y_i} = m_i y_i', \qquad \frac{\delta V}{\partial z_i} = m_i z_i' \qquad (i = 1, 2, \ldots, n) \tag{4.36}$$

and that

$$\frac{\delta V}{\delta a_i} = -m_i a_i', \qquad \frac{\delta V}{\delta b_i} = -m_i b_i', \qquad \frac{\delta V}{\partial c_i} = -m_i c_i' \qquad (i = 1, 2, \ldots, n) \tag{4.36'}$$

"and finally, the equation[14]

$$\frac{\delta V}{\delta H} = t." \tag{4.36''}$$

[14] Hamilton [1834], pp. 107–108.

Now if V is known, then Hamilton remarks that H can be eliminated between the $3n + 1$ equations (4.36) and (4.36″) "in order to obtain all the $3n$ intermediate integrals." Alternatively, it can be eliminated between (4.36′) and (4.36″) "to obtain all the $3n$ final integrals of the differential equations of motion; that is, ultimately, to obtain the $3n$ sought relations between the $3n$ varying coordinates and the time, involving also the masses and the $6n$ initial data . . . , the discovery of which relations would be (as we have said) the general solution of the general problems of dynamics. We have, therefore, at least reduced that general problem to the search and differentiation of a single function V."

Hamilton now proceeds to combine relations (4.35) and (4.36) into

$$\frac{1}{2}\sum\cdot\frac{1}{m}\left\{\left(\frac{\delta V}{\delta x}\right)^2+\left(\frac{\delta V}{\delta y}\right)^2+\left(\frac{\delta V}{\delta z}\right)^2\right\}=U+H \tag{4.37}$$

at the final point and

$$\frac{1}{2}\sum\cdot\frac{1}{m}\left\{\left(\frac{\delta V}{\delta a}\right)^2+\left(\frac{\delta V}{\delta b}\right)^2+\left(\frac{\delta V}{\delta c}\right)^2\right\}=U_0+H \tag{4.37′}$$

at the initial point. He remarks that his *characteristic function* V must satisfy these two partial differential equations.

At the close of his 1834 paper Hamilton introduced a new function S of paramount importance. He writes $V = tH + S$,

$$S=\int_0^t(T+U)\,dt \tag{4.38}$$

and notes that since S is a function of x, y, z, a, b, c and t,

$$\begin{aligned}&\frac{\delta S}{\delta t}=-H;\\ &\frac{\delta S}{\delta x_i}=m_i x_i', \qquad \frac{\delta S}{\delta y_i}=m_i y_i', \qquad \frac{\delta S}{\delta z_i}=m_i z_i';\\ &\frac{\delta S}{\delta a_i}=-m_i a_i', \qquad \frac{\delta S}{\delta x_i}=-m_i b_i', \qquad \frac{\delta S}{\delta c_i}=-m_i c_i'.\end{aligned} \tag{4.38′}$$

He then says that the "limits of the present essay do not permit us here to develope the consequences of these new expressions. We can only observe, that the auxiliary function S must satisfy the two following equations . . . "[15]

$$\frac{\delta S}{\delta t}+\sum\cdot\frac{1}{2m}\left\{\left(\frac{\delta S}{\delta x}\right)^2+\left(\frac{\delta S}{\delta y}\right)^2+\left(\frac{\delta S}{\delta z}\right)^2\right\}=U, \tag{4.39}$$

$$\frac{\delta S}{\delta t}+\sum\cdot\frac{1}{2m}\left\{\left(\frac{\delta S}{\delta a}\right)^2+\left(\frac{\delta S}{\delta b}\right)^2+\left(\frac{\delta S}{\delta c}\right)^2\right\}=U_0. \tag{4.39′}$$

[15] Hamilton [1834], pp. 160–161.

It seems clear from its position in the last section of the paper that he only came on S at the last moment and in the 1835 paper made it the chief function of his work. In that latter paper he calls S the *principal function* of motion of a system.[16]

To get at the paramount role of S, in his second paper Hamilton makes uses of an elegant result first deduced by Lagrange in his *Mécanique Analytique*,[17] where Lagrange changes from the variables (x_i, y_i, z_i), $(i = 1, 2, \ldots, n)$ to new ones $\eta_1, \eta_2, \ldots, \eta_{3n}$ and shows, without any reference to the calculus of variations, that the differential equations of motion now assume the form (in Hamilton's notation)

$$\frac{d}{dt}\frac{\delta T}{\delta \eta'_\alpha} - \frac{\delta T}{\delta \eta_\alpha} = \frac{\delta U}{\delta \eta_\alpha} \qquad (\alpha = 1, 2, \ldots, 3n). \tag{4.40}$$

(Here, of course, $2T = \sum \cdot m(x'^2 + y'^2 + z'^2)$.) Hamilton next sets

$$\frac{\delta T}{\delta \eta'_\alpha} = \omega_\alpha \qquad (\alpha = 1, 2, \ldots, 3n) \tag{4.41}$$

and considers "T (as we may) as a function of the following form,"

$$T = F(\omega_1, \omega_2, \ldots, \omega_{3n}, \eta_1, \eta_2, \ldots, \eta_{3n}). \tag{4.42}$$

Then it is not difficult to arrive at the relations

$$\frac{\delta F}{\delta \omega_\alpha} = \eta'_\alpha, \qquad \frac{\delta F}{\delta \eta_\alpha} = -\frac{\delta T}{\delta \eta_\alpha} \qquad (\alpha = 1, 2, \ldots, 3n), \tag{4.43}$$

"T being here considered as a function of the $6n$ quantities of the form η' and $\eta, \ldots$." To derive the equations due to Lagrange (4.40) above, Hamilton notes with the help of

$$m_i x_i'' = \frac{\delta U}{\delta x_i}, \qquad m_i y_i'' = \frac{\delta U}{\delta y_i}, \qquad m_i z_i'' = \frac{\delta U}{\delta z_i} \qquad (i = 1, 2, \ldots, n)$$

that, in his notation,

$$\begin{aligned} \frac{\delta U}{\delta \eta_i} &= \sum \cdot m\left(x''\frac{\delta x}{\delta \eta_i} + y''\frac{\delta y}{\delta \eta_i} + z''\frac{\delta z}{\delta \eta_i}\right) \\ &= \frac{d}{dt}\sum \cdot m\left(x'\frac{\delta x}{\delta \eta_i} + y'\frac{\delta y}{\delta \eta_i} + z'\frac{\delta z}{\delta \eta_i}\right) \\ &\quad - \sum \cdot m\left(x'\frac{d}{dt}\frac{\delta x}{\delta \eta_i} + y'\frac{d}{dt}\frac{\delta y}{\delta \eta_i} + z'\frac{d}{dt}\frac{\delta z}{\delta \eta_i}\right). \end{aligned} \tag{4.44}$$

Since

$$x' = \eta'_1\frac{\delta x}{\delta \eta_1} + \eta'_2\frac{\delta x}{\delta \eta_2} + \cdots + \eta'_{3n}\frac{\delta x}{\delta \eta_{3n}},$$

---

[16] Hamilton [1834], pp. 162–211. Notice that the integrand function $T + U$ is the so-called Lagrangian function and $T - U$ the famous hamiltonian function.

[17] Lagrange, *Mécanique Analytique*, 3rd ed., Vol. I, pp. 290–292.

etc., it is clear that

$$\sum \cdot m\left(x'\frac{\delta x}{\delta \eta_i}+y'\frac{\delta y}{\delta \eta_i}+z'\frac{\delta z}{\delta \eta_i}\right)$$
$$=\sum \cdot m\left(x'\frac{\delta x'}{\delta \eta_i'}+y'\frac{\delta y'}{\delta \eta_i'}+z'\frac{\delta z'}{\delta \eta_i'}\right)=\frac{\delta T}{\delta \eta_i'}$$

and also that

$$\sum \cdot m\left(x'\frac{d}{dt}\frac{\delta x}{\delta \eta_i}+y'\frac{d}{dt}\frac{\delta y}{\delta \eta_i}+z'\frac{d}{dt}\frac{\delta z}{\delta \eta_i}\right)$$
$$=\sum \cdot m\left(x'\frac{\delta x'}{\delta \eta_i}+y'\frac{\delta y'}{\delta \eta_i}+z'\frac{\delta z'}{\delta \eta_i}\right)=\frac{\delta T}{\delta \eta_i}.$$

Putting these together, Hamilton recreates Lagrange's equations (4.40).

Hamilton next introduces the hamiltonian function

$$H = F - U = F(\omega_1, \omega_2, \ldots, \omega_{3n}, \eta_1, \eta_2, \ldots, \eta_{3n}) - U(\eta_1, \eta_2, \ldots, \eta_{3n})$$

and observes that

$$\frac{d\eta_\alpha}{dt}=\frac{\delta H}{\delta \omega_\alpha}, \qquad \frac{d\omega_\alpha}{dt}=-\frac{\delta H}{\delta \eta_\alpha} \qquad (\alpha = 1, 2, \ldots, 3n). \tag{4.45}$$

(These are often called the *canonical equations* of motion and $\eta_\alpha, \omega_\alpha$ the *canonical variables*. Lagrange had introduced these equations to discuss perturbation theory; see *Mécanique Analytique*, 3rd ed., Vol. I, p. 310.) This system is then precisely equivalent to that of Lagrange (4.40). To find its solutions, Hamilton considers his function S in the form

$$S=\int_0^t\left(\sum \cdot \omega\frac{\delta H}{\delta \omega}-H\right)dt=\int_0^t S'\,dt \tag{4.46}$$

and forms the variation of S "without varying t or dt" (see p. 181 below or Hamilton [1835], p. 166). This gives him

$$\delta S=\int_0^t \delta S'\,dt, \qquad \delta S'=\sum\left(\omega\delta\frac{\delta H}{\delta \omega}-\frac{\delta H}{\delta \eta}\delta\eta\right).$$

But by equations (4.45), this implies that

$$\delta S'=\sum\left(\omega\delta\frac{d\eta}{dt}+\frac{d\omega}{dt}\delta\eta\right)=\frac{d}{dt}\sum \cdot \omega\,\delta\eta$$

and hence

$$\delta S=\sum(\omega\,\delta\eta-p\,\delta e). \tag{4.47}$$

where Hamilton uses e and p as the initial values of η and ω. This relation may then be decomposed into the equations

$$\omega_\alpha=\frac{\delta S}{\delta \eta_\alpha}, \qquad p_\alpha=-\frac{\delta S}{\delta e_\alpha} \qquad (\alpha = 1, 2, \ldots, 3n). \tag{4.47'}$$

This then reduces the solution of the equations of motion to finding a function S satisfying these $6n$ equations.[18] (Notice that the form of S given in (4.46) is the same as in (4.38). In other words, the same function has been used in both papers. To see this, note first with the help of (4.43) and (4.41) that since $2T$ is homogeneous of degree 2 in the η' and since $H = F(\omega, \eta) - U(\eta)$,

$$2T = 2F(\omega, \eta) = \sum \cdot \eta' \frac{\delta T}{\delta \eta'} = \sum \cdot \frac{\delta F}{\delta \omega} \omega = \sum \cdot \omega \frac{\delta H}{\delta \omega} .$$

Then

$$S' = \sum \cdot \omega \frac{\delta H}{\delta \omega} - H = 2F - (F - U) = F + U = T + U.)$$

Hamilton goes on to say that the function S when it is expressed in the form

$$S = \int_0^t (T + U)\, dt \tag{4.48}$$

has the property that the Euler equations found by minimizing S "are precisely the differential equations of motion . . . [(4.40)] under the forms assigned by Lagrange." (See also Section 2.4 above.)

He next proceeds to find the famous Hamilton–Jacobi partial differential equation. To do this, he notes that the variation of S in (4.47) omitted the variation of t, the time. To find $\delta S/\delta t$, he observes that

$$\frac{dS}{dt} = \frac{\delta S}{\delta t} + \sum \frac{\delta S}{\delta \eta} \frac{d\eta}{dt}$$

so that

$$\frac{\delta S}{\delta t} = S' - \sum \cdot \omega \frac{\delta H}{\delta \omega} = -H = U - F.$$

Moreover, H "is constant, so as not to alter during the motion of the system . . . ," since by (4.45)

$$\frac{dH}{dt} = \sum \left(\frac{\delta H}{\delta \eta} \frac{d\eta}{dt} + \frac{\delta H}{\delta \omega} \frac{d\omega}{dt} \right) = \sum \left(-\frac{d\omega}{dt} \frac{d\eta}{dt} + \frac{d\eta}{dt} \frac{d\omega}{dt} \right) = 0.$$

It then follows that

$$\frac{\delta S}{\delta t} + F\left(\frac{\delta S}{\delta \eta_1}, \frac{\delta S}{\delta \eta_2}, \ldots, \frac{\delta S}{\delta \eta_{3n}}, \eta_1, \eta_2, \ldots, \eta_{3n} \right) = U(\eta_1, \eta_2, \ldots, \eta_{3n}), \tag{4.49}$$

$$\frac{\delta S}{\delta t} + F\left(\frac{\delta S}{\delta e_1}, \frac{\delta S}{\delta e_2}, \ldots, \frac{\delta S}{\delta e_{3n}}, e_1, e_2, \ldots, e_{3n} \right) = U(e_1, e_2, \ldots, e_{3n}). \tag{4.50}$$

[18] Hamilton [1835], p. 167.

These last relations are what we need from Hamilton's paper. The former is the Hamilton–Jacobi partial differential equation; the latter was thrown out by Jacobi, as we see next. The problem of rendering the integral (4.48) for S an extremum is Hamilton's principle.

In 1838 Jacobi wrote a most important paper on this same subject.[19] In this paper he levels two criticisms at Hamilton's work as we shall see. He says:[20]

> It appears to me that Hamilton has presented his beautiful discovery in a false light, which both complicates and limits its usefulness unnecessarily. His theorem as he has stated it, has also the disadvantage of being obscure when one does not have his proof in front of one, since one can not define one function by two partial differential equations without first showing that such a function really exists. By the choice he has made of the special function S the arbitrary constants become the initial values of the coordinates and of the components of the velocities with respect to the coordinate axes; but this is not an advantage since the introduction of these constants ordinarily makes the integral equations more complicated, and since one can transform the integral equations to this form from any other form. It is perhaps because he has always to consider at the same time two partial differential equations that Hamilton has not applied to his theorem the general rules that Lagrange gives in his lectures on the calculus of functions for integrating a non-linear partial differential equation of the first order in three variables; and for this reason, as I shall show in another memoir, results of the greatest interest for mechanics have escaped him. Finally the requirement that the function S after having to satisfy the first partial differential equation satisfies also a second one leads to a restriction in that it excludes the case where the force-function U contains the time explicitly: for this case, in fact, the second partial differential equation is not valid.

Jacobi's criticisms are then first, that Hamilton does not know that his equations necessarily have a solution and second, that his second equation is unnecessary; it suffices to show that a function S exists which satisfies the first one, as we shall see.

Jacobi considers more or less the same dynamical problem as does Hamilton: he has n heavy particles of masses m_i, located at the points (x_i, y_i, z_i) with initial values (a_i, b_i, c_i) and initial components of velocity (a_i', b_i', c_i') $(i = 1, 2, \ldots, n)$. However, he permits the force function U to contain t explicitly, as well as the coordinates x_i, y_i, z_i and the initial values a_i, b_i, c_i. He is then able to introduce the time into U since he does not make use, as does Hamilton, of the principle of living force—conservation of energy.

Jacobi's main result is contained in the theorem:[21]

---

[19] Jacobi [1838′].

[20] Jacobi [1838′], pp. 73–74. The translation above is from the French version, pp. 76–77.

[21] Jacobi [1838′], pp. 71–72.

Let the differential equations of motion of a free [i.e., not constrained] system of n material particles be the following 3n second-order differential equations:

$$m_i \frac{d^2x_i}{dt^2} = \frac{\partial U}{\partial x_i}, \qquad m_i \frac{d^2y_i}{dt^2} = \frac{\partial U}{\partial y_1}, \qquad m_i \frac{d^2z_i}{dt^2} = \frac{\partial U}{\partial z_i}, \tag{4.51}$$

where U signifies a given function of the 3n coordinates $x_1, y_1, z_1, x_2, y_2, z_2, \ldots, x_n, y_n, z_n$, and the time t, and i takes on all values $1, 2, \ldots, n$. Further let S be some complete integral of the partial differential equation:

$$U = \frac{\partial S}{\partial t} + \frac{1}{2} \sum \frac{1}{m_i} \left[\left(\frac{\partial S}{\partial x_i} \right)^2 + \left(\frac{\partial S}{\partial y_i} \right)^2 + \left(\frac{\partial S}{\partial z_i} \right)^2 \right], \tag{4.52}$$

which, in addition to an additive constant, contains 3n other arbitrary constants $\alpha_1, \alpha_2, \ldots, \alpha_{3n}$. Then a complete and finite integral of the preceding 3n ordinary differential equations of the second order with 6n arbitrary constants is given by:

$$\frac{\partial S}{\partial \alpha_1} = \beta_1, \qquad \frac{\partial S}{\partial \alpha_2} = \beta_2, \ldots, \qquad \frac{\partial S}{\partial \alpha_{3n}} = \beta_{3n}, \tag{4.53}$$

where the quantities $\beta_1, \beta_2, \ldots, \beta_{3n}$ are 3n new arbitrary constants. Further the components of the velocity relative to the coordinate axes are:

$$x_i' = \frac{1}{m_i} \frac{\partial S}{\partial x_i}, \qquad y_i' = \frac{1}{m_i} \frac{\partial S}{\partial y_i}, \qquad z_i' = \frac{1}{m_i} \frac{\partial S}{\partial z_i}. \tag{4.54}$$

Notice that Jacobi, in essence, has shown the converse of what Hamilton did. His proof is quite straightforward. He starts by differentiating relations (4.53) with respect to t and finds

$$0 = \frac{d(\partial S / \partial \alpha_p)}{dt} = \frac{\partial^2 S}{\partial \alpha_p\, \partial t} + \sum \left(\frac{\partial^2 S}{\partial \alpha_p\, \partial x_i} x_i' + \frac{\partial^2 S}{\partial \alpha_p\, \partial y_i} y_i' + \frac{\partial^2 S}{\partial \alpha_p\, \partial z_i} z_i' \right)$$

$$(p = 1, 2, \ldots, 3n).$$

He remarks that these can be used to find the quantities x_i', y_i', z_i'. To do this, he differentiates the Hamilton–Jacobi relation (4.52) above with respect to α_p and finds the equivalent relations

$$0 = \frac{\partial^2 S}{\partial \alpha_p\, \partial t} + \sum \frac{1}{m_i} \left[\frac{\partial^2 S}{\partial \alpha_p\, \partial x_i} \cdot \frac{\partial S}{\partial x_i} + \frac{\partial^2 S}{\partial \alpha_p\, \partial y_i} \frac{\partial S}{\partial y_i} + \frac{\partial^2 S}{\partial \alpha_p\, \partial z_i} \frac{\partial S}{\partial z_i} \right]$$

$$(p = 1, 2, \ldots, 3n).$$

Then a comparison of these two systems shows that

$$x_i' = \frac{dx_i}{dt} = \frac{1}{m_i} \frac{\partial S}{\partial x_i}, \qquad y_i' = \frac{dy}{dt} = \frac{1}{m_i} \frac{\partial S}{\partial y_i}, \qquad z_i' = \frac{dz_i}{dt} = \frac{1}{m_i} \frac{\partial S}{\partial z_i},$$

as was to be shown. Next he differentiates these equations and obtains the new relations

$$m_i \frac{d^2x_i}{dt^2} = \sum_{k=1}^{n} \left[\frac{\partial^2 S}{\partial x_i\, \partial x_k} x'_k + \frac{\partial^2 S}{\partial x_i\, \partial y_k} y'_k + \frac{\partial^2 S}{\partial x_i\, \partial z_k} z'_k \right] + \frac{\partial^2 S}{\partial x_i\, \partial t},$$

$$m_i \frac{d^2y_i}{dt^2} = \sum_{k=1}^{n} \left[\frac{\partial^2 S}{\partial y_i\, \partial x_k} x'_k + \frac{\partial^2 S}{\partial y_i\, \partial y_k} y'_k + \frac{\partial^2 S}{\partial y_i\, \partial z_k} z'_k \right] + \frac{\partial^2 S}{\partial y_i\, \partial t},$$

$$m_i \frac{d^2z_i}{dt^2} = \sum_{k=1}^{n} \left[\frac{\partial^2 S}{\partial z_i\, \partial x_k} x'_k + \frac{\partial^2 S}{\partial z_i\, \partial y_k} y'_k + \frac{\partial^2 S}{\partial z_i\, \partial z_k} z'_k \right] + \frac{\partial^2 S}{\partial z_i\, \partial t}.$$

He substitutes the values of x'_i, y'_i, z'_i he just found into these; this gives him

$$m_i \frac{d^2x_i}{dt^2} = \sum_k \frac{1}{m_k} \left[\frac{\partial^2 S}{\partial x_i\, \partial x_k} \cdot \frac{\partial S}{\partial x_k} + \frac{\partial^2 S}{\partial x_i\, \partial y_k} \cdot \frac{\partial S}{\partial y_k} + \frac{\partial^2 S}{\partial x_i\, \partial z_k} \cdot \frac{\partial S}{\partial z_k} \right] + \frac{\partial^2 S}{\partial x_i\, \partial t},$$

$$m_i \frac{d^2y_i}{dt^2} = \sum_k \frac{1}{m_k} \left[\frac{\partial^2 S}{\partial y_k\, \partial x_k} \cdot \frac{\partial S}{\partial x_k} + \frac{\partial^2 S}{\partial y_i\, \partial y_k} \cdot \frac{\partial S}{\partial y_k} + \frac{\partial^2 S}{\partial y_i\, \partial z_k} \cdot \frac{\partial S}{\partial z_k} \right] + \frac{\partial^2 S}{\partial y_i\, \partial t},$$

$$m_i \frac{d^2z_i}{dt^2} = \sum_k \frac{1}{m_k} \left[\frac{\partial^2 S}{\partial z_i\, \partial x_k} \cdot \frac{\partial S}{\partial x_k} + \frac{\partial^2 S}{\partial z_i\, \partial y_k} \cdot \frac{\partial S}{\partial y_k} + \frac{\partial^2 S}{\partial z_i\, \partial z_k} \cdot \frac{\partial S}{\partial z_k} \right] + \frac{\partial^2 S}{\partial z_i\, \partial t}.$$

It is clear that the right-hand members of these equations are readily derivable from the Hamilton–Jacobi equation (4.52); e.g.,

$$\frac{\partial U}{\partial x_i} = \sum \frac{1}{m_k} \left[\frac{\partial S}{\partial x_k} \cdot \frac{\partial^2 S}{\partial x_i\, \partial x_k} + \frac{\partial S}{\partial y_k} \cdot \frac{\partial^2 S}{\partial x_i\, \partial y_k} + \frac{\partial S}{\partial z_k} \cdot \frac{\partial^2 S}{\partial x_i\, \partial z_k} \right] + \frac{\partial^2 S}{\partial x_i\, \partial t}.$$

This enables Jacobi to conclude that

$$m_i \frac{d^2x_i}{dt^2} = \frac{\partial U}{\partial x_i}, \qquad m_i \frac{d^2y_i}{dt^2} = \frac{\partial U}{\partial y_i}, \qquad m_i \frac{d^2z_i}{dt^2} = \frac{\partial U}{\partial z_i},$$

which are the equations of motion of his free system.

He has therefore shown that all solutions of the equations of motion of his dynamical system can be found once a complete integral of the Hamilton–Jacobi equation has been obtained. Furthermore, he has also shown that Hamilton's second partial differential equation is unnecessary. In fact, he says:[22]

> I do not understand why Hamilton in order to give the complete solution of the proposed differential equations, finds it necessary to seek a function S of $6n + 1$ variables, namely the $3n$ quantities x_i, y_i, z_i the $3n$ quantities a_i, b_i, c_i, and the quantity t, which at the same time satisfies

[22] Jacobi [1838′], p. 73.

both partial differential equations of the first order

$$\frac{\partial S}{\partial t} + \frac{1}{2}\sum \frac{1}{m_i}\left[\left(\frac{\partial S}{\partial x_i}\right)^2 + \left(\frac{\partial S}{\partial y_i}\right)^2 + \left(\frac{\partial S}{\partial z_i}\right)^2\right] = U,$$

$$\frac{\partial S}{\partial t} + \frac{1}{2}\sum \frac{1}{m_i}\left[\left(\frac{\partial S}{\partial a_i}\right)^2 + \left(\frac{\partial S}{\partial b_i}\right)^2 + \left(\frac{\partial S}{\partial c_i}\right)^2\right] = U_0,$$

since, as we have seen, it suffices to find one function of the $3n + 1$ quantities t, x_i, y_i, z_i, that satisfy the single equation

$$\frac{\partial S}{\partial t} + \frac{1}{2}\sum \frac{1}{m_i}\left[\left(\frac{\partial S}{\partial x_i}\right)^2 + \left(\frac{\partial S}{\partial y_i}\right)^2 + \left(\frac{\partial S}{\partial z_i}\right)^2\right] = U$$

and contains an additive arbitrary constant in addition to $3n$ other arbitrary constants.

Jacobi then sets $V = S + Ht = S + (T - U)t$ and observes that the variation of V is given by

$$\delta V = t\,\delta H + \sum m_i(x_i'\,\delta x_i + y_i'\,\delta y_i + z_i'\,\delta z_i) - \sum m_i(a_i'\,\delta a_i + b_i'\,\delta b_i + c_i'\,\delta c_i),$$

since $\partial S/\partial t = -H$ by (4.52), (4.38′) and

$$\delta S = -H\,\delta t + \sum m_i(x_i'\,\delta x_i + y_i'\,\delta y_i + z_i'\,\delta z_i) - \sum m_i(a_i'\,\delta a_i + b_i'\,\delta b_i + c_i'\,\delta c_i),$$

where

$$\tfrac{1}{2}\sum m_i(x_i'^2 + y_i'^2 + z_i'^2) - U = H.$$

Now he wishes to eliminate t from S by means of the equation

$$\frac{1}{2}\sum \frac{1}{m_i}\left[\left(\frac{\partial S}{\partial x_i}\right)^2 + \left(\frac{\partial S}{\partial y_i}\right)^2 + \left(\frac{\partial S}{\partial z_i}\right)^2\right] - U = H$$

so that both S and V become functions of H, x_i, y_i, z_i, and a_i, b_i, c_i $(i = 1, 2, \ldots, n)$. Thus

$$\frac{\partial V}{\partial H} = t, \qquad \frac{\partial V}{\partial x_i} = m_i x_i', \qquad \frac{\partial V}{\partial a_i} = -m_i a_i', \qquad \frac{\partial V}{\partial y_i} = m_i y_i',$$

$$\frac{\partial V}{\partial b_i} = -m_i b_i', \qquad \frac{\partial V}{\partial z_i} = m_i z_i', \qquad \frac{\partial V}{\partial c_i} = -m_i c_i'.$$

With the help of these values he finds that V satisfies the partial differential equation

$$\frac{1}{2}\sum \frac{1}{m_i}\left[\left(\frac{\partial V}{\partial x_i}\right)^2 + \left(\frac{\partial V}{\partial y_i}\right)^2 + \left(\frac{\partial V}{\partial z_i}\right)^2\right] = U + H, \tag{4.55}$$

where t is replaced by $\partial V/\partial H$ in case it occurs explicitly in U. However, if U does not contain t explicitly but depends only on the coordinates, as is often the case, then H can be regarded as a constant.

A complete solution of (4.55) contains $3n + 1$ constants, one additive, and $3n$ others, $\alpha_1, \ldots, \alpha_{3n}$. The $3n$ complete integrals of the ordinary differential equations of the second order

$$m_i \frac{d^2x_i}{dt^2} = \frac{\partial U}{\partial x_i}, \qquad m_i \frac{d^2y_i}{dt^2} = \frac{\partial U}{\partial y_i}, \qquad m_i \frac{d^2z_i}{dt^2} = \frac{\partial U}{\partial z_i}$$

can be found by setting

$$\frac{\partial V}{\partial \alpha_1} = \beta_1, \qquad \frac{\partial V}{\partial \alpha_2} = \beta_2, \ldots, \qquad \frac{\partial V}{\partial \alpha_{3n}} = \beta_{3n},$$

where $\beta_1, \beta_2, \ldots, \beta_{3n}$ are new arbitrary constants. Thus the solution of these $3n$ equations for the $3n$ quantities x_i, y_i, z_i contains $6n$ constants of integration $\alpha_1, \alpha_2, \ldots, \alpha_{3n}$; $\beta_1, \beta_2, \ldots, \beta_{3n}$. Moreover, Jacobi notes that $3n$ intermediate integrals containing $3n$ constants arise from the relations

$$\frac{\partial V}{\partial x_i} = m_i x_i', \qquad \frac{\partial V}{\partial y_i} = m_i y_i', \qquad \frac{\partial V}{\partial z_i} = m_i z_i',$$

and the value H appearing in these equations can be replaced by t by virtue of the fact that $\partial V/\partial H = t$.

To illustrate the use of the theory Jacobi discusses planetary motion once again, but it is probably not relevant here to follow up on this further. It should suffice to remark that these results of Hamilton and Jacobi were not only of great importance to classical mechanics but are of paramount importance to quantum mechanics, where the entire theory is permeated with these ideas. The generalization of the Hamilton–Jacobi equation to more general problems of the calculus of variations was suggested by Beltrami [1868].

4.6. Hesse's Commentary

In 1857 Hesse published in Vol. LIV of the *Journal für Mathematik* the last and best of the series of the commentaries on Jacobi's classic papers. In the two decades from V.-A. Lebesgue and Delaunay to Hesse there was considerable deepening of understanding, and the interested reader can see this in Hesse's paper. By this time, however, Clebsch had begun to examine more general problems of the calculus of variations by other means. In a sense, therefore, much of Hesse's paper is of only modest interest. I propose in this section to mention just the parts of his paper that are of some relevance to his successors. Much of this is contained in his ninth section, (Hesse [1857], pp. 255–260).

He considers in this section the simplest problem of the calculus of variations: to minimize the integral

$$\Delta = \int_a^b f(x, y, y')\, dx$$

among curves through fixed end-points. He makes use of a family of extremals $y = F(x, \alpha_1, \alpha_2)$ and writes the integrand of the second variation as

$$2\psi = a_{00}zz + 2a_{01}zz' + a_{11}z'z',$$

sets

$$\Psi(z) = \psi'(z) - \frac{d\psi'(z')}{dx},$$

and supposes that the coefficients in ψ are defined along some extremal of his family. He also supposes that along that extremal $a_{11} \neq 0$. [The expression for Ψ would now be written as $\partial\psi/\partial z - d(\partial\psi/\partial z')/dx$.] He can then write the second variation in the familiar form (see Section 4.1)

$$\Delta_2 = \int_a^b z\Psi(z)\, dx.$$

He expresses Ψ as

$$\Psi(z) = \mathfrak{A}_0 z - \frac{da_{11}z'}{dx}$$

and finds

$$u\Psi(z) - z\Psi(u) = -\frac{d\{a_{11}(uz' - zu')\}}{dx}$$

so that

$$\int \{u\Psi(z) - z\Psi(u)\}\, dx = -a_{11}(uz' - zu').$$

He now sets $u = a_1 r_1 + a_2 r_2$, $r_1 = \partial y/\partial\alpha_1$, and $r_2 = \partial y/\partial\alpha_2$ and discovers that

$$\Delta_2 = \int_a^b a_{11} \frac{(uz' - zu')^2}{u^2}\, dx.$$

[This is Legendre's relation (4.10) above.]

Hesse now takes up the Jacobi condition with the help of $u/a_2 = r_2 - mr_1$.

> If there is a value which the ratio r_2/r_1 does not assume between the limits a and b for x, then one can always assign the arbitrary constant m such a value that the expression
>
> $$\frac{u}{a_2} = r_1\left(\frac{r_2}{r_1} - m\right)$$

> never vanishes between the limits for x mentioned above .... If, on the other hand, that ratio assumes all values from $-\infty$ to $+\infty$ between the same limits for x, then that expression must necessarily vanish one time whatever value the arbitrary constant m takes on. For that reason we investigate the character of the ratio r_2/r_1.

He now notes that both r_1 and r_2 satisfy the Jacobi differential equation $\Psi(z)=0$ and thus $-a_{11}(r_1r_2'-r_2r_1')=C$, a constant, i.e.,

$$\frac{d(r_2/r_1)}{dx}=-\frac{C}{a_{11}r_1^2}\,.$$

The right-hand member of this equation does not change its sign since a_{11} does not, if the given extremal truly furnishes a maximum or minimum. It follows from this that r_2/r_1 is a monotone function which takes on all values between $-\infty$ and $+\infty$ "provided it takes on the same value for two different values of x. From this one perceives how necessary it is to restrict the limits of integration .... The limits of the integral must therefore only be extended to the point at which for the first time the equation

$$(r_2/r_1)_a=(r_2/r_1)_b$$

holds and this same point itself must surely be excluded." [He means, e.g., by $(r_2/r_1)_a$ the value of r_2/r_1 at $x=a$.]

Hesse now turns to the geometrical significance of the conjugate point. He calls his family of extremals $y=F(x,\alpha_1,\alpha_2)$ and asks that they all pass through the given point (a, y_a). He solves the equation $y_a=F(a,\alpha_1,\alpha_2)$ for α_2 as a function of α_1. Consider now the family $y=F[x,\alpha_1,\alpha_2(\alpha_1)]$ of curves through the first point and examine two of its curves defined by α_1 and $\alpha_1+\epsilon$. Then their intersection point satisfies the equations

$$y=F(x,\alpha_1,\alpha_2),\qquad y=F(x,\alpha_1,\alpha_2)+\left(\frac{\partial y}{\partial\alpha_1}+\frac{\partial y}{\partial\alpha_2}\,\frac{\partial\alpha_2}{\partial\alpha_1}\right)\epsilon,$$

for $|\epsilon|$ small. Hesse remarks that in general there is such an intersection point with $x\neq a$, which he calls (b, y_b). Then he has

$$\left(\frac{\partial y}{\partial\alpha_1}\right)_b+\left(\frac{\partial y}{\partial\alpha_2}\right)_b\frac{\partial\alpha_2}{\partial\alpha_1}=0$$

and also

$$\left(\frac{\partial y}{\partial\alpha_1}\right)_a+\left(\frac{\partial y}{\partial\alpha_2}\right)_a\frac{\partial\alpha_2}{\partial\alpha_1}=0,$$

where the subscripts a and b mean that the relevant expressions are evaluated for $x=a$ and b. Eliminating $\partial\alpha_2/\partial\alpha_1$, he finds

$$\frac{(r_2)_a}{(r_1)_a}=\frac{(r_2)_b}{(r_1)_b}\,,$$

which is his condition above. He discusses this and concludes that this second point is on the envelope of the family of extremals through the given first point.

He then proceeds to say:

> If the integral expression Δ shall be a maximum or minimum the second variation
>
> $$\Delta_2 = \int_a^b z\Psi(z)\,dx$$
>
> must keep the same sign for all functions z. It is however still permissible that it vanish for a designated function z. The vanishing of the second variation for a designated function z is nevertheless the limit of the extension of the limits for the integral Δ. For if one extends the limits of the integral further, then the principle of continuity offers at least the possibility that the second variation also will change its sign.

Hesse goes on to observe that the second variation vanishes along any solution of $\Psi(z) = 0$, and the most general such z is given by $z = a_1 r_1 + a_2 r_2$. He concludes by observing that if there is such a z, then

$$\left(\frac{r_2}{r_1}\right)_a = \left(\frac{r_2}{r_1}\right)_b.$$

5. Weierstrass

5.1. Weierstrass's Lectures

These lectures by Weierstrass on the calculus of variations were written up by a number of his students and were made available in the *Mathematische Verein* in Berlin and the *Mathematische Lesezimmer* in Göttingen. They were given by Weierstrass during the summer semesters of 1875, 1879, and 1882 and represent his contributions to the field.[1] The editor, Rothe, compiled his text principally from lecture notes prepared by Burckhardt based on the 1882 summer-semester lectures and from those by Schwarz. Since these lectures were not formally published, presumably they did not reach the entire mathematical community; as a result, much went on in our subject without reference to this monumental achievement of Weierstrass. In fact, Weierstrass's result became known in considerable measure through the dissertations of his students.

It is pointless to attempt to summarize the entire contents of Weierstrass's lectures. Instead I have decided to concentrate on his main contributions and to exhibit them. These include the famous Weierstrass–Erdmann corner condition and Weierstrass's necessary condition, his discovery of fields, his treatment of the four necessary conditions, and his treatment of the sufficient conditions, as well as his systematic treatment of the parametric problem. Also of importance was his insistence on rigor and clarity, which help to distinguish his work and that of his successors from their predecessors.

---

[1] K. Weierstrass, *VOR*. The first six chapters are concerned with the "Theory of Maxima and Minima of Functions of one and many Variables." According to Bolza, Weierstrass first started to lecture on the calculus of variations in 1865 and continued until 1890.

5.2. Formulation of the Parametric Problem

Throughout his lectures from at least 1872 on Weierstrass worked with a parametric formulation of the problem of the calculus of variations in the plane. He considered the integral

$$I = \int_{t_0}^{t_1} F(x, y, x', y')\, dt \tag{5.1}$$

where $x' = dx/dt$ and $y' = dy/dt$, and asked that it be a minimum (maximum) usually for fixed end-points. He assumes throughout that there is a region R of xy-space in which F is single-valued and regular—expansible in a convergent power series—for all $x', y' \neq 0, 0$.[2] He further needs the requirement on F of positive homogeneity, i.e., for all $\kappa > 0$

$$F(x, y, \kappa x', \kappa y') = \kappa F(x, y, x', y'). \tag{5.2}$$

To see how he uses this condition, let $\kappa = 1 + h$ and let $F(x, y, x', y')$ be written as F. Then by (5.2)

$$F(x, y, x' + hx', y' + hy') = F + hF$$

and by the regularity of F

$$F(x, y, x' + hx', y' + hy') = F + hx' \frac{\partial F}{\partial x'} + hy' \frac{\partial F}{\partial y'} + \cdots .$$

Equating coefficients of h, Weierstrass finds that

$$F = x' \frac{\partial F}{\partial x'} + y' \frac{\partial F}{\partial y'} . \tag{5.3}$$

It then follows by partial differentation that

$$x' \frac{\partial^2 F}{\partial x'^2} + y' \frac{\partial^2 F}{\partial x' \partial y'} = 0, \qquad x' \frac{\partial^2 F}{\partial x' \partial y'} + y' \frac{\partial^2 F}{\partial y'^2} = 0,$$

and consequently that

$$\frac{\partial^2 F}{\partial x'^2} : \frac{\partial^2 F}{\partial x' \partial y'} : \frac{\partial^2 F}{\partial y'^2} = y'^2 : -x'y' : x'^2.$$

There is thus a function $F_1(x, y, x', y')$ such that

$$\frac{\partial^2 F}{\partial x'^2} = F_1 y'^2, \qquad \frac{\partial^2 F}{\partial x' \partial y'} = -F_1 x'y', \qquad \frac{\partial^2 F}{\partial y'^2} = F_1 x'^2. \tag{5.4}$$

[2] One of Weierstrass's important contributions to mathematics was his insistence on sharpness of formulation, clarity, and rigor. In addition to his work in this direction, duBois–Raymond [1879] and Scheeffer [1886] contributed largely. Weierstrass emphasized this push toward rigor in his 1879 lectures. See also Zermelo's dissertation [1894].

For notational purposes, Weierstrass writes

$$G_1 = \frac{\partial F}{\partial x} - \frac{d}{dt}\frac{\partial F}{\partial x'} = 0, \qquad G_2 = \frac{\partial F}{\partial y} - \frac{d}{dt}\frac{\partial F}{\partial y'} = 0 \tag{5.5}$$

and

$$G = \frac{\partial^2 F}{\partial y\, \partial x'} - \frac{\partial^2 F}{\partial x\, \partial y'} - F_1\left(x'\frac{dy'}{dt} - y'\frac{dx'}{dt}\right) = 0. \tag{5.5'}$$

He then shows easily that the first variation δI is expressible as

$$\delta I = \int_{t_0}^{t_1} G\cdot(x'\eta - y'\xi)\, dt,$$

where ξ, η are the variations, and that $G_1 = -y'G$, $G_2 = x'G$.[3] He concludes from this that $G = 0$ is the first necessary condition.

In Chapter 11 Weierstrass works out his corner condition.[4] He states his result in this way: "*If the tangent direction of a given* [*minimizing*] *curve changes suddenly* [*discontinuously*] *in one or more places, then the quantities* $\partial F/\partial x'$ *and* $\partial F/\partial y'$ *do not undergo any break in their continuity, provided that there are only a finite number of such places in the interval* $(t_0 \ldots t_1)$."

To prove this, Weierstrass considers a value t' for which $t_0 < t' < t_1$ where the curve changes direction discontinuously. He further knows that there are values t'', t''' such that on the intervals $t'' < t < t'$, $t' < t < t'''$ there are no other discontinuities in the curve's tangent. Then he chooses variations $\xi = \kappa u$, $\eta = \kappa v$ with κ any constant and u and v arbitrary continuous [differentiable] functions of t which vanish at t'' and t'''. [They vanish outside the interval (t'', t''').] Then the first variation, which must vanish, has the form

$$\int_{t_0}^{t_1}\left\{\frac{\partial F}{\partial x}u + \frac{\partial F}{\partial y}v + \frac{\partial F}{\partial x'}\frac{du}{dt} + \frac{\partial F}{\partial y'}\frac{dv}{dt}\right\}dt.$$

After the usual integration by parts he has

$$\int_{t_0}^{t_1} G\cdot(x'v - y'u)\, dt + \left[\frac{\partial F}{\partial x'}u + \frac{\partial F}{\partial y'}v\right]_{t''}^{t'} + \left[\frac{\partial F}{\partial x'}u + \frac{\partial F}{\partial y'}v\right]_{t'}^{t'''} = 0.$$

Now $G = 0$ holds on $[t_0, t_1]$; and hence since u, v vanish at t'', t''',

$$u\left\{\left(\frac{\partial F}{\partial x'}\right)_{t'}^{-} - \left(\frac{\partial F}{\partial x'}\right)_{t'}^{+}\right\} + v\left\{\left(\frac{\partial F}{\partial y'}\right)_{t'}^{-} - \left(\frac{\partial F}{\partial y'}\right)_{t'}^{+}\right\} = 0.$$

It follows then that $\partial F/\partial x'$ and $\partial F/\partial y'$ are continuous at $t = t'$ since u and v are independent of each other. The symbols + and − are used to indicate right- and left-hand limits, respectively.

---

[3] Weierstrass, *VOR*, Chap. X. His analysis of the fundamental lemma is based on ideas of Heine and du Bois–Raymond, which we discuss in Chapter 6.

[4] See also Erdmann [1877′]. His work is independent of that of Weierstrass. Weierstrass worked out his condition during his summer semester lecture in 1865.

5.3. The Second Variation

In his 13th chapter Weierstrass begins his consideration of the second variation. (This was introduced into his lectures of 1872 or earlier.) He carries through the analysis of Legendre and Lagrange to show that the function F_1 of (5.4) satisfies the Legendre condition "that if the given integral is to be a minimum (or maximum), then the function F_1 necessarily must be never negative (or positive) inside the limits of integration." He proceeds in his next chapters to consider the Jacobi condition. To understand his analysis, let us first look at his form of the second variation. He starts by expressing this variation in the form

$$\begin{aligned}\delta^2 F = {} & \frac{\partial^2 F}{\partial x^2}\xi^2 + 2\frac{\partial^2 F}{\partial x\,\partial y}\xi\eta + \frac{\partial^2 F}{\partial y^2}\eta^2 \\ & + F_1\left\{ y'^2\left(\frac{\partial \xi}{dt}\right)^2 - 2x'y'\frac{d\xi}{dt}\frac{d\eta}{dt} + x'^2\left(\frac{d\eta}{dt}\right)^2 + 2y'\frac{dy'}{dt}\xi\frac{d\xi}{dt}\right. \\ & \left. + 2x'\frac{dx'}{dt}\eta\frac{d\eta}{dt} - 2x'\frac{dy'}{dt}\xi\frac{d\eta}{dt} - 2y'\frac{dx'}{dt}\eta\frac{d\xi}{dt}\right\} \\ & + 2\left\{ L\xi\frac{d\xi}{dt} + M\left(\xi\frac{d\eta}{dt} + \eta\frac{d\xi}{dt}\right) + N\eta\frac{d\eta}{dt}\right\},\end{aligned}$$

where

$$\begin{aligned}&\frac{\partial^2 F}{\partial x\,\partial x'} - y'\frac{dy'}{dt}F_1 = L,\\ &\frac{\partial^2 F}{\partial x\,\partial y'} + x'\frac{dy'}{dt}F_1 = \frac{\partial^2 F}{\partial y\,\partial x'} + y'\frac{dx'}{dt}F_1 = M,\\ &\frac{\partial^2 F}{\partial y\,\partial y'} - x\frac{dx'}{dt}F_1 = N.\end{aligned}$$

To simplify this expression further, he sets $x'\eta - x'\xi = w$ and

$$L\xi^2 + 2M\xi\eta + N\eta^2 = R.$$

This results in the new form

$$\delta^2 F = F_1\left(\frac{dw}{dt}\right)^2 + L_1\xi^2 + 2M_1\xi\eta + N_1\eta^2 + \frac{dR}{dt},$$

where

$$\begin{aligned}&\frac{\partial^2 F}{\partial x^2} - F_1\left(\frac{dy'}{dt}\right)^2 - \frac{dL}{dt} = L_1,\\ &\frac{\partial^2 F}{\partial x\,\partial y} + F_1\frac{dx'}{dt}\frac{dy'}{dt} - \frac{dM}{dt} = M_1,\\ &\frac{\partial^2 F}{\partial y^2} - F_1\left(\frac{dx'}{dt}\right)^2 - \frac{dN}{dt} = N_1.\end{aligned}$$

Weierstrass now remarks with the help of (5.3) that

$$\frac{\partial F}{\partial x} = \frac{\partial^2 F}{\partial x\,\partial x'}x' + \frac{\partial^2 F}{\partial x\,\partial y'}y', \qquad \frac{\partial F}{\partial y} = \frac{\partial^2 F}{\partial y\,\partial x'}x' + \frac{\partial^2 F}{\partial y\,\partial y'}y';$$

but since $G = 0$

$$\frac{\partial F}{\partial x} = \frac{\partial^2 F}{\partial x\,\partial x'}x' + \frac{\partial^2 F}{\partial y\,\partial x'}y' - x'y'F_1\frac{dy'}{dt} + y'^2F_1\frac{dx'}{dt},$$

$$\frac{\partial F}{\partial y} = \frac{\partial^2 F}{\partial y\,\partial y'}y' + \frac{\partial^2 F}{\partial x\,\partial y'}x' - x'y'F_1\frac{dx'}{dt} + x'^2F_1\frac{dy'}{dt}$$

and hence

$$\frac{\partial F}{\partial x} = Lx' + My', \qquad \frac{\partial F}{\partial y} = Mx' + Ny'.$$

He now differentiates these equations with respect to t and finds after some calculation that

$$L_1x' + M_1y' = 0, \qquad M_1x' + N_1y' = 0.$$

From these last two equations he deduces the existence of a function F_2 of x, y, x', y' such that

$$L_1 = y'^2F_2, \qquad M_1 = -x'y'F_2, \qquad N_1 = x'^2F_2. \tag{5.4'}$$

Moreover, $L_1\xi^2 + 2M_1\xi\eta + N_1\eta^2 = F_2w^2$, and so the second variation of F becomes

$$\delta^2F = F_1\left(\frac{dw}{dt}\right)^2 + F_2w^2 + \frac{dR}{dt}.$$

Thus Weierstrass has

$$\delta^2I = \int_{t_0}^{t_1}\left\{F_1\left(\frac{dw}{dt}\right)^2 + F_2w^2\right\}dt + \left[R\right]_{t_0}^{t_1}; \tag{5.6}$$

and since $R = L\xi^2 + 2M\xi\eta + N\eta^2$, the expression outside the integral sign vanishes at t_0 and t_1 for the fixed end-point case. Weierstrass notes that the total variation of the integral I can be represented as

$$\Delta I = \frac{1}{2!}\delta^2I + \frac{1}{3!}\delta^3I + \cdots,$$

when I is evaluated along an arc for which $\delta I = 0$. He remarks "*that in the case of a minimum for the integral I the second variation δ^2I must always possess a positive value, in the case of a maximum a negative value provided that it does not vanish.*" From this he concludes that

$$\int_{t_0}^{t_1}\left\{F_1\left(\frac{dw}{dt}\right)^2 + F_2w^2\right\}dt$$

must be always positive or always negative for all curves satisfying $G = 0$.

In his 14th chapter Weierstrass continues his analysis of the second variation. He divides the interval $(t_0 \ldots t_1)$ into pieces of length τ and approximates the integral above by the sum

$$\sum_{\nu}\left\{F_1^{(\nu)}\left(\frac{w_{\nu+1}-w_\nu}{\tau}\right)^2+F_2^{(\nu)}w_\nu^2\right\}\tau,$$

in which $w_\nu, F_1^{(\nu)}, F_2^{(\nu)}$ are the values of w, F_1, F_2 at $t = t_0 + \nu\tau$. This sum is a quadratic form in the w_ν and can therefore be expressed "as a sum of squares" in the form

$$\sum_{\nu} H^{(\nu)}v_\nu^2,$$

where the v_ν are linear functions of the w_ν, and all the $H^{(\nu)}$ possess the same sign; in the case of a minimum, they are all positive.

"This deliberation Lagrange carried out for the purpose of bringing the second variation into the form

$$\delta^2 I=\int_{t_0}^{t_1} Hv^2\,dt,$$

and he completed this transformation somewhat in the following manner."

Let v be any given function of t, and let the integral

$$\int_{t_0}^{t_1}\frac{d}{dt}(vw^2)\,dt=\left[vw^2\right]_{t_0}^{t_1}$$

be added to and subtracted from

$$\int_{t_0}^{t_1}\left\{F_1\left(\frac{dw}{dt}\right)^2+F_2w^2\right\}dt+\left[R\right]_{t_0}^{t_1}.$$

This gives Weierstrass the form

$$\delta^2 I=\int_{t_0}^{t_1}\left\{F_1\left(\frac{dw}{dt}\right)^2+F_2w^2+\frac{d}{dt}(vw^2)\right\}dt+\left[R-vw^2\right]_{t_0}^{t_1},$$

and the expression under the integral sign may be written as

$$F_1\left(\frac{dw}{dt}\right)^2+2vw\frac{dw}{dt}+\left(F_2+\frac{dv}{dt}\right)w^2.$$

For this to be a perfect square, v must satisfy the equation

$$v^2-F_1\left(F_2+\frac{dv}{dt}\right)=0.$$

For such a v, the second variation becomes

$$\delta^2 I=\int_{t_0}^{t_1}F_1\left\{\frac{dw}{dt}+\frac{vw}{F_1}\right\}^2dt+\left[R-vw^2\right]_{t_0}^{t_1},$$

and for fixed end-points the expression outside the integral sign vanishes. (In what Weierstrass has done it is evident that he has tacitly assumed $F_1 \neq 0$.)

Weierstrass now wishes to discuss the character of Legendre's differential equation above to see whether its solution v is everywhere finite and continuous. To do this, he replaces v with $u^* = vu$, where $u \neq 0$ in $(t_0 \ldots t_1)$. Then the equation for v is transformed into

$$u^*\left(u^* + F_1 \frac{du}{dt}\right) - uF_1\left(F_2 u + \frac{du^*}{dt}\right) = 0.$$

He now requires u and u^* to satisfy the pair of equations

$$u^* + F_1 \frac{du}{dt} = 0, \qquad F_2 u + \frac{du^*}{dt} = 0,$$

or equivalently

$$F_2 u - \frac{d}{dt}\left(F_1 \frac{du}{dt}\right) = 0. \tag{5.7}$$

(Recall that this is Jacobi's differential equation.)

Notice that

$$v = \frac{u^*}{u} = -\frac{F_1}{u}\frac{du}{dt},$$

and consequently that the second variation is expressible as

$$\delta^2 I = \int_{t_0}^{t_1} F_1 \left\{ \frac{dw}{dt} - \frac{w}{u}\frac{du}{dt} \right\}^2 dt.$$

This form has meaning only when (5.7) has a solution $u \neq 0$. Weierstrass, however, reasons nonetheless (see Section 5.4 below) that "*If the exhibited integral is to be a minimum* (*or maximum*), *the function* F_1 *necessarily must be never negative* (*or positive*) *between the limits of integration.*" His proof is essentially the one given in Section 3.5 above.

In his 15th chapter Weierstrass turns to Jacobi's investigations. He again notes that $G_1 = -y'G$, $G_2 = +x'G$ and reasons that

$$G_1 + \Delta G_1 = -\left(y' + \frac{d\eta}{dt}\right)(G + \Delta G),$$

$$G_2 + \Delta G_2 = \left(x' + \frac{d\xi}{dt}\right)(G + \Delta G),$$

when he varies x, y by ξ, η. From these it follows that

$$\delta G_1 = -y'\delta G - \frac{d\eta}{dt}G, \qquad \delta G_2 = x'\delta G + \frac{d\xi}{dt}G,$$

which imply that

$$\left(x'\frac{d\eta}{dt} - y'\frac{d\xi}{dt}\right)\delta G = \frac{d\xi}{dt}\delta G_1 + \frac{d\eta}{dt}\delta G_2.$$

Weierstrass then makes use of the definitions (5.5), (5.5′) of G_1, G_2 and G to calculate δG_1, δG_2, and δG. With the help of the definition $w = x'\eta - y'\xi$ and of (5.4) of F_1, he is able to show after considerable calculation that

$$\delta G_1 = -y'\left(F_2 w - \frac{d}{dt}\left(F_1 \frac{dw}{dt}\right)\right), \qquad \delta G_2 = x'\left(F_2 w - \frac{d}{dt}\left(F_1 \frac{dw}{dt}\right)\right)$$

and thus that

$$\delta G = F_2 w - \frac{d}{dt}\left(F_1 \frac{dw}{dt}\right)$$

provided that $x'\eta' - y'\xi' \neq 0$.

Weierstrass now wishes to show that "If the general solution of the differential equation $G = 0$ is known, the general solution of the differential equation . . . [(5.7)] is thereby also completely determined, and certainly only through differentiations." He does this with the help of a two-parameter family of solutions of $G = 0$ without developing any new results.

5.4. Conjugate Points

In his 16th chapter Weierstrass is concerned with an extremal arc along which $F_1 \neq 0, \neq \infty$. Consider with him two linearly independent integrals $\theta_1(t)$, $\theta_2(t)$ of the Jacobi equation (5.7) above, whose general solution is $c_1\theta_1(t) + c_2\theta_2(t)$. Then for a given value t' of t, the function

$$\Theta(t, t') = \theta_1(t)\theta_2(t') - \theta_1(t')\theta_2(t) \tag{5.8}$$

is a solution of (5.7) vanishing at $t = t'$. Consider now the zeros of Θ greater than t'. Weierstrass says "the function $\Theta(t, t')$ vanishes for no second value of t or it can be the case that $\Theta(t, t')$ has another zero in addition to t'. Then let t'' be the next zero after $t', \ldots, t'' > t'$. Two such consecutive zeros of $\Theta(t, t')$ determine points on the curve which will be said to be *conjugate* to each other."[5] With the help of this definition, he proves Jacobi's results on conjugate points: "*If inside the integration interval* $(t_0 \ldots t_1)$ *there are no two points conjugate to each other, then it is always possible to find a solution of the differential equation* . . . [5.7] *which never vanishes inside the given interval.*

If, on the contrary, two conjugate points are contained inside the integration interval or if the integration extends beyond a point conjugate to the initial point t_0, *then the integral under consideration cannot be a maximum or a minimum.*"

Weierstrass wishes to show that as t' moves along an extremal curve—a solution of $G = 0$—the point t'' conjugate to t' moves in the same direction. To do this, he shows that if $\Theta(t'', t') = 0$ and if $\Theta(t, t')$ has no zeros inside $(t' \ldots t'')$, then $\Theta(t, t' + \tau)$, with $\tau > 0$ and small, has no zeros inside $(t' + \tau \ldots t'')$. The proof depends on simple properties of the linear differential equation (5.7). Let it have two linearly independent solutions u_1, u_2

[5] Weierstrass, *VOR*, Chapter 16, p. 154.

and eliminate F_2 between the two resulting equations; then

$$0 = u_2 \frac{d}{dt}\left(F_1 \frac{du_1}{dt}\right) - u_1 \frac{d}{dt}\left(F_1 \frac{du_2}{dt}\right) = \frac{d}{dt}\left(F_1\left(u_2 \frac{du_1}{dt} - u_1 \frac{du_2}{dt}\right)\right)$$

or

$$F_1 p = C \tag{5.9}$$

where $p = (u_2 du_1/dt) - (u_1 du_2/dt)$. If now $u_1 \neq 0$, this implies that

$$\frac{d}{dt}\left(\frac{u_2}{u_1}\right) = -\frac{C}{F_1 u_1^2}$$

and consequently

$$u_2 = -Cu_1 \int_{t_0}^{t} \frac{dt}{F_1 u_1^2}.$$

Now recall that $\Theta(t, t')$ is a solution of (5.7) containing a parameter t'. Choose $u_1 = \Theta(t, t')$, $u_2 = \Theta(t, t' + \tau)$ and note that when t_0 is taken to be $t' + \tau$, the relation above for u_2 becomes

$$\Theta(t, t' + \tau) = -C\Theta(t, t') \int_{t'+\tau}^{t} \frac{dt}{F_1 \Theta(t, t')^2}.$$

By hypothesis, Weierstrass has $\Theta(t, t') \neq 0$ inside the interval $(t' \ldots t'')$ and $F_1 \neq 0$ on that interval; hence $\Theta(t, t' + \tau)$ cannot vanish for t between $t' + \tau$ and t''. Next he proceeds to show that $\Theta(t'', t' + \tau) \neq 0$. To do this, he notes that $\Theta(t'', t') = \theta_1(t'')\theta_2(t') - \theta_1(t')\theta_2(t'') = 0$; if now

$$\Theta(t'', t' + \tau) = \theta_1(t'')\theta_2(t' + \tau) - \theta_1(t' + \tau)\theta_2(t'')$$

were 0, then

$$\Theta(t', t' + \tau) = \theta_1(t')\theta_2(t' + \tau) - \theta_1(t' + \tau)\theta_2(t')$$

would also be 0, which is a contradiction since $t' + \tau$ lies between t' and t'', which are a pair of conjugate points; i.e., t'' is the first point after t' conjugate to t'. Thus, if $\Theta(t, t' + \tau)$ vanishes for any $t > t' + \tau$, the value t must be greater than t''. (It is worth noting that the constant $C \neq 0$. If the constant C in the relation above relating u_1 and u_2 were zero, then p would also be 0, and u_2 would be a constant times u_1, which is a contradiction of the fact that u_1, u_2 are linearly independent.)

Weierstrass goes on to show that for $\tau > 0$ sufficiently small there is a point greater than t'', which is conjugate to $t' + \tau$. Since Θ satisfies the Jacobi differential equation (5.7), it cannot vanish along with its derivative with respect to t at any point t without vanishing identically. Thus it must change sign in a neighborhood of each zero. This being so, let τ' be such that $\Theta(t'' + \tau', t' + \tau) = 0$, and let $\Theta_1(t, t'), \Theta_2(t, t')$ be the partial derivatives of Θ with respect to t and t'. Then by a series expansion in τ, τ' of the expression $\Theta(t'' + \tau', t' + \tau)$, Weierstrass reasons that

$$\Theta_1(t'', t')\tau' + \Theta_2(t'', t')\tau + [\tau, \tau']^2 = 0, \tag{5.10}$$

where Θ_1, Θ_2 are the partial derivatives of Θ with respect to t and t' and the third term in this relation is at least quadratic in τ and τ'. It is not difficult to show that $\Theta_1(t'', t')$ and $\Theta_2(t'', t')$ are not both zero; in fact, since $\Theta(t'', t') = -\Theta(t', t'')$, it is evident that $\Theta_1(t'', t') = -\Theta_2(t', t'')$. The quantity τ' can then be expanded for sufficiently small τ into a power series of the form

$$\tau' = a\tau + \tau^2 \mathfrak{P}(\tau),$$

where a is a nonzero coefficient. When this is substituted into equation (5.10) above, it can be seen with the help of (5.9) with $u_1 = \theta_1$, $u_2 = \theta_2$ and the fact that t'' is conjugate to t' that

$$a = -\frac{\Theta_2(t'', t')}{\Theta_1(t'', t')} = \left[\frac{\theta_1(t'')}{\theta_1(t')}\right]^2 \frac{F_1(t'')}{F_1(t')},$$

provided the terms of order 2 in τ are neglected. It is then clear that a is positive since F_1 does not change sign on the interval $(t_0 \ldots t_1)$. Thus τ' has the same sign as τ for small τ.

With the help of these facts, the first of Weierstrass's theorems above on conjugate points now follows easily. Suppose t_2 to be conjugate to t_0; then by hypothesis, it must $> t_1$. Now choose t' so near to t_0 that if t'' is conjugate to t', it is very near to t_2 and hence $> t_1$. The points then are arranged in this way t', t_0, t_1, t'', t_2. It follows that "*For suitable choice of the quantity t' the quantity $u[= \Theta(t, t')]$ neither vanishes inside the interval $(t_0 \ldots t_1)$ nor at the boundaries t_0 and t_1.*"

As a consequence, the second variation can be expressed in the form[6]

$$\delta^2 I = \int_{t_0}^{t_1} F_1 \left\{ \frac{dw}{dt} - \frac{w}{u}\frac{du}{dt} \right\}^2 dt.$$

In his 17th chapter Weierstrass continues his analysis of the Jacobi criterion. He starts with the form of the second variation

$$\delta^2 I = \int_{t_0}^{t_1} \left\{ F_1 \left(\frac{dw}{dt} \right)^2 + F_2 w^2 \right\} dt$$

and by integration by parts reduces this expression to the new form

$$\delta^2 I = \int_{t_0}^{t_1} \left\{ -\frac{d}{dt}\left(F_1 \frac{dw}{dt} \right) + F_2 w \right\} w \, dt$$

for the case of fixed end-points—w vanishes at t_0 and t_1. If w is taken to be $\Theta(t, t_0)$ and if Θ vanishes at t_1, but for no value in between t_0 and t_1, then it is easy to see that the second variation is zero for this choice of w. From this we begin to see the connection of conjugate points to the sign of the second variation, which was asserted by Jacobi.

Let $\varepsilon \neq 0$ be a constant having the same sign as F_1, and write the second

[6] Weierstrass, *VOR*, p. 161.

variation in the form

$$\delta^2 I = \int_{t_0}^{t_1} \left\{ (F_1 + \varepsilon)\left(\frac{dw}{dt}\right)^2 + (F_2 + \varepsilon)w^2 \right\} dt - \varepsilon \int_{t_0}^{t_1} \left(\left(\frac{dw}{dt}\right)^2 + w^2 \right) dt.$$

Weierstrass now chooses w so that the first integral in this relation will vanish. To do this, he applies the usual integration by parts technique and has

$$\delta^2 I = \int_{t_0}^{t_1} \left\{ -\frac{d}{dt}\left((F_1 + \varepsilon)\frac{dw}{dt}\right) + (F_2 + \varepsilon)w \right\} w\, dt$$
$$+ \left[w(F_1 + \varepsilon)\frac{dw}{dt} \right]_{t_0}^{t_1} - \varepsilon \int_{t_0}^{t_1} \left(\left(\frac{dw}{dt}\right)^2 + w^2 \right) dt. \tag{5.11}$$

To make the first integral in this expression vanish, he wishes to choose for w a solution of the differential equation

$$(F_2 + \varepsilon)u - \frac{d}{dt}\left((F_1 + \varepsilon)\frac{du}{dt}\right) = 0, \tag{5.7'}$$

which vanishes for $t = t_0$ and $t = t_1$. [Clearly, for $\varepsilon = 0$ this reduces to (5.7).]

If t'' is conjugate to t', then (5.7) has, by definition, a solution $\Theta(t, t')$ which vanishes at t' and t'' and is continuous. Weierstrass now shows by a continuity argument for linear differential equations that there is a solution $u = \Theta(t, t', \varepsilon)$ of (5.7′) with a root t''' near to t'' for $|\varepsilon|$ sufficiently small and ε of the same sign as F_1. If there is a point $t_0' < t_1$ conjugate to t_0, then for such ε there is a point $t_2 < t_1$ for which $w = \Theta(t, t_0, \varepsilon)$ vanishes at $t = t_0$ and t_2. When this w is introduced into (5.11) with $w = 0$ on $[t_2, t_1]$, there results for the second variation the expression

$$\delta^2 I = -\varepsilon \int_{t_0}^{t_2} \left(\left(\frac{d\Theta(t, t_0, \varepsilon)}{dt} \right)^2 + \Theta(t, t_0, \varepsilon)^2 \right) dt.$$

For this variation w it is evident that $\delta^2 I$ and thereby also the complete variation of I have the opposite sign to F_1. In this connection Weierstrass proves in his 18th chapter that $\delta^2 I$ has the same sign as the complete variation

$$\Delta I = \int_{t_0}^{t_1} \left(F\left(x + \xi, y + \eta, x' + \frac{d\xi}{dt}, y' + \frac{d\eta}{dt} \right) - F(x, y, x', y') \right) dt.$$

This shows that the integral cannot be either a maximum or a minimum if there is a point conjugate to t_0 inside the interval $(t_0 \dots t_1)$. If t_1 is conjugate to t_0, then these considerations are not applicable and the third variation needs to be considered. (In this case, $\delta^2 I = 0$, Erdmann [1877] calculated that $\delta^3 I \neq 0$ except in two special situations. He made this analysis for the nonparametric problem. These special situations occur when the envelope of the extremals through the initial point has a cusp at the point being considered or when it degenerates to a point.)

In the course of this analysis Weierstrass has, in effect, shown that if the Jacobi differential equation (5.7) has no solution u vanishing at any point of the closed interval from t_0 to t_1, then for ε sufficiently small the same is true of the perturbed equation (5.7′). This point is needed in what follows.

5.5. Necessary Conditions and Sufficient Conditions

In his 18th chapter Weierstrass precisely formulates the first three necessary conditions and then states and proves the first completely correct sufficiency theorem (cf. p. 177):[7]

> **First Condition**: *The coordinates x, y of an arbitrary point on the curve along which the integral is evaluated must, when viewed as functions of the integration variable t, satisfy the differential equation $G = 0$.*
>
> **Second Condition**: *Along such a curve the function F_1 must not be positive for a maximum, nor not negative for a minimum. The case when F_1 vanishes at some points or along the curve must, in view of this, be reserved for special consideration.*
>
> **Third Condition**: *If a maximum or a minimum is being considered, the integration interval must reach from its initial point at most to its conjugate point, but surely not beyond this.*

Prior to formulating his sufficiency theorem, Weierstrass has assumed (p. 175) that

(1) the three necessary conditions are satisfied along a given arc;

(2) F_1 does not vanish nor become infinite on the entire interval of integration; and

(3) there is no pair of conjugate points on the closed interval $[t_0, t_1]$.

Given these things, here is his first sufficiency theorem

> *Let the variations of the curve be restricted to the case where the distances between corresponding points on the given curve and the comparison curve are*

---

[7] The first published sufficiency proof was by Scheeffer [1886]. This is a fundamental paper and did much to clarify the basic ideas of our subject. See Section 5.11 below for a discussion of Scheeffer's methods. See also Bolza, *VOR*, pp. 92ff. In more recent times the fundamental necessary conditions are indicated as I, II, III, and IV or as the conditions of Euler, Weierstrass, Legendre, and Jacobi, respectively. The conditions II′ and III′, IV′, or the strengthened conditions of Weierstrass, Legendre, and Jacobi, are the conditions II, and III, and IV with the further restrictions that the Weierstrass $\mathcal{E}$-function $\neq 0$, that $F_1 \neq 0$, and that there is no point conjugate to the initial point on the closed interval. Condition II, the Weierstrass condition, is discussed in Section 5.7.

arbitrarily small and where the directions of the tangents to the two curves in these points deviate only a little from each other. Then the three conditions . . . [above]. . . are not only necessary but also sufficient for the exhibited integral to have a maximum provided the function F_1 is negative and, on the contrary, a minimum provided that F_1 is positive.

Weierstrass's proof is of real interest and depends on an analysis of the complete variation of the integral when evaluated along a curve satisfying $G = 0$, i.e., when $\delta I = 0$[8]:

$$\Delta I = \int_{t_0}^{t_1}\left(F\left(x+\xi, y+\eta, x'+\frac{d\xi}{dt}, y'+\frac{d\eta}{dt}\right) - F(x,y,x',y')\right)dt.$$

Since $\delta I = 0$, he can write this complete variation by means of Taylor's theorem in the form

$$\Delta I = \int_{t_0}^{t_1}\int_0^1 (1-\varepsilon)\Big(\sum_{\alpha,\beta} F_{\alpha\beta}(x+\varepsilon\xi_1, y+\varepsilon\xi_2, x'+\varepsilon\xi_3, y'+\varepsilon\xi_4)\Big)\xi_\alpha\xi_\beta \, d\varepsilon \, dt,$$

where α, β run over the integers 1, 2, 3, 4,

$$\xi_1 = \xi, \qquad \xi_2 = \eta, \qquad \xi_3 = \frac{d\xi}{dt}, \qquad \xi_4 = \frac{d\eta}{dt};$$

and the $F_{\alpha\beta}$ are the second partial derivatives of F with respect to ξ_α and ξ_β. He then expands the $F_{\alpha\beta}$ in powers of ε and notices that the first term in the right-hand side of the resulting expansion of ΔI is expressible in terms of the second variation of I. Thus

$$\Delta I = \frac{1}{2}\delta^2 I + \frac{1}{2}\int_{t_0}^{t_1}\sum_{\alpha\beta}\varepsilon_{\alpha\beta}\xi_\alpha\xi_\beta \, dt \qquad (\alpha, \beta = 1,2,3,4),$$

where the coefficients $\varepsilon_{\alpha\beta}$ are themselves functions of the ξ_α, which can be made arbitrarily small, uniformly in t. By this means Weierstrass has brought the problem of deciding on the sign of ΔI to that of deciding on the sign of the second variation $\delta^2 I$.

He chooses for ξ, η the values

$$\xi = -\frac{y'w}{x'^2 + y'^2}, \qquad \eta = \frac{x'w}{x'^2 + y'^2},$$

where w is an arbitrarily differentiable function of t such that w and dw/dt can be chosen as small as one desires on the interval from t_0 to t_1. (The quantities ξ and η are then continuous along with their derivatives ξ', η'.) Under the hypothesis that the original problem is one of fixed end-points, the second variation is, as was shown above, expressible as

$$\delta^2 I = \int_{t_0}^{t_1}\left(F_1\left(\frac{dw}{dt}\right)^2 + F_2 w^2\right)dt;$$

[8] As we shall soon see, Weierstrass later developed a more elegant way to handle sufficiency proofs, which was further refined by his successors.

and the quadratic form in the ξ_α appearing in ΔI goes over into such a form in w and dw/dt

$$fw^2 + 2gw\frac{dw}{dt} + h\left(\frac{dw}{dt}\right)^2,$$

where f, g, h are functions of w and dw/dt. Weierstrass now transforms this form by a rotation of axes into $f_1u_1^2 + f_2u_2^2$ with

$$u_1^2 + u_2^2 = w^2 + \left(\frac{dw}{dt}\right)^2;$$

here the functions f_1, f_2 have the same property as the $\varepsilon_{\alpha\beta}$; i.e., as functions of w they can be made small, uniformly in t. (Notice that f_1, f_2 are the roots of the quadratic equation $s^2 - (f+h)s + (fh - g^2) = 0$.)

He now chooses l as a value between f_1 and f_2 so that

$$f_1u_1^2 + f_2u_2^2 = l(u_1^2 + u_2^2) = l\left(w^2 + \left(\frac{dw}{dt}\right)^2\right)$$

and then can write

$$\begin{aligned}\Delta I &= \frac{1}{2}\int_{t_0}^{t_1}\left((F_1 + l)\left(\frac{dw}{dt}\right)^2 + (F_2 + l)w^2\right)dt,\\ &= \frac{1}{2}\int_{t_0}^{t_1}\left((F_1 - k)\left(\frac{dw}{dt}\right)^2 + (F_2 - k)w^2\right)dt\\ &\quad + \frac{1}{2}\int_{t_0}^{t_1}(l + k)\left(\left(\frac{dw}{dt}\right)^2 + w^2\right)dt,\end{aligned}$$

where k is a constant yet to be fixed. To do this, Weierstrass chooses k small and with the same sign as F_1 and chooses $w, dw/dt$ so that $l + k$ also has the same sign. This latter is always possible for sufficiently small values of w and dw/dt. He now knows that the second integral in the relation above has the same sign as F_1, and he needs to show that the first one does also. To do this, he must be sure that if the Jacobi differential equation (5.7) above has no continuous solution u vanishing at any point on the closed interval $(t_0 \ldots t_1)$, then the perturbed equation

$$\frac{d}{dt}\left((F_1 - k)\frac{d\bar{u}}{dt}\right) - (F_2 - k)\bar{u} = 0$$

has the same property. This follows from an examination of equation (5.7′) with $k = -\varepsilon$ and the remarks at the close of Section 5.4.

Knowing this, Weierstrass can then transform the first integral in the expression for ΔI above into

$$\int_{t_0}^{t_1}(F_1 - k)\left(\frac{dw}{dt} - \frac{w}{\bar{u}}\frac{d\bar{u}}{dt}\right)^2 dt$$

"and one sees that it has the same sign as $F_1 - k$, and therefore also as F_1. This is however what was to be proved."

He sums up in the theorem he first gave in his 1879 lectures:

> *Let the variations of the curve be restricted to the case where the distances between corresponding points on the original and on the varied curves are made arbitrarily small and also where the directions of the tangents to the curves at these points can only deviate arbitrarily little from each other. Then the three conditions set forth at the beginning of this chapter are not only necessary but also sufficient for the exhibited integral to be a maximum provided that the function F_1 is negative, and on the other hand a minimum provided F_1 is positive.*

In his 19th chapter Weierstrass establishes some function-theoretical theorems he needs. We will mention them only briefly in the next section since they are now so well understood. These theorems are essentially existence and embedding theorems.

5.6. Geometrical Considerations of Conjugate Points

In his 20th chapter Weierstrass begins to seek a geometrical significance to the notion of conjugate points; to do so, he examines a two-parameter family

$$x = \phi(t;\alpha,\beta), \qquad y = \psi(t;\alpha,\beta) \tag{5.12}$$

of curves satisfying $G = 0$. He proves that given one solution of $G = 0$ in the family, he can always find another passing through a given point A on the first and making an arbitrarily small angle with it at that point. To make this proof, Weierstrass first establishes an implicit-function theorem in Chapter 19. Let there be given a system of equations

$$y_\lambda = \sum_{\mu=1}^{n} (a_{\lambda\mu} + X_{\lambda\mu})x_\mu \qquad (\lambda = 1, 2, \ldots, n) \tag{5.13}$$

with $0 \neq A = \det a_{\lambda\mu}$ and with the $X_{\lambda\mu}$ continuous functions of x_1, $x_2, \ldots, x_n$, which vanish along with their arguments and which have finite-valued partial derivatives. It is now always possible to find constants $g_1, g_2, \ldots, g_n$ and $h_1, h_2, \ldots, h_n$ so that for $|y_\mu| < h_\mu$ there is one and only one set x_μ satisfying system (5.13) with $|x_\mu| < g_\mu$. This solution, moreover, is of the form

$$x_\lambda = \sum_{\mu=1}^{n} \left(\frac{A_{\lambda\mu}}{A} + Y_{\lambda\mu} \right) y_\mu \qquad (\lambda = 1, 2, \ldots, n),$$

where the $Y_{\lambda\mu}$ are continuous functions of $y_1, y_2, \ldots, y_n$, which tend to zero with their arguments.

Weierstrass now uses this result in considering the curve

$$\bar{x} = \phi(t+\tau;\alpha+\alpha',\beta+\beta'), \qquad \bar{y} = \psi(t+\tau;\alpha+\alpha',\beta+\beta'). \tag{5.12'}$$

He seeks to determine α', β', τ so that this new curve goes through the point A given by $t = t_0$ in equations (5.12) and makes with the given curve defined by α, β as small an angle as desired. If $\bar{x} = x + \xi, \bar{y} = y + \eta$, then he makes certain assumptions: first, that if $t = t_0$, then $\xi = 0$, $\eta = 0$; and second, that if

$$\frac{dx}{dt} = x' = \rho \cos \lambda, \qquad \frac{dy}{dt} = y' = \rho \sin \lambda,$$

$$\frac{d\bar{x}}{dt} = \bar{\rho} \cos \bar{\lambda}, \qquad \frac{d\bar{y}}{dt} = \bar{\rho} \sin \bar{\lambda},$$

then the determinant

$$x' \frac{d\bar{y}}{dt} - y' \frac{d\bar{x}}{dt} = x' \frac{d\eta}{dt} - y' \frac{d\xi}{dt} = \rho\bar{\rho} \sin(\lambda - \bar{\lambda}) = \kappa \tag{5.14}$$

must vanish with the angle $\bar{\lambda} - \lambda$.

Weierstrass expands the functions ξ, η, κ as power series in α', β', τ through first-order terms and then sets $t = t_0$. His conditions above then become

$$\begin{aligned}
0 &= (\phi'(t_0) + r_1)\tau + (\phi_1(t_0) + p_1)\alpha' + (\phi_2(t_0) + q_1)\beta', \\
0 &= (\psi'(t_0) + r_2)\tau + (\psi_1(t_0) + p_2)\alpha' + (\psi_2(t_0) + q_2)\beta', \\
\kappa &= (\phi'(t_0)\psi''(t_0) - \phi''(t_0)\psi'(t_0) + r)\tau \\
&\quad + (\phi'(t_0)\psi_1'(t_0) - \phi_1'(t_0)\psi'(t_0) + p)\alpha' \\
&\quad + (\phi'(t_0)\psi_2'(t_0) - \phi_2'(t_0)\psi'(t_0) + q)\beta',
\end{aligned} \tag{5.15}$$

where $\phi' = d\phi/dt$, $\psi' = d\psi/dt$, $\phi_1 = \partial\phi/\partial\alpha$, $\psi_1 = \partial\psi/\partial\alpha$, $\phi_2 = \partial\phi/\partial\beta$, and $\psi_2 = \partial\psi/\partial\beta$ and where $p, q, r, p_1, q_1, r_1, p_2, q_2, r_2$ are quantities that vanish with α', β', τ.

By Weierstrass's implicit-function theorem, to solve this system for α', β', τ, it suffices to examine the case when these nine last quantities are set to zero. Before doing this, he first seeks to simplify the relation for κ in (5.15) above. In the expression (5.14)

$$\frac{d}{dt}(\eta x' - \xi y') = \kappa + \eta \frac{dx'}{dt} - \xi \frac{dy'}{dt} = \kappa + \eta\phi''(t) - \xi\psi''(t), \tag{5.16}$$

he replaces ξ, η by their expansions in powers of τ, α', β' and introduces the functions

$$\begin{aligned}
\theta_0(t) &= \phi_2(t)\psi_1(t) - \psi_2(t)\phi_1(t), \\
\theta_1(t) &= \phi'(t)\psi_1(t) - \psi'(t)\phi_1(t), \\
\theta_2(t) &= \phi'(t)\psi_2(t) - \psi'(t)\phi_2(t).
\end{aligned} \tag{5.17}$$

Then the expression $\eta x' - \xi y'$ can be expanded into

$$\eta x' - \xi y' = s\tau + (\theta_1(t) + s_1)\alpha' + (\theta_2(t) + s_2)\beta',$$

where s, s_1, s_2 are quantities that vanish with τ, α', β'. Weierstrass now differentiates both members of this relation with respect to t at $t = t_0$ and has, with the help of (5.16),

$$\kappa = \sigma\tau + (\theta_1'(t) + \sigma_1)\alpha' + (\theta_2'(t) + \sigma_2)\beta',$$

where $\sigma, \sigma_1, \sigma_2$ are quantities that vanish with τ, α', β' since $\xi = \eta = 0$ at $t = t_0$.

Using this relation, Weierstrass returns to his system (5.15) above with $p, q, r, p_1, q_1, r_1, p_2, q_2, r_2$ set to zero and replaces the third equation there with the equivalent one

$$\kappa = \theta_1'(t_0)\alpha' + \theta_2(t_0)\beta'$$

obtained from the one above by setting $\sigma, \sigma_1, \sigma_2$ to zero. The new system has the determinant

$$D(t_0) = \begin{vmatrix} \phi'(t_0) & \phi_1(t_0) & \phi_2(t_0) \\ \psi'(t_0) & \psi_1(t_0) & \psi_2(t_0) \\ 0 & \theta_1'(t_0) & \theta_2'(t_0) \end{vmatrix} = \theta_1(t_0)\theta_2'(t_0) - \theta_1'(t_0)\theta_2(t_0).$$

We proved earlier with the help of equation (5.9) that this expression is not null, since it is equal to $-C/F_1$. This shows that the simplified system has the solutions

$$\tau = -\frac{\theta_0(t_0)}{D(t_0)}\kappa, \qquad \alpha' = -\frac{\theta_2(t_0)}{D(t_0)}\kappa, \qquad \beta' = \frac{\theta_1(t_0)}{D(t_0)}\kappa,$$

and hence the original system (5.15) has, for $|\kappa|$ sufficiently small, the solutions

$$\tau = \left(-\frac{\theta_0(t_0)}{D(t_0)} + k\right)\kappa, \qquad \alpha' = \left(-\frac{\theta_2(t_0)}{D(t_0)} + k_1\right)\kappa,$$

$$\beta' = \left(\frac{\theta_1(t_0)}{D(t_0)} + k_2\right)\kappa, \tag{5.18}$$

where k, k_1, k_2 are quantities vanishing with κ. Weierstrass then summarizes as follows: "*To any given curve satisfying the differential equation $G = 0$ there always corresponds a second one which intersects the first in a given point A at a given, sufficiently small angle*; *and that the second curve extends as far as the first*."

Weierstrass next asks whether the second curve (5.12′) has any other point in common with the first one, (5.12). To see this he considers the distance between them measured along the normal to the original curve. If X, Y are the coordinates of a point on the normal, the distance from the foot of the normal to the original curve is given by

$$\frac{x'(Y - y) - y'(X - x)}{\sqrt{x'^2 + y'^2}}.$$

(If there are no singular points on the first curve, the denominator is not zero.) In terms of the curve (5.12′) the numerator is expressible through terms of the first order as

$$\phi'(t)(\psi'(t)\tau + \psi_1(t)\alpha' + \psi_2(t)\beta') - \psi'(t)(\phi'(t)\tau + \phi_1(t)\alpha' + \phi_2(t)\beta')$$
$$= \theta_1(t)\alpha' + \theta_2(t)\,\beta',$$

where θ_1, θ_2 are given by definitions (5.17). In terms of the τ, α', β' of (5.18) and Θ of (5.8), this becomes

$$\left(\frac{-\theta_1(t)\theta_2(t_0) + \theta_2(t)\theta_1(t_0)}{D(t_0)} + (\kappa) \right)\kappa = \left(\frac{\Theta(t_0, t)}{D(t_0)} + (\kappa) \right)\kappa,$$

where (κ) is a quantity vanishing with κ. Thus the distance along the normal to the first curve from its base to where it cuts the second curve goes to zero with the angle of intersection at the initial point A. If this distance vanishes, then

$$\frac{\Theta(t_0, t)}{D(t_0)} + (\kappa) = 0. \tag{5.19}$$

This is only possible if $\Theta(t_0, t)/D(t_0)$ vanishes with κ. But this means that "the intersection point of the two curves must, with vanishing κ, become infinitely near to a point conjugate to t_0." If in (5.19) the value t is chosen not conjugate to t_0, then $\Theta(t_0, t) \neq 0$ and (5.13) cannot be satisfied for κ very small. Near such a point the curves cannot intersect.

On the other hand, if $t = t_0' + \tau$ is a value very near to a point t_0' conjugate to t_0, then (5.19) can be satisfied. To see this, recall that Θ changes sign near a root (see p. 198 above.) Thus there are positive constants τ_1, τ_2 such that the left-hand member of (5.19) has for $t = t_0' - \tau_1$ the opposite sign from the one for $t = t_0' + \tau_2$. Since it is a continuous function, it consequently must vanish for some intermediate value, such as $t = t_0' + \tau$.

Weierstrass summarizes in the result: "*If two solution curves of the differential equations $G = 0$ passing through an arbitrary initial point make a sufficiently small angle and if these curves have another point in common, then as the angle vanishes this point approaches a determined limiting position; this is then conjugate to the initial point.*"[9]

In concluding the chapter Weierstrass takes up the problem of determining whether there is a solution of the differential equation $G = 0$ through any two preassigned points A and B. He wishes to show that there is a circular neighborhood of B such that every point B' in this small circle can be joined to A by a solution of $G = 0$, which approaches the original curve AB as B' approaches B.

To do this, he considers his two-parameter family (5.12) and fixes one of

[9] Weierstrass, VII, p. 197. Notice that he has not yet got the idea of the envelope into his discussion. We will see more in Section 5.8, where we will also finish up a discussion of the remainder of the present, Weierstrass's 20th, chapter. This is his remarkable discovery of what he called a *plane strip* and what we call a *field of extremals.*

the parameters so that the curves go through the initial point A. "The second constant thereupon determines the initial direction, and it is now to be so chosen that the curve goes through the given second point, if this is possible." He supposes this is so and remarks that it can then be shown that there is a neighborhood of B such that each point B' in the neighborhood can be joined to A by a solution of $G = 0$, i.e., by an extremal. He asserts, moreover, that this solution viewed as a function of B' is continuous and that it converges to the original curve when B' approaches B. It is only necessary to assume that B is not conjugate to A.

To show these things Weierstrass returns to his curves (5.12) and (5.12′) and first requires that they pass through the point A for $t = t_0$. This means that

$$\phi(t_0 + \tau, \alpha + \alpha', \beta + \beta') - \phi(t_0, \alpha, \beta) = 0,$$
$$\psi(t_0 + \tau, \alpha + \alpha', \beta + \beta') - \psi(t_0, \alpha, \beta) = 0.$$

Suppose now that point B corresponds to t_1, B' to $t_1 + \tau_1$ and that B' has coordinates $x + \xi_1, y + \eta_1$. This means that

$$\xi_1 = \phi(t_1 + \tau_1, \alpha + \alpha', \beta + \beta') - \phi(t_1, \alpha, \beta),$$
$$\eta_1 = \psi(t_1 + \tau_1, \alpha + \alpha', \beta + \beta') - \psi(t_1, \alpha, \beta).$$

He expands the functions ϕ, ψ in powers of $\tau, \tau_1, \alpha', \beta'$ and keeps only first-order terms, finding thereby

$$\phi'(t_0)\tau + \phi_1(t_0)\alpha' + \phi_2(t_0)\beta' = 0,$$
$$\psi'(t_0)\tau + \psi_1(t_0)\alpha' + \psi_2(t_0)\beta' = 0,$$
$$\phi'(t_1)\tau_1 + \phi_1(t_0)\alpha' + \phi_2(t_0)\beta' = \xi_1,$$
$$\psi'(t_1)\tau_1 + \psi_1(t_1)\alpha' + \psi_2(t_0)\beta' = \eta_1.$$

The determinant of this system of equations in $\tau, \tau_1, \alpha', \beta'$ is clearly expressible as

$$D = \theta_1(t_1)\theta_2(t_0) - \theta_1(t_0)\theta_2(t_1) = \Theta(t_1, t_2).$$

If then the points A and B are not conjugate to each other, this determinant $D \neq 0$, and Weierstrass can find values for $\tau, \tau_1, \alpha', \beta'$. In fact, the system of equations above has the solution

$$\tau = \frac{G_1\xi_1 + H_1\eta_1}{D}, \qquad \tau_1 = \frac{G_2\xi_1 + H_2\eta_1}{D},$$
$$\alpha' = \frac{G_3\xi_1 + H_3\eta_1}{D}, \qquad \beta' = \frac{G_4\xi_1 + H_4\eta_1}{D}:$$

where the quantities $G_1, H_1, \ldots, G_4, H_4$ are all continuous functions of ξ_1, η_1 when $|\xi_1|, |\eta_1|$ are within determined bounds. This solution completely determines the location of the point B' in the theorem above.

However, when the point B approaches the point A, the argument above becomes questionable since the denominator D is approaching zero—recall that $\Theta(t_0, t_0) = 0$. Instead of discussing the behavior of the numerators in the solution for $\tau, \tau_1, \alpha', \beta'$, Weierstrass uses another approach, which is indicated below.

Let the given point A be $x = a, y = b$. Weierstrass then rotates axes with the help of the transformation

$$x = a + t\cos\lambda + u\sin\lambda, \qquad y = b + t\sin\lambda + u\cos\lambda.$$

The equation $G = 0$ for the extremals is transformed by this rotation into

$$\frac{\partial^2 f}{\partial u'^2}\frac{du'}{dt} + \frac{\partial^2 f}{\partial u\,\partial u'}u' + \frac{\partial^2 f}{\partial t\,\partial u'} - \frac{\partial f}{\partial u} = 0,$$

and

$$\frac{\partial^2 f}{\partial u'^2} = F_1 \cdot (y'\sin\lambda + x'\cos\lambda)^2 = F_1.$$

He now seeks a solution of the differential equation above with the initial conditions $u = 0$, $u' = 0$ and says that this is possible only when $F_1 \neq 0$. Since the point A is fixed this means the angle λ must be such that F_1 is different from zero in that direction. By the continuity of F_1 it must remain different from zero near to this direction. Weierstrass now has a pencil of curves emanating from the point A, which are all close together in direction.

The solutions of $G = 0$ in these directions are expressible as convergent power series

$$u = ct + \tfrac{1}{2}c't^2 + \tfrac{1}{6}c''t^3 + \cdots$$

for $|t|$ small and for all values c between two given bounds. Consider a small circle $u^2 + t^2 = \rho^2$, $\rho > 0$ about the point A. Weierstrass now asserts that the circle cuts these curves in distinct points within a given small sector determined by the condition that $F_1 \neq 0$. He writes the power-series expansion for ρ

$$\rho = \sqrt{1 + c^2}\, t + c_2t^2 + c_3t^3 + \cdots$$

and inverts this to find

$$t = \frac{1}{\sqrt{1 + c^2}}\rho + \cdots.$$

He then concludes that

$$u = \frac{c}{\sqrt{1 + c^2}}\rho + \cdots$$

is a convergent series for all sufficiently small ρ. He has then shown that for all permissible values of c and ρ, the series above give the points t, u of intersection with the circle; he has also shown that no two coincide. This completes his analysis of the situation near the point A:

Let AB be a piece of a curve satisfying the differential equation $G = 0$, that does not contain any point conjugate to A; then one can mark off on both sides of the curve a plane strip with the property that a curve, which always satisfies the differential equation $G = 0$, can be drawn from A to any other point B' in it, varying continuously with B' and coinciding with the original curve AB when B' approaches the point B.

It is not asserted that only one extremal can be drawn through the point B', nor that it must lie completely inside the strip. It is clear, however, that except possibly for point A the original curve AB does lie wholly inside the strip. Point A may lie on the boundary, but if the function $F_1(x_0, y_0, x', y')$ never vanishes nor becomes infinite for any set x', y', then the strip also contains the point A as an interior point. As he conceives of a strip, it is a set of points (x, y) at each of which he has a given direction $x' : y'$. The x', y' are proportional to the direction cosines of the extremal—solution of $G = 0$—from A through the point (x, y).

5.7. The Weierstrass Condition

This famous condition was called by Weierstrass the fourth necessary condition but is now referred to as the second one. He introduced it during his 1879 lectures. As is well known today, he realized that not all comparison arcs to a given arc need be close in direction as well as position. He therefore set out to determine what must follow if comparison arcs lie close to the given one in position but not direction. To do this, he first set up a geometric apparatus, and then in his 22nd chapter proceeded to formulate his condition.

Let (01) be an arc of a curve satisfying $G = 0$ (Figure 5.1), and let 2 be a point on this arc not conjugate to 0. Through 2 construct any curve, such as the one shown in Figure 5.1, through a nearby point 3—it need not satisfy $G = 0$—but close to (02) at every point; then let (03) be an arc of a curve satisfying $G = 0$, which passes through 0 and 3; let the coordinates of any point on (02) be x, y and the corresponding point on (03) be $x + \xi, y + \eta$.

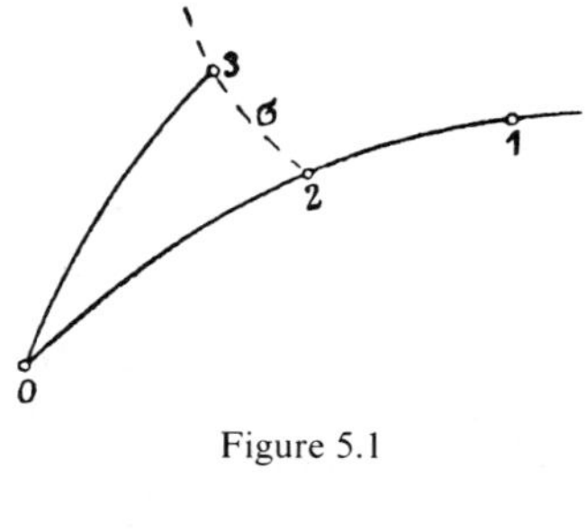

Figure 5.1

Finally, let the parameter value of t corresponding to 0 on (02) be t_0 and on (03), $t_0 + \tau'$; let the value corresponding to 2 be t_2 and the one corresponding to 3, $t_2 + \tau$.

If the family of extremals (5.12) is again $x = \phi(t, \alpha, \beta)$, $y = \psi(t, \alpha, \beta)$ with $\phi' = \partial\phi/\partial t$, $\psi' = \partial\psi/\partial t$, $\phi_1 = \partial\phi/\partial\alpha$, $\psi_1 = \partial\psi/\partial\alpha$, $\phi_2 = \partial\phi/\partial\beta$, and $\psi_2 = \partial\psi/\partial\beta$, it follows that the equations

$$\begin{aligned} \phi(t_0, \alpha, \beta) &= \phi(t_0 + \tau', \alpha + \alpha', \beta + \beta'), \\ \psi(t_0, \alpha, \beta) &= \psi(t_0 + \tau', \alpha + \alpha', \beta + \beta'), \end{aligned} \tag{5.20}$$

must be satisfied by the values $\tau, \tau', \alpha', \beta'$, where α, β are the parameter values corresponding to (02) and $\alpha + \alpha'$, $\beta + \beta'$ to (03). Moreover, since $x_2 + \xi_2, y_2 + \eta_2$ are the coordinates of point 3, the equations

$$\begin{aligned} \xi_2 &= \phi(t_2 + \tau, \alpha + \alpha', \beta + \beta') - \phi(t_2, \alpha, \beta), \\ \eta_2 &= \psi(t_2 + \tau, \alpha + \alpha', \beta + \beta') - \psi(t_2, \alpha, \beta) \end{aligned} \tag{5.20$'$}$$

must also hold for $\tau, \tau', \alpha', \beta'$. To solve these four equations, Weierstrass expands the indicated functions in power series in the variables and finds that through first-order terms, the equations for $\tau, \tau', \alpha', \beta'$ are

$$\begin{aligned} 0 &= \phi'(t_0)\tau' + * \quad +\phi_1(t_0)\alpha' + \phi_2(t_0)\beta', \\ 0 &= \psi'(t_0)\tau' + * \quad +\psi_1(t_0)\alpha' + \psi_2(t_0)\beta', \\ \xi_2 &= \quad * + \phi'(t_0)\tau + \phi_1(t_2)\alpha' + \phi_2(t_0)\beta', \\ \eta_2 &= \quad * + \psi'(t_0)\tau + \psi_1(t_2)\alpha' + \psi_2(t_0)\beta'. \end{aligned} \tag{5.21}$$

The last two of these equations give him the variations of the coordinates x_2, y_2 of point 2, so that the coordinates $x_2 + \xi_2, y_2 + \eta_2$ signify the coordinates of point 3.

Next Weierstrass introduces quantities $\bar{p}_2, \bar{q}_2$ proportional to the direction cosines of the tangent to the curve $(\overline{23})$ at point 2 and chooses σ to be arc-length along $(\overline{23})$ measured from 2 toward 3. Then he has

$$\xi_2 = \sigma(\bar{p}_2 + \sigma_1), \qquad \eta_2 = \sigma(\bar{q}_2 + \sigma_2), \tag{5.22}$$

where σ_1, σ_2 are quantities that vanish with σ. He has shown in his 20th chapter with the help of his implicit-function theorem how equations (5.21), (5.22) can be solved for $\tau, \tau', \alpha', \beta'$ provided that ξ_2, η_2, and σ are small. With the help of these solutions, he can then express the variations of the coordinates in the form

$$\xi = (X + \sigma')\sigma, \qquad \eta = (Y + \sigma'')\sigma, \tag{5.23}$$

where σ', σ'' vanish with σ and X, Y are given functions of t.

He designates by $I_{02}, I_{03}, \bar{I}_{32}$ the values of the integral

$$I = \int F(x, y, x', y')\, dt$$

evaluated along the arcs (02), (03), and $(\overline{32})$. With the help of (5.23), he calculates the value $I_{03} - I_{02}$ in powers of $\xi, \eta, d\xi/dt, d\eta/dt$ and finds

$$I_{03} = I_{02} + \int_{t_0}^{t_2} G\cdot(x'\eta - y'\xi)\,dt + \left[\frac{\partial F}{\partial x'}\xi + \frac{\partial F}{\partial y'}\eta\right]_{t_0}^{t_2} + \cdots,$$

in which the additional terms are regular functions of σ of degree at least 2. But since $G = 0$ along (02) and (03) and ξ, η vanish at $t = t_0$, he has

$$I_{03} = I_{02} + \left(\left(\frac{\partial F}{\partial x'}\right)_{(2)} \cdot \bar{p}_2 + \left(\frac{\partial F}{\partial y'}\right)_{(2)} \cdot \bar{q}_2\right)\sigma + S\sigma, \tag{5.24}$$

where S is a quantity vanishing with σ and the subscript (2) on $\partial F/\partial x'$ and $\partial F/\partial y'$ means that these quantities are evaluated for $t = t_2$.

To estimate the remaining integral $\bar{I}_{32}$, he again makes use of (5.22) and writes the coordinates of an arbitrary point on $(\overline{23})$ as

$$x = x_2 + (\bar{p}_2 + \sigma_1')\bar{\sigma}, \qquad y = y_2 + (\bar{q}_2 + \sigma_1'')\bar{\sigma},$$

where $\bar{\sigma}$ varies over the closed interval $[0, \sigma]$ as (x, y) goes from the point x_2, y_2 to x_3, y_3. This gives him the value

$$\bar{I}_{32} = -\int_{\sigma}^{0} F(x, y, x', y')\,d\bar{\sigma} = \int_0^{\sigma} F\,d\bar{\sigma} = \sigma\bar{F},$$

when $\bar{F}$ is a mean value of F on $0 \leqslant \bar{\sigma} \leqslant \sigma$. He now chooses σ so small that, by continuity considerations, he can write

$$\bar{F} = F(x_2, y_2, -\bar{p}_2, -\bar{q}_2) + (\sigma),$$

where he uses (σ) to indicate a quantity which goes to zero with σ. Combining these relations, he is able to assert that

$$I_{03} + \bar{I}_{32} - I_{02} = \left\{\left(\frac{\partial F}{\partial x'}\right)_{(2)} \cdot \bar{p}_2 + \left(\frac{\partial F}{\partial y'}\right)_{(2)} \cdot \bar{q}_2 + F(x_2, y_2, -\bar{p}_2, -\bar{q}_0)\right\}\sigma + S'\sigma,$$

where S' vanishes with σ.

Weierstrass now is able to remark that the broken arc $(0\overline{32})$ is a permissible variation of the arc (02) with the same end-points, which is near (02) in position but not in direction. He asserts[10]:

> If I_{02} is a minimum, in any case the left side of the equation above must have a determined positive value. One can then conclude that the quantities inside the vincula cannot take on negative values; since otherwise one could choose the quantity σ to be so small in absolute value that the entire variation of the integral I_{02} would be negative. Therewith the desired new necessary condition for the attainment of a minimum of the exhibited integral is certainly found.

He then proceeds to express the condition in its well-known form. To do

[10] Weierstrass, *VOR*, p. 213.

this he writes

$$\frac{\partial F}{\partial x'} = F^{(1)}, \qquad \frac{\partial F}{\partial y'} = F^{(2)}, \qquad -\bar{p}_2 = \bar{p}, \qquad -\bar{q}_2 = \bar{q} \tag{5.25}$$

and in place of x', y', he writes the direction cosines p, q of the curve (02). His condition is then expressible in terms of his function

$$\mathcal{E}(x, y;p,q;\bar{p},\bar{q}) = F(x, y, \bar{p},\bar{q}) - F^{(1)}(x, y, p,q)\bar{p} - F^{(2)}(x, y, p,q)\bar{q}; \tag{5.26}$$

and it states that for all x, y, p, q belonging to a curve C, satisfying $G = 0$ and $\bar{p}, \bar{q}$ arbitrary, the function $\mathcal{E} \geqslant 0$ if the curve C renders the integral a minimum and $\mathcal{E} \leqslant 0$ if a maximum.

He goes on to show that if $\kappa, \bar{\kappa}$ are positive constants, then

$$\mathcal{E}(x, y;\kappa p,\kappa q;\bar{\kappa}\bar{p},\bar{\kappa}\bar{q}) = \bar{\kappa}\mathcal{E}(x, y;p,q;\bar{p},\bar{q}),$$

and

$$\mathcal{E}(x, y;p,q;\kappa p,\kappa q) = 0.$$

Furthermore, by the homogeneity of F, he can write

$$\begin{aligned}\mathcal{E}(x, y;p,q;\bar{p},\bar{q}) = \big(&F^{(1)}(x, y, \bar{p},\bar{q}) - F^{(1)}(x, y, p,q)\big)\bar{p}\\ &+ \big(F^{(2)}(x, y, \bar{p},\bar{q}) - F^{(2)}(x, y, p,q)\big)\bar{q}.\end{aligned}$$

Next he wishes to exhibit "a close and deep connection" between $\mathcal{E}$ and F_1. To do this, he first uses the homogeneity property of F to note that $F^{(\alpha)}(x, y, \kappa x', \kappa y') = F^{(\alpha)}(x, y, x', y')$ $(\alpha = 1,2)$. If $\cos\phi = x'/(x' + y'^2)^{1/2}$, $\sin\phi = y'/(x'^2 + y'^2)^{1/2}$, then $F^{(\alpha)}(x, y, x', y') = F^{(\alpha)}(x, y, \cos\phi, \sin\phi)$. He next sets

$$\begin{aligned} p &= r\cos\chi, & q &= r\sin\chi,\\ \bar{p} &= \bar{r}\cos\bar{\chi}, & \bar{q} &= \bar{r}\sin\bar{\chi}. \end{aligned}$$

By the homogeneity properties of $\mathcal{E}$ just noted above he has

$$\mathcal{E}(x, y;p,q;\bar{p},\bar{q}) = \bar{r}\mathcal{E}(x, y;\cos\chi,\sin\chi;\cos\bar{\chi},\sin\bar{\chi});$$

and, by the result above, he has the expression

$$F^{(1)}(x, y, \bar{p},\bar{q}) - F^{(1)}(x, y, p,q),$$

which enters $\mathcal{E}$, represented as

$$\int_\chi^{\bar{\chi}} \frac{d}{d\phi} F^{(1)}(x, y, \cos\phi, \sin\phi)\, d\phi$$

with a corresponding formula for $F^{(2)}$. Weierstrass expresses these integrands in the form

$$\begin{aligned}\frac{d}{d\phi} F^{(1)}(x, y, \cos\phi, \sin\phi) &= -\sin\phi\,\frac{\partial^2 F}{\partial x'^2} + \cos\phi\,\frac{\partial^2 F}{\partial x'\,\partial y'},\\ \frac{d}{d\phi} F^{(2)}(x, y, \cos\phi, \sin\phi) &= -\sin\phi\,\frac{\partial^2 F}{\partial x'\,\partial y'} + \cos\phi\,\frac{\partial^2 F}{\partial y'^2}.\end{aligned} \tag{5.27}$$

However, as we saw earlier in definition (5.4) for F_1,

$$\frac{\partial^2 F}{\partial x'^2} = \sin^2\phi \cdot F_1, \qquad \frac{\partial^2 F}{\partial x' \partial y'} = -\cos\phi\sin\phi \cdot F_1,$$

$$\frac{\partial^2 F}{\partial y'^2} = \cos^2\phi \cdot F_1.$$

The right-hand members of (5.27) then become $-\sin\phi \cdot F_1$ and $\cos\phi \cdot F_1$, respectively, and Weierstrass has the result

$$\mathcal{E}(x, y; \cos\chi, \sin\chi; \cos\bar{\chi}, \sin\bar{\chi}) = \int_{\chi}^{\bar{\chi}} F_1(x, y, \cos\phi, \sin\phi)\sin(\bar{\chi} - \phi)\, d\phi.$$

He comments that $\sin(\bar{\chi} - \phi)$ takes on all values between $\sin(\bar{\chi} - \chi)$ and 0 and that $\bar{\chi}$ and χ are known up to multiples of 2π; hence $\sin(\bar{\chi} - \phi)$ cannot change sign inside the interval of integration. By means of the mean-value theorem of the integral calculus, he can then assert that

$$\mathcal{E}(x, y; \cos\chi, \sin\chi; \cos\bar{\chi}, \sin\bar{\chi}) = F_1(x, y, \cos\tilde{\chi}, \sin\tilde{\chi})(1 - \cos(\bar{\chi} - \chi)), \tag{5.28}$$

where $\tilde{\chi} = \chi + \theta(\bar{\chi} - \chi)$ with $0 < \theta < 1$ (Weierstrass gave this in his 1882 lectures). He concludes that "if the function $F_1(x, y, p, q)$ maintains the same sign not only for such values of the arguments p, q which are proportional to direction cosines along the original curve but also for arbitrary arguments p, q, then the same is true for the function $\mathcal{E}$, and it is clear that under this hypothesis the above formulated fourth necessary condition is itself fulfilled."

5.8. Sufficiency Arguments

Before taking up Weierstrass's discussion in his 23rd chapter, entitled "Proof that the fourth necessary condition is also sufficient," let us assume with him that $F_1 \neq 0$ throughout a given strip surrounding a solution curve AB of $G = 0$. This notion of a strip conceived by Weierstrass is what is now usually called a field or a field of extremals in two dimensions.

This concept ranks in importance with his introduction of the $\mathcal{E}$-function. As we shall see later, the extension of this notion to higher space is not a mere notational matter, but brings in a new idea. The extension was undertaken in 1905 by Mayer.[11]

The basic tool used by Weierstrass in his 23rd chapter is his relation (5.28) above, which he writes, for convenience, in the form

$$\mathcal{E}(x, y; p, q; \bar{p}, \bar{q}) = (1 - p\bar{p} - q\bar{q})\bar{F}_1,$$

[11] See Chapter 6 below.

where $\bar{F}_1$ is a value of $F_1(x, y, x', y')$ calculated for x' between p and $\bar{p}$ and y' between q and $\bar{q}$. He notes that if $\bar{F}_1$ does not vanish for any pair x', y' associated with his plane strip, then since it is a continuous function of these arguments its absolute value must have a positive lower bound on the strip. He concludes that the strip can be chosen so small that for every solution of $G = 0$ lying in it, either $\mathcal{E} \geqslant 0$ or $\mathcal{E} \leqslant 0$.

He now wishes to show that if the comparison arcs lie wholly inside a plane strip surrounding the given arc, then the value of the integral calculated along this arc is greater (less) than its value along the given one. The important thing is that he no longer needs to insist that the comparison arcs be near to the given arc in both position and direction, but only in position. In Figure 5.2 we find a reproduction of Weierstrass's illustration. In this figure 0 is the initial point, and 1 the terminal point of the arc (01) of the extremal being considered; he supposes that there is no point on it conjugate to 0. Inside the strip let $(\overline{01})$ be an arbitrary comparison arc joining 0 and 1, which is regular. Point 2 is an arbitrary point on $(\overline{01})$, and (02) is the extremal through points 0 and 2. (Weierstrass did not use the word *extremal*; he always said a solution of $G = 0$.) He consequently has $I_{02} + \bar{I}_{21} - I_{01} > 0$ provided that I_{01} is the minimum value of I. He designates the left-hand member of this inequality by $\phi(s)$, where s is arc-length along $(\overline{01})$ calculated from 0:

$$I_{02} + \bar{I}_{21} - I_{01} = \phi(s).$$

This function is continuous, and Weierstrass wishes to show it to be differentiable with a negative value for the derivative.

To do this, he now examines the value of the integral

$$\int_{t_0}^{t_1} F(x, y, x', y')\, dt$$

along a sufficiently small piece of a curve with arc-length σ measured from t_0, where the coordinates and the direction cosines of the curve are x_0, y_0, p_0, q_0. He then expresses an arbitrary point as

$$x = x_0 + (p_0 + (\sigma)_1)\sigma, \qquad y = y_0 + (q_0 + (\sigma)_2)\sigma,$$

where $(\sigma)_1, (\sigma)_2$ are functions of σ that vanish with it, and the function F as

$$F(x, y, x', y') = F(x_0, y_0, p_0, q_0) + (\sigma)_3;$$

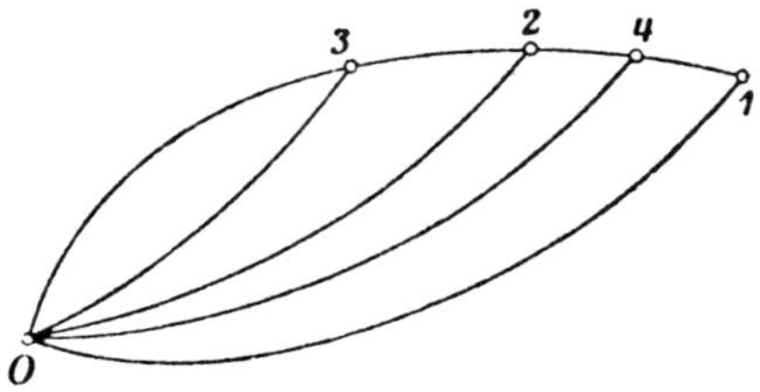

Figure 5.2

with $(\sigma)_3$ a function of σ vanishing with it. He can then assert that for $\sigma_1 = t_1 - t_0$, a sufficiently small value of σ,

$$\int_{t_0}^{t_1} F(x, y, x', y')\, dt = F(x_0, y_0, p_0, q_0)\sigma_1 + \bar{\sigma}_1\sigma_1,$$

where again $\bar{\sigma}_1$ is a small quantity vanishing with σ_1. But the expression $F(x_1, y_1, p_1, q_1) - F(x_0, y_0, p_0, q_0)$ is also a quantity that vanishes with σ, and so there is a quantity $\bar{\bar{\sigma}}_1$ such that the integral can also be expressed in the form

$$\int_{t_0}^{t_1} F(x, y, x', y')\, dt = F(x_1, y_1, p_1, q_1)\,\sigma_1 + \bar{\bar{\sigma}}_1\sigma_1,$$

where x_1, y_1 are the coordinates of the end-point defined by $t = t_1$.

Weierstrass now chooses on $(\overline{01})$ a point 3 between 0 and 2 and designates by σ the length of the arc $(\overline{32})$ so that

$$\phi(s - \sigma) = I_{03} + \bar{I}_{31} - I_{01},$$

and

$$\phi(s - \sigma) - \phi(s) = -I_{02} - \bar{I}_{21} + I_{01} + I_{03} + \bar{I}_{31} - I_{01},$$

i.e.,

$$\phi(s - \sigma) - \phi(s) = -I_{02} + \bar{I}_{32} + I_{03}.$$

Now when point 3 is so near to point 2 that the earlier considerations apply to the integral $\bar{I}_{32}$, then

$$\bar{I}_{32} = F(x_2, y_2, \bar{p}_2, \bar{q}_2)\sigma + \bar{\sigma}\sigma,$$

where $\bar{p}_2, \bar{q}_2$ are proportional to the direction cosines of the curve $(\overline{01})$ at point 2. For the integral I_{03}, Weierstrass makes use of relation (5.24) above with $-\bar{p}_2; -\bar{q}_2$, the direction cosines to $(\overline{32})$, and has

$$I_{03} - I_{02} = \left\{ -F^{(1)}(x_2, y_2, p_2, q_2)\bar{p}_2 - F^{(2)}(x_2, y_2, p_2, q_2)\bar{q}_2 + S \right\}\sigma.$$

Putting his various results together, Weierstrass finds with the help of his $\mathcal{E}$-function that

$$\phi(s - \sigma) - \phi(s) = \mathcal{E}(x_2, y_2; p_2, q_2; \bar{p}_2, \bar{q}_2)\sigma + (\sigma)\sigma,$$

and thus that the limit of $(\phi(s - \sigma) - \phi(s))/(-\sigma)$ is

$$-\mathcal{E}(x_2, y_2; p_2, q_2; \bar{p}_2, \bar{q}_2).$$

In a similar way he chooses a point 4 on $(\overline{01})$ between points 2 and 1 and again uses σ to designate the length of $(\overline{24})$. Then

$$\phi(s + \sigma) - \phi(s) = I_{04} + \bar{I}_{41} - I_{01} - I_{02} - \bar{I}_{21} + I_{01} = I_{04} - \bar{I}_{24} - I_{02},$$

and by (5.24)

$$I_{04} - I_{02} = \left\{ F^{(1)}(x_2, y_2, p_2, q_2)\bar{p}_2 + F^{(2)}(x_2, y_2, p_2, q_2)\bar{q}_2 + S' \right\}\sigma$$

as well as

$$\bar{I}_{24} = F(x_2, y_2, \bar{p}_2, \bar{q}_2)\sigma + \bar{\sigma}'\sigma,$$

by analogy with what we saw above. Then the limit of $(\phi(s+\sigma)-\phi(s))/\sigma$ is also $-\mathcal{E}(x_2, y_2; p_2, q_2; \bar{p}_2, \bar{q}_2)$ so that this value is in fact $d\phi(s)/ds$, where x_2, y_2 are the coordinates of an arbitrary point on $(\overline{01})$—any curve through points 0 and 1 and inside the plane strip; $\bar{p}_2, \bar{q}_2$ are the direction cosines of this curve at the point 2; and p_2, q_2 are the direction cosines at 2 of the given curve (01) satisfying the equation $G = 0$. (Weierstrass remarks that his proof makes use of the fact that the curve $(\overline{01})$ has a continuously varying tangent.) Integrating the quantity $\phi'(s)$, we find Weierstrass's famous theorem:

$$\Delta I = \int_{t_0}^{t_1} \mathcal{E}\, dt. \tag{5.29}$$

(This result also follows easily by the use of Hilbert's invariant integral, as we shall see later.The importance of this result lies in the fact that the total variation of the integral I is now related directly to the $\mathcal{E}$-function, and this makes sufficiency theorems easy to state and prove.)

If it is assumed that $\mathcal{E}(x_2, y_2; p_2, q_2; \bar{p}_2, \bar{q}_2)$ is positive for all values of its arguments under consideration, then the derivative $\phi'(s) < 0$ and ϕ is a monotonic decreasing function with the final value of 0; hence it is always positive and

$$I_{02} + \bar{I}_{21} > I_{01};$$

and if the point 2 coincides with 0, then

$$\bar{I}_{01} > I_{01}.$$

Weierstrass (*VOR*, p. 224) summarizes in the theorem

> *If the function $\mathcal{E}$ is not positive at any point of an arbitrary curve $(\overline{01})$ inside the plane strip and joining the points* 0 *and* 1 *and if it does not vanish at every point on the curve, then the integral in hand, evaluated along the curve* (01), *which satisfies the differential equation $G = 0$, has a greater value than it does evaluated along the curve $(\overline{01})$; and if the function $\mathcal{E}$ is never negative along the curve $(\overline{01})$ and also not zero at each point, then the integral along the original curve* (01) *is smaller than along the arbitrary curve* $(\overline{01})$.

He goes on to remark that the condition on the arbitrary curve $(\overline{01})$ that the tangent be continuously changing can be weakened to permit the curve to consist of a finite number of pieces, each of which has the property. This follows by a simple reconsideration of $\phi(s)$.

He then asks what happens if the function $\mathcal{E}$ vanishes along the curve $(\overline{01})$. Under the hypothesis, made earlier, that F_1 not only is different from zero along the curve (01) but also for x', y' arbitrary arguments in the plane strip, it is clear that $\mathcal{E}$ can only vanish when $1 - p\bar{p} - q\bar{q}$ does. But this is only possible when the directions determined by p, q and $\bar{p}, \bar{q}$ coincide, i.e., when $p\bar{q} - q\bar{p} = 0$. This means that the arbitrary curve $(\overline{01})$ must coincide with the extremal (02) at their intersection point 2 in both position and direction; but 2 is an arbitrary point, and hence they must coincide.

Weierstrass now considers the one-parameter family of solutions of $G = 0$, passing through the same initial point $t = t_0$ and having their initial directions near each other; let them be represented in the form

$$x = \phi(t, \kappa), \qquad y = \psi(t, \kappa).$$

Then if t, κ and $t + \tau', \kappa + \kappa'$ define the original curve x, y and a neighboring one $x + \xi, y + \eta$ at corresponding points, then for τ', κ' sufficiently small

$$\xi = \phi'(t)\tau' + \frac{\partial\phi}{\partial\kappa}\kappa' + (\tau', \kappa')_2, \qquad \eta = \psi'(t)\tau' + \frac{\partial\psi}{\partial\kappa}\kappa' + (\tau, \kappa')_2.$$

If the two curves have another point in common, then at that point $\xi = 0$, $\eta = 0$ and the determinant of these equations must vanish, i.e.,

$$\phi'(t)\frac{\partial\psi}{\partial\kappa} - \psi'(t)\frac{\partial\phi}{\partial\kappa} = 0.$$

Let $t_1 = t_1(\kappa)$ be the smallest root after t_0 of this equation as a function of κ. Then the locus of points conjugate to the initial point 0 is given by the equations $\bar{x} = \phi(t_1, \kappa)$ and $\bar{y} = \psi(t_1, \kappa)$, where $\bar{x}, \bar{y}$ are the coordinates of the conjugate points. It follows readily that

$$\frac{d\bar{x}}{d\kappa} = \frac{\partial\phi(t_1, \kappa)}{\partial t_1}\frac{dt_1}{d\kappa} + \frac{\partial\phi(t_1, \kappa)}{\partial\kappa} = \phi'(t_1)\frac{dt_1}{d\kappa} + \frac{\partial\phi(t_1, \kappa)}{\partial\kappa},$$

$$\frac{d\bar{y}}{d\kappa} = \frac{\partial\psi(t_1, \kappa)}{\partial t_1}\frac{dt_1}{d\kappa} + \frac{\partial\psi(t_1, \kappa)}{\partial\kappa} = \psi'(t_1)\frac{dt_1}{d\kappa} + \frac{\partial\psi(t_1, \kappa)}{\partial\kappa}.$$

Thus the equation above defining conjugate points becomes

$$\phi'(t_1)\frac{d\bar{y}}{d\kappa} - \psi'(t_1)\frac{d\bar{x}}{d\kappa} = 0.$$

Weierstrass notes that since $\phi'(t_1), \psi'(t_1)$ are proportional to the direction cosines of the tangent to the extremal arc through points 0 and 1 at point 1, this equation implies that the locus of conjugate points is tangent to the extremal at their intersection. He concludes that this proves the envelope of the one-parameter family of extremals through 0 is the locus of points conjugate to 0.

He now considers a curve $\bar{x} = f(u)$, $\bar{y} = g(u)$ through 0 lying totally inside the region determined by the envelope and not coinciding with any extremal through 0. For this curve to be tangent to such an extremal, it is necessary that

$$\phi(t, \kappa) = f(u), \qquad \psi(t, \kappa) = g(u), \qquad \frac{\partial\phi}{\partial t}\frac{dg}{du} - \frac{\partial\psi}{\partial t}\frac{df}{du} = 0.$$

Weierstrass remarks that the first two of these relations give t and u as functions of κ for $|\kappa|$ sufficiently small. It then results that

$$\frac{\partial\phi}{\partial t}\frac{dt}{d\kappa} + \frac{\partial\phi}{\partial\kappa} = \frac{df}{du}\frac{du}{d\kappa}, \qquad \frac{\partial\psi}{\partial t}\frac{dt}{d\kappa} + \frac{\partial\psi}{\partial\kappa} = \frac{dg}{du}\frac{du}{d\kappa};$$

combining these with the third relation above, he has

$$\frac{\partial\phi}{\partial t}\frac{\partial\psi}{\partial\kappa} - \frac{\partial\psi}{\partial t}\frac{\partial\phi}{\partial\kappa} = 0,$$

But this is the equation for determining the points conjugate to 0. This would require that the curve being considered be part of the envelope, which is contrary to hypothesis. Therefore, inside the region enclosed by the envelope there can be no curve of the sort described.

Weierstrass closes the chapter by noting that the Legendre condition, the third necessary condition, follows from the second one since when $\bar{p}, \bar{q}$ approach p, q, $\overline{F}_1$ goes over into $F_1(x, y, p, q)$; and by noting that the formula $\mathcal{E}(x, y; p, q; \bar{p}, \bar{q}) = (1 - p\bar{p} - q\bar{q})\overline{F}_1$ implies that F_1 has a fixed sign along the extremal arc (01). It follows, by continuity considerations, that $\overline{F}_1$ is positive along curve (01) for $\bar{p}, \bar{q}$ arbitrary and hence inside a suitably chosen surface strip.

In his 24th chapter Weierstrass considers a number of ancillary points which are by way of a conclusion to his discussion of the simplest problem. Perhaps it is not out of place to mention one example. If light moves in an isotropic medium whose density varies continuously, from a point A to another, then the time of transit is a minimum. The integral to be minimized is then $\int ds/\rho$, where ds is an element of arc length and ρ is the density. This problem was considered in detail by Kummer in the *Reports* of the Berlin Academy.

5.9. The Isoperimetric Problem

Weierstrass opens his 25th chapter with the observation that the problem he has just considered can be generalized in a number of ways.[12] The first is to consider not two, but n dependent variables. He remarks that there are no real difficulties with this case but says that there are greater difficulties connected with the case where the variables $x_1, x_2, \ldots, x_n$ satisfy various side-conditions. As long as these conditions do not involve the derivatives dx_i/dt, there is no real difficulty. For in this case he asserts that the problem can be reduced to one in fewer variables and without side-conditions, as is clear. If the side-conditions contain derivatives of the $x_1, x_2, \ldots, x_n$, he points out it is necessary only to consider the case when first derivatives enter. For if higher derivatives enter, they can be eliminated by appending additional variables such as $dx_1/dt = x_{n+1}$, $dx_2/dt = x_{n+2}, \ldots$. Instead of taking up such problems, Weierstrass turned his attention to the isoperimetric problem.

It had earlier been believed that such a problem was simply equivalent

[12] Weierstrass, *VOR*, pp. 242ff.

to the unrestricted problem of minimizing an integral

$$\int_{t_0}^{t_1} (F^0 - \lambda F^1)\, dt.$$

This was first shown to be false by Lundström.[13] A conjugate point for the isoperimetric problem can exist, beyond which no extremal can be found but which is not a conjugate point for the unrestricted problem above.

The problem at hand is to render the integral

$$I^0 = \int_{t_0}^{t_1} F^0(x, y, x', y')\, dt$$

a maximum or a minimum subject to the condition that at the same time another integral

$$I^1 = \int_{t_0}^{t_1} F^1(x, y, x', y')\, dt$$

has a prescribed value and the end-points are fixed.[14] Both integrals are to be understood as being evaluated along the same curve. To do this, Weierstrass notes that he must ensure that the variations given to the original curve will be such as to leave the integral I^1 unchanged in value. He considers variations of the form $\xi = \kappa u + \kappa_1 u_1$, $\eta = \kappa v + \kappa_1 v_1$, where κ and κ_1 are constants and u, u_1, v, v_1 are differentiable functions of t, that vanish for $t = t_0$ and t_1. He now goes to show that the parameters κ and κ_1 can be so chosen that I^1 remains unchanged in value when $|\xi|, |\eta|$ are sufficiently small. As we saw earlier (p. 192),

$$\Delta I^1 = \int_{t_0}^{t_1} G^1 \cdot (x'\xi - y'\eta)\, dt + \int_{t_0}^{t_1} \left[\xi, \eta, \frac{d\xi}{dt}, \frac{d\eta}{dt} \right]^2 dt.$$

In this expression he sets $x'v - y'u = w$, $x'v_1 - y'u_1 = w_1$ so that the total variation becomes

$$\Delta I^1 = \kappa \int_{t_0}^{t_1} G^1 w\, dt + \kappa_1 \int_{t_0}^{t_1} G^1 w_1\, dt + (\kappa, \kappa_1)_1 \kappa + (\kappa, \kappa_1)_2 \kappa_1,$$

where $(\kappa, \kappa_1)_1, (\kappa, \kappa_1)_2$ are quantities that vanish with κ, κ_1.[15]

If I^1 remains unchanged, its total variation $\Delta I^1 = 0$, and Weierstrass has a relation between κ and κ_1 of the form

$$\kappa\big(W^1 + (\kappa, \kappa_1)_1\big) + \kappa_1\big(W_1^1 + (\kappa, \kappa_1)_2\big) = 0,$$

where

$$\int_{t_0}^{t_1} G^1 w\, dt = W^1, \qquad \int_{t_0}^{t_1} G^1 w_1\, dt = W_1^1.$$

---

[13] Lundström [1869], and later Mayer [1877], p. 54.

[14] Weierstrass took up this subject in his lectures of 1879 and 1882.

[15] In his analysis Weierstrass makes use of *free* or *unrestricted* variations as well as *restricted* or *unfree* ones. If a curve under investigation has points or a segment on the boundary of the region of definition of the problem, then clearly the class of admissible variations for this curve must be restricted so that the resulting comparison curves will lie in the region. If the curve lies wholly interior to the region, no such restrictions apply.

If $W_1^1 \lessgtr 0$, then he can express κ_1 as

$$\kappa_1 = -\left(\frac{W^1}{W_1^1} + (\kappa)\right)\kappa.$$

If $W_1^1 = 0$ for all permissible choices of u, u_1, v, v_1, then $G^1 = 0$; and the given curve is an extremal for the integral I^1. He therefore postulates that the given curve is not an extremal for I^1, i.e., that $G^1 \not\equiv 0$ along that curve. According to this hypothesis he can, as we saw above, always find comparison arcs for which $\Delta I^1 = 0$ with the help of the relations

$$\xi = \kappa\left(u - u_1\frac{W^1}{W_1^1}\right) + (\kappa)\kappa, \qquad \eta = \kappa\left(v - v_1\frac{W^1}{W_1^1}\right) + (\kappa)\kappa.$$

The complete variation of I^0 is then expressible as

$$\Delta I^0 = \int_{t_0}^{t_1} G^0 \cdot (x'\eta - y'\xi)\,dt + \cdots = \kappa\int_{t_0}^{t_1} G^0 \cdot \left(w - w_1\frac{W^1}{W_1^1}\right)dt + (\kappa)\kappa.$$

When W^0, W_1^0 are given as

$$\int_{t_0}^{t_1} G^0 w\,dt = W^0, \qquad \int_{t_0}^{t_1} G^0 w_1\,dt = W_1^0,$$

then the total variation of I^0 is

$$\Delta I^0 = \kappa\left(W^0 - W_1^0\frac{W^1}{W_1^1}\right) + (\kappa)\kappa.$$

If the integral is to be a maximum or a minimum, then ΔI^0 must be either always negative or always positive; but this is only possible when the coefficient of the first power of κ vanishes, i.e., when

$$W^0 : W^1 = W_1^0 : W_1^1.$$

The left-hand member of this relation contains only the functions u, v and the right-hand only u_1, v_1. There must therefore exist a constant λ such that

$$W^0 = \lambda W^1, \qquad W_1^0 = \lambda W_1^1$$

or equivalently that

$$W^0 - \lambda W^1 = \int_{t_0}^{t_1} (G^0 - \lambda G^1) w\,dt = 0.$$

But this implies that

$$G = G^0 - \lambda G^1 = 0, \tag{5.30}$$

and Weierstrass sums up in the theorem: "*In general if there is a curve for which the integral I^0 assumes a greatest or least value while the integral I^1 maintains a prescribed value, then there must be a constant λ given so that the coordinates of an arbitrary point of the curve satisfy the differential equation* . . . $[G^0 - \lambda G^1 = 0]$."

Since this differential equation is in general of the second order, it must possess a two-parameter family $x = \phi(t, \alpha, \beta, \lambda)$, $y = \psi(t, \alpha, \beta, \lambda)$ of solutions. Thus there are three constants α, β, λ to be determined so that the integral I^1 has its preassigned value and the curve passes through both end-points. Weierstrass argues, as did Mayer ([1877], p. 65n), that even if the minimizing arc contains "restricted" or discontinuous segments or points, the "free" or continuous segments—those whose variation is not restricted—must satisfy equation (5.30) above with the same value for the constant λ.

If $F = F^0 - \lambda F^1$, then Weierstrass forms out of F the function $G = G^0 - \lambda G^1$ with the help of definition (5.5′). The equation $G = 0$ is then equivalent to the equations

$$\frac{\partial F}{\partial x} - \frac{d}{dt}\frac{\partial F}{\partial x'} = 0, \qquad \frac{\partial F}{\partial y} - \frac{d}{dt}\frac{\partial F}{\partial y'} = 0.$$

As was seen earlier, it follows that $\partial F/\partial x', \partial F/\partial y'$ must be continuous when evaluated along an extremal, even at corners.[16]

He goes on to indicate how the theory is also applicable when not one but many isoperimetric conditions must be satisfied. (Since this is not different from the simpler case, I do not include a discussion of it.)

Weierstrass, at the close of his 25th chapter, writes his total variations of I^0 and I^1 as

$$\Delta I^0 = \delta I^0 + \frac{1}{2}\int_{t_0}^{t_1}\left\{F_1^0\left(\frac{dw}{dt}\right)^2 + F_2^0 w^2\right\}dt + \cdots,$$

$$0 = \delta I^1 + \frac{1}{2}\int_{t_0}^{t_1}\left\{F_1^1\left(\frac{dw}{dt}\right)^2 + F_2^1 w^2\right\}dt + \cdots,$$

by analogy with the simplest problems, and concludes that

$$\Delta I^0 = \delta I^0 - \lambda\delta I^1 + \frac{1}{2}\int_{t_0}^{t_1}\left\{F_1\left(\frac{dw}{dt}\right)^2 + F_2 w^2\right\}dt + \cdots,$$

where F_1 and F_2 are formed, as before, from $F = F^0 - \lambda F^1$. But

$$\delta I^0 - \lambda\delta I^1 = \int_{t_0}^{t_1}(G^0 - \lambda G^1)w\,dt.$$

Along an arc satisfying $G = 0$, the right-hand member of this equation vanishes and consequently

$$\Delta I^0 = \frac{1}{2}\int_{t_0}^{t_1}\left\{F_1\left(\frac{dw}{dt}\right)^2 + F_2 w^2\right\}dt + \cdots.$$

As before, the first term in the right-hand member of this relation can be transformed into the expression

$$\frac{1}{2}\int_{t_0}^{t_1}F_1\left(\frac{dw}{dt} - \frac{w}{u}\frac{du}{dt}\right)^2 dt,$$

[16] Weierstrass, *VOR*, pp. 249–250. This is the Weierstrass–Erdmann corner condition.

provided that u is a solution of the differential equation

$$F_1 \frac{d^2u}{dt^2} + \frac{dF_1}{dt}\frac{du}{dt} - F_2 u = 0.$$

He summarizes by noting that for a maximum or a minimum of the integral I^0 "*the function* F_1, *on all parts of the curve which can be freely varied, must be non-positive in the first case and non-negative in the second one*."

5.10. Sufficient Conditions

Weierstrass now proceeds in his 27th chapter to establish his famous results on those conditions which are sufficient to ensure that a given arc is indeed an extremum. He has shown in his 25th chapter that a first necessary condition for the isoperimetric problem is that the curve in question must satisfy the differential equation

$$G = G^0 - \lambda G^1 = 0$$

for some constant λ. The second condition is that the function $F_1 \leqslant 0$ for a maximum and $\geqslant 0$ for a minimum. The equation $G = 0$ is in general a second-order differential equation containing a parameter λ and therefore has a three-parameter family of solutions. He starts his 27th chapter by making two assumptions, whose reasonableness he proposes to consider in the following chapter.

He first assumes that it is possible to determine the three constants so that through any two points A and B a solution of $G = 0$ passes for which the integral I^1 has its prescribed value. He then assumes that it is possible "*to join the two points A and B by an arbitrary regular curve so that along this curve the integral* I^1 *likewise has the prescribed value*; *in addition through every point C on this curve there can pass one and only one curve through A, which satisfies the differential equation* $G = 0$ *and for which the second integral* I^1, *evaluated from the point A to the point C, has the same value as when it is evaluated in the first place along the assumed arbitrary curve between the same two points*."[17]

[17] It might be well to note that Weierstrass here makes use of what is sometimes called an "improper field," or as he said a plane strip, about the given curve. It was first Schwarz ([1885], p. 225) and later Kneser (*LV*, p. 76) who introduced the present notation of a field for the simplest problem of the calculus of variations. They chose a point, say, 5, on the curve but to the left of the initial point 0 and near to 0 and used that instead of 0. (Bolza says that this idea was introduced by Zermelo [1894], pp. 87–88, but on p. 226 below we shall see that Weierstrass himself made use of this notion.) Thus they could define a field as a region of x, y space through every point of which passes a unique extremal of the family at hand. The initial point 0 is no longer an exceptional point.

He comments that it will be seen in the next chapter how easy it is to construct a plane strip about the given curve so that these properties are satisfied provided that inside the arc AB there is no point conjugate to A. He assumes that (01) is an arc of a solution of $G = 0$, satisfying the hypotheses made above: specifically, that along this arc the functions F^0 and F^1 are regular in their arguments x, y, x', y'; and that F_1 cannot vanish nor become infinite.

Weierstrass now chooses any point 2 on the given arc between 0 and 1 and a point 3 not on the arc but inside the above-mentioned plane strip. Weierstrass now considers the broken arc $(0\overline{32})$ made up of an arc (03) of a solution of $G = 0$ and an arc $(\overline{32})$ joining points 2 and 3 and in the strip. He finds then that $I^1_{02} = I^1_{03} + \bar{I}^1_{32}$. (Recall that $\bar{I}^1_{23}$, e.g., means the value of the integral I^1 evaluated along the curve $(\overline{23})$.) Now he compares the value of I^0 along the broken curve $(0\overline{32})$ with its value along (02) and finds, as before, that

$$\Delta I^0_{02} = \int_{t_0}^{t_2} G^0 w\, dt + \left[\frac{\partial F^0}{\partial x'} \xi_2 + \frac{\partial F^0}{\partial y'} \eta_2 \right]_{t_0}^{t_2} + \bar{F}^0[x_2, y_2, \bar{p}, \bar{q})\sigma + \cdots,$$

where $\bar{p}, \bar{q}$ signify the direction cosines of the curve $(\overline{32})$ at point 2 and ξ_2, η_2 are the variations of the coordinates of point 2; they are given, as we saw earlier in (5.22), by the relations

$$\xi_2 = (-\bar{p} + \sigma_1)\sigma, \qquad \eta_2 = (-\bar{q} + \sigma_2)\sigma.$$

Thus

$$\Delta I^s_{02} = \int_{t_0}^{t_2} G^s w\, dt + \left\{ F^2(x_2, y_2, \bar{p}, \bar{q}) - \left(\frac{\partial F^s}{\partial x'} \right)_{(2)} \bar{p} - \left(\frac{\partial F^s}{\partial y'} \right)_{(2)} \bar{q} \right\} \sigma + \cdots$$

$$(s = 0, 1).$$

Weierstrass now multiplies $\Delta I^1_{02} = 0$ by $-\lambda$, adds the result to ΔI^0, and finds

$$\Delta I^0_{02} = \mathcal{E}(x_2, y_2;\, p_2, q_2;\, \bar{p}, \bar{q})\sigma + \cdots,$$

$\mathcal{E}$ being formed from $F = F^0 - \lambda F^1$, where p_2, q_2 are the direction cosines of the extremal (02) at point 2 and the terms not shown in the relation above are regular functions of σ which contain a factor σ^2.

It follows that if ΔI^0_{02} does not change sign, then neither can the function $\mathcal{E}(x, y;\, p, q;\, \bar{p}, \bar{q})$ along the given curve for any two pairs p, q and $\bar{p}, \bar{q}$. Consider now in Figure 5.3 the arc (01) through points 0 and 1 and satisfying $G = 0$ and another curve $(\overline{01})$ through those points and having on it a point 2. Weierstrass now wishes to examine $\bar{I}^0_{02} - I^0_{02}$ as a function of point 2, and he denotes by σ arc-length along $(\overline{02})$. By the hypotheses made above, there is a solution of $G = 0$ through points 0 and 2 so that $I^1_{02} = \bar{I}^1_{02}$. He then examines the continuous function $f(s) = \bar{I}^0_{02} - I^0_{02}$ under the hy-

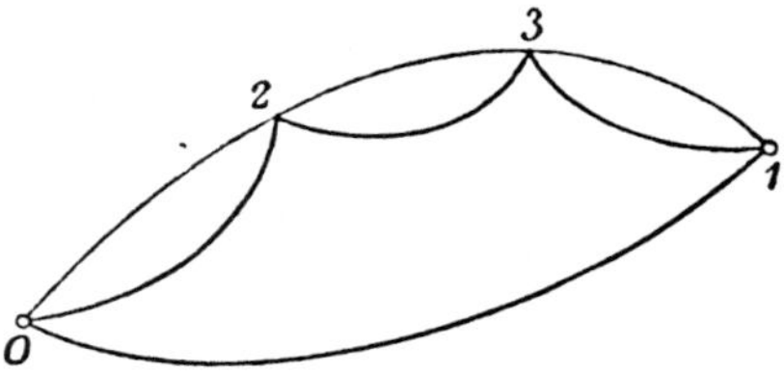

Figure 5.3

pothesis that (01) is a minimizing arc. Let 3 be a point between 0 and 2 and 4 a point between 2 and 1 so chosen that the distances along the arcs $(\overline{32})$ and $(\overline{24})$ have the same value σ, which will be understood to be small. Then

$$f(s-\sigma)=\bar{I}^0_{03}-I^0_{03}=\bar{I}^0_{02}-\bar{I}^0_{32}-I^0_{03}$$

and

$$f(s-\sigma)-f(s)=-\bar{I}^0_{32}-I^0_{03}+I^0_{02}=-\mathcal{E}(x_2,y_2;p_2,q_2;\bar{p},\bar{q})\sigma+\cdots.$$

Similarly,

$$f(s+\sigma)-f(s)=\bar{I}^0_{24}-I^0_{04}+I^0_{02}=\mathcal{E}(x_2,y_2;p_2,q_2;\bar{p},\bar{q})\sigma+\cdots$$

and hence

$$\frac{df(s)}{ds}=\mathcal{E}(x_2,y_2;p_2,q_2;\bar{p},\bar{q}),$$

provided only that the curve $(\overline{02})$ has a continuously turning tangent at the point 2.

Suppose now that $\mathcal{E}$ is positive for all values of its arguments; then $f(s)$ is an increasing function of the point 2, starting at $f(0)=0$. Thus when 2 coincides with 1, $f(s)$ is positive. Weierstrass summarizes in the theorem: "*If the function* $\mathcal{E}(x,y;p,q;\bar{p},\bar{q})$, *evaluated along the curve and for arbitrary* $\bar{p},\bar{q}$, *is always positive then* $\bar{I}^0_{01}>I^0_{01}$, *i.e., a minimum is surely exhibited.*"

In case $df/ds\equiv 0$, then f is a constant equal to 0 since $f(0)=0$, and hence $\bar{I}^0_{02}=I^0_{02}$ for every point 2 on $(\overline{01})$, in particular $\bar{I}^0_{01}=I^0_{01}$. But the curve $(\overline{01})$ does not satisfy the differential equation $G=0$ since by hypothesis only one solution (01) of $G=0$ can pass through points 0 and 1. Since $(\overline{01})$ is not a solution of $G=0$, a variation can always be found so that the corresponding integral will be smaller or larger than $\bar{I}^0_{01}=I^0_{01}$, and there can be no maximum or minimum.

Weierstrass remarks that the investigations above require continuity assumptions on the arbitrary curve $(\overline{01})$. Along the curve the quantities x',y' must vary continuously, and it must consist of a finite number of pieces along each of which the second derivatives are continuous in order that the complete variations ΔI can be expanded into power series.

If an arbitrary curve is given, it can always be replaced by a finite number of regular arcs for which the integral I^0 has nearly the same value. It follows that the value of the integral along the arbitrary curve is at least never less than its value along the original minimizing curve (01). To see this, consider in Figure 5.3 the minimizing arc (01) through points 0 and 1 and an arbitrary curve $(\overline{01})$ through the same points. Weierstrass chooses on the latter curve points 2 and 3 and draws the extremals (02), (23), and (31) for which the integral I^1 has the same values as on $(\overline{02})$, $(\overline{23})$, and $(\overline{31})$. But $\bar{I}^0_{02} \geqslant I^0_{02}$, $\bar{I}^0_{23} \geqslant I^0_{23}$, $\bar{I}^0_{31} \geqslant I^0_{31}$, and hence $\bar{I}^0_{0231} \geqslant I^0_{0231}$. But since points 2 and 3 do not lie on the minimizing arc, $I^0_{0231} > I^0_{01}$, and thus $\bar{I}^0_{0231} > I^0_{01}$, which is what Weierstrass wished to prove.

He again notes that when his necessary condition holds, so does the Legendre condition, but not conversely, and sums up his necessary condition in the theorem[18]: "*So long as the given curve is embedded in a plane strip with the properties described above the function* $\mathcal{E}(x, y; p, q; \bar{p}, \bar{q})$ *evaluated along the curve with* $\bar{p}, \bar{q}$ *arbitrary but different from* p, q *can assume only positive or only negative values.*"

Weierstrass now remarks on a point mentioned earlier in this chapter, which was later exploited by Zermelo and Kneser. He says "The considerations which have been set forth in this chapter also remain valid if on the same curve to which the arc (01) belongs, a point $\bar{0}$ sufficiently near to 0 is chosen, and a plane strip with respect to it constructed with the properties desired above." He goes on to note that this strip of field can be so chosen that the arc (01) is entirely in its interior. He now considers the arbitrary curve $(\overline{01})$ which is traced out by the point 2. By the hypotheses of this chapter there is one and only one arc $(\bar{0}2)$ satisfying $G = 0$ through that point 2 and $\bar{0}$. He assumes that

$$I^1_{\bar{0}2} = I^1_{\bar{0}0} + \bar{I}^1_{02},$$

i.e., that I^1 has the same value along the arc $(\bar{0}2)$ as along $(\bar{0}02)$. If in particular point 2 coincides with 1, then the value of the integral is $I^1_{\bar{0}01}$. As before, he can write the total variation in the form

$$\Delta I^0_{\bar{0}21} = \mathcal{E}(x_2, y_2; p_2, q_2; \bar{p}_2, \bar{q}_2)\sigma + \cdots,$$

where now p_2, q_2 are the direction cosines of the arc $(\bar{0}2)$ at point 2 and $\bar{p}_2, \bar{q}_2$ of the arc $(\overline{12})$ at the same point. Again he finds that

$$I^0_{\bar{0}1} < I^0_{\bar{0}0} + \bar{I}^0_{01}$$

and hence that $I^0_{01} < \bar{I}^0_{01}$, which means under his assumption that I^0_{01} is a minimum.

(Since he discovered the idea of moving the initial point slightly to the

[18] Weierstrass enunciated this in his 1879 lectures. It appears on p. 271.

left of 0 to $\bar{0}$, it would appear that Weierstrass also understood and conceived of the usual notion of a field and that others copied him.)

In closing the chapter Weierstrass says "From this result it is a possibility that the function $\mathcal{E}$ vanishes at some points on the curve or even along some arc of the curve without mattering. Only in the case that the function $\mathcal{E}$ vanishes along the entire curve must in addition a special investigation be employed."

In his 28th chapter Weierstrass discusses the meaning of the hypotheses he made above. To do this, he considers the family of solutions of $G = 0$, $x = \phi(t, \alpha, \beta, \lambda)$, $y = \psi(t, \alpha, \beta, \lambda)$, and supposes the given arc (01) is embedded in this family for given values of the parameters α, β, λ. He defines a function z of t with the help of the relation

$$z = \int_{t_0}^{t} F^1(x, y, x', y')\, dt$$

evaluated along the curve (01). Thus for $t = t_1$, he has $I^1 = z(t)$. He now considers a neighboring curve defined at corresponding points by the values t, α, β and $t + \tau, \alpha + \alpha', \beta + \beta', \lambda + \lambda'$ with end-points given by t_0, t_1 and $t_0 + \tau_0, t_1 + \tau_1$. He denotes by ξ, η, ζ the differences of corresponding coordinates x, y, z and finds that

$$\xi = \phi'(t)\tau + \phi_1(t)\alpha' + \phi_2(t)\beta' + \phi_3(t)\lambda' + \cdots,$$

$$\eta = \psi'(t)\tau + \psi'(t)\alpha' + \psi_2(t)\beta' + \psi_3(t)\lambda' + \cdots,$$

$$\zeta = \int_{t_0}^{t} G^1 \cdot (\phi'(t)\eta - \psi'(y)\xi)\, dt + \frac{\partial F^1}{\partial x'}\xi + \frac{\partial F^1}{\partial y'}\eta - \left(\frac{\partial F^1}{\partial x'}\right)_0 \xi_0 - \left(\frac{\partial F^1}{\partial y'}\right)_0 \eta_0 + \cdots,$$

where ξ_0, η_0 are the values of ξ, η at $t = t_0$ and $\phi_3 = \partial\phi/\partial\lambda$, $\psi_3 = \partial\psi/\partial\lambda$ with corresponding definitions for $\phi_1, \psi_1, \phi_2, \psi_2$. He gives the formula for ζ another form by setting

$$\theta_\nu(t) = \phi'(t)\psi_\nu(t) - \psi'(t)\phi_\nu(t) \qquad (\nu = 1, 2, 3)$$

and

$$\Theta_\nu(t, t_0) = \int_{t_0}^{t} G^1 \cdot \theta_\nu(t)\, dt \qquad (\nu = 1, 2, 3).$$

He finds that

$$\zeta - \frac{\partial F^1}{\partial x'}\xi - \frac{\partial F^1}{\partial y'}\eta + \left(\frac{\partial F^1}{\partial x'}\right)_0 \xi_0 + \left(\frac{\partial F^1}{\partial y'}\right)_0 \eta_0$$
$$= \Theta_1(t, t_0)\alpha' + \Theta_2(t, t_0)\beta' + \Theta_3(t, t_0)\lambda' + \cdots.$$

Combining these, Weierstrass finds for $t = t_0$ and t_1 that

$$\xi_0 = \phi'(t_0)\tau_0 + \phi_1(t_0)\alpha' + \phi_2(t_0)\beta' + \phi_3(t_0)\lambda' + \cdots,$$

$$\eta_0 = \psi'(t_0)\tau_0 + \psi_1(t_0)\alpha' + \psi_2(t_0)\beta' + \psi_3(t_0)\lambda' + \cdots,$$

$$\xi_1 = \phi'(t_1)\tau_1 + \phi_1(t_1)\alpha' + \phi_2(t_1)\beta' + \phi_3(t_1)\lambda' + \cdots,$$

$$\eta_1 = \psi'(t_1)\tau_1 + \psi_1(t_1)\alpha' + \psi_2(t_1)\beta' + \psi_3(t_1)\lambda' + \cdots,$$

$$\zeta_1 - \left(\frac{\partial F^1}{\partial x'}\right)_1 \xi_1 - \left(\frac{\partial F^1}{\partial y'}\right)_1 \eta_1 + \left(\frac{\partial F^1}{\partial x'}\right)_0 \xi_0 + \left(\frac{\partial F^1}{\partial y'}\right)_0 \eta_0$$
$$= \Theta_1(t_1, t_0)\alpha' + \Theta_2(t_1, t_0)\beta' + \Theta_3(t_1, t_0)\lambda' + \cdots, \tag{5.31}$$

since $\zeta_0 \equiv 0$. Now to solve this system of equations for sufficiently small values of $\tau_0, \tau_1, \alpha', \beta', \lambda'$, it suffices by his implicit-function theorem to be sure that the determinant

$$\Theta(t, t_0) = \begin{vmatrix} \phi'(t_0) & 0 & \phi_1(t_0) & \phi_2(t_0) & \phi_3(t_0) \\ \psi'(t_0) & 0 & \psi_1(t_0) & \psi_2(t_0) & \psi_3(t_0) \\ 0 & \phi'(t) & \phi_1(t) & \phi_2(t) & \phi_3(t) \\ 0 & \psi'(t) & \psi_1(t) & \psi_2(t) & \psi_3(t) \\ 0 & 0 & \Theta_1(t, t_0) & \Theta_2(t, t_0) & \Theta_3(t, t_0) \end{vmatrix}$$

$$= \begin{vmatrix} \theta_1(t) & \theta_2(t) & \theta_3(t) \\ \theta_1(t_0) & \theta_2(t_0) & \theta_3(t_0) \\ \Theta_1(t, t_0) & \Theta_2(t, t_0) & \Theta_3(t, t_0) \end{vmatrix} \tag{5.32}$$

does not vanish for $t = t_1$. He now says that "disregarding the case when the function $\Theta(t, t_0)$ vanishes identically for all values of t, *the first root of the equation being considered after* t_0 *as* t *moves across the interval* $(t_0 \ldots t_1)$ *determines the point conjugate to* t_0" (p. 275). (This is his generalization of conjugate points to the isoperimetric problem.)

Weierstrass now asserts that if there is no point conjugate to the point 0 on (01), it can be embedded in a plane strip which has the property that between 0 and each of its points passes one and only one curve satisfying the equation $G = G^0 - \lambda G^1 = 0$; this curve lies arbitrarily near to the original curve, and their points can be so ordered that the integral

$$\int_{t_0}^{t} F^1(x, y, x', y')\,dt$$

always has the same value on each of them; moreover, the value of λ on the new curve deviates only a little from its value on the original one. His proof is given later in the same chapter and will appear in due course.

He then states the fourth necessary condition: "*A maximum or minimum for the integral* I^0 *can not occur under the condition that the integral* I^1 *takes on a prescribed value if the arc contains in its interior a point conjugate to* t_0."

He does not attempt to demonstrate this until later. Instead, he carries out a certain amount of analysis he needs to characterize conjugate points further. To this end he wishes to consider the complete variation of G when $t, \alpha, \beta, \lambda$ vary. (The varied values are $t+\tau, \alpha+\alpha', \beta+\beta', \lambda+\lambda'$.) He writes this in the form

$$\Delta G = G^0(x+\xi, y+\eta) - G^0(x, y) - \lambda\big(G^1(x+\xi, y+\eta) - G^1(x, y)\big) - \lambda' G^1(x+\xi, y+\eta).$$

When $w = x'\eta - y'\xi$, it is not very difficult to show that

$$G^s(x+\xi, y+\eta) - G^s(x, y) = F_2^s \cdot w - \frac{d}{dt}\left(F_1^s \frac{dw}{dt}\right) + \cdots \qquad (s = 0, 1),$$

where the suppressed terms are at least second order. (The functions F_1 and F_2 appear in (5.4) and (5.4′) above.) He next expands w in powers of $\tau, \alpha', \beta', \lambda'$; to do this he aggregates all terms of first degree and calls the resulting expression $\overline{w}$. Then $\overline{w}$ has the form

$$\overline{w} = \theta_1(t)\alpha' + \theta_2(t)\beta' + \theta_3(t)\lambda',$$

and consequently

$$\Delta G = -\frac{d}{dt}\left(F_1 \frac{d\overline{w}}{dt}\right) + F_2\overline{w} - \lambda' G^1(x, y) + (\tau, \alpha', \beta', \lambda')_2.$$

Weierstrass now argues that since the varied curve must also satisfy $G = 0$ for sufficiently small $|\tau|, |\alpha'|, |\beta|, |\lambda'|$, the coefficients of $\alpha', \beta', \lambda'$ in the relation above must vanish. This implies that

$$F_1 \frac{d^2\theta_s}{dt^2} + \frac{dF_1}{dt}\frac{d\theta_s}{dt} - F_2\theta_s = 0, \qquad (s = 1, 2),$$

$$F_1 \frac{d^2\theta_3}{dt^2} + \frac{dF_1}{dt}\frac{d\theta_3}{dt} - F_2\theta_3 + G^1 = 0. \tag{5.33}$$

He expresses these in the equivalent form

$$\frac{d}{dt}\left(F_1\left(\theta_3 \frac{d\theta_s}{dt} - \theta_s \frac{d\theta_3}{dt}\right)\right) = \theta_s G^1 \qquad (s = 1, 2),$$

$$\frac{d}{dt}\left(F_1\left(\theta_1 \frac{d\theta_2}{dt} - \theta_2 \frac{d\theta_1}{dt}\right)\right) = 0,$$

or in the form

$$\left[F_1\left(\theta_3 \frac{d\theta_s}{dt} - \theta_s \frac{d\theta_3}{dt}\right)\right]_{t_0}^{t} = \Theta_s(t, t_0) \qquad (s = 1, 2),$$

$$F_1\left(\theta_1 \frac{d\theta_2}{dt} - \theta_2 \frac{d\theta_1}{dt}\right) = C, \tag{5.33′}$$

where C is an integration constant, which cannot be null. For if it were, then $\theta_1 = \text{const.}\cdot\theta_2$, which would mean by (5.33) that

$$\Theta_1(t, t_0) = const.\cdot\Theta_2(t, t_0);$$

this would imply with the help of the second determinantal form for $\Theta(t, t_0)$ in (5.32) that Θ vanished identically, which is contrary to Weierstrass's assumption.

He now goes on to show that $\Theta(t, t_0)$ changes sign as t passes through a zero. To do this he has from the definition of Θ above that

$$\frac{d\Theta\ (t, t_0\ )}{dt} = \begin{vmatrix} \theta_1(t) & \theta_2(t) & \theta_3(t) \\ \theta_1(t_0) & \theta_2(t_0) & \theta_3(t_0) \\ \dfrac{d\Theta_1}{dt} & \dfrac{d\Theta_2}{dt} & \dfrac{d\Theta_3}{dt} \end{vmatrix} + \begin{vmatrix} \theta_1'(t) & \theta_2'(t) & \theta_3'(t) \\ \theta_1(t_0) & \theta_2(t_0) & \theta_3(t_0) \\ \Theta_1(t, t_0) & \Theta_2(t, t_0) & \Theta_3(t, t_0) \end{vmatrix},$$

and from the expressions (5.33′) for Θ_ν that

$$\frac{d\Theta_\nu(t, t_0)}{dt} = G^1 \cdot \theta_\nu(t) \qquad (\nu = 1, 2, 3).$$

It then follows that

$$\frac{d\Theta(t, t_0)}{dt} = \begin{vmatrix} \theta_1'(t) & \theta_2'(t) & \theta_3'(t) \\ \theta_1(t_0) & \theta_2(t_0) & \theta_3(t_0) \\ \Theta_1(t, t_0) & \Theta_2(t, t_0) & \Theta_3(t, t_0) \end{vmatrix}.$$

To simplify notations, Weierstrass now defines f_1, f_2, f_3 with the help of the relations

$$\begin{aligned} \theta_2(t)\theta_3(t_0) - \theta_3(t)\theta_2(t_0) &= f_1(t), \\ \theta_3(t)\theta_1(t_0) - \theta_1(t)\theta_3(t_0) &= f_2(t), \\ \theta_1(t)\theta_2(t_0) - \theta_2(t)\theta_1(t_0) &= f_3(t). \end{aligned}$$

This enables him to write

$$\begin{aligned} \Theta(t, t_0) &= f_1(t)\Theta_1(t, t_0) + f_2(t)\Theta_2(t, t_0) + f_3(t)\Theta_3(t, t_0), \\ \frac{d}{dt}\Theta(t, t_0) &= f_1'(t)\Theta_1(t, t_0) + f_2'(t)\Theta_2(t, t_0) + f_3'(t)\Theta_3(t, t_0), \end{aligned}$$

since

$$f_1(t)\theta_1(t_0) + f_2(t)\theta_2(t_0) + f_3(t)\theta_3(t_0) = 0.$$

Combining these, he has the equation

$$\frac{d}{dt}\frac{\Theta(t, t_0)}{f_3(t)} = \frac{(f_3(t)f_1'(t) - f_1(t)f_3'(t))\Theta_1(t, t_0) + (f_3(t)f_2'(t) - f_2(t)f_3'(t))\Theta_2(t, t_0)}{f_3(t)^2}.$$

The numerator of the right-hand member of this equation is expressible as the product of F_1 and the square of $E = \theta_1'(t)f_1(t) + \theta_2'(t)f_2(t) + \theta_3'(t)f_3(t)$. To see this note the following. Since

$$\theta_2(t_0)E = f_1(t)f_3'(t) - f_3(t)f_1'(t), \qquad \theta_1(t_0)E = f_3(t)f_2'(t) - f_2(t)f_3'(t),$$

the numerator in question is representable as

$$(\theta_1(t_0)\Theta_2(t,t_0) - \theta_2(t_0)\Theta_1(t,t_0))E.$$

With the help of (5.33′), it follows that

$$\begin{aligned}\theta_1(t_0)\Theta_2(t,t_0) - \theta_2(t_0)\Theta_1(t,t_0) &= \left[F_1\theta_3(t)(\theta_1(t_0)\theta_2'(t) - \theta_2(t_0)\theta_1'(t))\right]_{t_0}^{t} \\ &\quad - F_1(t)\theta_3'(t)(\theta_1(t_0)\theta_2(t) - \theta_2(t_0)\theta_1(t)).\end{aligned}$$

The first term in the right-hand member of this equation—the one evaluated between t_0 and t—is expressible as

$$\begin{aligned}&\left[F_1(\theta_1(t)\theta_2'(t) - \theta_2(t)\theta_1'(t))\right]_{t_0}^{t}\theta_3(t_0) + F_1(t)\theta_3(t)(\theta_1(t_0)\theta_2'(t) - \theta_2(t_0)\theta_1'(t)) \\ &\quad - F_1(t)\theta_3(t_0)(\theta_1(t)\theta_2'(t) - \theta_2(t)\theta_1'(t)).\end{aligned}$$

The third relation in (5.33′) shows that the expression in brackets above is zero. When these facts are assembled, it results that

$$\theta_1(t_0)\Theta_2(t,t_0) - \theta_2(t_0)\Theta_1(t,t_0) = F_1E,$$

and consequently that

$$\frac{d}{dt}\,\frac{\Theta(t,t_0)}{f_3(t)} = \frac{F_1E^2}{f_3(t)^2}.$$

Weierstrass now supposes that t' is a root of $\Theta(t,t_0) = 0$ of order κ—the series expansion of Θ about t' begins with the term $(t - t')^{\kappa}$—and that $f_3(t') \neq 0$.[19] Accordingly, the expansion of the expression

$$f_3(t)\frac{d\Theta(t,t_0)}{dt} - f_3'(t)\Theta(t,t_0)$$

begins with the term $(t - t')^{\kappa-1}$. This expression is, as we just saw, equal to F_1E^2; since, by hypothesis, F_1 neither vanishes along the given curve nor is infinite, the expansion of F_1E^2 must begin with an even power of $t - t'$—it cannot change sign—and κ must be odd. This implies immediately that Θ changes sign at each root.

If $f_3(t') = 0$, then $f_3'(t') \neq 0$, because otherwise

$$\theta_1(t_0)\theta_2(t') - \theta_2(t_0)\theta_1(t') = 0 \qquad \text{and} \qquad \theta_1(t_0)\theta_2'(t') - \theta_2(t_0)\theta_1'(t') = 0,$$

which is possible only if $\theta_1(t_0) = \theta_2(t_0) = 0$ or $\theta_1(t')\theta_2'(t') - \theta_2(t')\theta_1'(t') = 0$.

---

[19] Notice that $f_3 = 0$ implies at once that

$$\theta_1(t)\theta_2(t_0) - \theta_2(t)\theta_1(t_0) = 0$$

and thus that t is a point conjugate to t_0 for the unrestricted problem, i.e., the problem without the isoperimetric condition (see Bolza, *VOR* pp. 478ff). We examine this case directly.

But by the third equation (5.33′), these equations are impossible since $C \neq 0$, $F_1 \neq 0$, $\neq \infty$ and hence not both $f_3(t')$ and $f_3'(t')$ vanish at the same point. From this Weierstrass infers that the expansions of $\Theta(t, t_0)$ and $f_3(t)$ are of the forms

$$\Theta(t, t_0) = a(t - t')^{\kappa} + \cdots, \qquad f_3(t) = b(t - t') + \cdots$$

and consequently that in this case the series expansion of

$$f_3(t)\frac{d\Theta(t, t_0)}{dt} - f_3'(t)\Theta(t, t_0)$$

begins with the term $ab(\kappa - 1)(t - t')^{\kappa}$. When $\kappa = 1$ his proof is still valid. However this is not necessary. See his example on pp. 301–302 or see Bolza, *VOR*, p. 478 for a discussion of the exceptional case. The condition $f_3 = 0$ determines the conjugate points for the unrestricted problem as mentioned in footnote 19.

Weierstrass turns now to a generalization of his notion of a strip for isoperimetrical problems. He again considers the family of solutions of $G = 0$, $x = \phi(t, \alpha, \beta, \lambda)$, $y = \psi(t, \alpha, \beta, \lambda)$, which contains the given arc (01) for the parameter values α, β, λ. He now appends to the coordinates x, y the third one

$$z = \int_{t_0}^{t} F^1(x, y, x', y')\,dt$$

evaluated along (01) and draws, as in Figure 5.4, the space curve (01′). Let the curve (01) vary in the family, and set up a correspondence between points on the original and the varied curve so that at corresponding points the values of z are the same.[20] Suppose that 2 is a point on (01) such that there is no point conjugate to 0 between 0 and 2, and suppose that $\bar{2}$ is the point corresponding to 2 on $(0\bar{2})$, a neighboring curve. He then constructs the space curve $(0\bar{2}')$ and notes that the space curves embedding (02′) fill a

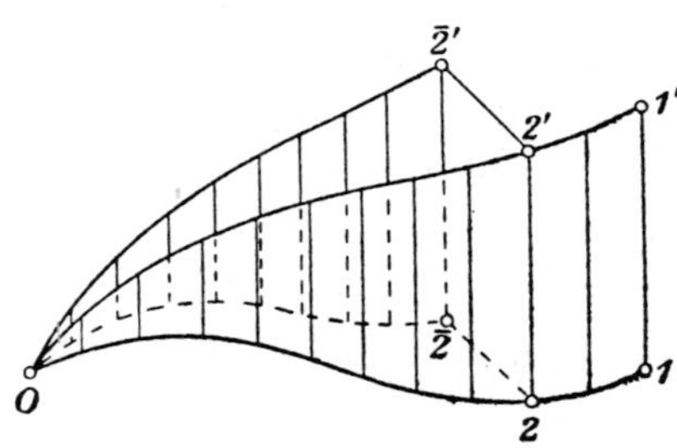

Figure 5.4

[20]The problem implicitly posed by Weierstrass is that of finding among all space curve passing through the fixed points x_0, y_0, 0 and x_1, y_1, $z_1 = \int_{t_0}^{t_1} F^1\,dt$ one which minimizes the integral $\int_{t_0}^{t_1} F^0\,dt$ subject to the side-condition $z' = F^1(x, y, x', y')$ and the end-conditions that x, y, z are fixed for $t = t_0$ and t_1. This is, in effect, the reason for his recourse to space curves. What he produces is the first example of a field in three-space.

portion of three-space. The projection of this region of three-space onto the plane embeds the original curve (02) with the exception of the point 0, which lies on the boundary of the projection. When he prolongs the curve (01) slightly backward to $\bar{0}$ and carries out the construction above with point 0 replaced by $\bar{0}$, then everything goes through as before, and (01) is completely inside the projection.

Weierstrass now wishes to examine the geometrical significance of his notion of conjugate points for the isoperimetric problem. He considers a space curve of the family, described by the parameter values α', β', λ', which cuts the original curve in a point (t''). By power series expansions he has with the help of equations (5.31) above

$$\begin{aligned}
0 &= \phi'(t_0)\tau_0 + \phi_1(t_0)\alpha' + \phi_2(t_0)\beta' + \phi_3(t_0)\lambda' + (\tau_0, \alpha', \beta', \lambda')_2,\\
0 &= \psi'(t_0)\tau_0 + \psi_1(t_0)\alpha' + \psi_2(t_0)\beta' + \psi_3(t_0)\lambda' + (\tau_0, \alpha', \beta', \lambda')_2,\\
0 &= \phi'(t'')\tau + \phi_1(t'')\alpha' + \phi_2(t'')\beta' + \phi_3(t'')\lambda' + (\tau, \alpha', \beta', \lambda')_2,\\
0 &= \psi'(t'')\tau + \psi_1(t'')\alpha' + \psi_2(t'')\beta' + \psi_3(t'')\lambda' + (\tau, \alpha', \beta', \lambda')_2,\\
0 &= * \qquad\quad + \Theta_1(t'', t_0)\alpha' + \Theta_2(t'', t_0)\beta' + \Theta_3(t'', t_0')\lambda' + (\alpha', \beta', \lambda')_2,
\end{aligned}$$

where $(\tau_0, \alpha', \beta', \lambda')_2$, etc. designate terms in the expansions of at least second degree. He goes to eliminate τ_0 and τ from the linearized equations; in doing this he finds that

$$\begin{aligned}
0 &= \theta_1(t_0)\alpha' + \theta_2(t_0)\beta' + \theta_3(t_0)\lambda' + (\alpha', \beta', \lambda')_2,\\
0 &= \theta_1(t'')\alpha' + \theta_2(t'')\beta' + \theta_3(t'')\lambda' + (\alpha', \beta', \lambda')_2,\\
0 &= \Theta_1(t, t_0)\alpha' + \Theta_2(t'', t_0)\beta' + \Theta_3(t'', t_0)\lambda' + (\alpha', \beta', \lambda')_2,
\end{aligned}$$

where the θ_ν $(\nu = 1, 2, 3)$ are defined on p. 227 above. From these he eliminates α' and β' and finds that $0 = \Theta(t'', t_0) + (\lambda')$, where (λ') vanishes with λ'.

If now $\Theta(t'', t_0) \neq 0$, then $|\lambda'|$ could be chosen so small that

$$|\Theta(t'', t_0)| > |(\lambda')|,$$

which contradicts the condition just found. If t' is a value of t'' for which $\Theta(t', t_0) \neq 0$, then by continuity considerations there is an interval $t' - \tau_1 \ldots t' + \tau_1$ inside which Θ remains different from zero. It consequently follows for $\Theta(t', t_0) \neq 0$ that no member of the family of space curves can, for small changes α', β', λ' in the original parameter values, cut the original curve near the point t'.

If t' is a zero of $\Theta(t, t_0) = 0$, then for some $\tau > 0$ the values $\Theta(t' + \tau, t_0)$ and $\Theta(t' - \tau, t_0)$ have opposite signs provided that $|\lambda'|$ is sufficiently small; and there is a value t'' in the interval $t' - \tau \ldots t' + \tau$ such that

$$\Theta(t'', t_0) + (\lambda') = 0.$$

Weierstrass (*VOR*, p. 284) states:

> With this it has been shown that when one delimits a very small region of the point conjugate to t_0 and puts a very small upper bound on the variations of the constants $\alpha', \beta', \lambda'$, one can always find among the admissible space curves some which pass through the point t_0 and intersect the original curve inside the region. All space curves through t_0 for which the $\alpha', \beta', \lambda'$ are situated beneath a fixed limit, even intersect the original curve in the region discussed. This limit for the quantities $\alpha', \beta', \lambda'$ will become infinitely small equally with the enlargement of the interval; one can therefore describe the point conjugate to t_0 as the limit point to which the intersections of the neighboring space curves (in the previously explained sense) with the original curve approach without bound.
>
> In a similar way one can show that to a point on the space curve, which is not conjugate to t_0, a very small region of space can be delimited so that the points conjugate to t_0 on neighboring space curves likewise do not lie in it; that on the other hand when one delimits the point conjugate to t_0 to a very small region of space, then the points conjugate to t_0 on the neighboring space curves all will be situated in its interior.

In concluding the chapter Weierstrass asserts that

$$\Theta(t, t_0 + \varepsilon) = \Theta(t, t_0) + (\varepsilon, t),$$

where (ε, t) vanishes with ε for every value of t. If t' is conjugate to t_0, $\Theta(t' - \tau, t_0)$ and $\Theta(t' + \tau, t_0)$ have opposite signs and so $\Theta(t, t_0 + \varepsilon)$ does also for $t = t' - \tau$, $t = t' + \tau$. It must therefore vanish for some value t between $t' - \tau$ and $t' + \tau$. The change in the point conjugate to t_0 caused by a small variation in t_0 is then arbitrarily small. These remarks apply when f_1, f_2, f_3 do not all vanish together.

In his 29th chapter he continues the discussion and returns to the form for the complete variation of I^1

$$\Delta I^1 = \kappa \int_{t_0}^{t_1} G^1 w\, dt + \kappa_1 \int_{t_0}^{t_1} G^1 w_1\, dt + (\kappa, \kappa_1)_2$$

that he established on p. 220 above and to the form for the complete variation of I^0

$$\Delta I^0 = \frac{1}{2}\kappa^2 \int_{t_0}^{t_1} \left\{ F_1\left(\frac{dw}{dt}\right)^2 + F_2 w^2 \right\} dt + (\kappa)_3.$$

He now rewrites this in the form

$$\Delta I^0 = \frac{1}{2}\varepsilon\kappa^2 \int_{t_0}^{t_1} w^2 dt + \frac{1}{2}\kappa^2 \int_{t_0}^{t_1} \left\{ F_1\left(\frac{dw}{dt}\right)^2 + (F_2 - \varepsilon) w^2 \right\} dt + (\kappa)_3,$$

for ε an arbitrary constant.

He asserts that when $|\varepsilon|$ is sufficiently small and when the integration extends beyond a point t' conjugate to t_0, a function $w \not\equiv 0$ can be so chosen that it vanishes for $t = t_0$ and $t = t_1$ and that[21]

$$W^1 = \int_{t_0}^{t_1} G^1 w\,dt = 0, \tag{5.34}$$

$$\int_{t_0}^{t_1}\left\{ F_1\left(\frac{dw}{dt}\right)^2 + (F_2 - \varepsilon)w^2 \right\} dt = 0. \tag{5.34'}$$

After this has been proved, the complete variation ΔI^0 for such a w reduces to

$$\frac{1}{2}\varepsilon\kappa^2 \int_{t_0}^{t_1} w^2\,dt;$$

and thus this variation has the same sign as ε, i.e., it can be made either positive or negative and hence there can be no extremum.

To make his proof, Weierstrass writes the relation (5.34′) with the help of (5.34) as

$$\int_{t_0}^{t_1}\left\{ -\frac{d}{dt}\left(F_1\frac{dw}{dt}\right) + (F_2 - \varepsilon)w - e_3 G^1 \right\} w\,dt = 0,$$

where e_3 is a quantity independent of t. He then considers the three linear differential equations

$$\frac{d}{dt}\left(F_2 \frac{d\theta_s(t,\varepsilon)}{dt}\right) - (F_2 - \varepsilon)\theta_s(t,\varepsilon) = 0, \qquad (s = 1,2),$$

$$\frac{d}{dt}\left(F_1 \frac{d\theta_3(t,\varepsilon)}{dt}\right) - (F_2 - \varepsilon)\theta_3(t,\varepsilon) + G^1 = 0. \tag{5.35}$$

(Compare these to the comparable equations (5.33).) If $F_1 \neq 0$ or ∞, then the solutions $\theta_\nu(t,\varepsilon)$ $(\nu = 1,2,3)$ can differ from the $\theta_\nu(t)$ only by quantities that vanish with ε.

Suppose again that t' is conjugate to t_0 and that $t'' < t_1$; then define

$$w = \begin{cases} e_1\theta_1(t,\varepsilon) + e_2\theta_2(t,\varepsilon) + e_3\theta_3(t,\varepsilon) & \text{on} \quad t_0 \ldots t'', \\ 0 & \text{on} \quad t'' \ldots t_1. \end{cases}$$

---

[21] The basic reason why the unrestricted problem of the calculus of variations and the isoperimetric one differ is that for the former one

$$\int_{t_0}^{t_1}\left\{ F_1\left(\frac{dw}{dt}\right)^2 + F_2 w^2 \right\} dt \geqslant 0$$

for every w vanishing at $t = t_0$ and t_1, whereas for the latter one w must satisfy the additional condition

$$\int_{t_0}^{t_1} G^1 w\,dt = 0.$$

It is *a priori* evident that the isoperimetric problem is, in general, more restrictive. The proper generalization for the isoperimetric problem of the notion of conjugate points was first discovered by Lundström [1869].

This function w is not identically zero since the solutions of the differential equations (5.35) could be linearly dependent only if the equation $G^1 = 0$ were satisfied. Weierstrass explicitly bars this case. Now the function w must clearly satisfy the differential equation

$$\frac{d}{dt}(F_1 w) - (F_2 - \varepsilon)w + e_3 G^1 = 0 \quad \text{on} \quad t_0 \ldots t'';$$

he wishes next to show that it vanishes at $t = t_0$ and t'' and is such that

$$\int_{t_0}^{t''} G^1 w\, dt = 0.$$

He has

$$\int_{t_0}^{t''} G^1 w\, dt = e_1 \int_{t_0}^{t''} G^1 \theta_1(t,\varepsilon)\, dt + e_2 \int_{t_0}^{t''} G^1 \theta_2(t,\varepsilon)\, dt + e_3 \int_{t_0}^{t''} G^1 \theta_3(t,\varepsilon)\, dt,$$

and he defines the quantities

$$\int_{t_0}^{t''} G^1 \theta_\nu(t,\varepsilon)\, dt = \Theta_\nu(t'', t_0, \varepsilon) \qquad (\nu = 1,2,3).$$

Then he needs to show that the conditions

$$\begin{aligned} &e_1\theta_1(t_0,\varepsilon) + e_2\theta_2(t_0,\varepsilon) + e_3\theta_3(t_0,\varepsilon) = 0, \\ &e_1\theta_1(t'',\varepsilon) + e_2\theta_2(t'',\varepsilon) + e_3\theta_3(t'',\varepsilon) = 0, \\ &e_1\Theta_1(t'',t_0,\varepsilon) + e_2\Theta_2(t'',t_0,\varepsilon) + e_3\Theta_3(t'',t_0,\varepsilon) = 0 \end{aligned} \tag{5.36}$$

are nontrivially satisfied. The determinant of this system differs from the determinant $\Theta(t'', t_0)$ of (5.32) by an amount which vanishes with ε.

It was proved above that $\Theta(t'', t_0)$ has opposite signs for $t'' = t' - \tau$ and $t'' = t' + \tau$; and consequently for $|\varepsilon|$ very small, the determinant above does also. This implies that the determinant must vanish for some value t'' between $t' - \tau$ and $t' + \tau$. For this value, system (5.36) must have a solution $e_1, e_2, e_3 \neq 0, 0, 0$; and Weierstrass has shown that w has the properties he asserted.

In his 27th chapter Weierstrass mentioned the special case when the $\mathfrak{E}$-function vanishes along a curve. Here he discusses the case and after a lengthy calculation concludes that this cannot occur in his spatial strip or field. It is not pointful to include it here.

In his 30th chapter (pp. 269ff) Weierstrass gives examples to illustrate his procedures. The first of these involves finding the surface of revolution containing a fixed volume which, when moving through the air, will encounter the least resistance. This is an isoperimetric version of Newton's famous problem discussed above. In his 31st and final chapter Weierstrass takes up briefly and tentatively two new classes of problems: those with the variable end-points and those with restricted variations, i.e., those where points or segments of the minimizing arc lie on the boundary of the region.

5.11. Scheeffer's Results

Before moving on to the work of Clebsch, Mayer, Lipschitz, and those others who investigated complex problems of the calculus of variations, it is not without interest to look briefly at the work of a young mathematician, Ludwig Scheeffer of Munich. He wrote three papers on the subject in 1885 and died before his last one appeared in the *Annalen* for 1886.[22] His papers are included here mainly to serve as a bridge to those of Clebsch, Mayer, and others and to record the first published sufficiency proof. Scheeffer gave that proof for weak minima in ignorance of Weierstrass's work. At the end of his 1885 paper he has a two-page note entitled "Remarks on the preceding paper." In it he says *inter alia*[23]:

> Two communications, for which I have to express thanks for the kindness of Messrs. A. Mayer and Weierstrass, cause me to make the following remarks.
>
> In §3 of the preceding paper I have shown that the Lagrange rule for handling the first variation for problems of relative minima in their generality is not properly established. In the meantime Mr. A. Mayer has found a completely satisfactory proof which covers the aforementioned general case . . . .
>
> The second remark concerns itself with the domain of validity of the aforementioned criteria for the occurrence of maximum or a minimum. I have called special attention to the fact that they are only valid in case one compares the given integral evaluated along some curve with such other integrals which are calculated along *neighboring* curves . . . . By a *neighboring* curve however one does not mean in general every curve, which lies inside of a small imbedding plane strip but only those in particular which remain uniformly nearly *parallel* to the given one over their entire domain; since the power series, whose terms of second order furnish the second variations, contains in addition to powers of the coordinate differences η also their derivatives η', these must thence also be under these given bounds, if the sign of the second variation is to be decisive.
>
> If one asks the question whether a maximum or minimum occurs in general in comparison to all curves which lie inside of a small plane strip imbedding the given curve without the condition of near parallelism having to be satisfied, then the preceding criteria arising out of a consideration of the second variation certainly are always *necessary* but no longer *sufficient*. In this case Mr. Weierstrass has found examples in which a minimum in this *extended* sense does not obtain even though the second variation is positive throughout.

After an example to show that an arc may make the first variation vanish and the second variation positive and yet not furnish a strong minimum, he goes on to discuss what he learned from a "generous oral

---

[22] Scheeffer [1885], [1885′], [1886].

[23] Scheeffer [1885′], pp. 594–595.

communication," presumably from Weierstrass. What he learned was that Weierstrass had known how to characterize "the maximum or minimum in this *extended* sense for some years and had lectured on this to his students." This obviously was a cause of great concern to Scheeffer since he felt called on to say "therefore I would like explicitly to emphasize that I got to know of the existence of Mr. Weierstrass's investigations only later; I had at my disposal only lecture notes for the year 1877 in which no indication of these ideas is found."

Inasmuch as Scheeffer's first paper is an excellent introduction to the papers of Clebsch and Mayer, it is perhaps worth our while to discuss briefly his results. To start he concerns himself with an integral

$$J = \int_{x_0}^{x_1} F(x, y_1, y_1', \ldots, y_n, y_n')\, dx$$

and assumes that the values of $y_1, y_2, \ldots, y_n$ are given at x^0 and x^1. He turns at once to a consideration of the second variation

$$\begin{aligned}\Delta^2 J &= \frac{1}{2}\int_{x^0}^{x^1} \Omega(\eta, \eta')\, dx \\ &= \frac{1}{2}\int_{x^0}^{x^1} \sum_{i,k=1}^{n} \left(\frac{\partial^2 F}{\partial y_i\, \partial y_k}\eta_i\eta_k + 2\frac{\partial^2 F}{\partial y_i\, \partial y_k'}\eta_i\eta_k' + \frac{\partial^2 F}{\partial y_i'\, \partial y_k'}\eta_i'\eta_k' \right) dx\end{aligned}$$

in which $\eta_1, \eta_2, \ldots, \eta_n$ are arbitrary continuous functions of x, which vanish at x^0 and x^1. The question of the sign of $\Delta^2 J$ for various η is then the topic under investigation, and Scheeffer first shows easily that if Ω does not depend on $\eta_1', \eta_2', \ldots, \eta_n'$ but only on the η_i, then it is necessary and sufficient for the second variation to be positive that $\Omega(\eta)$ never be negative at any point between x^0 and x^1.

Clebsch transformed the expression $\Omega(\eta, \eta')$ into a new form

$$\Omega_1(\eta_1\eta') + \frac{d}{dx}\Omega_2(\eta),$$

in which Ω_2 is a quadratic form in the n quantities η and Ω_1 is a quadratic form in η, η' which can be expressed as a quadratic form $\Omega_1(\omega)$ in n new variables ω. The actual form of Ω_1 is

$$\Omega_1(\omega) = \sum_{i,k=1}^{n} \frac{\partial^2 F}{\partial y_i'\, \partial y_k'}\omega_i\omega_k.$$

Clebsch then has $\Omega(\eta, \eta') = \Omega_1(\omega) + d\Omega_2(\eta)/dx$ and hence since Ω_2 vanishes at $x = x^0$ and x^1,

$$\Delta^2 J = \frac{1}{2}\int_{x^0}^{x^1} \Omega_1(\omega)\, dx.$$

Scheeffer also remarks that Ω_2 contains in it, as Clebsch and Mayer show, as a factor the reciprocal of the determinant

$$\Delta(x, x^0) = \begin{vmatrix} u_{11} & u_{12} & \cdots & u_{1,2n} \\ u_{21} & u_{22} & \cdots & u_{2,2n} \\ \cdots & \cdots & \cdots & \cdots \\ u_{n1} & u_{n2} & \cdots & u_{n,2n} \\ u_{11}^0 & u_{12}^0 & \cdots & u_{1,2n}^0 \\ u_{21}^0 & u_{12}^0 & \cdots & u_{2,2n}^0 \\ \cdots & \cdots & \cdots & \cdots \\ u_{n1}^0 & u_{n2}^0 & \cdots & u_{n,2n}^0 \end{vmatrix}$$

Here the u_{ij} are partial derivatives with respect to the parameters of a $2n$-parameter family of solutions of the Euler equations.

Scheeffer sums up in the theorem:

> *Let the determinant $\Delta(x, x^0)$ not vanish for any value of the variable x between x^0 and x^1; then the maximum or minimum of the integral J depends upon whether the quadratic form*
>
> $$\Omega_1(\omega) = \sum_{ik} \frac{\partial^2 F}{\partial y_i' \, \partial y'_k} \omega_i \omega_k$$
>
> *is always negative or always positive at every value x between x^0 and x^1 for arbitrary values of $\omega_1 \omega_2 \ldots \omega_n$ or whether it can be positive as well as negative.*

In his second paper Scheeffer [1886] considers the integral

$$J = \int_{x^0}^{x^1} F(x, y, y') \, dx$$

and asks that it be a minimum among the arcs in a given set joining two fixed points. He assumes that "I. All partial derivatives through the third order of the function $F(x, y, y')$ are finite and continuous on the interval $x^0 x^1$ and $\partial^2 F / \partial y'^2$ [$= F_{y'y'}$] is moreover *positive throughout*. II. A certain determinant $\Delta(x, x^0)$ is *different from zero* for all values of x between x^0 and x^1 and also for $x = x^1$ and is finite and continuous together with its first derivative."[24]

He goes on to define the function Δ in terms of two solutions u_1, u_2 of the accessory differential equation, (5.38) below, as

$$\Delta(x, x^0) = u_1(x_0) u_2(x) - u_2(x_0) u_1(x).$$

To make his proof of sufficiency, he expresses the integrand $F(x, y, y')$ with the help of a series expansion in the form

$$F(x, y + \Delta y, y' + \Delta y') = F(x, y, y') + g_1 + g_2 + r_3,$$

[24] Scheeffer [1886], p. 202.

where he has

$$g_1 = \frac{\partial F}{\partial y}\Delta y + \frac{\partial F}{\partial y'}\Delta y', \qquad 2g_2 = \frac{\partial^2 F}{\partial y^2}\Delta y^2 + 2\frac{\partial^2 F}{\partial y\,\partial y'}\Delta y \Delta y' + \frac{\partial^2 F}{\partial y'^2}\Delta y'^2,$$

and a corresponding third-order expression for r_3 where y, y' have been evaluated at some values $y + \theta\Delta y$, $y' + \theta\Delta y'$ $(0 < \theta < 1)$. It follows from this that since the integral of g_1, the first variation, vanishes along his minimizing curve, the total variation of his integral J is

$$\Delta J = \int_{x^0}^{x^1} F(x, y + \Delta y, y' + \Delta y')\,dx - \int_{x^0}^{x^1} F(x, y, y')\,dx = G_2 + R_3,$$

where G_2 and R_3 are the integrals from x^0 to x^1 of g_2 and r_3.

His plan is to show that R_3 is very small compared to G_2 provided that both $|\Delta y|$ and $|\Delta y'|$ are sufficiently small. From this he infers that the difference ΔJ will have the sign of the second variation. To do this he notes that Jacobi expressed the second variation in the form

$$G_2 = \frac{1}{2}\int_{x^0}^{x^1} \frac{\partial^2 F}{\partial y'^2} u^2 \left(\frac{d}{dx}\frac{\Delta y}{u} \right)^2 dx,$$

where u is a linear combination of the solutions u_1, u_2 above.[25] He asserts that if $F_{y'y'} > 0$ on $[x^0, x^1]$, it has a positive lower bound $2a^2$, and as a result he can infer that if $\zeta = \Delta y / u$, then

$$G_2 > a^2 \int_{x^0}^{x^1} \zeta'^2\,dx.$$

To bound R_3, he replaces Δy and $\Delta y'$ by $u\zeta$ and $u\zeta' + u'\zeta$ and finds that

$$|r_3| < c^2(|\zeta| + |\zeta'|)(\zeta^2 + \zeta'^2),$$

$$|R_3| < c^2(z + z')\int_{x^0}^{x^1} (\zeta^2 + \zeta'^2)\,dx,$$

where z and z' are upper bounds for ζ and ζ' on $[x^0, x^1]$. After some further discussion he concludes his last paper with the theorem:

> *Suppose for a given function y that the first variation of the integral $J = \int_{x^0}^{x^1} F(x, y, y')\,dx$ (in which the end-values of y for $x = x^0$ and $x = x^1$ are preassigned) is identically zero; suppose further that on the interval x^0x^1 the partial derivatives of $F(x, y, y')$ through the third order as well as the determinant $\Delta(x, x^0)$ are finite and continuous. If $\partial^2 F/\partial y'^2$ has a positive value on the interval x^0x^1 and if the determinant does not vanish either between x^0 and x^1 or at the place $x = x^1$ itself, then not only is the second variation always positive but also the integral J has a minimum in the sense that the difference*
>
> $$\Delta J = \int_{x^0}^{x^1} F(x, y + \Delta y, y' + \Delta y')\,dx - \int_{x^0}^{x^1} F(x, y, y')\,dx$$

---

[25] In his 1885 paper, Scheeffer has used $u = \Delta(x, x_0)$. This, or any other, choice is valid so long as there is no point on the closed interval $[x^0, x^1]$ conjugate to x_0.

is positive for every function Δy whose absolute value remains less than a given bound g while the absolute value of its derivative $\Delta y'$ throughout remains less than a second bound. If, on the contrary, $\partial^2 F/\partial y'^2$ is negative at any place on the interval x^0x^1 or if $\Delta(x, x^0)$ vanishes inside the interval x^0x^1, then the second variation is also capable of taking on a negative sign and for this reason no minimum can occur for the integral J.

Scheeffer remarks in his penultimate paragraph that the proof for the second part of his theorem is carried out in his 1885 paper. His proof that the second variation can take on either positive or negative values when there is a conjugate point inside the interval x^0x^1 is very similar to one due to Erdmann ([1878], pp. 365–368). Let us look at Scheeffer's proof in [1885], pp. 547ff.

He writes the integrand of the second variation in the usual form

$$\Omega(\eta, \eta') = F_{yy}\eta^2 + 2F_{yy'}\eta\eta' + F_{y'y'}\eta'^2$$

and shows without difficulty (p. 547) that, in his notation,

$$F_{y'^2}(u_1 u_2' - u_2 u_1') = C, \tag{5.37}$$

where u_1 and u_2 are two solutions of the accessory equation

$$\Omega_u - \frac{d}{dx}\Omega_{u'} = 0, \tag{5.38}$$

and he assumes that $F_{y'y'} \neq 0$ along the arc being considered. He points out that the constant C above cannot be zero when these solutions u_1, u_2 are taken to be the partial derivatives $\partial y/\partial c_1, \partial y/\partial c_2$ of a two-parameter family of solutions $y(x, c_1, c_2)$ of the Euler equation. His proof that $C \neq 0$ depends on writing an expansion for y about $x = x^0$. He has $y = c_1 + c_2(x - x^0) + \cdots$ and thus $\partial y/\partial c_1 = u_1 = 1 + \cdots$, $\partial y/\partial c_2 = u_2 = (x - x^0) + \cdots$. From this he has $[u_1u_2' - u_2u_1']_{x=x_0} = 1$ and $C = [\partial F/y'^2]_{x=x_0}$. Furthermore, since

$$u_1u_2' - u_2u_1' = u_1^2 \frac{d}{dx}\left(\frac{u_2}{u_1}\right),$$

it follows that u_2/u_1 is not a constant; and hence $\Delta(x, x^0) = u_2^0 u_1 - u_1^0 u_2$ is not identically zero. (The expressions u_s^0 $(s = 1, 2)$ mean u_s evaluated at $x = x^0$. In what follows I have slightly modified Scheeffer's notation to simplify matters a little.)

Scheeffer now considers the arc A^0A^1 in Figure 5.5 and fixes attention on two arbitrary points $A^2 : (x^2, y^2)$ and $A^3 : (x^3, y^3)$ on this arc and a point $A : (\xi, y(\xi) + \eta)$ between these points near to the given arc. He then forms

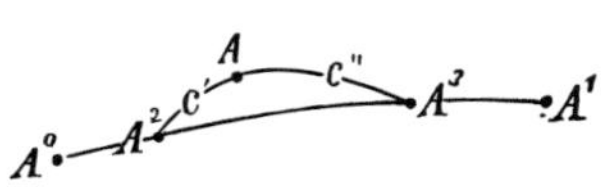

Figure 5.5

two arcs c', c'' neighboring to the arc A^2A^3 with the help of the variations

$$\bar{u}(x) = \frac{\Delta(x, x^2)}{\Delta(\xi, x^2)}\eta(\xi), \qquad \bar{v}(x) = \frac{\Delta(x, x^3)}{\Delta(\xi, x^3)}\eta(\xi), \tag{5.39}$$

which are linear combinations with constant coefficients of u_1 and u_2 and hence also solutions of the accessory differential equations (5.38). The variation defined by

$$u = \begin{cases} 0 & \text{on } [x^0, x^2] \\ \bar{u} & \text{on } [x^2, \xi] \\ \bar{v} & \text{on } [\xi, x^3] \\ 0 & \text{on } [x^3, x^1] \end{cases}$$

is then such that along it $\Delta^2 J = (\Delta^2 J)_{c'+c''}$. Clearly, this variation is continuous since $\bar{u}(\xi) = \bar{v}(\xi)$ but does not have a continuous derivative at $x = \xi$. Scheeffer then expresses the second variation evaluated along the arcs c' and c'' as

$$(\Delta^2 J)_{c'+c''} = (\Delta^2 J)_{c'} + (\Delta^2 J)_{c''} = \frac{1}{2}\int_{x^2}^{\xi}\Omega(\bar{u}, \bar{u}')\,dx + \frac{1}{2}\int_{\xi}^{x^3}\Omega(\bar{v}, \bar{v}')\,dx;$$

(this is clear since, e.g., $\bar{u}(x^2) = 0$, $\bar{u}(\xi) = \eta(\xi)$, $\bar{v}(\xi) = \eta(\xi)$, $\bar{v}(x^3) = 0$) and consequently he has, with the help of the accessory equation,

$$(\Delta^2 J)_{c'} = \tfrac{1}{4}\left[\Omega_{u'}(\bar{u}, \bar{u}')\bar{u}\right]_{x^2}^{\xi}, \qquad (\Delta^2 J)_{c''} = \tfrac{1}{4}\left[\Omega_{u'}(\bar{v}, \bar{v}')\bar{v}\right]_{\xi}^{x^3}.$$

He then finds easily that

$$\begin{aligned} (\Delta^2 J)_{c'+c''} &= \frac{1}{2}F_{y'y'}\cdot\frac{(u_1u_2' - u_2u_1')\Delta(x^3, x^2)}{\Delta(\xi, x^2)\cdot\Delta(\xi, x^3)}\eta^2(\xi) \\ &= \frac{1}{2}C\frac{\Delta(x^3, x^2)}{\Delta(\xi, x^2)\cdot\Delta(\xi, x^3)}\eta^2(\xi). \end{aligned} \tag{5.40}$$

He next shows that each of the expressions $\Delta(x^3, x^3)$, $\Delta(\xi, x^2)$, and $\Delta(\xi, x^3)$ has the same sign as $u_1u_2' - u_2u_1'$. To do this he notes, e.g., that

$$\frac{u_1u_2^3 - u_2u_1^3}{x^3 - x} = u_1\frac{u_2^3 - u_2}{x^3 - x} - u_2\frac{u_1^3 - u_1}{x^3 - x},$$

and consequently as A^3 approaches A that $\Delta(x, x^3)$ approaches $u_2u_1' - u_1u_2'$. By this means he sees from the first form of (5.40) that $(\Delta^2 J)$ must have the same sign as $F_{y'y'}$. This gives him a proof of the Legendre condition.

Scheeffer next assumes that $F_{y'y'} > 0$ along the given arc and discusses the behavior of the quotient u_2/u_1. With the help of (5.37) above, he finds that

$$\frac{d}{dx}\left(\frac{u_2}{u_1}\right) = \frac{C}{u_1^2F_{y'y'}};$$

since the right-hand member is always positive or negative, u_2/u_1 must be either monotone increasing or decreasing. (By proper choice of signs of u_1, u_2, he can assert that C is positive.) He then argues that the function u_2/u_1 increases monotonously toward infinity with the possibility that it may jump to $-\infty$ at points where $u_1 = 0$ or decreases in this way with a jump to $+\infty$.

If there is an $\bar{x}$ between x^0 and x^1 or at x^1 for which

$$\left(\frac{u_2}{u_1}\right)_{x=\bar{x}} = \left(\frac{u_2}{u_1}\right)_{x=x^0},$$

then there must have been a jump of the quotient u_2/u_1 at the value $\bar{\bar{x}}$ on $[x^0, x^1]$. (Recall this was Hesse's condition on conjugate points.) Moreover, on either side of the value $x = \bar{\bar{x}}$ the quotient u_2/u_1 must have opposite signs, and consequently $\Delta(\bar{\bar{x}}, x^2)$ and $\Delta(\bar{\bar{x}}, x^3)$ have opposite signs for x^2 and x^3 sufficiently near to $\bar{\bar{x}}$ and on opposite sides of it. Thus $(\Delta^2 J)_{c'+c''}$ can be made to have a negative value by choosing for ξ the value $\bar{\bar{x}}$ and for x^2, x^3 values as indicated above. If, on the contrary, ξ is taken to be a value not near to $\bar{\bar{x}}$, then x^2 and x^3 can be so chosen that $\Delta(\xi, x^2)$ and $\Delta(\xi, x^3)$ have the same sign and $(\Delta^2 J)_{c'+c''}$ can be made positive.

(This result is very close to one due to Sturm [1900], who studied the roots of second-order linear differential equations. He showed, in effect, that if u and v are independent solutions of such an equation, then between two successive roots of u there is one and only one root of v.)

Scheeffer (p. 551) is hence able to conclude that "*In order that* $(\Delta^2 J)_{c'+c''}$ *never be negative for arbitrary choices of the points* $A^2 A A^3$, *it is necessary that* (I) $F_{y'y'}$ *be positive except for certain places where it can vanish*; *and that* (II) $\Delta(x, x^0)$ *never vanish for* $x^0 < x < x^1$. *If* $\Delta(x, x^0)$ *also does not vanish for* $x = x^1$, *then* $(\Delta^2 J)_{c'+c''}$ *is always positive*; *if, on the contrary,* $\Delta(x^1, x^0) = 0$, *then* $(\Delta^2 J)_{c'+c''}$ *certainly can not take on both signs but can be made to vanish*."

From this he goes on to show how the conclusions about the sign of the second variation, evaluated along the special arc c′ + c″, imply the same conclusions for general comparison arcs (pp. 551–555). To see how he handles this, consider the arcs in Figure 5.6. He examines a comparison arc b joining A^0 and A^1 and defined by a function η which is of class C', i.e., continuous together with a continuous derivative on $x^0 \leqslant x \leqslant x^1$. The segment between A^0 and A he calls b' and between A and A^1, b''. The segments c' and c'' in Figure 5.6 are defined by the functions $\bar{u}, \bar{v}$ above with $A^2 = A^0, A^3 = A^1$.

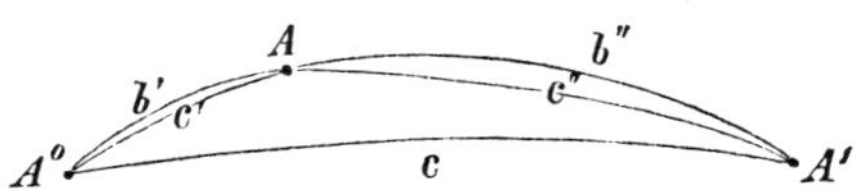

Figure 5.6

Scheefer now calculates the difference between the value of the second variation along b' and along c' and finds for it

$$(\Delta^2 J)_{b'} - (\Delta^2 J)_{c'} = \frac{1}{2}\int_{x^0}^{\xi} \frac{\partial^2 F}{\partial y' \partial y'}\left(\eta' - \frac{u_1' u_2^0 - u_2' u_1^0}{u_1 u_2^0 - u_2 u_1^0}\eta\right)^2 dx. \tag{5.41}$$

To make this estimation, he first notes that

$$\begin{aligned}(\Delta^2 J)_{c'} &= \frac{1}{2}\left[(F_{yy'}\bar{u} + F_{y'y'}\bar{u}')\bar{u}\right]_{x=x^0}^{x=\xi}\\ &= \frac{1}{2}\left[F_{yy'}\Delta(\xi, x^0) + F_{y'y'}(u_2^0 u_1' - u_1^0 u_2')\right]\frac{\eta^2}{u_2^0 u_1 - u_1^0 u_2}\\ &= \frac{1}{2}PQ,\end{aligned}$$

where P is the first and Q the second factor above and where $\bar{u}$ is defined in (5.39). He next writes

$$(\Delta^2 J)_{c'} = \int_{x^0}^{x} \frac{d}{dx}(\Delta^2 J)_{c'} dx = \frac{1}{2}\int_{x^0}^{x}\left(\frac{dP}{dx}Q + \frac{dQ}{dx}P\right)dx$$

and shows that for $u = \Delta(x, x^0)$,

$$\frac{dP}{dx}Q = F_{yy}\eta^2 + F_{yy'}\frac{u_2^0 u_1' - u_1^0 u_2'}{\Delta(x, x^0)}\eta^2,$$

provided that $\Delta \neq 0$. Further, he finds that

$$\begin{aligned}\frac{dQ}{dx}P = &- F_{yy'}\frac{u_2^0 u_1' - u_1^0 u_2'}{\Delta(x, x^0)}\eta^2 + 2F_{yy'}\eta\eta'\\ &- F_{y'y'}\left(\frac{u_2^0 u_1' - u_1^0 u_2'}{\Delta(x, x^0)}\right)^2\eta^2 + 2F_{y'y'}\frac{u_2^0 u_1' - u_1^0 u_2'}{\Delta(x, x^0)}\eta\eta'.\end{aligned}$$

Combining these, Scheeffer has his desired expression (5.41) above. He has then shown that $\Delta(x, x^0) \neq 0$ and $F_{y'y'} \geqslant 0$ on $[x^0, \xi]$ imply that the difference $(\Delta^2 J)_{b'} - (\Delta^2 J)_{c'}$ cannot be negative. In the same way

$$(\Delta^2 J)_{b''} - (\Delta^2 J)_{c''} \geqslant 0$$

for $\Delta(x, x^1) \neq 0$ and $F_{y'y'} \geqslant 0$ on $[\xi, x^1]$. He then combines these to show that $(\Delta^2 J)_b \geqslant (\Delta^2 J)_{c'+c''}$.

Scheeffer also remarks that if point A is taken to be A^1, then $(\Delta^2 J)_{c'} = 0$, $(\Delta^2 J)_{b'} = (\Delta^2 J)_b$ and relation (5.41) shows that when the necessary conditions, $F_{y'y'} \geqslant 0$ and $\Delta(x, x^0) \neq 0$ for $x^0 < x \leqslant x^1$, are fulfilled, the second variation can never be negative. He asks if it can ever be zero and shows that it can. To do this, he remarks that it will be whenever

$$\eta' - \frac{u_2^0 u_1' - u_1^0 u_2'}{\Delta(x, x^0)}\eta = 0.$$

The function $\eta = k \cdot \Delta(x, x^0)$, with k a constant, is clearly a solution of this equation which vanishes at x^0. This implies that $\eta \equiv 0$ except for the case that $\Delta(x^1, x^0) = 0$. "We see over again that only in this exceptional case the vanishing of the second variation without sign change is possible." This completes Scheeffer's discussion of the Jacobi amd Legendre necessary conditions. He then turns to more general problems whose analysis formed the major part of first Clebsch's and then Mayer's papers.

5.12. Schwarz's Proof of the Jacobi Condition

Recall that Jacobi formulated his necessary condition in 1837 but gave no proof. In his lectures Weierstrass discussed this condition, and Schwarz in his 1898/99 lectures simplified his proof. In 1900 Sommerfeld [1900], (Bolza, *VOR*, p. 83) showed how Schwarz's argument could be further simplified and even applied to double integral problems.

Suppose there is a point x' conjugate to x_0 on $x_0 < x < x_1$ and let $u = \Delta(x, x_0)$. Schwarz's proof, as given by Sommerfeld, is then concerned with a variation ζ given by

$$\zeta = \begin{cases} \Delta(x, x_0) + \varepsilon U = u + \varepsilon U & \text{on} \quad [x_0, x'] \\ \varepsilon U & \text{on} \quad [x', x_1], \end{cases}$$

where ε is a small constant and U is a function of class C'', vanishing at x_0 and x_1 but not zero at x'. This function ζ is then of class D'' on $[x_0, x_1]$; it is continuous, but its derivative has a discontinuity at $x = x'$. The second variation can now be expressed in the form

$$\begin{aligned} &\zeta\left[F_{yy'}\zeta + F_{y'y'}\zeta'\right]_{x'-0}^{x'+0} + \int_{x_0}^{x_1} \zeta\left[\left(F_{yy} - \frac{d}{dx}F_{yy'}\right)\zeta - \frac{d}{dx}\left(F_{y'y'}\frac{d\zeta}{dx}\right)\right] dx \\ &\quad = \zeta(x')F_{y'y'}\left[\zeta'(x'-0) - \zeta'(x'+0)\right] + \int_{x_0}^{x_1} \zeta\Psi(\zeta)\, dx, \end{aligned}$$

where $\Psi(\zeta) = (F_{yy} - dF_{yy'}/dx)\zeta - d(F_{y'y'}\,d\zeta/dx)/dx$, and $\zeta(x') = \varepsilon U(x')$, $\zeta'(x'-0) = u'(x') + \varepsilon U'(x')$, $\zeta'(x'+0) = \varepsilon U'(x')$. Thus the second variation is expressible in the form

$$J_2 = \varepsilon F_{y'y'}u'(x')U(x') + \int_{x_0}^{x'} (u + \varepsilon U)\Psi(u + \varepsilon U)\, dx + \int_{x'}^{x_1} \varepsilon U\Psi(\varepsilon U)\, dx.$$

But Ψ is a linear operator, $\Psi(u) = 0$, and

$$u\Psi(U) = -\frac{d}{dx}F_{y'y'}(uU' - u'U).$$

It follows then that

$$J_2 = 2\varepsilon F_{y'y'}u'(x')U(x') + \varepsilon^2\int_{x_0}^{x_1} U\Psi(U)\, dx.$$

Schwarz now has, by hypothesis, $F_{y'y'}$ different from zero on $[x_0, x_1]$, $u'(x') \neq 0$ since $u(x') = 0$, and $U(x') \neq 0$. Thus the coefficient of ε is different from zero, and by proper choice of ε, the second variation can be made negative (Bolza, *VOR*, pp. 84–85.). (Sommerfeld remarks that another proof appears in Kobb [1892/93], pp. 114ff.)

5.13. Osgood's Summary

At the turn of the century a number of men, who later became the leaders in American mathematics, went to Germany to study. Among these were Osgood and Hedrick, who made available in the United States the ideas of Weierstrass and Hilbert in the calculus of variations. In a very nice paper Osgood [1901″] gave a clear and succinct survey of Weierstrass's ideas, which I outline below. He also outlined Hilbert's and Kneser's ideas on sufficient conditions. (I discuss these ideas in Chapter 7.)

He tells us that Hilbert posed the problem of finding a function "y of the independent variable x which shall make the integral

$$I = \int_{x_0}^{x_1} F(x, y, y')\, dx$$

a maximum or minimum." The function $F(x, y, p)$ is supposed to be single-valued along with its partial derivatives of the first and second orders, continuous in the values x, y, p throughout the region

$$R : A_1 < x < B_1, \qquad A_2 < y < B_2, \qquad A_3 < p < B_3,$$

with $A_1 \leqslant x_0$, $x_1 \leqslant B_1$ and "with some or all the quantities $A_1, B_1, \ldots, B_3$ ... infinite." He further supposes fixed end-points and assumes that the functions $y(x)$ to be considered satisfy the

Conditions A

(1) $y(x)$ is a single-valued continuous function of x in the interval $x_0 \leqslant x \leqslant x_1$;

(2) y has a derivative y' at each point of the open interval $x_0 < x < x_1$, and y' is continuous on this interval;

(3) $y(x_0) = y_0$, $y(x_1) = y_1$, for given y_0, y_1;

(4) the point $(x_1\, y(x), y'(x))$ lies in the region R.

Osgood considers a curve $y(x)$, satisfying the conditions A, and a comparison or varied curve $Y(x) = y(x) + \eta(x)$ with $\eta(x_0) = \eta(x_1) = 0$, $|\eta(x)| < \varepsilon$ for x on $[x_0, x_1]$, ε an arbitrarily small positive constant and Y also satisfying conditions A. Then η is called a *strong variation*. If only those variations for which additionally $|\eta'(x)| < \varepsilon$ are considered, then η is termed a *weak variation*. A function y furnishes the integral I a *strong* or a *weak minimum* depending on the choice of variations permitted.

Consider now a function $y(x)$ satisfying conditions A and rendering the integral I a minimum; moreover, let η be any variation of y not identically zero. Then $\alpha\eta$ is also an admissible variation for $|\alpha| \leqslant 1$ and the function

$$\Delta I = f(\alpha) = \int_{x_0}^{x_1} \{ F(x, y + \alpha\eta, y' + \alpha\eta') - F(x, y, y') \} \, dx$$

is a minimum for $\alpha = 0$ and

$$f'(0) = \int_{x_0}^{x_1} \{ \eta F_y + \eta' F_{y'} \} \, dx.$$

It is necessary that $f'(0) = 0$, and hence the Euler–Lagrange condition follows.

At every point where $F_{y'y'} \neq 0$ the function y has a second derivative given by

$$y'' = \frac{F_y - F_{xy'} - y' F_{yy'}}{F_{y'y'}} . \tag{5.42}$$

Following Kneser, Osgood says that any function y satisfying (5.42) at all points inside an interval (x', x'') is called an *extremal* provided that all points $(x, y(x), y'(x))$ lie in R.[26] He also assumes that $F_{y'y'}(x, y, p) \neq 0$ at each point (x, y, p) of R.

Osgood then gives Schwarz's generalization of Weierstrass's notion of a strip or a *field* about the extremal C.[27] He assumes that there is a one-parameter family of extremals $y = \phi(x, \gamma)$, which contains C for $\gamma = \gamma_0$ and "which sweeps out the neighborhood of C just once; . . . ." More precisely, it is assumed that

(1) *a function* $\phi(x, \gamma)$ *exists which, together with its partial derivatives* $\phi_x = \phi', \phi_\gamma, \phi_{x\gamma} = \phi'_\gamma$, *is a continuous function of the two independent variables* (x, γ) *in the domain*

$$T : x_0 \leqslant x \leqslant x_1, \qquad \gamma_0 - \kappa \leqslant \gamma \leqslant \gamma_0 + \kappa,$$

where κ *is a suitably chosen positive constant; and that, for a constant value of* $\gamma, \phi(x, \gamma)$ *is an extremal, which for the special value* $\gamma = \gamma_0$, *coincides with* C;

(2) $\phi_\gamma(x, \gamma_0) \neq 0$ *when* $x_0 \leqslant x \leqslant x_1$.

He then goes to show that there is a neighborhood S of the curve C such that through each point of this neighborhood one and only one extremal

---

[26] Kneser, *LV*, p. 24 or *LV'*, p. 40.

[27] Schwarz, *GMA*, pp. 225–226. Here Schwarz actually uses a one-parameter family of minimal surfaces, not curves. Osgood ([1901] p. 168) gives a broader definition of a field than the one given here.

passes and that $\phi_\gamma(x,\gamma) \neq 0$ in this neighborhood.[28] His proof is discussed in Section 7.2. (Osgood remarks aptly that a field is very much like a cross section of the strata in a geological formation. In the two-dimensional case studied by Schwarz the surfaces are themselves the boundaries of the strata.) To show the simple coverage, Osgood uses an implicit function theorem due to Dini.

Osgood then takes up Weierstrass's sufficient condition for a minimum. Suppose that C is the curve under examination and that a field, as described below, exists about C. He goes on to show that under certain conditions the integral I has a smaller value along C than along any other curve $\overline{C}: y = Y(x)$ not identical to C. He supposes that the field about C is defined by a function $\phi(x,\gamma)$ which defines a one-parameter family of extremals passing through the point $A': (x_0', y_0')$ on the extremal C produced—$x_0' < x_0$—but very near to $A: (x_0, y_0)$.[29] He considers any point $P: (x_2, y_2)$ on $\overline{C}$, calls s the length of the arc AP along $\overline{C}$, and l the total length of $\overline{C}$ measured from A'; he then connects A' and P by the unique extremal $y = \phi(x, \gamma_2)$ of the family. Next he considers the integrals I_{02} and $\bar{I}_{21}$ defined as

$$I_{02} = \int_{x_0'}^{x_2} F(x, \phi(x,\gamma_2), \phi'(x,\gamma_2))\, dx, \qquad \bar{I}_{21} = \int_{x_2}^{x_1} F(x, Y(x), Y'(x))\, dx$$

and denotes their sum by $f(s)$. Thus $f(s) = I_{02} + \bar{I}_{21}$ and $f(l) = I' + J$, where I' is the value of I taken along C from (x_0', y_0') to (x_0, y_0) and J from (x_0, y_0) to (x_1, y_1); also $f(0) = I' + \bar{J}$, where now $\bar{J}$ is the value of I along $\overline{C}$ from (x_0, y_0) to (x_1, y_1). It follows that $f(l) - f(0) = -(\bar{J} - J)$.[30] Osgood now proceeds to establish the lemma:

> *The function $f(s)$ is a continuous function of s in the interval $0 \leqslant s \leqslant l$ and it has a derivative that is never positive and is sometimes negative, provided that the function*
>
> $$f_{y'y'}(x, y, p)$$
>
> *is positive at every point (x, y) of C when to the third argument p an arbitrary value is assigned.*

---

[28] A definition of a field used by Bliss, *LEC*, p. 44 and due, in essence, to Hilbert is this: A region F of xy space with a slope function $p(x, y)$ having the properties (1) p is single-valued and has continuous first partial derivatives in F; (2) the elements $(x, y, p(x, y))$ are in R; and (3) the Hilbert integral

$$I^* = \int [f(x, y, p(x, y))\, dx + (dy - p(x, y)\, dx) f_{y'}(x, y, p(x, y))]$$

is independent of the path for all curves in F with the same end-points.

[29] Recall that Weierstrass actually used both A and A'. This device considerably simplifies the analysis. As was mentioned above, it was Zermelo who, following Weierstrass, first published this ([1894], pp. 87–88), and Kneser who gave currency to the notion (*LV*, p. 59 or *LV'*, p. 76). Mayer [1903] then applied these ideas to more general problems.

[30] Recall how Weierstrass calculated the derivative $f'(s)$ on p. 216 above. Osgood asserts on p. 116n of [1901″] that his proof is superior to Weierstrass's. Actually both their methods are difficult and were replaced by Hilbert's method, which is simpler and more elegant.

This lemma shows that f is a monotone function which surely decreases in some part of the interval $[0, l]$ and consequently $f(l) < f(0)$, which means that $J < \bar{J}$. It remains then only to establish the lemma. Osgood's proof is tedious and not really to my mind an improvement on Weierstrass's. In Chapter 7 we shall see how Hilbert handles this problem in an elegant fashion.

The essential point is that $f'(s) = -\mathfrak{E}$, where Osgood defines $\mathfrak{E}$, the Weierstrass function, as

$$\cos\alpha\{F(x_2, y_2, M) - F(x_2, y_2, m) - (M - m)F_{y'}(x_2, y_2, m)\}$$
$$= \tfrac{1}{2}\cos\alpha(M - m)^2 F_{y'y'}(x_2, y_2, m + \theta(M - m)) \qquad (0 < \theta < 1);$$

in these expressions M is the slope of $\bar{C}$ at point P, m the slope of C at P, and $\alpha = \arctan M$ with $-\pi/2 < \alpha < \pi/2$.

Weierstrass's theorem is then the following:

A sufficient condition that a function y make the integral

$$I = \int_{x_0}^{x_1} F(x, y, y')\,dx$$

a minimum is

(1) *that the points* $(x_0, y_0), (x_1, y_1)$ *can be joined by an extremal C satisfying the Conditions A;*

(2) *that there exists a field about C . . . ;*

(3) *that* $F_{y'y'}(x, y, p)$ *be positive at all points of C, p being arbitrary in the case of a strong minimum, and equal to y′ in the case of a weak minimum.*

Osgood then takes up both Hilbert's and Kneser's methods. They are not germane here, so we close the discussion.

6. Clebsch, Mayer, and Others

6.1. Introduction

Much work went on in the calculus of variations without reference to that of Weierstrass's for quite a while. One reason for this was that Weierstrass's great accomplishments were generally made known to the mathematical community through the dissertations of his students. The other reason was the development of a school interested in generalizing the scope of the calculus of variations to more general problems than Weierstrass was considering.

In this chapter we start as an introduction with Clebsch's work in 1857 which, of course, preceded Weierstrass's; and we continue with an analysis of Mayer's papers which follow naturally out of Clebsch's. Another writer of this period is von Escherich, who followed closely in Clebsch's and Mayer's footsteps. However, his papers are so verbose and repititious of earlier writers that it seemed to me unreasonable to attempt a systematic study of them, but he did emphasize clearly the notion of abnormality (von Escherich [1899], p. 129) and gave an important form for the second variation on p. 1278 of that paper.

6.2. Clebsch's Treatment of the Second Variation

In 1857 Clebsch wrote a pioneering paper on the second variation. In spite of a number of technical shortcomings, the work is most important for several reasons: he found the necessary condition that today bears his name; he formulated, at least tentatively, what is now often called the *problem of Lagrange*; and he showed that his problem subsumed simply those earlier ones in which the integrand contained derivatives of order higher than the first.[1] It is unfortunate that Bolza in *VOR* attributes much of this work of Clebsch to von Escherich who wrote 40 years later. It is true that a good deal of Clebsch's analysis lacked the rigor one seeks today but nonetheless the basic ideas in these two papers, Clebsch [1858], [1858′], constitute a watershed in the calculus of variations. The work of all later writers is profoundly influenced by his ideas, and it is a pity that his contribution is so little regarded today, many of his ideas being attributed to Mayer or von Escherich.

Specifically, he considers a function f of $x, y = (y_1, y_2, \ldots, y_n)$, $y' = (y_1', y_2', \ldots, y_n')$ on the interval $a \leqslant x \leqslant b$ and asks that the integral

$$\int_a^b f(x, y, y')\,dx \tag{6.1}$$

be an extremum subject to the side-conditions

$$\phi_1 = 0, \qquad \phi_2 = 0, \ldots, \qquad \phi_\kappa = 0, \tag{6.2}$$

where each ϕ_ρ is dependent on x, y, y'. However, he does not explicitly make any statement about his end-conditions; presumably, he has $y(a) = y_0$, $y(b) = y_1$, i.e., fixed end-points. In his treatment of the second variation these end-conditions do not matter, as we shall see. Now he considers the expression

$$F = f + \sum_{\rho=1}^{\kappa} \lambda_\rho \phi_\rho,$$

where the λ_ρ are Lagrange's multipliers. Notice that Clebsch, like all his contemporaries, assumes the coefficient of f to be unity, i.e., that the extremal under consideration is *normal*; it was not until A. Mayer in 1886 and von Escherich in 1899 that the possibility of *abnormal* arcs was really considered.[2]

Be that as it may, Clebsch writes out the Euler–Lagrange equations which, he says, must be satisfied by an extremal

$$\begin{aligned} \frac{\partial F}{\partial y_i} = F_{y_i} = \frac{d}{dx}\frac{\partial F}{\partial y_i'} = \frac{d}{dx} F_{y_i'} \quad & (i = 1, 2, \ldots, n), \\ \phi_\rho = 0 \qquad\qquad\qquad & (\rho = 1, 2, \ldots, \kappa).^3 \end{aligned} \tag{6.3}$$

(In what follows I shall partially follow Clebsch's notations, but to simplify matters I use repeated indices for summation. The indices i, j range over $1, 2, \ldots, n$; ρ, σ, over $1, 2, \ldots, \kappa$; and r, s over $1, 2, \ldots, 2n$.) He now

---

[1] Clebsch [1858], pp. 254–273. In another paper ([1858′], pp. 335–355) in the same volume he goes beyond this, as we shall see below.

[2] Mayer [1886], pp. 79; von Escherich [1899], pp. 1290.

[3] Clebsch does not make clear a condition on his ϕ_ρ that is really needed. It is that the matrix of the $\phi_{\rho y_i'}$ is of rank $\kappa < n$ everywhere in the region being considered.

considers the functional

$$V = \int_a^b F(x, y, y', \lambda)\, dx$$

and replaces his minimizing arc y by $y + \epsilon w$, where the w are "arbitrary functions of x, ϵ however signifies a very small number." Then he notes that V can be expanded into $V + \epsilon V_1 + \epsilon^2 V_2$ and that V_1 vanishes along the extremal y. He concludes that for a minimum, V_2 must be positive (nonnegative) and for a maximum, negative (nonpositive). He does not in this place recognize that the variations w must satisfy the conditions

$$\phi_{\rho y_i} w_i + \phi_{\rho y_i'} w_i' = 0.$$

The integrand of V_2 is now the object of Clebsch's study. It is, as he says, a second-order homogeneous function and is expressible as

$$2\underline{F} = F_{y_i y_j} w_i w_j + 2F_{y_i y_j'} w_i w_j' + F_{y_i' y_j'} w_i' w_j'. \tag{6.4}$$

To transform the second variation V_2 by a complex integration by parts, Clebsch considers a $2n$-parameter family of solutions $y(x,c), \lambda(x,c)$, $c = (c_1, c_2, \ldots, c_{2n})$, of equations (6.3) and substitutes the family back into those equations so that they become identities not only in x, but also in c. He differentiates as to each c, sets $u_i = \partial y_i / \partial c$, and finds what we now call the *accessory differential equations* to be

$$\Omega_{u_i} = \frac{d}{dx} \Omega_{u_i'}, \qquad \Omega_{\mu_\rho} = 0, \qquad (i = 1, 2, \ldots, n; \rho = 1, 2, \ldots, \kappa), \tag{6.5}$$

where

$$2\underline{F}(u, u') = F_{y_i y_j} u_i u_j + 2F_{y_i y_j'} u_i u_j' + F_{y_i' y_j'} u_i' u_j',$$

$$2\Omega(u, \mu) = F_{y_i y_j} u_i u_j + 2F_{y_i y_j'} u_i u_j' + F_{y_i' y_j'} u_i' u_j' + 2\mu_\rho \Phi_\rho = 2\underline{F} + 2\mu_\rho \Phi_\rho,$$

$$\Phi_\rho(u, u') = \phi_{\rho y_i} u_i + \phi_{\rho y_i'} u_i'.$$

(Notice that the condition $\Omega_{\mu_\rho} = 0$ is equivalent to $\Phi_\rho = 0$. Recall also that the quantities $y, \zeta = \Omega_{u'}$ are called *canonical variables*.) Clebsch now takes linear combinations of the partial derivatives $y_{ic_r}, \lambda_{\rho c_r}$ to express

$$u_i = \gamma_r y_{ic_r}, \qquad \mu_\rho = \gamma_r \lambda_{\rho c_r}$$

and notes that they must satisfy the accessory equations (6.5) for all γ_r. This is Clebsch's generalization of Jacobi's result, which we noted above.

In what follows he writes the integrand 2Ω as

$$2\Omega = a_{ij} u_i u_j + 2b_{ij} u_i u_j' + c_{ij} u_i' u_j' + 2\mu_\rho (p_{\rho i} u_i + q_{\rho i} u_i')$$

where $a_{ij} = F_{y_i y_j}$, $b_{ij} = F_{y_i y_j'}$, $c_{ij} = F_{y_i' y_j'}$, $p_{\rho i} = \phi_{\rho y_i}$, $q_{\rho i} = \phi_{\rho y_i'}$. He then can write

$$\Omega_{u_j} = u_i a_{ij} + b_{ji} u_i' + \mu_\rho p_{\rho j}, \qquad \Omega_{u_j'} = u_i b_{ij} + u_i' c_{ij} + \mu_\rho q_{\rho j}.$$

He considers any two sets u, μ and v, ν of solutions of the accessory equations (6.5) (this is not his notation) and proceeds to show that

$$u_i\Omega_{v_i'} - v_i\Omega_{u_i'} = \text{const.} \tag{6.6}$$

(This is often called *Clebsch's relation*.) To do this he notes with the help of (6.5) that

$$\begin{aligned} u_i\Omega_{v_i} + u_i'\Omega_{v_i'} &= u_i \frac{d}{dx}\Omega_{v_i'} \quad + u_i'\Omega_{v_i'} \\ &= \frac{d}{dx}(u_i\Omega_{v_i'}) \end{aligned}$$

and with the help of the homogeneity of Ω that

$$\frac{d}{dx}(v_i\Omega_{u_i'}) = v_i\Omega_{u_i} + v_i'\Omega_{u_i'} = u_i\Omega_{v_i} + u_i'\Omega_{v_i'} = \frac{d}{dx}(u_i\Omega_{v_i'}). \tag{6.7}$$

He now desires to consider only those u, μ for which the constant in (6.6) is zero. (In more modern parlance, he chooses a *conjugate* set of solutions of the accessory equations; such sets exist, as we shall see later.) The maximal number of linearly independent conjugate sets can be shown to be n, and later we shall see how they can be found. Clebsch implicitly assumes he has such a set. (These matters received more thorough discussion by Clebsch [1858′], Mayer [1886], and von Escherich [1898], as we shall see below.) In his notation such a set is $u^{(r)}$, $\mu^{(r)}$; but for notational simplicity I prefer to write u_{ik}, $\mu_{\rho k}$. He asserts that there is a symmetric matrix β_{ij} such that

$$\Omega_{u_{ik}'} = \beta_{ij}u_{jk}, \qquad \beta_{ik} = \beta_{ki} \qquad (i,k = 1,2,\ldots,n). \tag{6.8}$$

It is easy to see that for such a matrix $\beta = (\beta_{ij})$, the solutions u_{ik} form a conjugate set since

$$u_{ik}\Omega_{u_{il}'} - u_{il}\Omega_{u_{ik}'} = u_{il}(\beta_{ji} - \beta_{ij})u_{jk} = 0.$$

He also asserts that there exists a matrix $\alpha = (\alpha_{ij})$ such that

$$u_{ik}' = \alpha_{ji}u_{jk} \qquad (i,k = 1,2,\ldots,n) \tag{6.9}$$

and a matrix $M = (M_{\rho k})$ such that

$$\mu_{\rho k} = M_{\rho j}u_{jk} \qquad (\rho = 1,2,\ldots,\kappa; k = 1,2,\ldots,n). \tag{6.9′}$$

Later he exhibits the form of the α_{ji} by inference as quotients of determinants, but not the form of the $M_{\rho j}$. (He writes the matrices $\alpha_{ji}, M_{\rho j}$ as $\alpha_j^{(i)}$ and M_ρ, respectively, with no other index on M.) In terms of these matrices Clebsch now has some relations he needs shortly. For $u_j = u_{jk}$, $\mu_\rho = \mu_{\rho k}$,

$$\Omega_{u_{jk}} = u_{ik}a_{ij} + b_{ji}u_{ik}' + \mu_{\rho k}p_{\rho j} = (a_{lj} + b_{ji}\alpha_{li} + p_{\rho j}M_{\rho l})u_{lk}, \tag{6.10}$$

$$\Omega_{u_{jk}'} = u_{ik}b_{ij} + u_{ik}'c_{ij} + \mu_{\rho k}q_{\rho j} = (b_{lj} + c_{ij}\alpha_{li} + q_{\rho j}M_{\rho l})u_{lk}. \tag{6.11}$$

Clebsch now wishes to relate the matrices which he has heuristically obtained. First, he shows with the help of (6.10), (6.11), and the defining relations (6.9) and (6.9′) for α and M that equations (6.8) become

$$\beta_{ik} = b_{ik} + \alpha_{ij}c_{jk} + M_{\rho i}q_{\rho k} \qquad (i,k = 1,2, \ldots, n). \tag{6.12}$$

Second, he differentiates equations (6.8) with respect to x, makes use of the accessory differential equations (6.5), and concludes that

$$\beta'_{ik} + \beta_{ij}\alpha_{kj} = a_{ik} + b_{ij}\alpha_{kj} + p_{\rho i}M_{\rho k}.$$

He, shows moreover, that

$$p_{\rho l} = -q_{\rho i}\alpha_{li} \tag{6.12′}$$

since the equations $\Omega_{\mu_\rho} = 0$ imply that $p_{\rho i}u_{ik} + q_{\rho i}u'_{ik} = 0$ and hence that $p_{\rho l}u_{lk} = -q_{\rho i}\alpha_{li}u_{lk}$. He asserts that the $n^2 + n\kappa$ equations (6.12) and (6.12′) suffice to find the matrices M and α. When these values are substituted into the conditions above for $\beta'_{i\kappa}$, he finds the differential equations

$$\beta'_{ik} = a_{ik} - \alpha_{il}c_{lj}\alpha_{kj} + p_{\rho i}M_{\rho k} + p_{\rho k}M_{\rho i}, \tag{6.13}$$

which define the matrix β.

Clebsch now turns to his transformation of the second variation, which Bolza calls *Escherich's fundamental form* (Bolza, *VOR*, pp. 628–632). To do this, he multiplies the equation (6.13) by $w_i w_k$ and sums as to i and k, finding

$$w_i a_{ik} w_k - w_i\alpha_{il}c_{lj}\alpha_{kj}w_k + 2p_{\rho i}w_i M_{\rho k}w_k = w_i\beta'_{ik}w_k. \tag{6.14}$$

He next multiplies equation (6.12) by $w_i w'_k$ and sums as to i and k, finding this time that

$$w_i\beta_{ik}w'_k = w_i b_{ik}w'_k + w_i\alpha_{ij}c_{jk}w'_k + M_{\rho i}w_i q_{\rho k}w'_k;$$

because of the symmetry of c_{ij} he can rewrite twice this expression as

$$2w_i b_{ik}w'_k + c_{jk}(w'_j\alpha_{ik}w_i + w'_k\alpha_{ij}w_i) + 2q_{\rho k}w'_k M_{\rho i}w_i = 2w_i\beta_{ik}w'_k. \tag{6.14′}$$

He now combines equations (6.14), (6.14′), notes that both β and c are symmetric matrices, and finds as a result that

$$2\underline{F} - W_i c_{ij} W_j + 2M_{\rho i}w_i(p_{\rho k}w_k + q_{\rho k}w'_k) = \frac{d}{dx}w_i\beta_{ik}w_k, \tag{6.15}$$

where $\underline{F}$ is defined by (6.4) and

$$W_i = w'_i - w_j\alpha_{ji}. \tag{6.16}$$

This yields for him the result he desires, namely:

$$\underline{F} = \overline{F} + \frac{dB}{dx} + \Psi_p\Phi_\rho, \tag{6.17}$$

where

$$2\overline{F} = W_i c_{ij} W_j, \qquad 2B = w_i\beta_{ij}w_j,$$
$$-\Psi_\rho = M_{\rho i}w_i, \qquad \Phi_\rho = p_{\rho i}w_i + q_{\rho i}w'_i. \tag{6.17′}$$

Moreover, by the relation just after (6.14) and (6.12′), the Φ can be expressed as linear functions of the W

$$0 = \Phi_\rho = q_{\rho i}W_i \tag{6.18}$$

since

$$\Phi_\rho = -q_{\rho j}\alpha_{ij}w_i + q_{\rho j}w_j' = q_{\rho j}(w_j' - w_i\alpha_{ij}).$$

(This elegant result is due to Clebsch.) Clebsch now has his transformation of the second variation in relations (6.17), (6.17′), (6.18). This is a generalization of Legendre's form and also of Jacobi's. From it his necessary condition follows directly, as we see from the fact that for variations w vanishing at the end-points and satisfying (6.18), Clebsch has shown that

$$V_2 = \int \underline{F}\,dx = \int \overline{F}\,dx.$$

He then investigates the form of the W. He has with the help of (6.9) and (6.16), the relation

$$W_i = \frac{1}{R}\begin{vmatrix} w_i' & w_1 & w_2 & \cdots & w_n \\ u_{i1}' & u_{11} & u_{21} & \cdots & u_{n1} \\ u_{i2}' & u_{12} & u_{22} & \cdots & u_{n2} \\ \cdot & \cdot & \cdot & \cdots & \cdot \\ u_{in}' & u_{1n} & u_{2n} & \cdots & u_{nn} \end{vmatrix}$$

where R, the determinant of u_{ik}, must be different from 0 for each x.

Clebsch now proceeds to show how his method is applicable to the problem discussed by Jacobi in which the integrand function f is of the form

$$f(x, y, y', y'', \ldots, y^{(n)})$$

and which caused so many commentaries to be written. Clebsch is concerned with minimizing or maximizing the integral

$$V = \int_a^b f\left(y, \frac{dy}{dx}, \ldots, \frac{d^{n-1}y}{dx^{n-1}}, \frac{d^n y}{dx^n}\right) dx.$$

To handle this, he sets

$$\phi_1 = \frac{dy}{dx} - y_1 = 0, \qquad \phi_2 = \frac{dy_1}{dx} - y_2 = 0, \ldots, \qquad \phi_{n-1} = \frac{dy_{n-2}}{dx} - y_{n-1} = 0.$$

(In what follows, the range of ρ is from 1 to $\kappa = n - 1$.) He has for the Euler–Lagrange equations, in his notation,

$$F = f + \lambda_\rho\phi_\rho, \qquad f_y = \lambda_1', \qquad f_{y_\sigma} - \lambda_\sigma = \lambda_{\sigma+1}' \qquad (\sigma = 1, 2, \ldots, n-2),$$

$$f_{y_{n-1}} - \lambda_{n-1} = \frac{d}{dx} f_{y_{n-1}'}.$$

When he eliminates the λ_ρ from these equations, he finds the well-known result

$$f_y - \frac{d}{dx} f_{y_1} + \frac{d^2}{dx^2} f_{y_2} - + \cdots = 0.$$

He has, also in his notation,

$$2\underline{F} = \frac{\partial^2 f}{\partial y^2} u^2 + 2\frac{\partial^2 f}{\partial y\,\partial y_1} u_1 u + \frac{\partial^2 f}{\partial y_1^2} u_1^2 + \cdots$$

$$+ \frac{\partial^2 f}{\partial(dy_{n-1}/dx)\partial(dy_{n-1}/dx)}\left(\frac{du_{n-1}}{dx}\right)^2;$$

and he transforms the second variation into the form

$$V_2 = \frac{1}{2}\int_a^b \frac{\partial^2 f}{\left[\partial(dy_{n-1}/dx)\right]^2}\cdot\frac{W^2}{R^2}\,dx = \frac{1}{2}\int_a^b \frac{\partial^2 f}{\left[\partial(d^n y/dx^n)\right]^2}\cdot\frac{W^2}{R^2}\,dx,$$

where

$$R = \begin{vmatrix} u^{(1)} & u_1^{(1)} & u_2^{(1)} & \cdots & u_{n-1}^{(1)} \\ u^{(2)} & u_1^{(2)} & u_2^{(2)} & \cdots & u_{n-1}^{(2)} \\ \cdot & \cdot & \cdot & \cdot & \cdot \\ u^{(n)} & u_1^{(n)} & u_2^{(n)} & \cdots & u_{n-1}^{(n)} \end{vmatrix} = \begin{vmatrix} u^{(1)} & \frac{du^{(1)}}{dx} & \cdots & \frac{d^{n-1}u^{(1)}}{dx^{n-1}} \\ u^{(2)} & \frac{du^{(2)}}{dx} & \cdots & \frac{d^{n-1}u^{(2)}}{dx^{n-1}} \\ \cdot & \cdot & \cdot & \cdot \\ u^{(n)} & \frac{du^{(n)}}{dx} & \cdots & \frac{d^{n-1}u^{(n)}}{dx^{n-1}} \end{vmatrix}$$

and

$$W = \begin{vmatrix} w & w_1 & w_2 & \cdots & \frac{dw_{n-1}}{dx} \\ u^{(1)} & u_1^{(1)} & u_2^{(1)} & \cdots & \frac{du_n^{(1)}}{dx} \\ \cdot & \cdot & \cdot & \cdot & \cdot \\ u^{(n)} & u_1^{(n)} & u_2^{(n)} & \cdots & \frac{du_n^{(n)}}{dx} \end{vmatrix}$$

$$= \begin{vmatrix} w & \frac{dw}{dx} & \frac{d^2w}{dx^2} & \cdots & \frac{d^nw}{dx^n} \\ u^{(1)} & \frac{du^{(1)}}{dx} & \frac{d^2u^{(1)}}{dx^2} & \cdots & \frac{d^nu^{(1)}}{dx^n} \\ \cdot & \cdot & \cdot & \cdot & \cdot \\ u^{(n)} & \frac{du^{(n)}}{dx} & \frac{d^2u^{(n)}}{dx^2} & \cdots & \frac{d^nu^{(n)}}{dx^n} \end{vmatrix}.$$

In these results the $u^{(i)}$ are a set of n conjugate solutions. The paper then closes with a discussion of multiple integral problems.

In closing our discussion of the first of Clebsch's papers in the 1858 volume of *Crelle's Journal*, it is worth stating formally the Clebsch condition for the problem of Lagrange. As stated by Bolza,[4] it is

> *For a minimum of the integral J with side-conditions $\phi_\beta = 0$ it is . . . necessary that at each point of an extremal arc*
>
> $$\sum_{i,k} \frac{\partial^2 F}{\partial y_i' \, \partial y_k'} \zeta_i \zeta_k \geqslant 0$$
>
> *for all systems of quantities $\zeta_1, \zeta_2, \ldots, \zeta_n$ which satisfy the equations*
>
> $$\sum_i \frac{\partial \phi_\beta}{\partial y_i'} \zeta_i = 0.$$

Notice that Clebsch did not quite prove this condition explicitly in that he did not show how, given the W, he could find the corresponding w which satisfy the differential equations (6.16) as well as the relations (6.18). Nonetheless, he did show the elegant way to handle Jacobi's general problem.

6.3. Clebsch, Continuation

In his second paper [1858′] Clebsch continues (pp. 335–355) to exploit his insight into more general problems of the calculus of variations than were considered by Jacobi. In this paper he proceeds to generalize the Hamilton–Jacobi theory to what are now called problems of Lagrange and of Mayer, and he shows how isoperimetric problems also fall under his general theory.[5] It is unfortunate that Clebsch's name is not today associated with any of the general problems. Perhaps this is the case since his analyses are lacking in a number of ways; e.g., some of his discussions are quite flawed. It would be wrong, however, not to recognize in Clebsch a man with an exceptional and fertile mind, who made possible a great advance in our subject. His originality is conspicuous when one considers that his 1858 papers came just a year later than Hesse's; there Hesse was still busy proving assertions by Jacobi, now outmoded by Clebsch's work. His great difficulty is his general inability to organize his analysis and to give clearer proofs. He probably furnished many basic ideas to Mayer.

---

[4] Bolza, *VOR*, p. 608.

[5] In his 1838 paper Jacobi [1838], pp. 79ff, said that the Hamilton–Jacobi theory is applicable to isoperimetric problems in which the integrand contains only the first derivative of the dependent variable; i.e., the integrand is of the form $f(x, y, y')$. Clebsch was able trivially to generalize this to any integrand by his method of adjoining differential equations as side-conditions.

He chooses as the most general single integral problem of the calculus of variations the following:

Let the integral

$$J = \int_{x^0}^{x^1} F\,dx \tag{6.19}$$

be a minimum or maximum, while at the same time a given set of differential equations

$$\phi_1 = 0, \qquad \phi_2 = 0, \ldots, \qquad \phi_\kappa = 0 \tag{6.20}$$

is satisfied, where κ *is smaller than* n.

He points out that the problem of minimizing a function

$$V(x^1, y_1, y_2, \ldots, y_n)$$

subject to the conditions (6.20) can be transformed into the problem just mentioned by setting

$$J = \int_{x^0}^{x^1} \frac{dV}{dx}\,dx.$$

This is a version of the problem now called the *problem of Mayer*, and Clebsch's general problem is now called the *problem of Lagrange*.

He reverts to the problem of Lagrange, as he formulated it, and sets

$$\Omega = F + \lambda_1\phi_1 + \lambda_2\phi_2 + \cdots + \lambda_\kappa\phi_\kappa; \tag{6.21}$$

he then writes out the Euler–Lagrange equations in the form

$$\frac{d}{dx}\frac{\partial\Omega}{\partial(dy_i/dx)} = \frac{\partial\Omega}{\partial y_i} \qquad (i = 1,2, \ldots, n),$$

$$\phi_\rho = 0 \qquad (\rho = 1,2, \ldots, \kappa).^6 \tag{6.22}$$

To make Clebsch's arguments reasonable, consider an n-parameter family of extremals

$$y_i = Y_i(x; x^0, y^0, a), \qquad \lambda_\rho = \Lambda_\rho(x; x^0, y^0, a) \tag{6.23}$$

all of which pass through the given initial point $(x^0, y_1^0, \ldots, y_n^0)$ with $a = (a_1, a_2, \ldots, a_n)$ arbitrary, i.e.,

$$y_i^0 = Y_i(x^0; x^0, y^0, a). \tag{6.23$'$}$$

He then defines a function

$$U(x; x^0, y^0, a) = \int_{x^0}^{x} F(x, Y, Y')\,dx = \int_{x^0}^{x} \Omega(x, Y, Y', \Lambda)\,dx$$

---

[6]Clebsch's discussion in his Sections 2 and 3 is flawed. However, the germs of several important ideas are in there, and I have therefore felt it reasonable to replace his discussion by a correctly formulated one. I have made this discussion conform as closely as possible to Clebsch's despite the fact that there are much more elegant ways to handle the problem in terms of Hilbert's invariant integral.

since $\phi_\rho(x, Y, Y') \equiv 0$. It is not difficult to calculate the partial derivatives of U with respect to its variables:

$$\frac{\partial U}{\partial x} = F(x, Y, Y'),$$

$$\frac{\partial U}{\partial x^0} = -F\Big|^{x_0} + \int_{x_0}^{x} \sum_i \left\{ \frac{\partial \Omega}{\partial y_i} \frac{\partial Y_i}{\partial x^0} + \frac{\partial \Omega}{\partial y_i'} \frac{\partial Y_i'}{\partial x^0} \right\} dx$$

$$= -F\Big|^{x_0} + \sum_i \frac{\partial \Omega}{\partial y_i'} \frac{\partial Y_i}{\partial x^0}\Big|_{x^0}^{x},$$

$$\frac{\partial U}{\partial y_j^0} = \sum_i \frac{\partial \Omega}{\partial y_i'} \frac{\partial Y_i}{\partial y_j^0}\Big|_{x^0}^{x}, \qquad \frac{\partial U}{\partial a_j} = \sum_i \frac{\partial \Omega}{\partial y_i'} \frac{\partial Y_i}{\partial a_j}\Big|^{x},$$

since $\partial \Omega / \partial \lambda_\rho = 0$ and $\partial Y_i / \partial a_j$ at $x = x^0$ is 0. However, it is clear from (6.23′) that

$$\left(Y_i' + \frac{\partial Y_i}{\partial x^0} \right)\Big|^{x^0} = 0, \qquad \frac{\partial Y_i}{\partial y_j^0}\Big|^{x^0} = \delta_{ij},$$

and hence that

$$\frac{\partial U}{\partial x} = F\Big|^{x}, \qquad \frac{\partial U}{\partial x^0} = \sum_i \frac{\partial \Omega}{\partial y_i'} \frac{\partial Y_i}{\partial x^0}\Big|^{x} + \left(\sum_i \frac{\partial \Omega}{\partial y_i'} Y_i' - \Omega \right)\Big|^{x^0},$$

$$\frac{\partial U}{\partial y_i^0} = \sum_i \frac{\partial \Omega}{\partial y_i'} \frac{\partial Y_i}{\partial y_j^0}\Big|^{x} - \frac{\partial \Omega}{\partial y_i'}\Big|^{x_0}, \qquad \frac{\partial U}{\partial a_j} = \sum_i \frac{\partial \Omega}{\partial y_i'} \frac{\partial Y_i}{\partial a_j}\Big|^{x}.$$

Clebsch, in effect, now fixes the constants $a_1, a_2, \ldots, a_n$ so that the curves of family (6.23) pass through another point $(\xi, \eta_1, \eta_2, \ldots, \eta_n)$. (This can be done if the initial- and end-points are not conjugate, as we see later.) Suppose then that for $a_i = A_i(x^0, y^0, \xi, \eta)$, the family (6.23) has the desired property, i.e., that

$$\eta_i = Y_i\big[\xi; x^0, y^0, A(x^0, y^0; \xi, \eta)\big].$$

In terms of this family, define another function $V(x^0, y^0; \xi, \eta)$ by the relation

$$V(x^0, y^0; \xi, \eta) = U\big[\xi; x^0, y^0, A(x^0, y^0; \xi, \eta)\big].$$

We then have

$$\frac{\partial V}{\partial \xi} = \left(\frac{\partial U}{\partial x} \right) + \sum_i \left(\frac{\partial U}{\partial a_i} \right) \frac{\partial A_i}{\partial \xi},$$

$$\frac{\partial V}{\partial \eta_j} = \sum_i \left(\frac{\partial U}{\partial a_i} \right) \frac{\partial A_i}{\partial \eta_j},$$

where Clebsch uses parentheses around a function to indicate that it is evaluated for $a_i = A_i$. We see also that

$$Y_i' + \sum_j \frac{\partial Y_i}{\partial a_j}\frac{\partial A_j}{\partial \xi} = 0, \qquad \sum_j \frac{\partial Y_i}{\partial a_j}\frac{\partial A_j}{\partial \eta_k} = \delta_{ik},$$

and as a result that

$$\frac{\partial V}{\partial \xi} = \left(\frac{\partial U}{\partial x}\right)\Bigg|^{\xi} - \sum_i \frac{\partial \Omega}{\partial y_i'} Y_i'\Bigg|^{\xi},$$

$$\frac{\partial V_i}{\partial \eta_j} = \frac{\partial \Omega}{\partial y_j'}\Bigg|^{\xi}.$$

He combines these to form the relation

$$\frac{\partial V}{\partial \xi} + \sum_i \frac{\partial V}{\partial \eta_i} Y_i' = F,$$

which is his statement of the Hamilton–Jacobi equation.

Clebsch then notes in passing his generalization of a very elegant theorem due to Jacobi. Suppose that $V(x, y, a)$ is an n-parameter solution of the Hamilton–Jacobi equation (he needs to assume also that the determinant $|V_{a_i y_i}| \neq 0$)

$$V_x + H(x, y, V_y) = 0. \tag{6.24}$$

Then the general solution of the canonical equations

$$\frac{dy_i}{dx} = \frac{\partial H}{\partial v_i}, \qquad \frac{dv_i}{dx} = -\frac{\partial H}{\partial y_i} \tag{6.25}$$

can be found from the equations

$$\frac{\partial V}{\partial y_i} = v_i, \qquad \frac{\partial V}{\partial a_i} = \alpha_i. \tag{6.26}$$

By solving the second set and substituting the solutions into the first set, he has

$$y = y_i(x, a, \alpha), \qquad v = v_i(x, a, \alpha)$$

with $a = (a_1, a_2, \ldots, a_n)$, $\alpha = (\alpha_1, \alpha_2, \ldots, \alpha_n)$. They form the general solutions for equations (6.25). The proof, which Clebsch does not give, is simple.

Suppose that the solutions $y_i(x, a, \alpha), v_i(x, a, \alpha)$ are substituted into the Hamilton–Jacobi equation above and the resulting identities differentiated with respect to a_j and y_j. There results the identities

$$V_{xa_j} + \sum_i H_{v_i} V_{y_i a_j} = 0, \qquad V_{xy_j} + \sum_i H_{v_i} V_{y_i y_j} = -H_{y_j}. \tag{6.26$'$}$$

It also follows from the relations (6.26) by differentiating with respect to x that

$$V_{xa_j} + \sum_i y_i' V_{y_i a_j} = 0, \qquad v_j' = V_{xy_j} + \sum_i y_i' V_{y_i y_j}.$$

When we compare these sets of equations with (6.26′), we see that the canonical equations (6.25) result. This is, in essence, what Clebsch remarks at the end of Section 2.

In Section 4 Clebsch turns to a discussion of the second variation. To this end he considers the two functions (in his notation)

$$\Omega_1 = \sum_i u_i \frac{\partial \Omega}{\partial y_i} + \sum_i \frac{du_i}{dx} \frac{\partial \Omega}{\partial (dy_i/dx)} + \sum_i \mu_i \frac{\partial \Omega}{\partial \lambda_i},$$

$$2\Omega_2 = \sum_i u_i \frac{\partial \Omega_1}{\partial y_i} + \sum_i \frac{du_i}{dx} \frac{\partial \Omega_1}{\partial (dy_i/dx)} + \sum_i \mu_i \frac{\partial \Omega_1}{\partial \lambda_i}.$$

He remarks that if $\delta y_i, \delta\lambda_i$ are substituted for u_i, μ_i, these become the first and second variations of Ω. (Notice his unhappy use of the index i to range over both the sets $1, 2, \ldots, n$ and $1, 2, \ldots, \kappa$.) In his earlier paper [1858] he had shown that

$$\delta^2 J = \frac{1}{2} \int_{x^0}^{x^1} \sum_i \sum_k \frac{\partial^2 \Omega}{\partial (dy_i/dx)\, \partial (dy_k/dx)} \frac{R_i R_k}{R^2} dx + B^1 - B^0,$$

where the R_i are linear functions of δy_i and $\delta dy_i/dx$ and B^0, B^1 are the values at $x = x^0$ and x^1 of a homogeneous function B of second order. The R_i are related by the equations

$$\sum_i R_i \frac{\partial \phi_\rho}{\partial (dy_i/dx)} = 0 \qquad (\rho = 1, 2, \ldots, \kappa).$$

Moreover, Clebsch observes that the R, R_i, B depend on the accessory differential equations for the problem of minimizing the second variation.

He writes the solutions of those differential equations with the help of the $2n$-parameter family (6.23) of solutions of the Euler–Lagrange equations (6.22). He expresses the general solution of the accessory equations as

$$u_i = \sum_k A_k \left[\frac{\partial y_i}{\partial a_k} \right], \qquad \mu_\rho = \sum_k A_k \left[\frac{\partial \lambda_\rho}{\partial a_k} \right], \tag{6.27}$$

where the $A = (A_1, A_2, \ldots, A_{2n})$ are new arbitrary constants. (He uses square brackets to indicate the functions in the brackets are evaluated on the family (6.23) $y(x, a_1, \ldots, a_{2n}), \lambda(x, a_1, \ldots, a_{2n})$.) He again chooses a conjugate set of solutions u_i^r $(i, r = 1, 2, \ldots, n)$ of the accessory equations so that

$$\sum_i \left\{ u_i^r \left(\frac{\partial \Omega_2}{\partial (du_i/dx)} \right)^s - u_i^s \left(\frac{\partial \Omega_2}{\partial (du_i/dx)} \right)^r \right\} = 0 \qquad (r, s = 1, 2, \ldots, n). \tag{6.28}$$

The quantity R is the determinant of the u_i^r, and the R_i are of the form

$$R_i = R\delta \frac{dy_i}{dt} - \sum_h \sum_k \delta y_h \frac{du_i^k}{dx} \cdot \frac{\partial R}{\partial u_h^k},$$

where h and k are summed from 1 through n; the function B is given by

$$2B = \sum_i \sum_k \beta_{ik}\, \delta y_i\, \delta y_k, \tag{6.29}$$

where the β_{ij} $(i, j = 1, 2, \ldots, n)$ are defined with the help of

$$\sum_k \beta_{ki} u_k^r = \left(\frac{\partial \Omega_2}{\partial(du_i/dx)} \right)^r \qquad (i, r = 1, 2, \ldots, n).$$

He now wishes to simplify matters with the help of an n-parameter family of solutions $V(x, y, a_1, \ldots, a_n)$ of the Hamilton–Jacobi equation. To this end (Section 5, pp. 343ff), Clebsch first observes that

$$\frac{\partial \Omega_2}{\partial(du_i/dx)} = \frac{\partial \Omega_1}{\partial(dy_i/dx)}$$

and hence that

$$\frac{\partial \Omega_2}{\partial(du_i/dx)} = \sum_h u_h \frac{\partial}{\partial y_h} \cdot \frac{\partial \Omega}{\partial(dy_i/dx)} + \sum_h \frac{du_h}{dx} \cdot \frac{\partial}{\partial(dy_h/dx)} \cdot \frac{\partial \Omega}{\partial(dy_i/dx)}$$
$$+ \sum_\rho \mu_\rho \frac{\partial}{\partial \lambda_\rho} \cdot \frac{\partial \Omega}{\partial(dy_i/dx)} \cdot$$

(I have slightly altered his notation to make his relations unambiguous.)[7]

He now substitutes for u, μ their values from (6.27) into this last relation and writes

$$\frac{\partial \Omega_2}{\partial(du_i/dx)} = \sum_k A_k \left[\frac{\partial}{\partial a_k} \cdot \frac{\partial \Omega}{\partial(dy_i/dx)} \right]$$

or equivalently

$$\frac{\partial \Omega_2}{\partial(du_i/dx)} = \sum_k A_k \left[\frac{\partial}{\partial a_k} \left(\frac{\partial V}{\partial y_i} \right) \right], \tag{6.29$'$}$$

as can be seen with the help of (6.25). (In these and in the following relations k and m range from 1 through $2n$.) When these are substituted into equations (6.28), there results

$$0 = \sum_k \sum_m A_k^r A_m^s \sum_i \left(\left[\frac{\partial y_i}{\partial a_k} \right] \left[\frac{\partial}{\partial a_m} \left(\frac{\partial V}{\partial y_i} \right) \right] - \left[\frac{\partial y_i}{\partial a_m} \right] \left[\frac{\partial}{\partial a_k} \left(\frac{\partial V}{\partial y_i} \right) \right] \right);$$

and by an interchange of k and m,

$$0 = -\sum_k \sum_m A_m^r A_k^s \sum_i \left(\left[\frac{\partial y_i}{\partial a_k} \right] \left[\frac{\partial}{\partial a_m} \left(\frac{\partial V}{\partial y_i} \right) \right] - \left[\frac{\partial y_i}{\partial a_m} \right] \left[\frac{\partial}{\partial a_k} \left(\frac{\partial V}{\partial y_i} \right) \right] \right).$$

---

[7] In what follows Clebsch uses symbols such as $(\partial V/\partial y_i)$ or $(\partial V/\partial a_i)$ to mean the appropriate partial derivatives of $V(x, y, a)$. However, when he writes $[\partial(\partial V/\partial y_i)/\partial a_k]$ he means that $(\partial V/\partial y_i)$ is formed, evaluated along the family of extremals $y = y(x, a_1, \ldots, a_{2n})$ and then the resulting expression is differentiated with respect to a_k. In this connection note that $[\partial V/\partial a_k] = \sum_i (\partial V/\partial y_i)[\partial y_i/\partial a_k] + (\partial V/\partial a_k)$.

(The A^r, A^s are the constants in (6.27) associated with u^r and u^s, respectively.) Clebsch thus has

$$0 = \sum_k \sum_m \left\{ (A_k^r A_m^s - A_m^r A_k^s) \sum_i \left(\left[\frac{\partial y_i}{\partial a_k} \right] \left[\frac{\partial}{\partial a_m} \left(\frac{\partial V}{\partial y_i} \right) \right] - \left[\frac{\partial y_i}{\partial a_m} \right] \left[\frac{\partial}{\partial a_k} \left(\frac{\partial V}{\partial y_i} \right) \right] \right) \right\}. \tag{6.30}$$

He proceeds to assert that this relation (6.30) can represent at most one equation between the A, and thus that the quantity inside the sum as to i must be a constant. To see this note that

$$\sum_i \left(\left[\frac{\partial y_i}{\partial a_k} \right] \left[\frac{\partial}{\partial a_m} \left(\frac{\partial V}{\partial y_i} \right) \right] - \left[\frac{\partial y_i}{\partial a_m} \right] \left[\frac{\partial}{\partial a_k} \left(\frac{\partial V}{\partial y_i} \right) \right] \right)$$
$$= \left[\frac{\partial}{\partial a_m} \sum_i \frac{\partial y_i}{\partial a_k} \left(\frac{\partial V}{\partial y_i} \right) \right] - \left[\frac{\partial}{\partial a_k} \sum_i \frac{\partial y_i}{\partial a_m} \left(\frac{\partial V}{\partial y_i} \right) \right], \tag{6.31}$$

since the additional terms introduced cancel each other. Clebsch further notes that

$$\sum_i \left[\frac{\partial y_i}{\partial a_k} \right] \left(\frac{\partial V}{\partial y_i} \right) = \frac{d(V)}{da_k} - \frac{\partial(V)}{\partial a_k} = \left[\frac{\partial V}{\partial a_k} \right] - \left(\frac{\partial V}{\partial a_k} \right).$$

(He means here that $d(V)/da_k = \sum V_{y_i} y_{ia_k} + V_{a_k}$.) It therefore follows that

$$\sum_i \left(\left[\frac{\partial y_i}{\partial a_k} \right] \left[\frac{\partial}{\partial a_m} \left(\frac{\partial V}{\partial y_i} \right) \right] - \left[\frac{\partial y_i}{\partial a_m} \right] \left[\frac{\partial}{\partial a_k} \left(\frac{\partial V}{\partial y_i} \right) \right] \right)$$
$$= \left[\frac{\partial}{\partial a_k} \left(\frac{\partial V}{\partial a_m} \right) \right] - \left[\frac{\partial}{\partial a_m} \left(\frac{\partial V}{\partial a_k} \right) \right]. \tag{6.31'}$$

He now chooses the constants $a_1, a_2, \ldots, a_{2n}$ so that they are the sets $a_1, a_2, \ldots, a_n, \alpha_1, \alpha_2, \ldots, \alpha_n$ in (6.26). He then has $V_{a_i} = (V)_{a_i} = \alpha_i$ $(i = 1, 2, \ldots, n)$. Notice also that $(V)_{a_i \alpha_j} = 0$ since V does not contain α explicitly. Equation (6.30) then reduces to the relation

$$\sum_{k=1}^{n} (A_k^r A_{n+k}^s - A_{m+k}^s A_k^r) = 0.$$

Thus expression (6.31) has the value 1 for $k = 1, 2, \ldots, n$ and $m = n + k$ and the value 0 otherwise.

Clebsch now divides the constants $A_1, A_2, \ldots, A_{2n}$ into two sets: for the first n he writes $\underline{A}_1, \underline{A}_2, \ldots, \underline{A}_n$, but for the second n he writes $A_1, A_2, \ldots, A_n$. In this notation he rewrites (6.30) in the form

$$0 = \sum_k (\underline{A}_k^r A_k^s - \underline{A}_k^s A_k^r), \tag{6.32}$$

where "the sum now only extends from 1 to n."[8] He goes on to say these are the equations that the constants $\underline{A}, A$ must satisfy and that he will show they are actually identities. To do this, he says that, in general, one can always find n^2 quantitites c so that the equations

$$A_m^i = \sum_j c_{jm} \underline{A}_j^i \qquad (i, m = 1, 2, \ldots, n) \tag{6.33}$$

are satisfied. He substitutes these into (6.32) and finds that

$$\begin{aligned} 0 &= \sum_k \sum_h (\underline{A}_k^r \underline{A}_h^s - \underline{A}_h^r \underline{A}_k^s) c_{hk} \\ &= \sum_k \sum_h (\underline{A}_h^r \underline{A}_k^s - \underline{A}_k^r \underline{A}_h^s) c_{kh} \end{aligned}$$

and hence that

$$0 = \sum_k \sum_h (\underline{A}_k^r \underline{A}_h^s - \underline{A}_h^r \underline{A}_k^s)(c_{hk} - c_{kh}).$$

From these $n(n-1)/2$ relations he concludes that relation (6.32) can be replaced by the condition that $C = (c_{hk})$ is a symmetric matrix; or, as he expresses the matter,

"*The arbitrary constants $\underline{A}, A$ are related to each other by an arbitrary system of linear equations with a symmetric determinant.*" He concludes the section with the remark that the $2n^2$ interdependent constants $\underline{A}_k^r, A_k^r$ can be replaced by the $n^2 + n(n+1)/2$ linearly independent constants $\underline{A}_k^r$, c_{hk}.[9]

In Section 6 Clebsch proceeds to his first form of the second variation. To this end, he inquires as to how the choice of the constants influences the quantities R and R_i. He substitutes the values for A_m^i in (6.33) into relations (6.27), finding, in his notation,

$$u_i^r = \sum_{k=1}^{k=n} \underline{A}_k^r \left\{ \left[\frac{\partial y_i}{\partial a_k} \right] + c_{1k} \left[\frac{\partial y_i}{\partial \alpha_1} \right] + c_{2k} \left[\frac{\partial y_i}{\partial \alpha_2} \right] + \cdots + c_{nk} \left[\frac{\partial y_i}{\partial \alpha_n} \right] \right\},$$

which implies that the u are of the form

$$u_i^r = \underline{A}_1^r v_i^1 + \underline{A}_2^r v_i^2 + \cdots + \underline{A}_n^r v_i^n,$$

where

$$v_i^k = \left[\frac{\partial y_i}{\partial a_k} \right] + c_{1k} \left[\frac{\partial y_i}{\partial \alpha_1} \right] + c_{2k} \left[\frac{\partial y_i}{\partial \alpha_2} \right] + \cdots + c_{nk} \left[\frac{\partial y_i}{\partial \alpha_n} \right]. \tag{6.34}$$

He concludes that the determinant R of u_i^r is expressible as

$$R = A \cdot S,$$

where A is the determinant of $\underline{A}_i^r$ and S of v_i^r.

He then turns to the form of the R_i given above just after (6.28). The

[8] Clebsch [1858′], p. 344.

[9] Clebsch [1858′], p. 345.

coefficient of $\delta y_i'$ is R and of δy_h is

$$\frac{du_i^1}{dx}\cdot\frac{\partial R}{\partial u_h^1}+\frac{du_i^2}{dx}\cdot\frac{\partial R}{\partial u_h^2}+\cdots+\frac{\partial u_i^n}{dx}\cdot\frac{\partial R}{\partial u_h^n}.$$

This he remarks "is a determinant of the same nature as R itself, only that in place of the v_k the functions dv_i/dx appear." He notes, moreover, that this expression transforms into

$$A\left\{\frac{dv_i^1}{dx}\frac{\partial S}{\partial v_h^1}+\frac{dv_i^2}{dx}\cdot\frac{\partial S}{\partial v_h^2}+\cdots+\frac{dv_i^n}{dx}\frac{\partial S}{\partial v_h^n}\right\},$$

so that the R_i are of the form

$$R_i = A\cdot S_i, \tag{6.35}$$

where

$$S_i = S\delta\frac{dy_i}{dx}-\sum_k\sum_h \delta y_h\frac{dv_i^k}{dx}\cdot\frac{\partial S}{\partial v_h^k}. \tag{6.36}$$

Notice, however, that in the second variation only the ratio R_i/R appears, and thus the factor A does not appear. Clebsch now has

$$\delta^2 J=\frac{1}{2}\int_{x^0}^{x^1}\sum_i\sum_k\frac{\partial^2\Omega}{\partial(dy_i/dx)\,\partial(dy_k/dx)}\cdot\frac{S_iS_k}{S^2}dx+B^1-B^0. \tag{6.37}$$

From this he asserts that

"*Instead of using the particular integrals u it suffices to use the particular integrals v which involve a lesser number (only $n(n+1)/2$) of arbitrary constants without restricting the generality of the discussion.*"[10] He goes on to remark that before this theorem can be accepted, it is necessary to show that the S_i satisfy the equations

$$\sum_i S_i\frac{\partial\phi_\rho}{\partial(dy_i/dx)}=0 \qquad (\rho=1,2,\ldots,k). \tag{6.38}$$

This is easy to see since equations (6.35), together with the equations

$$\sum_i R_i\frac{\partial\phi_\rho}{\partial(dy_i/dx)}=0 \qquad (\rho=1,2,\ldots,k),$$

imply the result at once.

He now turns to function B and asks about coefficients β_{ik} in (6.29). With the help of (6.29′), Clebsch expresses the situation as follows:

$$\left(\frac{\partial\Omega_2}{\partial(du_i/dx)}\right)^r=\underline{A}_1^r\Omega_i^1+\underline{A}_2^r\Omega_i^2+\cdots+\underline{A}_n^r\Omega_i^n,$$

[10] Clebsch [1858′], p. 347.

where

$$\Omega_i^k = \left[\frac{\partial}{\partial a_k} \left(\frac{\partial V}{\partial y_i} \right) \right] + c_{i1} \left[\frac{\partial}{\partial a_1} \left(\frac{\partial V}{\partial y_i} \right) \right] + c_{i2} \left[\frac{\partial}{\partial a_2} \left(\frac{\partial V}{\partial y_i} \right) \right] + \cdots$$
$$+ c_{in} \left[\frac{\partial}{\partial a_n} \left(\frac{\partial V}{\partial y_i} \right) \right].$$

He concludes from this with the help of the relation following (6.29) that

$$0 = \sum_k \underline{A}_k^r \{ \beta_{1i} v_1^k + \beta_{2i} v_2^k + \cdots + \beta_{ni} v_n^k - \Omega_i^k \}$$

for $r = 1, 2, \ldots, n$; since he has assumed that the determinant of $\underline{A}_k^r \neq 0$, he has

$$\beta_{1i} v_1^k + \beta_{2i} v_2^k + \cdots + \beta_{ni} v_n^k = \Omega_i^k. \tag{6.39}$$

A comparison of this result with relations (6.31′) and (6.29) shows that $\underline{A}_k^r = 0$ for $k \neq r$ and $\underline{A}_r^r = 1$ and thus

$$\left(\frac{\partial \Omega_2}{\partial (dv_i/dx)} \right)^r = \Omega_i^r. \tag{6.40}$$

Relations (6.34), (6.36), (6.37), (6.38), (6.39), and (6.40) completely define the new form of the second variation.[11]

In Section 7 Clebsch proceeds to his second and final form for the second variation, calculating both the parts under the integral sign and outside. To do this, he proceeds to examine the expressions $[\partial y_i / a_k]$, $[\partial y_i / \partial \alpha_k]$. Recall that

$$\left(\frac{\partial V}{\partial a_i} \right) = \alpha_i \qquad (i = 1, 2, \ldots, n)$$

and differentiate both members with respect to a_k. He clearly has

$$\left(\frac{\partial^2 V}{\partial a_i \, \partial y_1} \right) \left[\frac{\partial y_1}{\partial a_k} \right] + \left(\frac{\partial^2 V}{\partial a_i \, \partial y_2} \right) \left[\frac{\partial y_2}{\partial a_k} \right] + \cdots$$
$$+ \left(\frac{\partial^2 V}{\partial a_i \, \partial y_n} \right) \left[\frac{\partial y_n}{\partial a_k} \right] + \left(\frac{\partial^2 V}{\partial a_i \, \partial a_k} \right) = 0 \qquad (i, k = 1, 2, \ldots, n), \tag{6.41}$$

$$\left(\frac{\partial^2 V}{\partial a_i \, \partial y_1} \right) \left[\frac{\partial y_1}{\partial \alpha_k} \right] + \left(\frac{\partial^2 V}{\partial a_i \, \partial y_2} \right) \left[\frac{\partial y_2}{\partial \alpha_k} \right] + \cdots + \left(\frac{\partial^2 V}{\partial a_i \, \partial y_n} \right) \left[\frac{\partial y_n}{\partial \alpha_k} \right] = \delta_{ik}. \tag{6.41′}$$

To solve these equations, Clebsch designates by p the determinant of the $(\partial^2 V / \partial a_k \partial y_i)$ and by p_i^k, the signed minor of p corresponding to the indices

[11] Clebsch [1858′], p. 348.

i, k. It then follows that equations (6.41) can be solved in the form

$$-p\left[\frac{\partial y_i}{\partial a_k}\right] = p_i^1\left(\frac{\partial^2 V}{\partial a_1\,\partial a_k}\right) + p_i^2\left(\frac{\partial^2 V}{\partial a_2\,\partial a_k}\right) + \cdots + p_i^n\left(\frac{\partial^2 V}{\partial a_n\,\partial a_k}\right). \tag{6.42}$$

and equations (6.41′) in the form

$$p\left[\frac{\partial y_i}{\partial \alpha_k}\right] = p_i^k.$$

With the help of these relations, Clebsch finds a new form for the functions v. By (6.34), he has

$$pv_i^k = p_i^1 w_{1k} + p_i^2 w_{2k} + \cdots + p_i^n w_{nk}, \tag{6.43}$$

where it is easy to see that

$$w_{hk} = c_{hk} - \left(\frac{\partial^2 V}{\partial a_h\,\partial a_k}\right).$$

He comments that the matrix of the w is symmetric and proceeds to examine the forms assumed by S and S_i, where S is the determinant $|v_i^r|$ and S is given in (6.36). He first shows directly that

$$S = \frac{1}{p}\cdot W, \qquad W = \sum \pm w_{11}w_{22}\ldots w_{nn},$$

i.e., W is the determinant of the w_{hk}. He then goes to discuss the forms of the S_i by considering the expression

$$\delta y_1 \frac{\partial S}{\partial v_1^k} + \delta y_2 \frac{\partial S}{\partial v_2^k} + \cdots + \delta y_n \frac{\partial S}{\partial v_n^k}. \tag{6.44}$$

This is also expressible as a determinant by replacing v^k by δy in S. He says that if he can give the δy the form

$$p\,\delta y_h = w_1 p_h^1 + w_2 p_h^2 + \cdots + w_n p_h^n, \tag{6.45}$$

then (6.44) will go over into the new form

$$\frac{1}{p}\left(w_1 \frac{\partial W}{\partial w_{1k}} + w_2 \frac{\partial W}{\partial w_{2k}} + \cdots + w_n \frac{\partial W}{\partial w_{nk}}\right).$$

He then proceeds to show that the w can always be so chosen. They are, in fact, given by

$$\begin{aligned} w_k &= \left(\frac{\partial^2 V}{\partial a_k\,\partial y_1}\right)\delta y_1 + \left(\frac{\partial^2 V}{\partial a_k \partial y_2}\right)\delta y_2 + \cdots + \left(\frac{\partial^2 V}{\partial a_k\,\partial y_n}\right)\delta y_n \\ &= \delta\left(\frac{\partial V}{\partial a_k}\right) = \delta\alpha_k \end{aligned}$$

since $(\partial V/\partial a_k) = \alpha_k$. Expression (6.44) then transforms into

$$\frac{1}{p}\left(\frac{\partial W}{\partial w_{1k}} \delta\alpha_1 + \frac{\partial W}{\partial w_{2k}} \delta\alpha_2 + \cdots + \frac{\partial W}{\partial w_{nk}} \delta\alpha_n \right).$$

But when this expression is substituted into the definition of S_k in (6.36) together with the values in (6.45) for δy_h and the values in (6.43) for v_i^k, there results

$$S_i = \frac{1}{p}\left\{ W \frac{d}{dx} \frac{p_i^1\,\delta\alpha_1 + p_i^2\,\delta\alpha_2 + \cdots + p_i^n\,\delta\alpha_n}{p} \right.$$

$$- \sum_k \left(\frac{\partial W}{\partial w_{1k}} \delta\alpha_1 + \frac{\partial W}{\partial w_{2k}} \delta\alpha_2 + \cdots + \frac{\partial W}{\partial w_{nk}} \delta\alpha_n \right)$$

$$\left. \times \frac{d}{dx} \frac{w_{1k}p_i^1 + w_{2k}p_i^2 + \cdots}{p} \right\}.$$

Clebsch now notes that the coefficient of $d(p_i^r/p)/dx$ in this relation is, apart from the factor $1/p$,

$$W\delta\alpha_r - \sum_k \left(\frac{\partial W}{\partial w_{1k}} \delta\alpha_1 + \frac{\partial W}{\partial w_{2k}} \delta\alpha_2 + \cdots + \frac{\partial W}{\partial w_{nk}} \delta\alpha_n \right) w_{rk},$$

which vanishes identically. The S_i then assume the form

$$S_i = \frac{1}{p^2} \left(p_i^1 T_1 + p_i^2 T_2 + \cdots + p_i^n T_n \right),$$

where

$$T_h = W \frac{d\delta\alpha_h}{dx} - \sum_k \sum_m \delta\alpha_m \frac{dw_{hk}}{dx} \cdot \frac{\partial W}{\partial w_{mk}}.$$

From this it is not hard to see with Clebsch that T_h is expressible as the determinant

$$T_h = \begin{vmatrix} \dfrac{d\delta\alpha_h}{dx} & \delta\alpha_1 & \delta\alpha_2 & \cdots & \delta\alpha_n \\ \dfrac{dw_{1h}}{dx} & w_{11} & w_{12} & \cdots & w_{1n} \\ \dfrac{dw_{2h}}{dx} & w_{21} & w_{22} & \cdots & w_{2n} \\ \cdot & \cdot & \cdot & \cdots & \cdot \\ \dfrac{dw_{nh}}{dx} & w_{n1} & w_{n2} & \cdots & w_{nn} \end{vmatrix}$$

Moreover, when these are introduced into the second variation (6.37), the expression under the integral sign is a homogeneous function in T_h/W of the second order, and the coefficients themselves are the expressions

$$\sum_i \sum_k \frac{\partial^2 \Omega}{\partial(dy_i/dx)\,\partial(dy_k/dx)} \frac{p_i^h p_i^r}{p^2}.$$

He now proceeds in a somewhat unclear way to find a form for these coefficients and for the function B of (6.29). Since it is not essential to follow the details, I prefer to skip over this material and proceed to Mayer's work (see p. 275 below).

6.4. Mayer's Contributions

We come now to the first of the figures after Weierstrass who have shaped the modern theory of the calculus of variations and whose imprints are still on the subject in a number of ways, as we shall see. Unlike Clebsch, Mayer proceeded in a neat and fairly rigorous way. We shall see in his papers the genesis of much of the work of later mathematicians. In his earlier papers he has reconstructed Clebsch's results by careful means and put them on sound footings. All this work has been rendered superfluous by the observation that the Clebsch condition follows trivially from the Weierstrass condition. For this reason Mayer's early work is largely obsolete and irrelevant. I have reproduced his analysis largely to show how laborious was his direct approach to the problem. It helps make us sharply aware of how elegant the approach of Weierstrass really was.

In conformity with Clebsch, Mayer considers in his 1868 paper the problem of minimizing

$$V = \int_{x_0}^{x_1} f(x\, y_1 y_1' y_2 y_2' \ldots y_n y_n')\, dx \tag{6.46}$$

subject to the side conditions

$$\phi_1 = 0, \qquad \phi_2 = 0, \ldots, \qquad \phi_m = 0, \tag{6.47}$$

where the ϕ_k are differential equations of the first order and $m < n$. He also states that the values x_0, x_1 as well as the end-values of y are specified.[12] He then proceeds to determine the y and λ so that the integral

$$J = \int_{\lambda_0}^{x_1} \Omega\, dx \tag{6.48}$$

will be a maximum or a minimum, where

$$\Omega = f + \lambda_1\phi_1 + \lambda_2\phi_2 + \cdots + \lambda_m\phi_m.$$

[12] Mayer [1868].

He replaces y_h by $y_h + \epsilon z_h$ and calculates $\delta^2 J$ as well as

$$\delta\phi_k = \sum_1^n {}_h\left\{ \frac{\partial\phi_k}{\partial y_h} z_h + \frac{\partial\phi_k}{\partial y_h'} z_h' \right\} = 0; \tag{6.49}$$

he also points out that inasmuch as the end-values of the y are fixed, the z must vanish at x_0 and x_1. He then, as usual, finds the Euler–Lagrange equations

$$\frac{\partial\Omega}{\partial y_h} = \frac{d}{dx}\frac{\partial\Omega}{\partial y_h'}, \qquad \phi_k = 0. \tag{6.50}$$

He now assumes explicitly that the determinant

$$R = \begin{vmatrix} \frac{\partial^2\Omega}{\partial y_1'\partial y_1'} & \cdots & \frac{\partial^2\Omega}{\partial y_n'\partial y_1'} & \cdots & \frac{\partial\phi_1}{\partial y_1'} & \cdots & \frac{\partial\phi_m}{\partial y_1'} \\ \cdot & \cdot & \cdot & \cdot & \cdot & \cdot & \cdot \\ \frac{\partial^2\Omega}{\partial y_1'\partial y_n'} & \cdots & \frac{\partial^2\Omega}{\partial y_n'\partial y_n'} & \cdots & \frac{\partial\phi_1}{\partial y_n'} & \cdots & \frac{\partial\phi_m}{\partial y_n'} \\ \frac{\partial\phi_1}{\partial y_1'} & \cdots & \frac{\partial\phi_1}{\partial y_n'} & \cdots & 0 & \cdots & 0 \\ \cdot & \cdot & \cdot & \cdot & \cdot & \cdot & \cdot \\ \frac{\partial\phi_m}{\partial y_1'} & \cdots & \frac{\partial\phi_m}{\partial y_n'} & \cdots & 0 & \cdots & 0 \end{vmatrix} \tag{6.51}$$

is not identically zero, so that he can solve the system

$$\frac{\partial\Omega}{\partial y_h'} = v_h, \qquad \phi_k = 0 \tag{6.52}$$

for the $n + m$ quantitites y' and λ as functions of y and v. He then considers the canonical equations

$$\frac{dy_h}{dx} = \frac{\partial H}{\partial v_h}, \qquad \frac{dv_h}{dx} = -\frac{\partial H}{\partial y_h} \tag{6.53}$$

in which H designates the function of y and v, which is found from

$$\sum_1^n {}_h y_h' v_h - f \tag{6.54}$$

by replacing y' by its value above in terms of y and v.[13]

To express the complete integral of (6.53) or (6.50), Mayer writes

$$y_h = [y_h], \qquad \lambda_k = [\lambda_k]$$

---

[13] See, e.g. Bliss, *LEC*, pp. 65ff. This is the hamiltonian, which Hamilton called his *principal function*, as we saw above on p. 179. See also Mayer [1886′].

and hence

$$v_h = \left[\frac{\partial \Omega}{\partial y_h'} \right] = [v_h].$$

These solutions contain $2n$ integration constants $a_1, a_2, \ldots, a_{2n}$. They are to be determined so that the n functions $[y_h]$ assume for $x = x_0$ and $x = x_1$ the given values y_{h0}, y_{h1}. He also assumes that the determinant R evaluated for the functions $[y_h], [\lambda_k]$ is different from zero for all values of the constants. He does this, of course, to guarantee the existence of a $2n$-parameter family of extremals.

In Section 2 after these preliminaries he turns to the second variation, the coefficient of $\epsilon^2/2$ in the expansion of J in terms of ϵ when y is replaced by $[y_h] + \epsilon z_h$. He writes it as

$$\delta^2 J = \int_{x_0}^{x_1} 2\delta^2 \Omega \, dx$$

and proceeds to find a simpler form for $2\delta^2\Omega$. To this end he first considers the function, in his notation,

$$2\Omega_2 = 2\sum_1^m {}_k \mu_k \sum_1^n {}_h \left\{ \left[\frac{\partial \phi_k}{\partial y_h} \right] z_h + \left[\frac{\partial \phi_k}{\partial y_h'} \right] z_h' \right\}$$

$$+ \sum_1^n {}_h \sum {}_i \left\{ \left[\frac{\partial^2 \Omega}{\partial y_h \, \partial y_i} \right] z_h z_i + 2\left[\frac{\partial \Omega}{\partial y_h \, \partial y_i'} \right] z_h z_i' + \left[\frac{\partial^2 \Omega}{\partial y_h' \, \partial y_i'} \right] z_h' z_i' \right\}, \quad (6.55)$$

where the quantities y, λ are replaced by $[y], [\lambda]$. He then has

$$2\delta^2\Omega = 2\Omega_2 - 2\sum_i^m {}_k \, \mu_k \delta\phi_k. \quad (6.56)$$

He now proceeds, as did Clebsch, to find the accessory differential equations, which he expresses as

$$\frac{\partial \Omega_2((\partial[y]/\partial a_i), (\partial[\lambda]/\partial a_i))}{\partial(\partial[y_h]/\partial a_i)} = \frac{d}{dx} \frac{\partial \Omega_2((\partial[y]/\partial a_i), (\partial[\lambda]/\partial a_i))}{\partial(\partial[y_h]'/\partial a_i)},$$

$$\frac{\partial \Omega_2((\partial[y]/\partial a_i), (\partial[\lambda]/\partial a_i))}{\partial(\partial[\lambda_k]/\partial a_i)} = 0.$$

He does this by differentiating equations (6.50) as to a_i; i.e., he replaces in (6.50) the y, λ by $[y], [\lambda]$ and then differentiates the resulting identities. He then sets

$$u_h = \sum_1^{2n} {}_i \gamma_i \frac{\partial[y_h]}{\partial a_i}, \qquad r_k = \sum_1^{2n} {}_i \gamma_i \frac{\partial[\lambda_k]}{\partial a_i}, \quad (6.57)$$

where the γ_i are $2n$ arbitrary constants. Then the accessory equations above become

$$\frac{\partial\Omega_2(u,r)}{\partial u_h} = \frac{d}{dx}\frac{\partial\Omega_2(u,r)}{\partial(du_h/dx)}, \qquad \frac{\partial\Omega_2(u,r)}{\partial r_k} = 0. \tag{6.58}$$

With the help of these equations, Mayer now simplifies the form of $2\Omega_2$ and finds by its homogeneity that

$$2\Omega_2(u,r) = \frac{d}{dx}\sum_1^n u_h \frac{\partial\Omega_2(u,r)}{\partial(du_h/dx)},$$

and for two different sets u,r and w,τ of solutions of the accessory equations (6.58) that

$$\sum_1^n {}_h\left\{ w_h \frac{\partial\Omega_2(u,r)}{\partial u_h} + \frac{dw_h}{dx}\frac{\partial\Omega_2(u,r)}{\partial(du_h/dx)} \right\} + \sum_1^n {}_k\tau_k \frac{\partial\Omega_2(u,r)}{\partial r_k}$$
$$= \frac{d}{dx}\sum_1^n {}_k w_k \frac{\partial\Omega_2(u,r)}{\partial(du_k/dx)}.$$

He then finds Clebsch's result (6.6) that

$$\sum_1^n {}_h\left\{ w_h \frac{\partial\Omega_2(u,r)}{\partial(du_h/dx)} - u_h \frac{\partial\Omega_2(w,\tau)}{\partial(dw_h/dx)} \right\} = \text{const.} \tag{6.59}$$

Again in imitation of Clebsch he chooses a complete solution of the accessory equations (6.58)

$$u_h^\sigma = \sum_1^{2n} \gamma_i^\sigma \frac{\partial[y_h]}{\partial a_i}, \qquad r_k^\sigma = \sum_1^{2n} \gamma_i^\sigma \frac{\partial[\lambda_k]}{\partial a_i}. \tag{6.60}$$

He also chooses n arbitrary functions g_σ and writes

$$z_h = \sum_1^n {}_\sigma g_\sigma u_h^\sigma, \qquad \mu_k = \sum_1^n {}_\sigma g_\sigma r_k^\sigma; \tag{6.61}$$

from these he finds directly that

$$\frac{dz_h}{dx} = \zeta_h + \eta_h, \qquad \zeta_h = \sum_1^n {}_\sigma g_\sigma \cdot \frac{du_h^\sigma}{dx}, \qquad \eta_h = \sum_1^n {}_\sigma \frac{dg_\sigma}{dx}\cdot u_h^\sigma. \tag{6.62}$$

For notational purposes, he calls Ω_2^0 the part of Ω_2 formed when the values in (6.60) for u,r are substituted into Ω_2 but the value ζ is put in place of $dz/dx = \zeta + \eta$.

Mayer now remarks that if the g_σ were constants, then the quantities z, μ of (6.61) would be solutions of the accessory differential equation and thus

$$2\Omega_2^0 = \frac{d}{dx}\sum_1^n {}_h\left\{ \sum_1^n {}_\sigma g_\sigma u_h^\sigma \sum_1^n {}_\rho g_\rho \frac{\partial\Omega_2(u^\rho, r^\rho)}{\partial(du_h^\rho/dx)} \right\};$$

He goes on to observe that since $2\Omega_2^0$ does not contain the derivatives of the g, this is indeed the form of $2\Omega_2^0$ whether the g are functions of x or not, provided that the terms containing the dg/dx are removed. He thus has

$$
\begin{aligned}
2\Omega_2^0 = \frac{d}{dx}\sum_{h=1}^{n}\left\{\sum_{\sigma=1}^{n} g_\sigma u_h^\sigma \sum_{\rho=1}^{n} g_\rho \frac{\partial\Omega_2(u^\rho, r^\rho)}{\partial(du_h^\rho/dx)}\right\} \\
-\sum_{h=1}^{n}\left\{\sum_{\sigma=1}^{n}\frac{dg_\sigma}{dx} u_h^\sigma \sum_{\rho=1}^{n} g_\rho \frac{\partial\Omega_2(u^\rho, r^\rho)}{\partial(du_h^\rho/dx)}\right\} \\
-\sum_{h=1}^{n}\left\{\sum_{\sigma=1}^{n} g_\sigma u_h^\sigma \sum_{\rho=1}^{n} \frac{dg_\rho}{dx} \frac{\partial\Omega_2(u^\rho, r^\rho)}{\partial(du_h^\rho/dx)}\right\}.
\end{aligned}
$$

This means that

$$
2\Omega_2 = 2\Omega_2^0 + 2\sum_{h=1}^{n}\frac{\partial\Omega_2^0}{\partial\zeta_h}\eta_h + \underset{1}{\overset{n}{\overline{\sum_h\sum_i}}}\ \frac{\partial^2\Omega_2^0}{\partial\zeta_h\,\partial\zeta_i}\eta_h\eta_i,
$$

and by the definition of Ω_2^0

$$
\frac{\partial\Omega_2^0}{\partial\zeta_h} = \sum_{\rho=1}^{n} g_\rho \frac{\partial\Omega_2(u^\rho, r^\rho)}{\partial(du_h^\rho/dx)}.
$$

This enables him to write

$$
\begin{aligned}
2\Omega_2^0 + 2\sum_{h=1}^{n}\frac{\partial\Omega_2^0}{\partial\zeta_h}\eta_h = \frac{d}{dx}\sum_{h=1}^{n} z_h \sum_{\rho} g_\rho \frac{\partial\Omega_2(u^\rho, r^\rho)}{\partial(du_h^\rho/dx)} \\
+\sum_{h=1}^{n}\left\{\sum_{\sigma=1}^{n}\frac{dg_\sigma}{dx} u_h^\sigma \sum_{\rho=1}^{n} g_\rho \frac{\partial\Omega_2(u^\rho, r^\rho)}{\partial(du_h^\rho/dx)}\right\} \\
-\sum_{h=1}^{n}\left\{\sum_{\sigma=1}^{n} g_\sigma u_h^\sigma \sum_{\rho=1}^{n} \frac{dg_\rho}{dx} \frac{\partial\Omega_2(u^\rho, r^\rho)}{\partial(du_h^\rho/dx)}\right\}.
\end{aligned}
$$

The last two terms combine into the expression

$$
\underset{1}{\overset{n}{\overline{\sum_h\sum_i}}}\ g_\rho \frac{dg_\sigma}{dx}\sum_{h=1}^{n}\left\{u_h^\sigma \frac{\partial\Omega_2(u^\rho, r^\rho)}{\partial(du_h^\rho/dx)} - u_h^\rho \frac{\partial\Omega_2(u^\sigma, r^\sigma)}{\partial(du_h^\sigma/dx)}\right\}.
$$

He now, again following Clebsch, chooses the constants γ_n^σ that determine the u_h^σ so that the constant in (6.59) is zero, and thus this expression vanishes. In other words, he chooses a conjugate set.

He proceeds to calculate that

$$
\frac{\partial^2\Omega_2^0}{\partial\zeta_h\,\partial\zeta_i} = \frac{\partial^2\Omega_2}{\partial z_h'\,\partial z_i'} = \left[\frac{\partial^2\Omega}{\partial y_h'\,\partial y_i'}\right]
$$

and hence that

$$
2\Omega_2 = \frac{d}{dx}\sum_{h=1}^{n} z_h \sum_{\rho=1}^{n} g_\rho \frac{\partial\Omega_2(u^\rho, r^\rho)}{\partial(du_h^\rho/dx)} + \underset{1}{\overset{n}{\overline{\sum_h\sum_i}}}\left[\frac{\partial^2\Omega}{\partial y_h'\,\partial y_i'}\right]\eta_h\eta_i.
$$

He has now transformed $2\delta^2\Omega$ into the form

$$2\delta^2\Omega = \overline{\sum_1^n {}_h \sum {}_i} \left[\frac{\partial^2\Omega}{\partial y_h' \, \partial y_i'} \right] \eta_h \eta_i + \frac{d}{dx} \sum_1^n z_h \sum_1^n g_\rho \frac{\partial \Omega_2(u^\rho, r^\rho)}{\partial (du_h^\rho / dx)}$$
$$- 2 \sum_1^m {}_k \mu_k \delta\phi_k,$$

where

$$z_h = \sum_1^n {}_\sigma u_h^\sigma g_\sigma, \qquad \mu_k = \sum_1^n {}_\sigma r_k^\sigma g_\sigma, \qquad \eta_h = \frac{dz_h}{dx} - \sum_1^n {}_\sigma g_\sigma \frac{du_h^\sigma}{dx} .$$

The expressions for the $\delta\phi$ have the form

$$\delta\phi_k = \sum_1^n {}_\sigma g_\sigma \sum_1^n \left\{ \left[\frac{\partial \phi_k}{\partial y_h} \right] u_h^\sigma + \left[\frac{\partial \phi_k}{\partial y_h'} \right] \frac{du_h^\sigma}{dx} \right\} + \sum_1^n {}_h \left[\frac{\partial \phi_k}{\partial y_h'} \right] \eta_h;$$

and Mayer notes that the coefficient of g_σ is $\partial\Omega_2(u^\sigma, r^\sigma)/\partial r_k^\sigma$ and thus is zero, as is seen by equations (6.58). This gives the form

$$\delta\phi_k = \sum_1^n {}_h \left[\frac{\partial \phi_k}{\partial y_h'} \right] \eta_h .$$

He now goes to find expressions for the quantities that enter into the second variation. To do this, he eliminates the g_σ between the equations

$$z_i = u_i^1 g_1 + u_i^2 g_2 + \cdots + u_i^n g_n,$$
$$-\eta_h + \frac{dz_h}{dx} = \frac{du_h^1}{dx} g_1 + \frac{du_h^2}{dx} g_2 + \cdots + \frac{du_h^n}{dx} g_n$$

and obtains the determinantal result

$$U \cdot \eta_h = \begin{vmatrix} \dfrac{dz_h}{dx} & \dfrac{du_h^1}{dx} & \cdots & \dfrac{du_h^n}{dx} \\ z_1 & u_1^1 & \cdots & u_1^n \\ \cdot & \cdot & \cdot & \cdot \\ z_n & u_n^1 & \cdots & u_n^n \end{vmatrix} = U_h, \tag{6.63}$$

where U is the determinant of the u_h^σ. In an analogous way he finds that

$$U \sum_1^n {}_\rho g_\rho \sum_1^n {}_h z_h \frac{\partial \Omega_2(u^\rho, r^\rho)}{\partial (du_h^\rho / dx)}$$

$$= - \begin{vmatrix} 0 & \sum_1^n {}_h z_h \dfrac{\partial \Omega_2(u^1, r^1)}{\partial (du_h^1 / dx)} & \cdots & \sum_1^n {}_h z_h \dfrac{\partial \Omega_2(u^n, r^n)}{\partial (du_h^n / dx)} \\ z_1 & u_1^1 & \cdots & u_1^n \\ \cdot & \cdot & \cdot & \cdot \\ z_n & u_n^1 & & u_n^n \end{vmatrix} = -2B. \tag{6.64}$$

Finally he finds that

$$U \cdot \mu_k = -\begin{vmatrix} 0 & r_k^1 & \cdots & r_k^n \\ z_1 & u_1^1 & \cdots & u_1^n \\ \cdot & \cdot & \cdot & \cdot \\ z_n & u_n^1 & \cdots & u_n^n \end{vmatrix} = -R_k. \tag{6.65}$$

This yields him the transformation of Clebsch

$$2\delta^2\Omega = \overline{\overline{\sum_h \sum_i}}_1^n \left[\frac{\partial^2\Omega}{\partial y_h' \partial y_i'} \right] \frac{U_h U_i}{U^2} - \frac{d}{dx} \frac{2B}{U} + 2\sum_1^m {}_k \frac{R_k}{U} \delta\phi_k, \tag{6.66}$$

with

$$\delta\phi_k = \sum_1^n {}_h \left[\frac{\partial \phi_k}{\partial y_h'} \right] \frac{U_h}{U}.^{14} \tag{6.67}$$

In Section 3 Mayer points out that Clebsch's relations (6.66) and (6.67) hold identically for all $2n^2$ constants γ_i^σ which satisfy the $n(n-1)/2$ conjugacy conditions

$$\sum_1^n {}_h \left\{ u_h^\sigma \frac{\partial \Omega_2(u^\rho, r^\rho)}{\partial(du_h^\rho/dx)} - u_h^\rho \frac{\partial \Omega_2(u^\sigma, r^\sigma)}{\partial(du_h^\sigma/dx)} \right\} = 0 \tag{6.68}$$

and for which the determinant U is not identically zero. As he notes, Clebsch used the most general γ_i^σ, but he finds that this is unnecessary. He now proceeds to find a special system of values γ_i^σ. To do this, he recalls that the $[y_h], [v_h]$ are a complete solution of the canonical equations (6.53) containing the $2n$ constants $a_1, a_2, \ldots, a_{2n}$. He can therefore solve the equations

$$y_h = [y_h], \qquad v_h = [v_h] = \left[\frac{\partial \Omega}{\partial y_h'} \right] \tag{6.69}$$

for the a as functions of the y and v. He calls these functions

$$(a_1), (a_2), \ldots, (a_{2n}),$$

and he calculates that

$$\begin{aligned} \frac{\partial y_h}{\partial y_\sigma} &= \sum_1^{2n} {}_i \frac{\partial [y_h]}{\partial a_i} \left[\frac{\partial (a_i)}{\partial y_\sigma} \right], \\ 0 &= \sum_1^{2n} {}_i \frac{\partial [y_h]}{\partial a_i} \left[\frac{\partial (a_i)}{\partial v_\sigma} \right], \\ 0 &= \sum_1^{2n} {}_i \frac{\partial [v_h]}{\partial a_i} \left[\frac{\partial (a_i)}{\partial y_\sigma} \right], \\ \frac{\partial v_h}{\partial v_\sigma} &= \sum_1^{2n} {}_i \frac{\partial [v_h]}{\partial a_i} \left[\frac{\partial (a_i)}{\partial v_\sigma} \right], \end{aligned} \tag{6.70}$$

where clearly $\partial y_h/\partial y_\sigma$ and $\partial v_h/\partial v_\sigma$ are Kronecker deltas.

[14] Mayer [1868], p. 248.

It follows from the definition of u_h^σ in (6.60) that

$$\frac{\partial\Omega_2(u^\sigma,r^\sigma)}{\partial(du_h^\sigma/dx)}=\sum_1^{2n}\gamma_i^\sigma\frac{\partial}{\partial a_i}\left[\frac{\partial\Omega}{\partial y_h'}\right]=\sum_1^{2n}{}_i\gamma_i^\sigma\frac{\partial[v_h]}{\partial a_i}.$$

Mayer now picks a value $x=x_\omega$ and defines the γ_i^σ by the relations

$$\gamma_i^\sigma=\left[\frac{\partial(a_i)}{\partial v_\sigma}\right]_\omega; \tag{6.70$'$}$$

it then follows from (6.70) that at $x=x_\omega$ he has

$$u_h^\sigma=0\quad\text{and}\quad\frac{\partial\Omega_2(u^\sigma,r^\sigma)}{\partial(du_h^\sigma/dx)}=0\text{ or }1,$$

according as $\sigma\neq h$ or $\sigma=h$. By this choice of the γ he has satisfied the conjugacy conditions (6.68) at x_ω; but since they are constant, he has satisfied them for all x.

He now moves ahead into new ground, leaving Clebsch behind. He formulates what Bolza calls *Mayer's determinant* and investigates some of its properties. He first defines the determinant

$$\Delta(x,x_\omega)=\begin{vmatrix}\dfrac{\partial[y_1]}{\partial a_1} & \dfrac{\partial[y_1]}{\partial a_2} & \cdots & \dfrac{\partial[y_1]}{\partial a_{2n}}\\ \cdot & \cdot & \cdot & \cdot\\ \dfrac{\partial[y_n]}{\partial a_1} & \dfrac{\partial[y_n]}{\partial a_2} & \cdots & \dfrac{\partial[y_n]}{\partial a_{2n}}\\ \dfrac{\partial[y_1]_\omega}{\partial a_1} & \dfrac{\partial[y_1]_\omega}{\partial a_2} & \cdots & \dfrac{\partial[y_1]_\omega}{\partial a_n}\\ \cdot & \cdot & \cdot & \cdot\\ \dfrac{\partial[y_n]_\omega}{\partial a_1} & \dfrac{\partial[y_n]_\omega}{\partial a_2} & \cdots & \dfrac{\partial[y_n]_\omega}{\partial a_n}\end{vmatrix} \tag{6.71}$$

in which the subscript ω attached to an expression means that the expression is to be evaluated at $x=x_\omega$. He also defines a determinant A, which he writes, as was common in his era,

$$A=\sum\pm\left[\frac{\partial(a_1)}{\partial v_1}\right]\cdots\left[\frac{\partial(a_n)}{\partial v_n}\right]\left[\frac{\partial(a_{n+1})}{\partial y_1}\right]\cdots\left[\frac{\partial(a_{2n})}{\partial y_n}\right]$$

and A_ω, which is A at $x=x_\omega$. He then concludes with the help of identities (6.70) that

$$U(x,x_\omega)=A_\omega\cdot\Delta(x,x_\omega),$$

where $U(x,x_\omega)$ is the value of $U=|u_h^\sigma|$ with the values in (6.70′) for γ substituted into it. He also has

$$\nabla_\omega U(x,x_\omega)=\Delta(x,x_\omega),$$

where ∇_ω is the reciprocal of A_ω and is thus given by

$$\nabla = \sum \pm \frac{\partial[v_1]}{\partial a_1} \cdots \frac{\partial[v_n]}{\partial a_n} \frac{\partial[y_1]}{\partial a_{n+1}} \cdots \frac{\partial[y_n]}{\partial a_{2n}}.$$

In passing, he observes from these relations why it is important that $U \neq 0$. He observes that $U = 0$ for some x is equivalent to $A_\omega = 0$ or $\nabla_\omega = \infty$, either of which is impossible since both A and ∇ are independent of x. To show this, he differentiates ∇ with respect to x and writes

$$\frac{d\nabla}{dx} = \sum_1^n {}_h \sum_1^{2n} {}_i \left\{ \frac{\partial \nabla}{\partial[\partial(v_h)/\partial a_i]} \frac{d}{dx} \frac{\partial[v_h]}{\partial a_i} + \frac{\partial \nabla}{\partial(\partial[y_h]/\partial a_i)} \frac{d}{dx} \frac{\partial[y_h]}{\partial a_i} \right\}$$

To simplify this, he substitutes the functions $[y], [\lambda]$ from (6.69) into the canonical equations (6.53). By differentiating the resulting identities, he finds at once that

$$\frac{d}{dx} \frac{\partial[y_h]}{\partial a_i} = \sum_1^n {}_k \left\{ \left[\frac{\partial^2 H}{\partial v_h \partial y_k} \right] \frac{\partial[y_k]}{\partial a_i} + \left[\frac{\partial^2 H}{\partial v_h \partial v_k} \right] \frac{\partial[v_k]}{\partial a_i} \right\},$$

$$\frac{d}{dx} \frac{\partial[v_h]}{\partial a_i} = -\sum_1^n {}_k \left\{ \left[\frac{\partial^2 H}{\partial y_h \partial y_k} \right] \frac{\partial[y_k]}{\partial a_i} + \left[\frac{\partial^2 H}{\partial y_h \partial v_k} \right] \frac{\partial[v_k]}{\partial a_i} \right\}.$$

It follows from these, after some calculations, that

$$\frac{d\nabla}{dx} = \nabla \sum_1^n {}_h \left\{ \left[\frac{\partial^2 H}{\partial v_h \partial y_h} \right] - \left[\frac{\partial^2 H}{\partial y_h \partial v_h} \right] \right\} = 0.$$

He has thus shown that ∇ is independent of x; hence it cannot vanish if the v, y are a complete solution of the canonical equations.

He sums up in this result:

(31.) *If the $2n^2$ constants γ_i^σ are given the values*

$$\gamma_i^\sigma = \left[\frac{\partial(a_i)}{\partial v_\sigma} \right]_{x = x_\omega},$$

in which x_ω designates an arbitrarily chosen value of x, then the conditions [(6.68)] . . . *are identically fulfilled and it is identically so that*

$$U = C \cdot \Delta(x, x_\omega)$$

where C is a non-null constant independent of x_ω. (Mayer [1868], p. 252)

Mayer now goes forward to formulate a generalization of Jacobi's condition concerning conjugate points. What he does is note that for a problem with fixed end-points the transformation (6.66), (6.67) of Clebsch implies immediately that

$$\delta^2 J = \int_{x_0}^{x_1} 2\delta^2 \Omega \, dx = \int_{x_0}^{x_1} dx \, \overline{\sum_1^n {}_h \sum {}_i} \left[\frac{\partial^2 \Omega}{\partial y_h' \partial y_i'} \right] \frac{U_h U_i}{U^2}, \tag{6.72}$$

subject to the conditions

$$\sum_1^n {}_h \left[\frac{\partial \phi_k}{\partial y'_h} \right] U_h = 0 \tag{6.73}$$

since the function B, evaluated at x_0 and x_1, vanishes along with the variations z. He then concludes that

> I. *So long as the arbitrary constants* γ_i^σ *can be so chosen that the determinant* U *never vanishes inside* [*or at*] *the limits* x_0 *and* x_1, *the form . . .* [(6.72)] *is valid for all variatioms z, in which the constants are given such values as satisfy the conditions above* [*i.e.*, (6.73)].[15]
>
> II. *If the arbitrary constants* γ_i^σ *can be so determined that the determinant* U *does not vanish inside* [*or at*] *the limits* x_0 *and* x_1, *and if, moreover, the homogeneous function of the second order*
>
> $$2W = \overline{\underline{\sum_h \sum_i}} \left[\frac{\partial^2 \Omega}{\partial y'_h \, \partial y'_i} \right] U_h U_i,$$
>
> *whose n arbitrary arguments* U_h *satisfy the m linear conditions* (6.73) *does not change its sign inside* [*or at*] *these limits, then the second variation* $\delta^2 J$ *can neither change its sign nor vanish, and it follows that the functions* [y] *surely furnish a* [*weak*] *maximum or a minimum.*[16]

Mayer's argument is as follows: The second variation cannot change sign or become infinite on the interval $[x_0, x_1]$ unless $2W$ vanishes identically for some U. To see that this is so, observe that $2W$ can always be transformed into

$$2W = \rho_1 V_1^2 + \rho_2 V_2^2 + \cdots + \rho_{n-m} V_{n-m}^2, \tag{6.74}$$

$$\sum_1^n {}_h U_h^2 = V_1^2 + V_2^2 + \cdots + V_{n-m}^2, \tag{6.74'}$$

i.e., the symmetric matrix of $2W$ can be reduced, for each x, to diagonal

---

[15] I have been forced to introduce the condition $U \neq 0$ on $[x_0, x_1]$, the *closed* interval, to make Mayer's subsequent results valid.

[16] Mayer's theorem does not seem quite precise unless he means by the expression "innerhalb der Grenzen" inside or at the limits. This is not absolutely clear, but later on p. 258 he does use "zwischen x_0 und x'" to mean "inside." It is known that a necessary and sufficient condition for the second variation to be positive definite for all nonidentically zero variations U_h which satisfy the conditions in (6.73) and vanish at x_0 and x_1 is that the determinant $|U_{ik}| \neq$ on $[x_0, x_1]$ (Bliss, *LEC*, p. 254). Moreover, if the determinant R in (6.51) is different from zero everywhere on $[x_0, x_1]$, then the given extremal is said to be *nonsingular*, and the condition that $2W \geqslant 0$ for all U_h satisfying (6.73) becomes $2W > 0$ for all nonidentically zero variations U_h satisfying (6.73); i.e., $2W$ is then positive definite (see Bolza, *VOR*, pp. 608–609). This is the Clebsch condition.

form by rotations. The ρ in (6.74) are the roots of the equation[17]

$$\begin{vmatrix} \left[\dfrac{\partial^2\Omega}{\partial y_1' \partial y_1'}\right] - \rho, & \cdots & \left[\dfrac{\partial^2\Omega}{\partial y_n' \partial y_1'}\right], & \left[\dfrac{\partial \phi_1}{\partial y_1'}\right], & \cdots & \left[\dfrac{\partial \phi_m}{\partial y_1'}\right] \\ \cdot & \cdots & \cdot & \cdot & \cdots & \cdot \\ \left[\dfrac{\partial^2\Omega}{\partial y_1' \partial y_n'}\right], & \cdots & \left[\dfrac{\partial^2\Omega}{\partial y_n' \partial y_n'}\right] - \rho, & \left[\dfrac{\partial \phi_1}{\partial y_n'}\right], & \cdots & \left[\dfrac{\partial \phi_m}{\partial y_n'}\right] \\ \left[\dfrac{\partial \phi_1}{\partial y_1'}\right], & \cdots & \left[\dfrac{\partial \phi_1}{\partial y_n'}\right], & 0, & \cdots & 0 \\ \cdot & \cdots & \cdot & \cdot & \cdots & \cdot \\ \left[\dfrac{\partial \phi_m}{\partial y_1'}\right], & \cdots & \left[\dfrac{\partial \phi_m}{\partial y_n'}\right], & 0, & \cdots & 0 \end{vmatrix} = 0.$$

The part of this equation which is free of ρ is exactly the determinant $[R]$, which is, by hypothesis, not identically zero. Thus none of the ρ can vanish identically. Since $2W$ cannot change sign for $x_0 \leqslant x \leqslant x_1$, the ρ must therefore be always positive, or always negative; and $2W$ can, for a given x, vanish only when the corresponding $n - m$ quantities V_i all vanish for that x, and this can only happen when all the U_h do also, as we see from (6.74′).

The second variation can then only vanish for variations z_h which, for some x, satisfy the n linear differential equations of the first order

$$U_1 = 0, \qquad U_2 = 0, \ldots, U_n = 0. \tag{6.75}$$

and vanish at the end-points x_0 and x_1. It is clear from (6.63) that equations (6.75) are linear differential equations of the first order and have as solutions the u_h^σ, so that their general solution is expressible as $z_h = c_1 u_h^1 + c_2 u_h^2 + \cdots + c_n u_h^n$ with the c arbitrary constants. The only way that such a solution z can avoid being zero at a given x is for the determinant U to vanish for that x. But by hypothesis, U is different from zero on the closed interval $[x_0, x_1]$, and so there is no set of variations z on (x_0, x_1) which makes the second variation vanish. This completes the proof.

Mayer now remarks that by the hypotheses of Theorem II, i.e., that $U \neq 0$ on the closed interval, the condition that $2W$ be of one sign viewed as a function of U for each x on $[x_0, x_1]$ is both necessary and sufficient for a minimum. (Note here and elsewhere that Mayer considers only weak minima.) He has just proved it is sufficient. To prove it necessary, suppose that $2W$ could change sign for a particular x on $[x_0, x_1]$; then clearly one of

[17] Mayer [1868], p. 254 or Bolza, *VOR*, pp. 608–609.

the ρ would have to change sign, and in this case the second variation would also, contrary to hypothesis.[18]

He concludes the section with a result originally stated by Richelot:

> III. *So long as the arbitrary constants* γ_i^σ *can be so determined that the determinant* U *is never zero on the interval* $[x_0, x_1]$, *then also the second variation can not vanish.*

He then returns (Mayer [1868], p. 256) to his determinant (6.71) and takes up the question of what it means for $\Delta(x, x_0)$ to vanish. He considers a value x' which is the first value of $x > x_0$ for which the function $\Delta(x, x_0) = 0$; he also assumes that $x_1 \geqslant x'$. The most general solution of the accessory differential equations

$$\frac{\partial \Omega_2}{\partial z_h} = \frac{d}{dx}\frac{\partial \Omega_2}{\partial z_h'}, \qquad \frac{\partial \Omega_2}{\partial \mu_k} = 0 \tag{6.76}$$

is of the form

$$z_h = \gamma_1 \frac{\partial[y_h]}{\partial a_1} + \gamma_2 \frac{\partial[y_h]}{\partial a_2} + \cdots + \gamma_{2n} \frac{\partial[y_h]}{\partial a_{2n}}, \tag{6.77}$$

$$\mu_k = \gamma_1 \frac{\partial[\lambda_k]}{\partial a_1} + \gamma_2 \frac{\partial[\lambda_k]}{\partial a_2} + \cdots + \gamma_{2n} \frac{\partial[\lambda_k]}{\partial a_{2n}}. \tag{6.77'}$$

Since $\Delta(x', x_0) = 0$, the $2n$ constants γ_i can be so chosen that the z_h in (6.77) vanish both at $x = x_0$ and $x = x'$. Mayer now chooses for z_h, μ_k the values given by (6.77) and (6.77′) on $[x_0, x']$ and the values zero on $[x', x_1]$. Then

$$\delta^2 J = \left[\sum_1^n {}_h z_h \frac{\partial \Omega_2}{\partial z_h}\right]_{x_0}^{x'} = 0$$

since

$$2\delta^2\Omega = 2\Omega_2 = \frac{d}{dx}\sum_1^n {}_h z_h \frac{\partial \Omega_2}{\partial z_h'}$$

for all z which satisfy the accessory equations (6.76). Mayer concludes from the hypothesis of Theorem III, $U \neq 0$ on $[x_0, x_1]$, that the second variation cannot vanish and hence necessarily the value $x' > x_1$.

In Section 7 he considers the converse problem. If the upper end-value x_1 lies between x_0 and x', then the γ_i^σ can always be so determined that the determinant $U \neq 0$ on $[x_0, x_1]$. Moreover, it follows by the methods described in Section 6 that if $\Delta(x_0', x_1') = 0$ for two different values x_0', x_1' on

[18] It is shown in Bliss, *LEC*, p. 254 that the second variation is positive definite for all variations z satisfying $\delta\phi_k = 0$, $z_h(x_0) = z_h(x_1) = 0$, if and only if there is conjugate system u_h^σ with $U \neq 0$ on the interval $[x_0, x_1]$. Bolza curiously does not credit Mayer with this result.

$[x_0, x_1]$, then the second variation can be made to vanish on $[x_0, x_1]$. This is perfectly clear. He therefore concludes that

> IV. *If it is possible to choose the constants γ_i^σ so that the determinant U is different from zero on $[x_0, x_1]$, then the determinant $\Delta(x, x_1)$ can not vanish for any value of x which is smaller than x_1 and greater than or equal to x_0.*

To proceed Mayer appeals to his Theorem (31) on p. 277 above to show that

$$U = C \cdot \Delta(x, x_\omega)$$

with C a constant independent of x_ω; and he chooses a value x^0 so that $x_0 < x^0 < x_1 < x'$, where x' is the first root of $\Delta(x, x_0)$ after x_0. Then for $x_\omega = x_0$, he has $U = C \cdot \Delta(x, x_0) \neq 0$ for $x^0 \leqslant x < x_1$; moreover, $\Delta(x_1, x_0) \neq 0$ since $\Delta(x_0, x_1) = (-1)^n \Delta(x_1, x_0)$.

He now remarks that so long as $x_0 < x_1 < x_1 + \epsilon < x'$, then $\Delta(x, x_1 + \epsilon) \neq 0$ on the interval $[x_0, x_1]$. From this he concludes that

> V. *So long as the larger end-point x_1 lies between x_0 and the next root x' after x_0 of the equation $\Delta(x, x_0) = 0$, the given integral will be a [weak] maximum or minimum for those functions y, which make the first variation vanish provided that the homogeneous function*
>
> $$2W = \overline{\underline{\sum_h \sum_i}}_1^n \left[\frac{\partial^2 \Omega}{\partial y_h' \, \partial y_i'} \right] U_h U_i,$$
>
> *whose n arbitrary arguments U_h satisfy the m conditions*
>
> $$\sum_1^n{}_h \left[\frac{\partial \phi_k}{\partial y_h'} \right] U_h = 0,$$
>
> *never changes its sign on the interval $[x_0, x_1]$; on the contrary, in general, neither a maximum nor minimum will be found as soon as $x_1 \geqslant x'$.*

In Mayer's final section he considers the curves $[y_h]$ passing through the fixed end-points (x_0, y_0) and (x_1, y_1), i.e., having

$$y_{h0} = [y_h]_0, \qquad y_{h1} = [y_h]_1;$$

and he supposes that these equations hold for $a_1, a_2, \ldots, a_{2n}$ and for $a_1 + \delta a_1, a_2 + \delta a_2, \ldots, a_{2n} + \delta a_{2n}$ with $\delta a_1, \delta a_2, \ldots, \delta a_{2n}$ restricted by the conditions

$$0 = \frac{\partial [y_h]_0}{\partial a_1} \delta a_1 + \frac{\partial [y_h]_0}{\partial a_2} \delta a_2 + \cdots + \frac{\partial [y_h]_0}{\partial a_{2n}} \delta a_{2n},$$

$$0 = \frac{\partial [y_h]_1}{\partial a_1} \delta a_1 + \frac{\partial [y_h]_1}{\partial a_2} \delta a_2 + \cdots + \frac{\partial [y_h]_1}{\partial a_{2n}} \delta a_{2n}.$$

These equations have a nonzero solution if and only if $\Delta(x, x_0) = 0$; thus Mayer begins to get an inkling of the character of conjugate points. When we reach Kneser, we will see this work carried out much more completely.

Mayer's paper closes with a few remarks on how to treat the variable end-point case.

6.5. Lagrange's Multiplier Rule

When Lagrange devised his multiplier rule, he gave at best a very cursory proof. Mayer suggests that a proof for the case where the side-conditions are differential equations had never been given.[19] Moreover, he points out that no counterexample is known. It is therefore reasonable for him to give a proof of this fundamental result. To this end he considers the following problem of Lagrange:

> *Among all functions $y_1, y_2, \ldots, y_n$ of x which satisfy the $m < n$ preassigned side conditions*
>
> $$\phi_k(x, y_1, \ldots, y_n, y_1', \ldots, y_n') = 0 \qquad (k = 1, 2, \ldots, m), \tag{6.78}$$
>
> *which have given values at the end-points x_0 and x_1, and which are continuous along with their derivatives, to find that one which furnishes the given integral*
>
> $$V = \int_{x_0}^{x_1} f(x, y_1, \ldots, y_n, y_1', \ldots, y_n')\, dx$$
>
> *its largest or least value.*

He assumes that the determinant

$$\sum \pm \frac{\partial \phi_1}{\partial y_1'} \frac{\partial \phi_2}{\partial y_2'} \cdots \frac{\partial \phi_m}{\partial y_m'} \tag{6.79}$$

does not vanish; thus the side-conditions (6.78) do not vanish identically and can be solved for the first m functions y in terms of the others.

He assumes that he has found the solution of his problem and concludes, without any real justification that along this solution curve the identity

$$\delta V \equiv \int_{x_0}^{x_1} dx \sum_1^n {}_i \left(\frac{\partial f}{\partial y_i} \delta y_i + \frac{\partial f}{\partial y_i'} \delta y_i' \right) = 0 \tag{6.80}$$

must hold for all continuous functions $\delta y_1, \ldots, \delta y_n$ of x, which vanish at

[19] Mayer [1886], p. 74.

$x = x_0$ and x_1 and which satisfy the relations

$$\sum_1^n {}_i\left(\frac{\partial \phi_\kappa}{\partial y_i} \delta y_i + \frac{\partial \phi_\kappa}{\partial y_i'} \delta y_i' \right) = 0.^{20} \tag{6.81}$$

Mayer multiplies these conditions by (as yet undertermined) multipliers μ_k and adds the result into expression (6.80) for δV, finding, after the usual integration by parts, that

$$\int_{x_0}^{x_1} dx \sum_1^n {}_i\left(\frac{\partial F}{\partial y_i} - \frac{d}{dx} \frac{\partial F}{\partial y_i'} \right) \delta y_i = 0 \tag{6.82}$$

where $f + \mu_1\phi_1 + \cdots + \mu_m\phi_m = F$. He now views the equations

$$\frac{\partial F}{\partial y_r} - \frac{d}{dx} \frac{\partial F}{\partial y_r'} = 0 \qquad (r = 1, 2, \ldots, m) \tag{6.83}$$

as m linear differential equations of the first order for the multipliers $\mu_1, \ldots, \mu_m$. (The functions y and y' that enter are those of the minimizing arc.) These equations can be solved since the determinant (6.79) is $\neq 0$.

With the help of these μ_k, conditions (6.82) become

$$U_{\delta y} \equiv \int_{x_0}^{x_1} dx \sum_{m+1}^n {}_s\left(\frac{\partial F}{\partial y_s} - \frac{d}{dx} \frac{\partial F}{\partial y_s'} \right) \delta y_s = 0. \tag{6.84}$$

Mayer now turns to an examination of the δy_s $(s = m + 1, \ldots, n)$. To do this, he multiplies conditions (6.81) by "new factors ν_κ and adds up the products." The sum is then expressible as

$$\sum_1^n {}_i\delta y_i \sum_1^m {}_\kappa\left(\nu_\kappa \frac{\partial \phi_\kappa}{\partial y_i} - \frac{d}{dx} \nu_\kappa \frac{\partial \phi_\kappa}{\partial y_i'} \right) + \frac{d}{dx}\left(\sum_1^n {}_i\delta y_i \sum_1^m {}_\kappa \nu_\kappa \frac{\partial \phi_\kappa}{\partial y_i'} \right) = 0.$$

With him, let ν_κ^ρ $(\kappa, \rho = 1, 2, \ldots, m)$ be "the linearly independent system of solutions $\nu_1, \nu_2, \ldots, \nu_m$ of the m abridged linear differential equations of the first order:

$$\sum_1^m {}_\kappa\left(\nu_\kappa \frac{\partial \phi_\kappa}{\partial y_r} - \frac{d}{dx} \nu_\kappa \frac{\partial \phi_\kappa}{\partial y_r'} \right) = 0 \qquad (r = 1, 2, \ldots, m)." \tag{6.85}$$

[20] There is an essential gap here in Mayer's reasoning. He has not shown that the given arc $y = y(x)$ can be embedded in a family of arcs $y_i = y_i(x, b)$ satisfying the side-conditions (6.78) and having

$$y_i(x, 0) = y_i(x), \qquad \frac{\partial y_i(x, 0)}{\partial b} = \delta y_i,$$

for any preassigned continuous variation δy_i such that $\delta y_i(x_0) = 0 = y_i(x_1)$. This gap was first filled by Kneser, *LV*, pp. 161ff, and by Hilbert [1905], [1905′]. See also Bolza, *VOR*, pp. 566–569 and Bolza [1907], pp. 370.

It then follows that conditions (6.84) become

$$\frac{d}{dx}\left(\sum_1^n{}_i\delta y_i\sum_1^m{}_\kappa \nu_\kappa^\rho\frac{\partial\phi_\kappa}{\partial y_i'}\right)+\sum_{m+1}^n{}_s\delta y_s\sum_1^m{}_\kappa\left(\nu_\kappa^\rho\frac{\partial\phi_\kappa}{\partial y_s}-\frac{d}{dx}\nu_\kappa^\rho\frac{\partial\phi_\kappa}{\partial y_s'}\right)=0.$$

Mayer integrates both members of this equation from x_0 to x, and since $\delta y = 0$ for $x = x_0$, he has

$$\sum_1^m{}_r\delta y_r\sum_1^m{}_\kappa \nu_\kappa^\rho\frac{\partial\phi_\kappa}{\partial y_r'}=-\sum_{m+1}^n{}_s\delta y_s\sum_1^m \nu_\kappa^\rho\frac{\partial\phi_\kappa}{\partial y_s'}$$
$$-\int_{x_0}^x dx\sum_{m+1}^n{}_s\delta y_s\sum_1^m{}_\kappa\left(\nu_\kappa^\rho\frac{\partial\phi_\kappa}{\partial y_s}-\frac{d}{dx}\nu_\kappa^\rho\frac{\partial\phi_\kappa}{\partial y_s'}\right). \quad (6.86)$$

He now views these as m linear equations in the m unknowns $\delta y_1, \ldots, \delta y_m$. The determinant of the system has been chosen by Mayer so that it does not vanish at x_0 or x_1. It is

$$\sum \pm \nu_1^1\nu_2^2\cdots\nu_m^m\sum\pm\frac{\partial\phi_1}{\partial y_1'}\frac{\partial\phi_2}{\partial y_2'}\cdots\frac{\partial\phi_m}{\partial y_m'}.$$

Mayer then observes that since all variations δy must vanish at $x = x_1$,

$$W_{\delta y}^\rho\equiv\int_{x_0}^{x_1}dx\sum_{m+1}^n\delta y_s\sum_1^m{}_\kappa\left(\nu_\kappa^\rho\frac{\partial\phi_\kappa}{\partial y_s}-\frac{d}{dx}\nu_\kappa^\rho\frac{\partial\phi_\kappa}{\partial y_s'}\right)=0. \quad (6.87)$$

These are then m conditions that the δy_s must fulfill. To exhibit the δy_s, he writes

$$\delta y_s = z_s+\sum_1^m{}_\sigma\alpha_\sigma u_s^\sigma \qquad (s = m+1, \ldots, n), \quad (6.88)$$

where the u_s^σ are arbitrary, as are the z_s, except that the u_s^σ and z_s must vanish at x_0 and x_1. Then the W in (6.87) must be such that

$$W_z^\rho+\sum_1^m{}_\sigma\alpha_\sigma W_{u^\sigma}^\rho=0. \quad (6.89)$$

He now desires the $m(n-m)$ functions u_s^σ to be so chosen that the determinant

$$\sum\pm W_{u^1}^1W_{u^2}^2,\ldots,W_{u^m}^m \quad (6.90)$$

does not vanish, and he picks the α_σ so that equations (6.89) are satisfied, i.e.,

$$\alpha_\sigma=\sum_1^m{}_\rho\,\beta_\rho^\sigma W_z^\rho$$

with β_ρ^σ dependent only on the u_s^σ.

Mayer next substitutes the values (6.88) for δy_s into formula (6.84) for $U_{\delta y}$ and finds that

$$U_z+\sum_1^m\alpha_\sigma U_{u^\sigma}=0$$

or equivalently that

$$U_z + \sum_1^m {}_\rho \gamma_\rho W_z^\rho = 0,$$

with

$$\gamma_\rho = \sum_1^m {}_\sigma \beta_\rho^\sigma U_{u^\sigma}.$$

This may be expressed in the form

$$\int_{x_0}^{x_1} dx \sum_{m+1}^{n} {}_s z_s \left\{ \frac{\partial F}{\partial y_s} - \frac{d}{dx}\frac{\partial F}{\partial y_s'} + \sum_1^m {}_\rho \gamma_\rho \left(\nu_\kappa^\rho \frac{\partial \phi_\kappa}{\partial y_s} - \frac{d}{dx} \nu_\kappa^\rho \frac{\partial \phi_\kappa}{\partial y_s'} \right) \right\} = 0,$$

where the functions z are arbitrary, except that they must vanish at x_0 and x_1. But this implies that the coefficients of the z_s must vanish.[21] That is, since by (6.85)

$$\sum_1^m {}_\kappa \left(\nu_\kappa \frac{\partial \phi_\kappa}{\partial y_r} - \frac{d}{dx} \nu_\kappa \frac{\partial \phi_\kappa}{\partial y_r'} \right) = 0 \qquad (r = 1, 2, \ldots, n), \tag{6.91}$$

for

$$\nu_\kappa = \sum_1^m {}_\rho \gamma_\rho \nu_\kappa^\rho,$$

it follows from (6.83) that not only must the minimizing arc satisfy

$$\frac{\partial F}{\partial y_r} - \frac{d}{dx}\frac{\partial F}{\partial y_r'} = 0 \qquad (r = 1, 2, \ldots, m), \tag{6.92}$$

but also the equations

$$\frac{\partial F}{\partial y_s} - \frac{d}{dx}\frac{\partial F}{\partial y_s'} + \sum_1^m \left(\nu_\kappa \frac{\partial \phi_\kappa}{\partial y_s} - \frac{d}{dx} \nu_\kappa \frac{\partial \phi_\kappa}{\partial y_s'} \right) = 0 \qquad (s = m+1, \ldots, n), \tag{6.93}$$

where the ν_κ satisfy equations (6.91).

Mayer now points out that if corresponding members of equations (6.91) and (6.83) are added, equations (6.92) and (6.93) can be combined into

$$\frac{\partial f}{\partial y_i} + \sum_1^m {}_\kappa \lambda_\kappa \frac{\partial \phi_\kappa}{\partial y_i} - \frac{d}{dx}\left(\frac{\partial f}{\partial y_i'} + \sum_1^m {}_\kappa \lambda_\kappa \frac{\partial \phi_\kappa}{\partial y_i'} \right) = 0 \tag{6.94}$$

with $\mu_\kappa + \nu_\kappa = \lambda_\kappa$. This is Lagrange's multiplier rule.

Mayer now goes on to the first discussion of abnormality in the history of our subject. He notes that his proof requires the determinant of the W

---

[21] This is the essence of Mayer's proof. He has shown correctly that there are no restrictions other than continuity on the z, except that they vanish at the ends. Hence the fundamental lemma of the calculus of variations is applicable. (See du Bois-Raymond [1879], Bertrand [1842] and Stegmann, *LV*.) The proof due to Bertrand is the first given but is not very rigorous. It is, in fact, a proof of Euler's rule for isoperimetric problems (see Section 6.6 below).

(6.90) to be not zero for the set u_s^σ vanishing at x_0 and x_1. However, if there is a solution $\nu_1, \ldots, \nu_m$ of the m differential equations (6.85) which also satisfies the $n - m$ equations

$$\sum_1^m \kappa\left(\nu_\kappa \frac{\partial \phi_\kappa}{\partial y_s} - \frac{d}{dx}\nu_\kappa \frac{\partial \phi_\kappa}{\partial y_s'}\right) = 0 \qquad (s = m+1, \ldots, n), \tag{6.95}$$

then clearly the W in (6.87) will vanish. In this case there are constants $l_0 = 0$, λ_κ such that for

$$F = l_0 f + \lambda_1\phi_1 + \lambda_2\phi_2 + \cdots + \lambda_m\phi_m$$

the equations

$$\frac{\partial F}{\partial y_i} - \frac{d}{dx}\frac{\partial F}{\partial y_1'} = 0 \qquad (i = 1, 2, \ldots, n)$$

are satisfied. Not all the λ_κ can be identically zero. It was Hahn who called this the *abnormal* case.[22]

In Section 2 Mayer turns to the case where his new side-conditions are

$$\psi_\alpha(x, y_1, \ldots, y_n) = 0 \qquad (\alpha = 1, 2, \ldots, p), \tag{6.96}$$

$$\phi_\beta(x, y_1, \ldots, y_n, y_1', \ldots, y_n') = 0 \qquad (\beta = 1, 2, \ldots, q) \tag{6.97}$$

with $p + q < n$. He further requires that only $n - p$ values of y can be specified at the lower and upper limits.

He assumes the determinant

$$\sum \pm \frac{\partial \psi_1}{\partial y_1}\frac{\partial \psi_2}{\partial y_2} \cdots \frac{\partial \psi_p}{\partial y_p} \tag{6.98}$$

is not zero and solves the equations (6.96) for

$$y_\kappa = y_\kappa(x, y_{p+1}, \ldots, y_n) \qquad (\kappa = 1, 2, \ldots, p) \tag{6.99}$$

He substitutes these into the integrand f and into the ϕ_β, obtaining the new functions

$$f = \chi(x, y_{p+1}, \ldots, y_n, y_{p+1}', \ldots, y_n'),$$

$$\phi_\beta = \chi_\beta(x, y_{p+1}, \ldots, y_n, y_{p+1}', \ldots, y_n').$$

By this means he has reduced the new problem to that of solving the problem

$$\int_{x_0}^{x_1} \chi\, dx = \text{Max., Min.}, \chi_\beta = 0,$$

which is the form of the problem in Section 1. That is, he has

$$\frac{\partial \Phi}{\partial y_s} - \frac{d}{dx}\frac{\partial \Phi}{\partial y_s'} = 0 \qquad (s = p+1, \ldots, n), \tag{6.100}$$

$$\Phi = \chi + \lambda_1\chi_1 + \cdots + \lambda_q\chi_q.$$

[22] Hahn [1904], p. 152.

In terms of the original variables, he has

$$F \equiv f + \lambda_1\phi_1 + \cdots + \lambda_q\phi_q = \Phi.$$

For notational purposes, Mayer now considers n arbitrary functions $y_1, \ldots, y_n$ and relates them to $x, u_1, \ldots, u_\mu$ by the relation

$$F(x, y_1, \ldots, y_n, y_1', \ldots, y_n') = \Phi(x, u_1, \ldots, u_\mu, u_1', \ldots, u_\mu').$$

He then sees that

$$\frac{\partial \Phi}{\partial u_h} - \frac{d}{dx}\frac{\partial \Phi}{\partial u_h'} = \sum_1^n {}_i\left(\frac{\partial F}{\partial y_i} - \frac{d}{dx}\frac{\partial F}{\partial y_i'}\right)\frac{\partial y_i}{\partial u_h}.$$

With the help of equations (6.99), equations (6.100) become identically

$$\frac{\partial \Phi}{\partial y_s} - \frac{d}{dx}\frac{\partial \Phi}{\partial y_s'} = \sum_1^p {}_\kappa\left(\frac{\partial F}{\partial y_\kappa} - \frac{d}{dx}\frac{\partial F}{\partial y_\kappa'}\right)\frac{\partial y_\kappa}{\partial y_s} + \frac{\partial F}{\partial y_s} - \frac{d}{dx}\frac{\partial F}{\partial y_s'}$$

$$(s = p + 1, \ldots, n).$$

Mayer also has, as a consequence of the side-conditions $\psi_\alpha(x, y_1, \ldots, y_p) = 0$, that

$$0 = \sum_1^p {}_\kappa \frac{\partial \psi_\alpha}{\partial y_\kappa}\frac{\partial y_\kappa}{\partial y_s} + \frac{\partial \psi_\alpha}{\partial y_s}$$

and hence that

$$\frac{\partial \Phi}{\partial y_s} - \frac{d}{dx}\frac{\partial \Phi}{\partial y_s'} = \sum_1^p {}_\kappa\left(\frac{\partial F}{\partial y_\kappa} - \frac{d}{dx}\frac{\partial F}{\partial y_\kappa'} + \sum_1^p {}_\alpha \mu_\alpha \frac{\partial \psi_\alpha}{\partial y_\kappa}\right)\frac{\partial y_\kappa}{\partial y_s}$$

$$+ \frac{\partial F}{\partial y_s} - \frac{d}{dx}\frac{\partial F}{\partial y_s'} + \sum_1^p {}_\alpha \mu_\alpha \frac{\partial \psi_\alpha}{\partial y_s},$$

where $\mu_1, \ldots, \mu_p$ are arbitrary multipliers. But by the assumption that the determinant (6.98) $\neq 0$ "these p multipliers can be so determined that in the last formula the coefficient of each $\partial y_\kappa / \partial y_s$ vanishes." This reduces the situation to equations (6.96), (6.97) together with

$$\frac{\partial \Psi}{\partial y_i} - \frac{d}{dx}\frac{\partial \Psi}{\partial y_i'} = 0 \qquad (i = 1, 2, \ldots, n).$$

This is the multiplier rule in the present case, where

$$\Psi = f + \lambda_1\phi_1 + \cdots + \lambda_q\phi_q + \mu_1\psi_1 + \cdots + \mu_p\psi_p.$$

6.6. Excursus on the Fundamental Lemma and on Isoperimetric Problems

One of the basic problems in the calculus of variations is to go from the fact that the first variation must vanish along a minimizing arc to the Euler–Lagrange equations. This transition is effected by the so-called fundamental lemma. It is certainly not as self-evident as Lagrange seemed to feel it to be. The first proof seems to have been given by Stegmann in 1854.[23] The lemma has several possible formulations, one of which is that of Bolza:

Lemma. *If M is a continuous function of x on $[x_1, x_2]$ and if*

$$\int_{x_1}^{x_2} \eta M\, dx = 0 \tag{6.101}$$

for all functions η which vanish at x_1 and x_2 and possess a continuous derivative on $[x_1, x_2]$, then

$$M \equiv 0$$

on $[x_1, x_2]$.

Recall that for the case where the integrand is $f(x, y, y')$, the first variation is expressible as

$$\delta I = \int_{x_0}^{x_1} \left[\frac{\partial f}{\partial y} \eta + \frac{\partial f}{\partial y'} \eta' \right] dx.$$

Now if $df_{y'}/dx$ exists and is integrable, then the first variation is expressible as

$$\delta I = \int_{x_0}^{x_1} \left[f_y - \frac{d}{dx} f_{y'} \right] \eta\, dx,$$

for all η such that $\eta(x_0) = \eta(x_1) = 0$ and for which η' has some additional property such as continuity on $[x_0, x_1]$. The form of $df_{y'}/dx$ is such that

$$\frac{d}{dx} f_{y'} = f_{y'x} + f_{y'y} \cdot y' + f_{y'y'} \cdot y'',$$

and thus it was usual to assume that y'' existed and was continuous.

However, it is perfectly possible to express the first variation in the form

$$\delta I = \int_{x_0}^{x_1} \left[f_{y'} - \int_{x_0}^{x} f_y\, dx \right] \eta'\, dx = \int_{x_0}^{x_1} N \eta'\, dx \tag{6.102}$$

provided that $f_{y'}$ and f_y are integrable. This form, due to du Bois-Raymond, is applicable even when y'' does not necessarily exist.[24]

Let us examine the proofs due to Stegmann and du Bois-Raymond. The former's argument is not sound but is easily alterable into a correct form. He chooses η so that

$$\eta(x) = \frac{(x - x_0)^{\mu}(x_1 - x)^{\nu}}{M(x)}$$

with positive exponents μ, ν. He then finds that the integral condition

[23] Stegmann, *LV*, pp. 90–91 or Bolza, *VOR*, p. 26n.

[24] du Bois-Raymond [1879], p. 313.

(6.101) reduces to

$$\int_{x_0}^{x_1} (x - x_0)^{\mu}(x_1 - x)^{\nu}\, dx = 0,$$

which is impossible. Thus $M \equiv 0$.[25]

In connection with the fundamental lemma, we might also notice a short paper by Bertrand in 1842 in which he attempts to show that if

$$\int_a^b \omega u\, dx = 0$$

for all ω for which

$$\int_a^b \omega v\, dx = 0,$$

then $u = cv$ for some constant c.[26] The connection is this: let $u = N$, $v = 1$, and $\omega = \eta'$ in (6.102). Then Bertrand has shown that $N \equiv C$. Thus the fundamental lemma may be viewed as a special case of an isoperimetric problem.[27] Bertrand's proof is not very convincing or complete.

Another proof was given by Heine in 1870 in a brief note.[28] He wrote the first variation in the form

$$[P]_a^b + \int_a^b f(x, y, y', \ldots,)\, \delta y\, dx = 0, \tag{6.103}$$

in which "P represents a linear function of δy and its $\mu - 1$ derivatives with respect to x." He chooses for δy the function

$$\delta y = (x - a)^{\mu}(x - b)^{\mu}(x - \alpha_1)(x - \alpha_2) \ldots,$$

where he has assumed that f is analytic and that its sign changes at the places $\alpha_1, \alpha_2, \ldots$. Then the expression outside the integral sign in (6.103) vanishes, and he argues that the function f must now vanish also. His method is not free of objections and requires more differentiability on the minimizing arc than is needed.

The standard proofs of the fundamental lemma are in a paper by P. du Bois-Raymond.[29] This paper is of some interest, and I review parts of it below. However, he does devote a considerable amount of space to results on (Riemann) integrability and rectifiability, which now hardly seem germane. I have omitted these and start with Section 8 of his paper.

Here du Bois-Raymond establishes the fundamental lemma in several

---

[25] To correct Stegmann's argument, suppose that $M \not\equiv 0$ and that $M(x') \neq 0$ for $x_0 < x' < x_1$. Then in a small interval $x' - \delta$, $x' + \delta$, the function $M \neq 0$. Choose

$$\eta(x) = (x - x' + \delta)(x' + \delta - x)/M(x)$$

on this interval and zero outside.

[26] Bertrand [1842], pp. 55–58. This is a proof of Euler's rule for the isoperimetric problem.

[27] du Bois-Raymond [1879].

[28] Heine [1870].

[29] du Bois-Raymond [1879].

ways. First he considers the condition

$$\int_{x_0}^{x_1} dx\, \lambda(x) F'(x) = 0 \tag{6.104}$$

for all $\lambda(x)$ of class $C^{(n)}$ on $[x_0, x_1]$ which vanish at x_0 and x_1, i.e., all such functions which, along with their first n derivatives, are continuous. (In what follows he assumes F' is bounded and Riemann integrable.) He chooses for λ the function defined in terms of two arbitrary values α, β such that $x_0 < \alpha < \beta < x_1$; he sets

$$\lambda(x) = 0 \quad \text{for } x_0 \leqslant x < \alpha \text{ and } \beta < x \leqslant x_1,$$

$$\lambda(x) = \left[(x - \alpha)(\beta - x) \right]^{n+1} \quad \text{for } \alpha \leqslant x \leqslant \beta.$$

Then condition (6.104) reduces to

$$F'(\xi) \int_\alpha^\beta \lambda(x)\, dx = 0$$

with ξ some value between α and β. Thus $F'(\xi) = 0$ and by simple continuity arguments $F'(x) = 0$ at every point x at which F' is continuous since α and β are arbitrary.

In Section 10 du Bois-Raymond considers condition (6.104) as holding for all λ which vanish at a and b and possess continuous derivatives of all orders. To do this, he rewrites his condition (6.104) as

$$\int_a^b dx\, \lambda(x) f(x) = 0.$$

He divides up the interval $a \ldots b$ into five parts $a \ldots \alpha \ldots \alpha' \ldots \beta' \ldots \beta \ldots b$, and he seeks to determine λ so that on $a \leqslant x \leqslant \alpha$, $\beta \leqslant x \leqslant b$ it vanishes and is unity on $\alpha' \leqslant x \leqslant \beta'$. On the remaining intervals $\alpha < x < \alpha'$, $\beta' < x < \beta$ the function λ varies smoothly between 0 and 1. He sets

$$\lambda(x) = \frac{1}{\pi} \left\{ \arctan \rho(x) + \frac{\pi}{2} \right\},$$

$$\rho(x) = \left(x - \frac{\alpha + \alpha'}{2} \right)^{1/(x-\alpha)(\alpha' - x)}.$$

(The reader can satisfy himself that this function λ has the desired differentiability properties; in Figures 6.1 and 6.2 he can see the function and its limiting position as α', β' approach α, β.)

With the help of this function λ, he finds that

$$0 = \int_a^b dx\, \lambda(x) f(x)$$

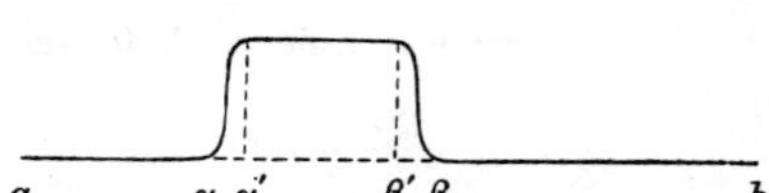

Figure 6.1

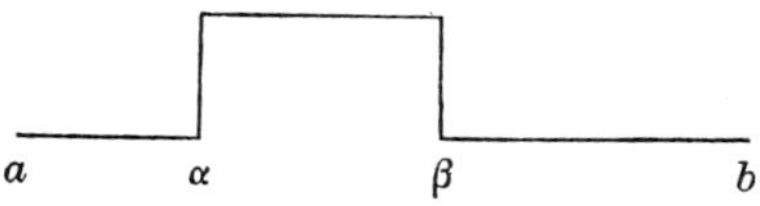

Figure 6.2

becomes

$$0=\int_{\alpha}^{\beta} f(x)\,dx+\int_{\alpha}^{\alpha'} dx\,(\lambda(x)-1)f(x)+\int_{\beta'}^{\beta} dx\,(\lambda(x)-1)f(x).$$

He writes this as

$$\int_{\alpha}^{\beta} f(x)\,dx=\int_{\alpha}^{\alpha'} dx\,(1-\lambda(x))f(x)+\int_{\beta'}^{\beta} dx\,(1-\lambda(x))f(x)$$

and notes the monotonic property of $1-\lambda(x)$. Then in the limit as $\alpha'\to\alpha$, $\beta'\to\beta$

$$\int_{\alpha}^{\beta} f(x)\,dx=0$$

for α, β arbitrary. He then concludes that $\int f\,dx$ vanishes on every subinterval of a,b.

On pp. 307–309 and 312–313 du Bois-Raymond, in effect, replaces his original form of the fundamental lemma by the one that appears in (6.102). He considers the lemma as an isoperimetric problem in the form

$$\int_{x_0}^{x_1} dx\,\lambda'(x)f(x)=0$$

for all δy such that

$$\int_{x_0}^{x_1} \lambda'(x)\,dx=0.$$

This implies the existence of a constant σ such that $f(x)+\sigma$ is a constant.

In Sections 16 and 17 he gives two proofs of the isoperimetric rule or, as it is called, Euler's rule. The "first is an application to the ordinary case of an idea of Herr R. Reiff, the other I beg permission to introduce."[30] In the first method he expresses the variations of his integrals in the form

$$\Big|_a^b U''+\int_a^b dx\,U'\delta y=0,\qquad \Big|_a^b V''+\int_a^b dx\,V'\delta y=0.$$

(Presumably there is no particular significance to the expressions U'',V'' other than the fact they represent what comes out from under the integral signs after the indicated integration by parts.) du Bois-Raymond assumes that U'',V'' vanish at a and b, and he sets δy to be null except in two equal

[30] Reiff's idea was contained in his inaugural dissertation at Tübingen [1879]. du Bois-Raymond's proof is reproduced not only in his paper, but also in Jordan, *COURS*. The problem at hand is to render the integral $\int_a^b U\,dx$ a maximum or a minimum while $\int_a^b V\,dx$ has a predetermined value.

but small subintervals $\alpha_0 \ldots \beta_0$, $\alpha_1 \ldots \beta_1$ in each of which δy is a constant, δy_0 and δy_1. Then the first integral becomes

$$\delta y_0 \int_{\alpha_0}^{\beta_0} dx\, U' + \delta y_1 \int_{\alpha_1}^{\beta_1} dx\, U' = 0$$

and by the mean-value theorem,

$$\delta y_0\, U_0' + \delta y_1\, U_1' = 0$$

since $\beta_0 - \alpha_0 = \beta_1 - \alpha_1$, where U_0', U_1' are values of U' somewhere on the intervals α_0, β_0 and α_1, β_1. In the same way

$$\delta y_0\, V_0' + \delta y_1\, V_1' = 0.$$

From these it follows that $U_0'V_1' - U_1'V_0' = 0$. Now du Bois-Raymond fixes $\alpha_0 = \beta_0$, $\alpha_1 = \beta_1$ and allows $x = \alpha_1 = \beta_1$ to be variable. (This is, in effect, John Bernoulli's uniformity principle.) Then

$$U' + \sigma V' = 0,$$

where $\sigma = -U_0'/V_0'$.

If there are two isoperimetric conditions

$$\delta \int dx\, V = 0, \qquad \delta \int dx\, W = 0$$

to be satisfied, he chooses $\delta y = 0$, except on three equal subintervals $\alpha_0 \ldots \beta_0$, $\alpha_1 \ldots \beta_1$, $\alpha_2 \ldots \beta_2$ in each of which δy is a constant $\delta y_0, \delta y_1, \delta y_2$. Then he has

$$\begin{aligned} \delta y_0\, U_0' + \delta y_1\, U_1' + \delta y_2\, U_2' &= 0, \\ \delta y_0\, V_0' + \delta y_1\, V_1' + \delta y_2\, V_2' &= 0, \\ \delta y_0\, W_0' + \delta y_1\, W_1' + \delta y_2\, W_2' &= 0, \end{aligned}$$

and he takes $x_0 = \alpha_0 = \beta_0$, $x_1 = \alpha_1 = \beta_1$ as fixed and $x = \alpha_2 = \beta_2$ as variable. From this he concludes that

$$U' + \sigma_1 V' + \sigma_2 W' = 0,$$

"where the σ_1 and σ_2 depend only on the fixed points x_0 and $x_1, \ldots$."[31]

du Bois-Raymond next takes up his own proof. To do this, he considers the relations

$$0 = \int \delta U\, dx = \int dx \left\{ \frac{\partial U}{\partial y} \delta y + \frac{\partial U}{\partial y'} \delta y' + \cdots \right\},$$

$$0 = \int \delta V\, dx = \int dx \left\{ \frac{\partial V}{\partial y} \delta y + \frac{\partial V}{\partial y'} \delta y' + \ldots \right\},$$

where again the second equation arises from the isoperimetric condition. He now needs what he calls *unbounded* variations. (These are variations which need not be of class $C^{(n)}$ as before.) Let $\delta_1 y, \delta_2 y$ be two such variations and consider $\delta y = \delta_1 y + C\delta_2 y$, where C is to be determined so

[31] Recall the problem Euler had with this case; see p. 100 above.

that the relation

$$0=\int dx\left\{\frac{\partial V}{\partial y}(\delta_1 y+C\delta_2 y)+\frac{\partial V}{\partial y'}(\delta_1 y'+C\delta_2 y')+\dots\right\}$$

is satisfied. This implies that for $\delta_1 y, \delta_2 y$ arbitrary,

$$C=-\frac{\int dx\{(\partial V/\partial y)\delta_1 y+(\partial V/\partial y')\delta_1 y'+\dots\}}{\int dx\{(\partial V/\partial y)\delta_2 y+(\partial V/\partial y')\delta_2 y'+\dots\}},$$

and hence that

$$\begin{aligned}0&=\int dx\left\{\frac{\partial}{\partial y}(U+\sigma V)\delta_1 y+\frac{\partial}{\partial y'}(U+\sigma V)\delta_1 y'+\dots\right\}\\&=\int dx\,\delta_1(U+\sigma V),\end{aligned}$$

where σ is a constant independent of $\delta_1 y$. It follows that this holds for $\delta_1 y$ arbitrary. He says that the variation $\delta_1 y$ must be allowed to have discontinuities if the minimizing arc does not have a continuous derivative everywhere.

In a second paper in the same journal du Bois-Raymond ([1879′], pp. 564–576) carries out a further analysis of the fundamental lemma and in the process coins the expression "extremum" to cover the two cases: "maximum" and "minimum."

In a paper in 1877 Mayer gives a very nice discussion of necessary conditions for isoperimetric problems by reducing such problems to special cases of the general problem of Lagrange.[32] It is in this paper that Mayer exhibits his famous reciprocity theorem.

The problem posed by Mayer is this:

> II. *To determine functions $y_1, \dots, y_n$ of x which are subject to the m isoperimetric conditions*
>
> $$\int_{x_0}^{x_1} f_\kappa(x, y_1, \dots, y_n y_1', \dots, y_n')\,dx = l_\kappa \qquad (\kappa = 1, 2, \dots, m)$$
>
> *and moreover which take on given values at x_0 and x_1, so that at the same time the integral*
>
> $$V=\int_{x_0}^{x_1} f(x, y_1, \dots, y_n y_1', \dots, y_n')\,dx$$
>
> *becomes a relative maximum or minimum.*

To solve this problem, Mayer sets

$$u_\kappa=\int f_\kappa\,dx, \quad \phi_\kappa\equiv f_\kappa-u_\kappa' \qquad (\kappa=1,2,\dots,m)$$

[32] Mayer [1877]. In his discussion Mayer calls attention to a paper by Lündstrom [1869] on this problem.

and observes that his problem now becomes

> III. *To determine the* $n + m$ *functions* $y_1, \ldots, y_n, u_1, \ldots, u_m$ *interrelated by the m given conditions*
> $$\phi_\kappa = f_\kappa - u'_\kappa = 0$$
> *so that for the given end-conditions on* $x, y_1, \ldots, y_n, u_1, \ldots, u_m$ *the integral V becomes a relative maximum or minimum.*

He lets $F(x, y, y', \lambda), \Omega(x, y, y', u', \lambda)$ be given as

$$F = f + \lambda_1 f_1 + \cdots + \lambda_m f_m, \qquad \Omega = f + \lambda_1 \phi_1 + \cdots + \lambda_m \phi_m$$

and concludes that

$$\frac{\partial F}{\partial y_h} = \frac{d}{dx}\frac{\partial F}{\partial y'_h}, \qquad 0 = -\frac{d\lambda_\kappa}{dx}, \qquad f_\kappa - u'_\kappa = 0.$$

Thus the λ_κ are constants and

$$u_\kappa = c_\kappa + \int f_\kappa \, dx.$$

He further finds that Clebsch's form of the second variation reduces to

$$2W = \sum_{h=1}^{h=n} \sum_{i=1}^{i=n} \frac{\partial^2 F}{\partial y'_h \, \partial y'_i} U_h U_i$$

since F does not involve the u'_κ explicitly. The U, moreover, satisfy the conditions

$$\sum_{h=1}^{h=n} \frac{\partial f_\kappa}{\partial y'_h} U_h = V_\kappa. \tag{6.105}$$

The V_κ are, along with the U_i, variable quantities, but unlike the latter, they do not appear in the expression for $2W$. They therefore can be chosen as one wishes, and conditions (6.105) are thus always satisfied.[33] If the determinant of $\partial^2 F / \partial y'_h \, \partial y'_i \neq 0$, the equations

$$\frac{\partial F}{\partial y_h} = \frac{d}{dx}\frac{\partial F}{\partial y'_h}$$

have solutions y and u which are functions of $2(n + m)$ constants which Mayer calls $a_1, \ldots, a_{2n}, \lambda_1, \ldots, \lambda_m, c_1, \ldots, c_m$. Then the condition that Mayer's determinant vanishes is

$$\sum \pm \frac{\partial y_1}{\partial a_1} \cdots \frac{\partial y_n}{\partial a_n} \frac{\partial y_{10}}{\partial a_{n+1}} \cdots \frac{\partial y_{n0}}{\partial a_{2n}} \frac{\partial u_1}{\partial \lambda_1} \cdots \frac{\partial u_m}{\partial \lambda_m} \frac{\partial u_{10}}{\partial c_1} \cdots \frac{\partial u_{m0}}{\partial c_m} = 0;$$

[33] See, e.g., Pascal *VR*, pp. 95–99 and also Scheefer [1885], p. 584.

but he knows that

$$\frac{\partial y_h}{\partial c_\kappa} = \frac{\partial y_{h0}}{\partial c_\kappa} = \frac{\partial u_h}{\partial c_\kappa} = \frac{\partial u_{h0}}{\partial c_\kappa} = 0, \qquad \frac{\partial u_\kappa}{\partial c_\kappa} = \frac{\partial u_{\kappa 0}}{\partial c_\kappa} = 1$$

and hence the determinantal equation simplifies to

$$\sum \pm \frac{\partial y_1}{\partial a_1} \cdots \frac{\partial y_n}{\partial a_n} \frac{\partial y_{10}}{\partial a_{n+1}} \cdots \frac{\partial y_{n0}}{\partial a_{2n}} \frac{\partial v_1}{\partial \lambda_1} \cdots \frac{\partial v_m}{\partial \lambda_m} = 0,$$

where

$$v_k = u_k - u_{k0} = \int_{x_0}^{x} f_k \, dx.$$

He concludes that

IV. *The Problem* II *is solvable with the help of the n differential equations*

$$\frac{\partial F}{\partial y_h} = \frac{d}{dx} \frac{\partial F}{\partial y_h'},$$

in which

$$F = f + \lambda_1 f_1 + \lambda_2 f_2 + \cdots + \lambda_m f_m,$$

and the λ *signify undetermined constants. The complete solution of these equations yields* $y_1, \ldots, y_n$ *as functions of* x*, the* m *isoperimetric constants* $\lambda_1, \ldots, \lambda_m$ *and* $2n$ *integration constants* $a_1, \ldots, a_{2n}$ *provided that the second derivatives* $y_1'', \ldots, y_n''$ *cannot be eliminated. If these* $2n + m$ *constants are determined by the* m *isoperimetric and the* $2n$ *end-conditions, it follows . . . that the functions* $y_1, \ldots, y_n$ *so determined afford the integral* V *a true relative maximum or minimum provided that the homogeneous function of the second order*

$$2W = \sum_{h=1}^{h=n} \sum_{i=1}^{i=n} \frac{\partial^2 F}{\partial y_h' \partial y_i'} U_h U_i$$

always has the same sign on the interval of integration and provided moreover that the upper limit [of this interval] x_1 *lies between* x_0 *and the next root after* x_0 *of the equation*

$$\Delta(xx_0) = \sum \pm \frac{\partial y_1}{\partial a_1} \cdots \frac{\partial y_n}{\partial a_n} \frac{\partial y_{10}}{\partial a_{n+1}} \cdots \frac{\partial y_{n0}}{\partial a_{2n}} \frac{\partial v_1}{\partial \lambda_1} \cdots \frac{\partial v_m}{\partial \lambda_m} = 0;$$

in this equation the functions v_κ *are found by the quadratures*

$$v_\kappa = \int_{x_0}^{x} f_\kappa \, dx.$$

If the first condition is not satisfied, then there is neither a maximum or minimum and in general this is also true if x_1 *reaches or exceeds the given root.*

Mayer now takes up his reciprocity theorem in Section 2 of his paper.

He formulates two problems:

$$V=\int_{x_0}^{x_1} f\,dx = M_a^i, \qquad \int_{x_0}^{x_1} f_1\,dx = l_1,\ \ldots,\ \int_{x_0}^{x_1} f_m\,dx = l_m \tag{α}$$

and

$$V_1=\int_{x_0}^{x_1} f_1\,dx = M_a^i, \qquad \int_{x_0}^{x_1} f\,dx = l,$$

$$\int_{x_0}^{x_1} f_2\,dx = l_2,\ \ldots,\ \int_{x_0}^{x_1} f_m\,dx = l_m. \tag{β}$$

In problem (α) it is V that is to be minimized, and in (β) it is V_1. He assumes the given end-points, end-conditions, and isoperimetric conditions to be the same, i.e., the x_0, x_1, l, l_k, and sets

$$\lambda_k = \frac{\mu_k}{\mu}, \qquad \mu F \equiv \mu f + \mu_1 f_1 + \cdots + \mu_m f_m = \Phi.$$

Note that the determinant $\Delta(xx_0)$ only changes by a constant factor if the $a_1, \ldots, a_{2n}, \lambda_1, \ldots, \lambda_m$ are replaced by $2n + m$ independent functions of these constants.

The solution of the problem (α) of Mayer can now be found by integrating the differential equations

$$\frac{\partial \Phi}{\partial y_h} = \frac{d}{dx}\frac{\partial \Phi}{\partial y_h'}. \tag{6.106}$$

The extremals are functions of $2n$ constants of integration $a_1, \ldots, a_{2n}$ and the ratios of the $m+1$ isoperimetric constants $\mu, \mu_1, \ldots, \mu_m$. These are to be fixed by the $2n$ end-conditions and the m isoperimetric conditions

$$\int_{x_0}^{x_1} f_1\,dx = l_1, \qquad \int_{x_0}^{x_1} f_2\,dx = l_2,\ \ldots,\ \int_{x_0}^{x_1} f_m\,dx = l_m.$$

The resulting arc is then a true maximum or minimum for problem (α) so long as the upper end-point x_1 is less than the next root after x_0 of the equation

$$\Delta_\alpha(xx_0) = \sum \pm \frac{\partial y_1}{\partial a_1} \cdots \frac{\partial y_n}{\partial a_n}\frac{\partial y_{10}}{\partial a_{n+1}} \cdots \frac{\partial y_{n0}}{\partial a_{2n}}\frac{\partial v_1}{\partial \mu}\frac{\partial v_2}{\partial \mu_1} \cdots \frac{\partial v_m}{\partial \mu_m} = 0$$

provided that, moreover, the second-order homogeneous function

$$2W_\alpha = \frac{1}{\mu}\sum_{h=1}^{h=n}\sum_{i=1}^{i=n}\frac{\partial^2 \Phi}{\partial y_h' \partial y_i'} U_h U_i$$

always has the same sign for x on the interval $[x_0, x_1]$.

For problem (β) equations (6.106) remain the same, but the isoperimetric conditions change insofar as

$$\int_{x_0}^{x_1} f_1\,dx = l_1$$

is replaced by

$$\int_{x_0}^{x_1} f\,dx = l.$$

Mayer now supposes that the maximal or the minimal value in problem (α) of the integral V is κ. Then problem (β) with $l = \kappa$ has the same Euler–Lagrange equations (6.106) as (α) and has

$$\int_{x_0}^{x_1} f\,dx = \kappa, \qquad \int_{x_0}^{x_1} f_1\,dx = \kappa_1, \ldots, \qquad \int_{x_0}^{x_1} f_m\,dx = \kappa_m.$$

The Mayer determinant for problem (β) is now

$$\Delta_\beta(xx_0) = \sum \pm \frac{\partial y_1}{\partial a_1} \cdots \frac{\partial y_n}{\partial a_n} \frac{\partial y_{10}}{\partial a_{n+1}} \cdots \frac{\partial y_{n0}}{\partial a_{2n}} \frac{\partial v}{\partial \mu} \frac{\partial v_2}{\partial \mu_2} \cdots \frac{\partial v_m}{\partial \mu_m} = 0,$$

where y_κ and v_κ are functions of the same a_i, μ_κ as in (α). The difference is that v_1 has been replaced by v, i.e., by

$$v = \int_{x_0}^{x} f\,dx.$$

It is easy to see that

$$\mu v + \mu_1 v_1 + \cdots + \mu_m v_m = \int_{x_0}^{x} \Phi\,dx.$$

Now let any one of the constants $a_1, \ldots, a_{2n}, \mu, \mu_1, \ldots, \mu_m$ be designated by c; then

$$\mu \frac{\partial v}{\partial c} + \mu_1 \frac{\partial v_1}{\partial c} + \cdots + \mu_m \frac{\partial v_m}{\partial c} = \int_{x_0}^{x_1} dx \sum_{h=1}^{h=n} \left(\frac{\partial \Phi}{\partial y_h} \frac{\partial y_h}{\partial c} + \frac{\partial \Phi}{\partial y_h'} \frac{\partial y_h'}{\partial c} \right)$$

and hence

$$\frac{\partial v}{\partial c} = -\frac{\mu_1}{\mu} \frac{\partial v_1}{\partial c}$$
$$- \frac{1}{\mu} \left\{ \sum_{k=2}^{k=m} \mu_k \frac{\partial v_k}{\partial c} - \sum_{h=1}^{h=n} \left(\frac{\partial \Phi}{\partial y_h'} \frac{\partial y_h}{\partial c} - \sum_{h=1}^{h=n} \left[\frac{\partial \Phi}{\partial y_h'} \right]_0 \frac{\partial y_{h0}}{\partial c} \right) \right\},$$

as can be seen with the help of the Euler–Lagrange equations (6.106). When this value of $\partial v/\partial c$ is substituted into the determinant $\Delta_\beta(xx_0)$, all terms drop out except the one in $\partial v_1/\partial c$, and therefore

$$\Delta_\beta(xx_0) = -\frac{\mu_1}{\mu} \Delta_\alpha(xx_0).$$

Mayer also shows that

$$2W_\beta = \frac{\mu}{\mu_1} 2W_\alpha = \frac{1}{\lambda_1} 2W_\alpha,$$

and he summarizes in the theorem

V. *Let the isoperimetric problem*

$$V=\int_{x_0}^{x_1} f\,dx = M_a^i, \qquad \int_{x_0}^{x_1} f_1\,dx = l_1, \ldots, \qquad \int_{x_0}^{x_1} f_m\,dx = l_m$$

be solved by the complete integration of the n differential equations

$$\frac{\partial F}{\partial y_h} = \frac{d}{dx}\frac{\partial F}{\partial y_h'}$$

in which

$$F \equiv f + \lambda_1 f_1 + \lambda_2 f_2 + \cdots + \lambda_m f_m$$

and the λ are undetermined constants whose values are to be determined by the m isoperimetric conditions; and let the maximum or minimum value of the integral V be

$$V = l.$$

Then the solution thus found for this problem is equally the solution of the reciprocal problem:

$$V_1 \equiv \int_{x_0}^{x_1} f_1\,dx = M_a^i, \quad \int_{x_0}^{x_1} f\,dx = l, \quad \int_{x_0}^{x_1} f_2\,dx = l_2, \ldots, \int_{x_0}^{x_1} f_m\,dx = l_m,$$

provided that the end-conditions are the same for both problems.

If, further,

$$V = l$$

in problem α) is a true maximum or minimum of the integral V, then at the same time in problem β)

$$V_1 = l_1$$

is a true maximum or minimum of the integral V_1 if $\lambda_1 < 0$; if, on the contrary, $\lambda_1 > 0$ it is a minimum or maximum for V_1. Finally if in problem α) the obtained value l of V is neither a maximum nor a minimum, the same is also true in problem β) for the value l_1 of V_1.

Mayer closes the paper with the classical example: "*To find the curve of given length and given end-points, whose center of gravity is the lowest.*"

Let the z axis be vertical, and let the *xy*-plane pass through the initial point on the curve. Then the problem is this:

$$\int_{x_0}^{x_1} z\sqrt{1 + y'^2 + z'^2}\,dx = \text{Min}, \qquad \int_{x_0}^{x_1} \sqrt{1 + y'^2 + z'^2}\,dx = l_1.$$

Thus $F = (z + \lambda_1)(1 + y'^2 + z'^2)^{1/2}$, and the extremals are

$$y = \frac{a_1}{a_2}x + a_3, \qquad z + \lambda_1 = \frac{1}{2}\sqrt{a_1^2 + a_2^2}\,\{e^{(x-a_4)/a_2} + e^{-(x-a_4)/a_2}\}.$$

He sets

$$v_1 = \int_{x_0}^{x} \sqrt{1 + y'^2 + z'^2}\,dx$$

and has

$$v_1 = \frac{1}{2}\sqrt{a_1^2 + a_2^2}\,\{e^{(x-a_4)/a_2} - e^{-(x-a_4)/a_2} - e^{(x_0-a_4)/a_2} + e^{-(x_0-a_4)/a_2}\}.$$

The five constants $a_1, a_2, a_3, a_4, \lambda$ are now to be determined by the end-conditions and the given arc-length l_1. Mayer fixes the sign of a_2 to be the same as that of $(a_1^2 + a_2^2)^{1/2}$ so that dv_1/dx is positive.

It is easy to see that

$$2W = \frac{z + \lambda_1}{\left(\sqrt{1 + y'^2 + z'^2}\right)^3} \left\{ U_1^2 + U_2^2 + (z'U_1 - y'U_2)^2 \right\},$$

and therefore it has the same sign as $(a_1^2 + a_1^2)^{1/2}$. For a minimum, he therefore has to choose the positive value for this square root. This corresponds to choosing for the extremal arc in question the one whose convex side hangs down. Then since $x_0 = 0$, it follows that $\lambda_1 > 0$.

Now Mayer's determinant is easy to calculate, and his determinantal equation becomes

$$\begin{vmatrix} \dfrac{\partial(y - y_0)}{\partial a_1} & \dfrac{\partial(y - y_0)}{\partial a_2} & 0 \\ \dfrac{\partial(z - z_0)}{\partial a_1} & \dfrac{\partial(z - z_0)}{\partial a_2} & \dfrac{\partial(z - z_0)}{\partial a_4} \\ \dfrac{\partial v_1}{\partial a_1} & \dfrac{\partial v_1}{\partial a_2} & \dfrac{\partial v_1}{\partial a_4} \end{vmatrix} = 0.$$

To simplify notations, Mayer sets

$$\frac{x - a_4}{a_2} = \xi, \qquad \frac{1}{2}(e^{\xi} + e^{-\xi}) = p, \qquad \frac{1}{2}(e^{\xi} - e^{-\xi}) = q$$

and has

$$y - y_0 = a_1(\xi - \xi_0), \qquad z - z_0 = \sqrt{a_1^2 + a_2^2}\,(p - p_0),$$

$$v_1 = \sqrt{a_1^2 + a_2^2}\,(q - q_0).$$

In terms of these, his equation above becomes

$$\Theta\Psi(\Theta) = 0,$$

where

$$\xi - \xi_0 = \frac{x - x_0}{a_2} = \Theta, \qquad \Psi(\Theta) = e^{\Theta} + e^{-\Theta} - \frac{\Theta}{2}(e^{\Theta} - e^{-\Theta}) - 2.$$

This equation has only one real root $\Theta = 0$, i.e., $x = x_0$, and therefore imposes no restriction on the second end-point x_1. He thus has the result that "*Among all curves of given length and end-points the catenary has always the lowest center of gravity.*" Moreover, since $\lambda_1 > 0$, he also has the reciprocal result that "*Among all curves with given end-points, whose centers of gravity lie in one and the same horizontal plane, the catenary always has the shortest length.*"

6.7. The Problem of Mayer

In two papers [1878, 1895] Mayer formulated what he deemed to be the most general problem of the calculus of variations. It is in a certain sense, but we shall see later how more general problems were formulated by allowing the end-points of the arcs being examined to be variable.

To understand the problem of Mayer consider first his formulation of the problem of Lagrange:

> A. *We seek to determine the dependent variables* $y_1, y_2, \ldots, y_n$ *as functions of* x, *satisfying the differential equations of the first order,*
>
> $$\phi_1 = 0, \phi_2 = 0, \ldots, \phi_m = 0 \qquad (m < n)$$
>
> *so that the integral*
>
> $$V = \int_{x_0}^{x_1} f(x, y_1, \ldots, y_n, y_1', \ldots, y_n')\, dx$$
>
> *becomes a maximum or minimum.*

He further postulates the end-conditions "*that all* n *functions* $y_1, y_2, \ldots, y_n$ *have fixed values at the given limits* x_0 *and* x_1."

He then formulates [1878] a general problem:

> I. *Between the independent variable* x *and the* n *unknown functions* $y_1, y_2, \ldots, y_n$ *there are* m *first-order differential equations*
>
> $$\phi_1 = 0, \phi_2 = 0, \ldots, \phi_m = 0 \qquad (m < n). \tag{6.107}$$
>
> *The problem at hand is to determine these functions so that while the functions* $y_2, \ldots, y_n$ *have preassigned values for two given values of* x, x_0 *and* x_1 $(> x_0)$, *the function* y_1 *for* $x = x_0$ *has a preassigned value and for* $x = x_1$ *becomes a maximum or minimum.*

Mayer points out that Problem A is a special case of I. He does this by, in effect, defining $y_1' = f$ and fixing y_1 at $x = x_0$. Then the problem becomes one of finding an arc $y_1, y_2, \ldots, y_n$ for which $y_1(x_1)$ is an extremum. Conversely, however, not every problem I is reducible to the form A because of the fixed end-conditions.[34]

Let us look at Mayer's 1878 treatment of the problem. He finds for a minimum that $\delta y_1(x_1) = \delta y_{11}$ must be 0 and that $\delta^2 y_{11}$ must have one sign. He also notes that the usual accessory side-conditions

$$\sum_{i=1}^{i=n} \left\{ \frac{\partial \phi_k}{\partial y_i} \delta y_i + \frac{\partial \phi_k}{\partial y_i'} \delta y_i' \right\} = 0 \tag{6.108}$$

---

[34] However, the problems are equivalent if one does not insist on fixed end-points because in this case

$$V = \int_{x_0}^{x_1} y_1'\, dx$$

can be chosen as the integral to be minimized, and the problem of Mayer becomes one of Lagrange.

must be satisfied by the variations $\delta y_1, \delta y_2, \ldots, \delta y_n$, and he defines a function Ω:

$$\lambda_1\phi_1 + \lambda_2\phi_2 + \cdots + \lambda_m\phi_m = \Omega.$$

In terms of this function, he rewrites the accessory side-conditions in the form

$$\begin{aligned} 0 &= \sum_{i=1}^{i=n} \left\{ \frac{\partial \Omega}{\partial y_i}\delta y_i + \frac{\partial \Omega}{\partial y_i'}\delta y_i' \right\} \\ &= \sum_{i=1}^{i=n}\left(\frac{\partial \Omega}{\partial y_i} - \frac{d}{dx}\frac{\partial \Omega}{\partial y_i'} \right)\delta y_i + \frac{d}{dx}\sum_{i=1}^{i=n}\frac{\partial \Omega}{\partial y_i'}\delta y_i, \end{aligned} \tag{6.109}$$

and he imposes the condition on the λ that

$$\frac{\partial \Omega}{\partial y_1} - \frac{d}{dx}\frac{\partial \Omega}{\partial y_1'} = 0.$$

Since $\delta y_2, \ldots, \delta y_m$ vanish for $x = x_0$ and $x = x_1$ and δy_1 also vanishes for $x = x_0$, relation (6.109) becomes, on integration with respect to x,

$$\left[\frac{\partial \Omega}{\partial y_1'} \right]_1 \delta y_{11} = -\int_{x_0}^{x_1} dx \sum_{i=2}^{i=n}\left(\frac{\partial \Omega}{\partial y_i} - \frac{d}{dx}\frac{\partial \Omega}{\partial y_i'} \right)\delta y_i.$$

This gives him an expression for δy_{11} in terms of the quantities $\delta y_2, \ldots, \delta y_n$ provided that $[\partial\Omega/\partial y_1'] \neq 0$ at $x = x_1$. He concludes, with the help of this expression and (6.109), that the equations

$$\frac{\partial \Omega}{\partial y_i} - \frac{d}{dx}\frac{\partial \Omega}{\partial y_i'} = 0 \qquad (i = 1, 2, \ldots, n), \tag{6.110}$$

as well as the side-conditions (6.107) must hold, since the variations $\delta y_2, \ldots, \delta y_n$ are arbitrary inside the interval x_0x_1. The complete solution of this system of $m + n$ equations yields y and λ as functions of x and some arbitrary constants. To ensure that the system is of order $2n$, Mayer assumes that the functional determinant

$$R = \begin{vmatrix} \dfrac{\partial^2\Omega}{\partial y_1' \partial y_1'} & \cdots & \dfrac{\partial^2\Omega}{\partial y_1' \partial y_n'} & \dfrac{\partial \phi_1}{\partial y_1'} & \cdots & \dfrac{\partial \phi_m}{\partial y_1'} \\ \cdot & \cdot & \cdot & \cdot & \cdot & \cdot \\ \dfrac{\partial^2\Omega}{\partial y_n' \partial y_1'} & \cdots & \dfrac{\partial^2\Omega}{\partial y_n' \partial y_n'} & \dfrac{\partial \phi_1}{\partial y_n'} & \cdots & \dfrac{\partial \phi_m}{\partial y_n'} \\ \dfrac{\partial \phi_1}{\partial y_1'} & \cdots & \dfrac{\partial \phi_1}{\partial y_n'} & 0 & \cdots & 0 \\ \cdot & \cdot & \cdot & \cdot & \cdot & \cdot \\ \dfrac{\partial \phi_m}{\partial y_1'} & \cdots & \dfrac{\partial \phi_m}{\partial y_n'} & 0 & \cdots & 0 \end{vmatrix}$$

is not identically zero. Note that equations (6.108) do not contain the λ, but that equation (6.109) contains the functions λ, λ' linearly and homogeneously. Hence one integration constant—Mayer takes it to be a_{2n}—is an amplitude factor for λ. Thus the y contain the constants $a_1, a_2, \ldots, a_{2n-1}$ —just enough to ensure that there is an extremal satisfying the end-conditions—and the λ contain the constants $a_1, a_2, \ldots, a_{2n}$ in the form $a_{2n}\lambda_k(x, a_1, a_2, \ldots, a_{2n-1})$.

He then turns to a consideration of the second variation. To this end he writes the second variation of the side-conditions (6.108) in the form

$$\sum_{i=1}^{i=n}\left(\frac{\partial\phi_k}{\partial y_i}\delta^2 y_i + \frac{\partial\phi_k}{\partial y_i'}\delta^2 y_i'\right) + 2\delta^2\phi_k = 0,$$

where the $2\delta^2\phi_k$ are homogeneous functions of the second order in $z_i = \delta y_i$ and $z_i' = \delta y_i'$. Multiplying these equations by the quantitites λ_k and adding, he finds the relation

$$\sum_{i=1}^{i=n}\left(\frac{\partial\Omega}{\partial y_i}\delta^2 y_i + \frac{\partial\Omega}{\partial y_i'}\delta^2 y_i'\right) + 2\delta^2\Omega = 0.$$

With the help of the Euler equations (6.110) this relation becomes

$$-2\delta^2\Omega = \frac{d}{dx}\sum_{i=1}^{i=n}\frac{\partial\Omega}{\partial y_i'}\delta^2 y_i$$

or when integrated from x_0 to x_1,

$$\delta^2 y_{11} = -\frac{1}{\left[(\partial\Omega/\partial y_1)\right]_1}\int_{x_0}^{x_1} 2\delta^2\Omega\,dx \tag{6.111}$$

since $\delta^2 y_2, \ldots, \delta^2 y_n$ all vanish at $x = x_0$ and x_1 and $\delta^2 y_1$ vanishes at $x = x_0$ since $y_2, \ldots, y_n$ have fixed end-values.

For an extremum, it is necessary that $\delta y_{11} = 0$ and that $\delta^2 y_{11}$ must be negative or positive for all $z_1, z_2, \ldots, z_n$ which satisfy the accessory side-conditions

$$\sum_{i=1}^{i=n}\left(\frac{\partial\phi_k}{\partial y_i} z_i + \frac{\partial\phi_k}{\partial y_i'} z_i'\right) = 0, \tag{6.112}$$

which are continuous on the interval from x_0 to x_1, and for which $z_2, \ldots, z_n$ vanish at those points. Also $z_1 = \delta y_1$ must satisfy $z_{11} = 0$. As a consequence of these conditions (6.112), the expression (6.111) for $\delta^2 y_{11}$ becomes

$$\delta^2 y_{11} = -\frac{1}{\left[(\partial\Omega/\partial y_i')\right]_1}\int_{x_0}^{x_1} 2\Omega_2\,dx,$$

when $2\Omega_2$ is defined as

$$2\Omega_2 = 2\delta^2\Omega + 2\sum_{k=1}^{k=m}\mu_k\sum_{i=1}^{i=n}\left\{\frac{\partial\phi_k}{\partial y_i} z_i + \frac{\partial\phi_k}{\partial y_i} z_i'\right\},$$

with μ_k arbitrary. Recall that equations (6.108) are expressible as

$$\frac{\partial\Omega_2}{\partial\mu_k} = 0. \tag{6.113}$$

The quantity $\delta^2 y_{11}$ can then be exhibited in the form

$$\delta^2 y_{11} = -\frac{1}{\left[(\partial\Omega/\partial y_i')\right]_1}\int_{x_0}^{x_1} dx \sum_{h=1}^{h=n}\left(\frac{\partial\Omega_2}{\partial z_h} - \frac{d}{dx}\frac{\partial\Omega_2}{\partial z_h'}\right)\delta z_h .$$

Mayer now writes the complete solution of equations (6.108) and (6.110) as functions of $2n$ constants $a_1, a_2, \ldots, a_{2n}$ and sets

$$z_i = \sum_{h=1}^{h=2n-1} \gamma_h \frac{\partial y_i}{\partial a_h}, \qquad \mu_k = \sum_{h=1}^{h=2n} \gamma_h \frac{\partial \lambda_k}{\partial a_h}.$$

It follows by Jacobi's argument that the z_i, μ_k are a complete solution of the system

$$\frac{\partial\Omega_2}{\partial z_h} - \frac{d}{dx}\frac{\partial\Omega_2}{\partial z_h'} = 0, \qquad \frac{\partial\Omega_2}{\partial \mu_k} = 0.$$

This implies that the second variation $\delta^2 y_{11}$ will vanish if the $2n-1$ constants $\gamma_1, \gamma_2, \ldots, \gamma_{2n-1}$ can be so determined that the z_i defined above vanish at any two values x_0', x_1' on the interval $[x_0, x_1]$ without vanishing identically.

Mayer observes that the z_i satisfy the equation

$$\sum_{i=1}^{i=n}\left\{\frac{\partial\Omega}{\partial y_i} z_i + \frac{\partial\Omega}{\partial y_i'} z_i'\right\} = 0$$

and consequently the equation

$$\frac{d}{dx}\sum_{i=1}^{i=n}\frac{\partial\Omega}{\partial y_i'} z_i = 0. \tag{6.114}$$

He concludes this with a result relating the second variations δ_{11}^2 to his determinant $\Delta(x, x_0)$ (in his form of Δ notice that y_1 does not appear for $x = x_1$):

II. *The second variation $\delta^2 y_{11}$ can always be made to vanish whenever the determinant*

$$\Delta(x, x_1) = \begin{vmatrix} \frac{\partial y_1}{\partial a_1} & \frac{\partial y_1}{\partial a_2} & \cdots & \frac{\partial y_1}{\partial a_{2n-1}} \\ \frac{\partial y_2}{\partial a_1} & \frac{\partial y_2}{\partial a_2} & \cdots & \frac{\partial y_2}{\partial a_{2n-1}} \\ \cdot & \cdot & \cdots & \cdot \\ \frac{\partial y_n}{\partial a_1} & \frac{\partial y_n}{\partial a_2} & \cdots & \frac{\partial y_n}{\partial a_{2n-1}} \\ \frac{\partial y_{21}}{\partial a_1} & \frac{\partial y_{21}}{\partial a_2} & \cdots & \frac{\partial y_{21}}{\partial a_{2n-1}} \\ \cdot & \cdot & \cdots & \cdot \\ \frac{\partial y_{n1}}{\partial a_1} & \frac{\partial y_{n1}}{\partial a_2} & \cdots & \frac{\partial y_{n1}}{\partial a_{2n-1}} \end{vmatrix}$$

vanishes for any x, different from x_1, between the limits x_0 and x_1.

Mayer notes with the help of equation (6.114) that

$$\sum_{i=1}^{i=n} \frac{\partial \Omega}{\partial y_i'} \frac{\partial y_i}{\partial a_h} = \text{const.} = b_h,$$

and therefore that

$$\frac{\partial \Omega}{\partial y_i'} \Delta(x, x_1) = \pm \left[\frac{\partial \Omega}{\partial y_i'} \right]_1 \Delta(x_1, x).$$

He now applies Clebsch's transformation to the second variation and writes

$$2W = -\frac{1}{\left[(\partial \Omega / \partial y_i')\right]_1} \sum_{h=1}^{h=n} \sum_{i=1}^{i=n} \frac{\partial^2 \Omega}{\partial y_h' \, \partial y_i'} U_h U_i,$$

when the n arguments $U_1, U_2, \ldots, U_n$ satisfy the m conditions

$$\sum_{i=1}^{i=n} \frac{\partial \phi_k}{\partial y_i'} U_i = 0.$$

For an extremum, $2W$ must be of fixed sign. He chooses a family u_i^σ, r_k^σ so that the determinant

$$\sum \pm u_1^1 u_2^2 \cdots u_n^n \neq 0$$

and for which[35]

$$\sum_{i=1}^{i=n} \left\{ u_i^\sigma \frac{\partial \Omega_2(u^\rho r^\rho)}{\partial (du_i^\rho / dx)} - u_i^\rho \frac{\partial \Omega_2(u^\sigma r^\sigma)}{\partial (du_i^\sigma / dx)} \right\} = 0.$$

The u_i^σ, r_k^σ are determined by choosing $2n^2$ constants γ_k^σ in the expressions

$$u_i^\sigma = \sum_{\tau=1}^{\tau=2n-1} \gamma_\tau^\sigma \frac{\partial y_i}{\partial a_\tau}, \qquad r_k^\sigma = \sum_{\tau=1}^{2n} \gamma_\tau^\sigma \frac{\partial \lambda_k}{\partial a_\tau} \qquad (\sigma = 1, 2, \ldots, n).$$

Mayer closes the paper with a discussion of the relation of Δ to the determinant of the u_i^σ (see pp. 276ff above) and the theorem

> III. *The Problem* I *is solvable with the help of the* $n + m$ *differential equations*
>
> $$\frac{\partial \Omega}{\partial y_i} - \frac{d}{dx} \frac{\partial \Omega}{\partial y_i'} = 0, \qquad \phi_k = 0$$
>
> *in which*
>
> $$\Omega = \lambda_1 \phi_1 + \lambda_2 \phi_2 + \cdots + \lambda_m \phi_m.$$
>
> *Under the assumption that the problem is in general solvable the complete integration of these equations produces* $y_1, \ldots, y_n, \lambda_1, \ldots, \lambda_m$ *as functions of* x *and of* $2n$ *arbitrary constants. Of these constants, however, those that enter into the solutions* y *always reduce to* $2n - 1$ *constants* $a_1, a_2, \ldots, a_{2n-1}$; *the* $2n$-th *constant* a_{2n} *enters only as a common factor of all*

[35] Recall the discussion of this on pp. 253 above. This is a conjugate set.

the solutions λ. The constants $a_1, a_2, \ldots, a_{2n-1}$ are then to be so determined that $y_1, y_2, \ldots, y_n$ for $x = x_0$ and $y_2, \ldots, y_n$ for $x = x_1$ have pre-assigned values. If such a system, satisfying the end-conditions, has been found and if for x_1 such a value has been chosen that the expression $\partial\Omega/\partial y_1'$ for $x = x_1$ is not null, then for the corresponding solutions $y_1, y_2, \ldots, y_n$ a true maximum or minimum for the problem is obtained provided that x_0 lies between x_1 and the next smaller root of the end equation

$$\Delta(xx_1) = 0;$$

assuming, moreover, that the homogeneous function of the second order

$$2W = -\frac{1}{[(\partial\Omega/\partial y_1')]_1} \sum_{h=1}^{h=n} \sum_{i=1}^{i=n} \frac{\partial^2\Omega}{\partial y_h' \partial y_i'} U_h U_i$$

between these limits [i.e., for all x on $[x_0, x_1]$] is always negative or always positive for all n arbitrary arguments that satisfy the conditions

$$\sum_{i=1}^{i=n} \frac{\partial\Omega}{\partial y_i'} U_i = 0.$$

On the contrary neither a maximum nor a minimum will occur if the last hypothesis is not fulfilled, and the same is valid in general also if x_0 is less then or equal to the given root [the first root of $\Delta(xx_1) = 0$ below x_1].

In his 1895 paper Mayer takes up the task of establishing the multiplier rule for his general problem of Mayer.[36] The procedure is not very different from that discussed above, but perhaps it is worthwhile to give a quick sketch of the method.

The problem posed here is to find among all continuous functions $y_0, y_1, \ldots, y_n$ of the independent variable x, which satisfy identically $r + 1$ given differential equations

$$\phi_k(x, y_0, y_1, \ldots, y_n, y_0', y_1', \ldots, y_n') = 0 \qquad k = 0, 1, \ldots, r < n,$$

and which are such that $y_0, \ldots, y_n$ take on given values for $x = x_0$ and $y_1, \ldots, y_n$ given values for $x = x_1$, one which renders y_0 at x_1 a maximum or minimum. Mayer assumes that the determinant

$$\sum \pm \phi_{0y_0'}\phi_{1y_1'} \cdots \phi_{ry_r'} \neq 0. \tag{6.115}$$

His variations δy_i must therefore satisfy the equations

$$\sum_0^n (\phi_{ky_i} \delta y_i + \phi_{ky_i'} \delta y_i') = 0$$

and be such that $\delta y_0, \ldots, \delta y_n$ vanish at $x = x_0$ and $\delta y_1, \ldots, \delta y_n$ vanish at $x = x_1$.[37] Mayer then rewrites these equations with the help of multipliers λ_k

---

[36] Mayer [1895]. There is also a very different proof in a paper by Turksma [1896], which is a part of his doctoral thesis.

[37] Mayer's notation in this paper is not very good. For instance, he writes the equations above in the form $\sum i\ (\phi_k' y_i\ \delta y_i + \phi_k' y_i'\ \delta y_i') = 0$, which makes it difficult to understand precisely what is meant, and he writes determinant (6.115) as $\sum \pm \phi_0' y_0' \phi_1' y_1' \cdots \phi_r' y_r'$.

in the form

$$\frac{d}{dx}\left(\sum_0^n {}_i\delta y_i \sum_0^r {}_k\lambda_k\phi_{ky_i'}\right) + \sum_0^n {}_i\delta y_i \sum_0^r {}_k\left(\lambda_k\phi_{ky_i} - \frac{d\lambda_k\phi_{ky_i'}}{dx}\right) = 0. \qquad (6.116)$$

Since determinant (6.115) is different from zero, the equations

$$\sum_0^r {}_k\left(\lambda_k\phi_{ky_\rho} - \frac{d\lambda_k\phi_{ky_\rho'}}{dx}\right) = 0 \qquad (\rho = 0, 1, \ldots, r) \qquad (6.117)$$

are solvable for the λ_k'; these solutions form a system of $r+1$ linear, first-order differential equations, whose solutions he writes as

$$\lambda_0 = \lambda_0^\sigma, \lambda_1 = \lambda_1^\sigma, \ldots, \lambda_r = \lambda_r^\sigma \qquad (\sigma = 0, 1, \ldots, r).$$

With the help of equation (6.116), Mayer finds by integration that

$$\sum_0^r {}_\rho\delta y_\rho \sum_0^r {}_k\lambda_k^\sigma\phi_{ky_\rho'} = -\sum_{r+1}^n {}_\tau\delta y_\tau \sum_0^r {}_k\lambda_k^\sigma\phi_{ky_\tau'}$$
$$-\int_{x_0}^x dx \sum_{r+1}^n {}_\tau\delta y_\tau \sum_0^r {}_k\left(\lambda_k^\sigma\phi_{ky_\tau} - \frac{d\lambda_k^\sigma\phi_{ky_\tau'}}{dx}\right),$$

since all δy vanish at $x = x_0$. These equations imply that for $x = x_1$,

$$\sum_0^r {}_\rho\delta y_{\rho 1}\left[\sum_0^r {}_k\lambda_k^\sigma\phi_{ky_\rho'}\right]_{x=x_1} = -\int_{x_0}^{x_1} dx \sum_{r+1}^n {}_\tau\delta y_\tau \sum_0^r {}_k\left(\lambda_k^\sigma\phi_{ky_\tau} - \frac{d\lambda_k^\sigma\phi_{ky_\tau'}}{dx}\right).$$

Moreover, since the determinant

$$\sum \pm \lambda_0^0\lambda_1^1 \cdots \lambda_r^r \sum \pm \phi_{0y_0'}\phi_{1y_1'} \cdots \phi_{ry_r'} \neq 0,$$

it follows that those equations can be solved for the $\delta y_{\rho 1}$ in the form

$$\delta y_{\rho 1} = -\int_{x_0}^{x_1} dx \sum_{r+1}^n {}_\tau\delta y_\tau \sum_0^r {}_k\left(\mu_k^\rho\phi_{ky_\tau} - \frac{d\mu_k^\rho\phi_{ky_\tau'}}{dx}\right),$$

where

$$\mu_k^\rho \equiv \sum_0^r {}_\sigma c_\rho^\sigma\lambda_k^\sigma \qquad (\rho, k = 0, 1, \ldots, r)$$

and the c are constants. Mayer remarks that

$$\lambda_0 = \mu_0^\rho, \qquad \lambda_1 = \mu_1^\rho, \ldots, \lambda_r = \mu_r^\rho$$

are again solutions of (6.117).

Now $\delta y_0 = 0$ along with $\delta y_1, \delta y_2, \ldots, \delta y_n$ at $x = x_1$, and hence Mayer has

$$W_{\delta y}^0 \equiv \int_{x_0}^{x_1} dx \sum_{r+1}^n {}_\tau\delta y_\tau \sum_0^r {}_k\left(\mu_k^0\phi_{ky_\tau} - \frac{d\mu_k^0\phi_{ky_\tau'}}{dx}\right) = 0, \qquad (6.118)$$

for all $\delta y_{r+1}, \ldots, \delta y_n$ which vanish at x_0 and x_1 and satisfy the equations

$$W^{\rho}_{\delta y} \equiv \int_{x_0}^{x_1} dx \sum_{r+1}^{n} \delta y_\tau \sum_{0}^{r} {}_k\left(\mu^{\rho}_k \phi_{k y_\tau} - \frac{d\mu^{\rho}_k \phi_{k y'_\tau}}{dx} \right) = 0 \qquad (\rho = 1, 2, \ldots, r). \tag{6.119}$$

He sets

$$\delta y_\tau = z_\tau + \sum_{1}^{r} {}_\sigma \alpha_\sigma u^\sigma_\tau \qquad (\tau = r+1, \ldots, n).$$

and proceeds to determine the $\alpha_\sigma, u^\sigma_\tau$ so that equations (6.118) and (6.119) are satisfied but the z_τ are arbitrary functions vanishing at x_0 and x_1. This is done as before (see pp. 284ff). Equation (6.118) becomes

$$W^0_z + \sum_{1}^{r} {}_\sigma \alpha_\sigma W^0_{u^\sigma} = 0, \tag{6.118$'$}$$

and equations (6.119) become

$$W^\rho_z + \sum_{1}^{r} {}_\sigma \alpha_\sigma W^\rho_{u^\sigma} = 0. \tag{6.119$'$}$$

The problem is to find whether the u^σ_τ can be so chosen that the determinant

$$\Delta_r = \sum \pm W^1_{u^1} W^2_{u^2} \cdots W^r_{u^r} \neq 0.$$

In this case equation (6.118′) is a consequence of (6.119′) and the z are not restricted. For each set $z_{r+1}, \ldots, z_n$, a system of constants $\alpha_1, \ldots, \alpha_n$ can be found satisfying relations (6.119′). Equations

$$W^0_z \equiv \beta^1 W^2_z + \cdots + \beta^r W^r_z$$

must then hold in which the β are constants independent of the z.

Mayer now sets

$$-\mu^0_k + \sum_{1}^{r} {}_\rho \beta^\rho \mu^\rho_k \equiv \nu_k$$

and combines the relations (6.118) and (6.119) in the form

$$\int_{x_0}^{x_1} dx \sum_{r+1}^{n} {}_\tau z_\tau \sum_{0}^{r} {}_k\left(\nu_k \phi_{k y_\tau} - \frac{d\nu_k \phi_{k y'_\tau}}{dx} \right) \equiv 0.$$

Since the z_τ are arbitrary, the sums as to k must vanish and by (6.117)

$$\sum_{0}^{r} {}_k\left(\lambda_k \phi_{k y_\rho} - \frac{d\lambda_k \phi_k y'_\rho}{dx} \right) = 0 \qquad (\rho = 0, 1, \ldots, r), \tag{6.120}$$

where $\lambda_0 = \nu_0, \lambda_1 = \nu_1, \ldots, \lambda_r = \nu_r$.

"So long as the determinant Δ_r does not vanish for all continuous functions u^σ_τ vanishing at both limits there must necessarily exist solutions

$\lambda_0, \lambda_1, \ldots, \lambda_r$ of the $r+1$ equations (6.120) which also satisfy the $n-r$ equations

$$\sum_0^r {}_k \left(\lambda_k \phi_{ky_\tau} - \frac{d\lambda_k \phi_{ky'_\tau}}{dx} \right) = 0 \qquad (\tau = r+1, \ldots, n).\text{''} \qquad (6.120')$$

If Δ_r does vanish for such u_τ^σ, then the $W_{u^\sigma}^\rho$ in (6.118) may each vanish, and this requires that

$$\sum_0^r {}_k \left(\mu_k^\rho \phi_{ky_\tau} - \frac{d\mu_k^\rho \phi_{ky'_\tau}}{dx} \right) = 0.$$

Mayer ([1895], p. 136) next considers the case when $\Delta_r = 0$ and there is a $p > 1$ and $\leqslant r$ such that all subdeterminants of the pth degree vanish but that one subdeterminant of $(p-1)$st degree, say,

$$\Delta_{p-1} \equiv \sum \pm W_{u^1}^1 W_{u^2}^2 \cdots W_{u^{p-1}}^{p-1}$$

is not null for all u_τ^σ. Mayer then chooses a set

$$u_\tau^1, u_\tau^2, \ldots, u_\tau^{p-1} \qquad \tau = r+1, \ldots, n$$

of continuous functions vanishing at x_0 and x_1 for which $\Delta_{p-1} \neq 0$, chooses u_τ^p arbitrary, and he expands

$$\Delta_p \equiv \sum \pm W_{u^1}^1 W_{u^2}^2 \cdots W_{u^p}^p$$

with respect to the elements $W_{u^p}^\rho$, obtaining

$$\Delta_p = \sum_1^p {}_\rho c_\rho W_{u^p}^\rho,$$

where the coefficients (one of which, $\Delta_{p-1}, \neq 0$)

$$c_\rho \equiv \frac{\partial \Delta_p}{\partial W_{u^p}^\rho}$$

are independent of the $W_{u^p}^\rho$ ($\rho = 1, 2, \ldots, p$) and u_τ^p ($\tau = r+1, \ldots, n$). He then sets

$$\sum_1^p {}_\rho c_\rho \mu_k^\rho \equiv \pi_k$$

and notes that Δ_p is expressible as

$$\Delta_p = \int_{x_0}^{x_1} dx \sum_{r+1}^n {}_\tau u_\tau^p \sum_0^r {}_k \left(\pi_k \phi_{ky_\tau} - \frac{d\pi_k \phi_{ky'_\tau}}{dx} \right).$$

But by hypothesis, Δ_p is null for all $u_\tau^p, z_\tau (\tau = r+1, \ldots, n)$ which vanish at x_0 and x_1 and therefore

$$\lambda_0 = \pi_0, \qquad \lambda_1 = \pi_i, \ldots, \lambda_r = \pi_r$$

must satisfy equations (6.120) and (6.120′). He summarizes in the theorem: "*In each case the solutions $y_0, y_1, \ldots, y_n$ of our problem must possess the* *property of providing common solutions* $\lambda_0, \lambda_1, \ldots, \lambda_r$ *of the* $n+1$ *differential equations*

$$\sum_0^r {}_k\left(\lambda_k \phi_{ky_i} - \frac{d\lambda_k \phi_{ky_i'}}{dx}\right) = 0 \qquad (i = 0, 1, \ldots, n)." \tag{6.121}$$

Mayer now sets

$$\Omega = \lambda_0\phi_0 + \lambda_1\phi_1 + \cdots + \lambda_r\phi_r$$

and rewrites equations (6.121) in the more usual form

$$\frac{d}{dx}\frac{\partial\Omega}{\partial y_i'} = \frac{\partial\Omega}{\partial y_i} \qquad (i = 0, 1, \ldots, n). \tag{6.121$'$}$$

He notes[38] that the system of differential equations (6.107) and (6.121′) "is then and only then of order $2n+2$ when the $n+r+2$ equations

$$\frac{\partial\Omega}{\partial y_i'} = v_i, \qquad \phi_k = 0 \tag{6.122}$$

determine the $n+r+2$ unknowns

$$y_0', y_1', \ldots, y_n', \lambda_0, \lambda_1, \ldots, \lambda_r."$$

He writes the solutions of (6.122) in the form

$$y_i' = (y_i'), \qquad \lambda_k = (\lambda_k)$$

and uses parentheses in other connections to indicate that these functions have been substituted. Equations (6.107) and (6.121′) can then be replaced by the $2n+2$ first-order differential equations

$$\frac{dy_i}{dx} = (y_i'), \qquad \frac{dv_i}{dx} = \left(\frac{\partial\Omega}{\partial y_i}\right) \tag{6.123}$$

between the $y_0, y_1, \ldots, y_n, v_1, \ldots, v_n$ and x and by the $r+1$ finite equations

$$\lambda_k = (\lambda_k). \tag{6.123$'$}$$

Clearly, the complete solution of system (6.123) furnishes the solutions y of (6.107) and (6.121′) and then by proper substitution the λ are found from (6.123′).

Mayer proceeds in Section 2 to the Hamilton–Jacobi theory for the problem at hand. He defines H as a function of $x, y_0, y_1, \ldots, y_n, v_0, v_1, \ldots, v_n$ through the equation

$$H \equiv \sum_0^n {}_i v_i(y_i') - (\Omega) \equiv \sum_0^n {}_i v_i(y_i')$$

[38] Mayer [1895], p. 138. Note that the y' and λ are now functions of x, y and v which he expresses as $(y_i'), (\lambda_k)$.

and also notes that

$$\left(\frac{\partial\Omega}{\partial y_i'}\right) \equiv v_i, \qquad (\phi_k) \equiv 0.$$

(Recall that the expressions in parentheses are functions of x, y, v.) On this basis he calculates that

$$\delta H \equiv \sum_0^n {}_i\left\{ (y_i')\,\delta v_i - \left(\frac{\partial\Omega}{\partial y_i'}\right)\delta y_i \right\};$$

or equivalently

$$\frac{\partial H}{\partial v_i} \equiv (y_i'), \qquad \frac{\partial H}{\partial y_i} \equiv -\left(\frac{\partial\Omega}{\partial y_i}\right),$$

which he writes in the canonical form

$$\frac{dy_i}{dx} = \frac{\partial H}{\partial v_i}, \qquad \frac{dv_i}{dx} = -\frac{\partial H}{\partial y_i}. \tag{6.124}$$

He thus expresses the Hamiltonian as

$$H \equiv \sum_0^n {}_i v_i \frac{\partial H}{\partial v_i}; \tag{6.124'}$$

i.e., H is a homogeneous function of order 1 in $v_0, v_1, \ldots, v_n$. Mayer then writes

$$\frac{v_h}{v_0} \equiv -p_h$$

and asserts that H is of the form

$$H \equiv v_0 F(x, y_0, y_1, \ldots, y_n, p_1, p_2, \ldots, p_n).$$

From these it follows that

$$\frac{dp_h}{dx} = -\frac{1}{v_0}\left(\frac{dv_h}{dx} + p_h\frac{dv_0}{dx}\right),$$

and the canonical equations reduce to

$$\frac{dy_0}{dx} = F - \sum_1^n {}_h p_h \frac{\partial F}{\partial p_h},$$

$$\frac{dy_h}{dx} = -\frac{\partial F}{\partial p_h}, \qquad \frac{dp_h}{dx} = \frac{\partial F}{\partial y_h} + p_h\frac{\partial F}{\partial y_0}, \tag{6.125}$$

The quantity v_0 is found with the help of the expression

$$\log v_0 = -\int \frac{\partial F}{\partial y_0}\,dx + \text{const.}$$

Thus the solutions y of the differential equations do not contain more than $2n + 1$ constants. However, these equations can be replaced by the Hamilton–Jacobi partial differential equation:

$$\frac{\partial y_0}{\partial x} = F\left(x, y_0, y_1, \ldots, y_n, \frac{\partial y_0}{\partial y_1}, \frac{\partial y_0}{\partial y_2}, \ldots, \frac{\partial y_0}{\partial y_n}\right).$$

Mayer now says "Then the system (6.125) is equivalent to the partial differential equation

$$\frac{\partial y_0}{\partial x} = F\left(x, y_0, y_1, \ldots, y_n, \frac{\partial y_0}{\partial y_1}, \frac{\partial y_0}{\partial y_2}, \ldots, \frac{\partial y_0}{\partial y_n}\right). \tag{6.126}$$

If

$$y_0 = Y_0(x, y_1, y_2, \ldots, y_n, \alpha, \alpha_1, \ldots, \alpha_n) \tag{6.127}$$

is a complete solution of it, then the $2n + 1$ equations

$$y_0 = Y_0, \qquad \frac{\partial Y_0}{\partial \alpha_h} = \beta_h \frac{\partial Y_0}{\partial \alpha}, \qquad p_h = \frac{\partial Y_0}{\partial y_h}, \tag{6.128}$$

containing the $2n + 1$ arbitrary constants $\alpha, \alpha_1, \ldots, \alpha_n, \beta_1, \ldots, \beta_n$, form the complete integral of the system (6.125)."

What he means by this is very elegant and is derived from Jacobi's work on the Hamilton–Jacobi equation.[39] Let us follow Jacobi's analysis to derive Mayer's result, modifying notations as needed. Suppose that the Y_0 in (6.127) satisfies (6.126). Jacobi notes that

$$\frac{dY_0}{dx} = \frac{\partial Y_0}{\partial x} + \sum_{k=1}^{n} \frac{\partial Y_0}{\partial y_k} \cdot \frac{dy_k}{dx} = F + \sum_{k=1}^{n} p_k \frac{dy_k}{dx}$$

and hence that

$$\begin{aligned}\delta \frac{dY_0}{dx} &= \sum_{k=1}^{n} \frac{\partial F}{\partial y_k} \cdot \delta y_k + \frac{\partial F}{\partial y_0} \cdot \delta Y_0 \\ &\quad + \sum_{k=1}^{n} \left(\frac{dy_k}{dx} + \frac{\partial F}{\partial p_k} \right) \delta p_k + \sum_{k=1}^{n} p_k \, \delta \frac{dy_k}{dx} .\end{aligned} \tag{6.129}$$

Jacobi also has for δY_0 the value

$$\begin{aligned}\delta Y_0 &= \sum_{k=1}^{n} \frac{\partial Y_0}{\partial y_k} \delta y_k + \frac{\partial Y_0}{\partial \alpha} \delta \alpha + \sum_{k=1}^{n} \frac{\partial Y_0}{\partial \alpha_k} \delta \alpha_k \\ &= \sum_{k=1}^{n} p_k \, \delta y_k + \frac{\partial Y_0}{\partial \alpha} \left(\delta \alpha + \sum_{k=1}^{n} \beta_k \, \delta \alpha_k \right),\end{aligned}$$

as can be seen with the help of the equations (6.128). It follows *inter alia*

[39] Mayer refers to Jacobi, *NACH*, p. 291; this is a profound paper on the Hamilton–Jacobi theory in Jacobi's *Nachlass* and is well worth examining.

that

$$\frac{d\delta Y_0}{dx} = \sum_{k=1}^{n} \frac{dp_k}{dx}\delta y_k + \sum_{k=1}^{n} p_k \frac{d\delta y_k}{dx} + \frac{d\partial Y_0/\partial\alpha}{dx}\left(\delta\alpha + \sum_{k=1}^{n} \beta_k\,\delta\alpha_k\right). \tag{6.129'}$$

By equating the two expressions (6.129) and (6.129′), he finds the resulting relation

$$0 = \sum_{k=1}^{n}\left(\frac{dy_k}{dx} + \frac{\partial F}{\partial p_k}\right)\delta p_k - \sum_{k=1}^{n}\left(\frac{dp_k}{dx} - \frac{\partial F}{\partial y_k} - p_k\frac{\partial F}{\partial y_0}\right)\delta y_k$$
$$+ \sum_{k=1}^{n}\left(\frac{\partial F}{\partial y_0}\frac{\partial Y_0}{\partial\alpha} - \frac{d\partial Y_0/\partial\alpha}{dx}\right)\left(\delta\alpha + \sum_{k=1}^{n}\beta_k\,\delta\alpha_k\right). \tag{6.129''}$$

Jacobi now argues that the δp_h, δy_h and $\Delta \equiv \delta\alpha + \sum \beta_k\,\delta\alpha_k$ are $2n+1$ quantities determined by the $2n+1$ "arbitrary and independent parameters" $\delta\alpha, \delta\alpha_1, \ldots, \delta\alpha_n, \delta\beta_1, \ldots, \delta\beta_n$.[40] He asserts that since the $\delta\alpha$, $\delta\alpha_1, \ldots, \delta\alpha_n, \delta\beta_1, \ldots, \delta\beta_n$ are independent of each other, so are $\delta p_h, \delta y_h, \Delta$. Their coefficients in (6.129″) then vanish, and he has relations (6.125):

$$\frac{dy_h}{dx} = -\frac{\partial F}{\partial p_h}, \qquad \frac{dp_h}{dx} = \frac{\partial F}{\partial y_h} + p_h\frac{\partial F}{\partial y_0},$$
$$\frac{\partial F}{\partial y_0}\cdot\frac{\partial Y_0}{\partial\alpha} - \frac{d(\partial Y_0/\partial\alpha)}{dx} = 0;$$

and, as we saw above,

$$\frac{dY_0}{dx} = F + \sum_{k=1}^{n} p_k\frac{dy_k}{dx} = F - \sum_{k=1}^{n} p_k\frac{\partial F}{\partial p_k}.$$

(In closing this discussion Jacobi remarks that the equation in $\partial Y_0/\partial\alpha$ also holds with α replaced by α_h.)

Mayer now returns to the form of F and shows, with the help of (6.122), (6.124′), the equations $v_h/v_0 = -p_h$, and $H = v_0F$, that F can be directly expressed as

$$F = y_0' - \sum_{1}^{n}{}_h\, p_h y_h',$$

provided that the $n+r+1$ unknowns $y_0', y_1', \ldots, y_n'; \lambda_0 : \lambda_1 : \ldots : \lambda_r$ are determined by the $n+r+1$ equations

$$\frac{\partial\Omega/\partial y_h'}{\partial\Omega/\partial y_0'} = -p_h, \qquad \phi_k = 0$$

[40] Jacobi, *NACH*, p. 293. Recall that, e.g.,

$$\delta p_h = \sum_{k=1}^{n}\frac{\partial p_h}{\partial\alpha_k}d\alpha_k + \frac{\partial p_h}{\partial\alpha}d\alpha + \sum_{k=1}^{n}\frac{\partial p_h}{\partial\beta_k}d\beta_k.$$

and their values substituted into F. He then closes with the following rule which also occurs in an earlier paper (Mayer [1878], p. 20):

Solve the $n + r + 1$ equations

$$\frac{\partial \Omega}{\partial y_h'} + \frac{\partial \Omega}{\partial y_0'} \frac{\partial y_0}{\partial y_h} = 0, \qquad \phi_k = 0 \qquad (h = 1, 2, \ldots, n;\ k = 0, 1, \ldots, r)$$

for the $n + r + 1$ unknowns $y_0', y_1', \ldots, y_n'; \lambda_0 : \lambda_1 : \cdots : \lambda_r$ and substitute the solutions y' into the equation

$$\frac{\partial y_0}{\partial x} = y_0' - \sum_1^n {}_h \frac{\partial y_0}{\partial y_h} y_h'$$

so that it becomes a partial differential equation of first order between the unknown function y_0 and the $n + 1$ independent variables $x, y_1, \ldots, y_n$.

Given some complete solution

$$y_0 = Y_0(x, y_1, \ldots, y_n, \alpha, \alpha_1, \ldots, \alpha_n)$$

of this partial differential equation, the $n + 1$ unknowns $y_0, y_1, \ldots, y_n$ found by solving the $n + 1$ equations

$$y_0 = Y_0, \qquad \frac{\partial Y_0}{\partial \alpha_h} = \beta_n \frac{\partial Y_0}{\partial \alpha},$$

with $2n + 1$ arbitrary constants $\alpha, \alpha_1, \ldots, \alpha_n, \beta_1, \ldots, \beta_n$, immediately furnish the complete solutions y of the differential equations [(6.107), (6.121′)] *where the ratios of the multipliers λ can be determined by a simple quadrature of some of the equations* [(6.121′)].

7. Hilbert, Kneser, and Others

7.1. Hilbert's Invariant Integral

In his world-famous presentation of mathematical problems at the 1900 International Mathematical Congress, Hilbert ([1900], p. 473) said the following by way of introduction to his discussion of the calculus of variations: "Nevertheless, I should like to close with a general problem, namely the indication of a branch of mathematics repeatedly mentioned in this lecture—which, in spite of the considerable advancement lately given it by Weierstrass, does not receive the general appreciation which, in my opinion, is its due—I mean the calculus of variations."

During the period 1899–1901 Hilbert lectured at Göttingen on the subject, and a very brief account has found its way into papers of Osgood, Hedrick, and others of his students.[1] Moreover, in his 1900 lecture Hilbert outlined his work on the so-called Hilbert invariant integral. This elegant device makes possible a considerable simplification of much work that preceded, as we shall see, particularly in the construction of fields of extremals and in the proofs of sufficiency theorems. In fact, the discoveries of this integral and of the so-called existence theory stemming from his initial work in the subject, in a sense closed the last great conceptual gaps in the basic structure of the calculus of variations before Morse's remarkable work. With the exception of Morse's ideas, the subject had very nearly assumed its theoretical shape with the help of Hilbert's contributions, except for many essential but not revolutionary discoveries made during the remainder of the twentieth century such as Kneser's work on variable end-points or his entire theory of the calculus of variations, which was in marked contrast to Weierstrass's[2]; the works of Mayer and then of Bliss on

---

[1] Osgood [1901″], Hedrick [1902]. There are also interesting results stemming from Hilbert's ideas in theses of his students. See, e.g. Bliss, *LEC*, p. 287 for several references.

[2] It is amusing to read Hedrick's not wholly unjustified criticism of Kneser's work. (Hedrick [1902], p. 24). He says, "In conclusion the author desires to enter protest against the extreme complication recently introduced in some quarters into the essentially simple subject of the

general problems; or the elegant efforts of Bolza to give a clear and succinct treatment to the entire subject. This is certainly not a categorical list of remaining accomplishments, nor is it intended as a denigration of these discoveries, but rather as an attempt to delimit the field. In doing this I should, of course, make mention of the remarkable renascence of the subject as a result of the developments in control theory in the past 25 years. In fact, the later developments of Morse theory and of Control theory have clearly moved the subject into beautiful new areas which are quite beyond the ken of this book.

Hilbert ([1900], p. 473) considers the problem of "finding a function y of a variable x such that the definite integral

$$J = \int_a^b F(y_x, y; x)\, dx, \qquad y_x = \frac{dy}{dx}$$

assumes a minimum value as compared with the values it takes when y is replaced by other functions of x with the same initial and final values." His *invariant integral* is then given by the expression

$$J^* = \int_a^b \{ F + (y_x - p) F_p \}\, dx,$$

where

$$\left[F = F(p, y; x),\ F_p = \frac{\partial F(p, y{:}x)}{\partial p} \right].$$

He continues, "*Now we inquire how p is to be chosen as a function of x, y in order that the value of this integral J* shall be independent of the path of integration, i.e., of the choice of the function y of the variable x.*"

To determine this, Hilbert notes that J^* is of the form

$$J^* = \int_a^b \{ A y_x - B \}\, dx,$$

where A and B are independent of y_x. If this integral is to be path-independent, i.e., have the same value for all curves through the same end-points, then its first variation must vanish and hence

$$\frac{\partial A}{\partial x} + \frac{\partial B}{\partial y} = 0.$$

This means that the first-order partial differential equation

$$\frac{\partial F_p}{\partial x} + \frac{\partial (pF_p - F)}{\partial y} = 0. \tag{7.1}$$

---

calculus of variations. In the case of the only modern text-book [Kneser, *Variationsrechnung*] . . . , this condition is so exaggerated as to essentially mar the usefulness of the book, in that many who would otherwise interest themselves in the subject are repelled by style and treatment." Osgood ([1901″], p. 105n) says of Kneser's book, "This is a work of high scientific merit; in point of style, however it is not all that could be desired."

must be satisfied. Hilbert now remarks that this last equation is closely related to the Euler differential equation

$$\frac{dF_{y_x}}{dx} - F_y = 0. \tag{7.2}$$

To see this connection, he writes the first variation of J^* in the form

$$\begin{aligned} \delta J^* &= \int_a^b \{ F_y \,\delta y + F_p \,\delta p + (\delta y_x - \delta p) F_p + (y_x - p)\,\delta F_p \}\, dx \\ &= \int_a^b \{ F_y \,\delta y + \delta y_x \, F_p + (y_x - p)\,\delta F_p \}\, dx \\ &= \delta J + \int_a^b (y_x - p)\,\delta F_p \, dx. \end{aligned}$$

He now makes an elegant leap in the following way. Suppose that a one-parameter family of solutions of equation (7.2) is given and that each of these solutions also satisfies the first-order differential equation

$$y_x = p(x, y). \tag{7.3}$$

He concludes that $p(x, y)$ is "always an integral of the partial differential equation" of the first order (7.1); "and conversely, if $p(x, y)$ denotes any solution of the partial differential equation [(7.1)], all the nonsingular integrals of the ordinary differential equation" (7.3) of the first order are simultaneously integrals of the second-order differential equation (7.2). That is, if $y_x = p(x, y)$ is an integral of the first order of the second-order equation (7.2), then p is also an integral of the partial differential equation (7.1) and conversely. This means, e.g., that the integral curves of the Euler equation—the extremals—are also characteristics of the first-order partial differential equation (7.1).[3]

Hilbert now represents equations (7.1) and (7.2) in the forms

$$(p_x + pp_y)F_{pp} + pF_{py} + F_{px} - F_y = 0, \tag{7.1$'$}$$

$$y_{xx}F_{y_x y_x} + y_x F_{y_x y} + F_{y_x x} - F_y = 0. \tag{7.2$'$}$$

It is evident from these relations how equation (7.3) relates these equations as stated by Hilbert. He goes on to note that if J^* is independent of the path, if $\bar{y}$ is a solution of (7.3) [i.e., $\bar{y}_x = p(x, \bar{y})$] and if y is any other curve

---

[3]It is not hard to relate this function $p(x, y)$ to the notion of a field. In fact, p is precisely what is called the *slope-function* of a field—a region F of xy-space with a slope-function $p(x, y)$ is said to be a field if p has continuous first partial derivatives in F, if $(x, y, p(x, y))$ are all in the region R of definition of F, and if the Hilbert integral J^* above is independent of the path (see Bliss, *LEC*, p. 85). It can be shown "that through every point of the field F there passes one and but one solution of the differential equation of the field $dy/dx = p(x, y)$ and that every solution of this equation is an extremal." We see from this how close Hilbert was to giving a formal definition of a field in terms of his integral J^*. In fact, we see below that he really did get to the root of the matter.

through the same end-points, then the fundamental relation

$$\int_a^b \{F(p) + (y_x - p)F_p(p)\}\,dx = \int_a^b F(\bar{y}_x)\,dx$$

obtains directly. (Notice his notation.) This may now be rewritten slightly, but significantly, as

$$\int_a^b F(y_x)\,dx - \int_a^b F(\bar{y}_x)\,dx = \int_a^b E(y_x, p)\,dx,$$

where E is Weierstrass's function, which Hilbert writes in the form[4]

$$E(y_x, p) = F(y_x) - F(p) - (y_x - p)F_p(p).$$

Hilbert went on to say "Since, therefore, the solution depends only on finding an integral $p(x, y)$ which is single-valued and continuous in a certain neighborhood of the integral curve $\bar{y}$, which we are considering, the developments just indicated lead immediately—without introduction of the second variation, but only by the application of the polar process to the differential equation [(7.2)] . . . to the expression of Jacobi's condition and to the answer to the question: How far this condition of Jacobi's in conjunction with Weierstrass's condition $E > 0$ is necessary and sufficient for the occurence of a minimum."[5]

Hilbert then proceeds to show how his remarks apply equally to multiple integrals (p. 476). However, we will not take up this topic.

7.2. Existence of a Field

Weierstrass gave only indications of how to show the existence of a strip or field about a given extremal, and it remained for Osgood to give what seems to be the first detailed proof that such a field exists under reasonable

[4] Let the curve defined by $\bar{y}$ be designated as E_{ab}, and the one defined by y as C_{ab}. Then the relation above says that

$$I(C_{ab}) - I(E_{ab}) = \int_a^b E\,dx.$$

This is Weierstrass's fundamental relation (see p. 217 above). This relation is clearly basic for any consideration of sufficiency; it expresses at once the relation derived rather painfully by Weierstrass. Bolza suggests that the result for fields was first given by Schwarz in his 1898/99 lectures.

[5] In 1868 Beltrami [1868] had already worked out, in effect, relation (7.1) above; but its importance and utility went unnoticed until Hilbert's time. For this reason Bolza in *VOR* (p. 108) refers to J^* as the Hilbert invariant integral and the following as the Beltrami–Hilbert independence theorem: "*The integral J^* is independent of the path C and depends only on the positions of the two end-points P_0, P_1 provided that $p(x, y)$ is the slope-function of a field.*"

If we have a one-parameter family of extremals $y(x, \gamma)$, containing the one under consideration for $\gamma = \gamma_0$ and passing through the first end-point P_0, and if $y_\gamma(x, \gamma_0) \neq 0$, then y_γ is a solution of the Jacobi equation and the zeros of $y_\gamma(x, \gamma_0)$ determine the points conjugate to the initial point P_0. Hilbert also clearly noticed from this that no conjugate point could exist between the P_0 and P_1.

hypotheses. This was followed by Bliss's and Bolza's proofs. Osgood and Bolza first show that under certain conditions a field can be constructed in the neighborhood of each point on the extremal in question and then go on to demonstrate the result globally. Osgood thus says "the lower limit . . . , as (x', y') describes C [the given extremal], will be a positive quantity, as can be shown at once by a well-known method of reasoning in higher analysis." Bliss's proof is quite different, as we shall see below.[6]

Osgood uses the definition of a field given above in Section 5.13 involving a family $\varphi(x,\gamma)$ of extremals. He wishes to show that there is a neighborhood S of the given curve C such that through each point of it there passes one and only one of the family of extremals, $\varphi(x,\gamma)$, and that $\varphi_\gamma(x,\gamma) \neq 0$ throughout S. From this he can conclude that the extremal family $\varphi(x,\gamma)$ exactly fills a region in which C is embedded.

He first shows the result "not for the neighborhood S about C but for the neighborhood of an arbitrary point $(x,', y')$ of C." To do this, he notes that since $\varphi_\gamma(x',\gamma_0) \neq 0$, there is a neighborhood $|x - x'| < h'$, $|\gamma - \gamma_0| < \kappa'$ about (x',γ_0) such that $\varphi_\gamma(x,\gamma) \neq 0$ in that neighborhood by continuity. He then can solve the equation $y = \varphi(x,\gamma)$ for γ as a function of x, y in the neighborhood $|x - x'| < h'$, $|y - y'| < h'$. Osgood's proof of this follows from an implicit-function theorem of Dini's which was given in lectures at Pisa and appears in the first of Dini's two volumes, *Analisi Infinitesimale*, 1877–1878, p. 163. (The printed version is in Peano–Genocchi [1884], Section 110; Jordan, *COURS*, 2nd ed., Vol. 1, Paris, 1893, Sections 91, 92; *Osgood*, *LEHR*, pp. 47–57; or Bliss, *PRINCE*, pp. 7–21. An historical survey of implicit-function theory can be found in Osgood, *Encyclopädie der Mathematischen Wissenschaft*, II.B.1, Section 44 and footnote 30.)

Dini's theorem as stated by Osgood [1901″] is this:

> *Let $\Phi(x, y,\gamma)$ be a continuous function of the three independent variables (x, y,γ) throughout the neighborhood D of the point (x', y',γ_0) and let*
>
> $$\Phi(x', y',\gamma_0) = 0;$$
>
> *let the first partial derivatives of Φ exist at each point of D and be continuous, and let*
>
> $$\Phi_\gamma(x', y',\gamma_0) \neq 0;$$
>
> *then for every point (x, y) of the neighborhood d of (x', y') the equation*
>
> $$\Phi(x, y,\gamma) = 0$$
>
> *has one and only one root γ that lies in the neighborhood of the point $\gamma = \gamma_0$, and thus defines γ as a single valued function of (x, y):*
>
> $$\gamma = \psi(x, y).$$
>
> *The function $\psi(x, y)$ is continuous throughout d and has continuous first partial derivatives there, given by the usual formulas for the differentiation of*

[6] See Osgood [1901″], pp. 113ff; Bliss [1904], pp. 113ff; and Bolza, *LEC*, pp. 80–81.

implicit functions:

$$\Phi_x + \Phi_\gamma \frac{\partial \gamma}{\partial x} = 0, \qquad \Phi_y + \Phi_\gamma \frac{\partial \gamma}{\partial y} = 0.$$

By the use of this theorem with $\Phi = \varphi - y$, Osgood now finds his desired neighborhood. He then makes the remark mentioned earlier about a "well-known method of reasoning in higher analysis" to extend his result.

The extension of Dini's theorem to arcs from points was first carried out in detail by Bolza [1906], p. 247 or *VOR*, pp. 160–163. His theorem concerns a transformation $y_i = f_i(x_1, \ldots, x_n)$ $(i = 1, 2, \ldots, n)$ between two regions of n-space and is this:

(A) *The functions* $f_i(x_1, \ldots, x_n)$ *are of class* C' *in some region* A.
(B) C *is a bounded and closed point set in the interior of* A.
(C) *The transformation* [*above*] *always carries two different points* (x'), (x'') *of* C *into two different points* $(y'), (y'')$.
(D) *The functional determinant*

$$\Delta(x_1, \ldots, x_n) = \frac{\partial(f_1, \ldots, f_n)}{\partial(x_1, \ldots, x_n)}$$

is different from zero in C.

Then a positive constant ρ *can be chosen so small that the transformation* [*above*] *defines a one-to-one relation between the neighborhood* $(\rho)_C$ *and its map* S_ρ *in* (y)*-space; or expressed otherwise, for each* (y) *in* S_ρ *the transformation* [*above*] *has one and only solution in* $(\rho)_C$.

Notice that $(\rho)_C$ means the neighborhood of the curve C given by the relation $|\bar{y}(x) - y(x)| < \rho$ for $x_0 \leqslant x \leqslant x_1$.[7]

Bliss's proof of the existence of a field is quite different, depending on a theorem of Picard's on solutions of second-order differential equations, and not on solutions of implicit equations. He does not presuppose the existence of a minimizing arc joining the given end-points and then show that it can be embedded in a family of extremals. Instead, he shows that there is a minimizing arc joining the end-points, if they are close enough together. Picard, in his *Traité d'Analyse*, ([1891], Vol. III, pp. 94–100), has shown that the equation

$$\eta'' = f(\xi, \eta, \eta')$$

"has a solution $\eta = \eta(\xi)$ joining two given points p and q, provided that f satisfies certain continuity restrictions and that q lies in a properly chosen region in the vicinity of the point p. The constants of integration are then the coordinates of p and q."[8] Bliss chooses to apply this theorem to the

---

[7] Another and different proof of Bolza's extension of the fundamental theorem was given by Bliss and Mason [1910].

[8] Bliss [1904′], p. 113. See Weierstrass's result on p. 207 above.

equation $1/\rho = G(x, y, x', y')$ with $1/\rho = (x'y'' - x''y')/(x'^2 + y'^2)^{3/2}$ and

$$G(x, y, x', y') = \frac{F_{x'y} - F_{xy'}}{(x'^2 + y'^2)^{3/2} F_1},$$

$$F_1 = \frac{1}{y'^2} F_{x'x'} = -\frac{1}{x'y'} F_{x'y'} = \frac{1}{x'^2} F_{y'y'} > 0.$$

Here F is positively homogeneous—$F(x, y, \kappa x, \kappa y') = \kappa F(x, y, x', y')$ for $\kappa > 0$—and continuous with continuous first, second, and third partial derivatives "for all (x, y) in a finite closed region $S_1 \ldots$, and for any values of x', y' which are not both zero." (Note that the equation $1/\rho = G$ is the Euler equation in parametric form. Problems for which $F_1 > 0$ were called *regular* by Hilbert [1900], p. 469.)

Bliss can then conclude from his use of Picard's method that any two points P and R in a region S such that the distance PR is suitably small can be joined by an extremal. For a regular problem, he can also prove a very nice local existence theorem for parametric problems of the calculus of variations with fixed end-points. It is this:

> *The function F of the integral I is supposed to have the properties described at the beginning of* [*Bliss's Section* 3], *when (x, y) is in a finite closed region S_1 of the (x, y)-plane and for all values of x', y' which are not both zero. S is a closed region interior to S_1.*
>
> *Then a positive quantity δ can be determined such that any two points P and R of S, whose distance is less than δ, can be joined by one and only one continuous solution of Euler's equation which has a continuously turning tangent and lies entirely in a circle of radius δ about the point P. This solution gives to I a smaller value than any other continuous curve which* 1) *joins P and R,* 2) *lies in the circle, and* 3) *consists of a finite number of pieces, each of which has a continuously turning tangent.*

In Section 7.9 below we shall see, by way of contrast, how Hilbert and his student Noble attacked frontally the problem of showing under quite general hypotheses that a minimizing arc actually exists.[9]

---

[9] In a remarkable paper Hilbert [1899], as in his lectures in 1900, outlined his method for showing the existence of a minimizing arc joining two given points; and Noble ([1901], Sections 5–14) in his dissertation carried Hilbert's results somewhat further. (However, Noble's paper lacks rigor in some places. Bolza says *inter alia*, "In particular the developments in §§9, 10 and 13 are imperfect.") As Hedrick ([1902], p. 240) says: "It was the original purpose of Hilbert's proof (merely) to show that, in this most favorable case of a regular problem, a curve must exist which renders our integral an unlimited strong minimum, compared with all continuous comparison curves; and that the minimizing curve is composed of a finite number of pieces of extremals. Hilbert's existence theorem may therefore properly be called a theorem in differential equations."

Hilbert ([1899], p. 184) said "Each problem of the calculus of variations has a solution as soon as suitable assumptions regarding the nature of the given end-conditions are fulfilled, and, if need be, the notion of a solution has undergone an obvious generalization." We say more on Hilbert's existence theory below.

Before leaving the topic of the existence of fields, we should perhaps establish a result of Bolza's concerning a family of extremals

$$y = \varphi(x, a) \tag{7.4}$$

which contains a given arc E_0 for $a = a_0$ and for which the functions φ and φ' are of class C' for

$$X_1 \leqslant x \leqslant X_2, \qquad |a - a_0| \leqslant d_0, \qquad X_1 < x_1, \qquad x_2 < X_2.$$

Then he shows that

"*If*

$$\varphi_a(x, a_0) \neq 0 \quad on \ [x_1, x_2],$$

one can choose a positive constant k so small that the family of arcs

$$y = \varphi(x, a), \qquad x_1 \leqslant x \leqslant x_2, \qquad |a - a_0| \leqslant k$$

forms a field embedding the arc E_0."

Bolza says, in accord with Schwarz and Kneser, that a family (7.4) of extremals joining two curves $x_1(a), x_2(a)$, $a_1 \leqslant a \leqslant a_2$ [with x_1, x_2 continuous, with $x_1(a) < x_2(a)$, and with φ, φ' of class C' for $a_1 \leqslant a \leqslant a_2$, $x_1(a) \leqslant x \leqslant x_2(a)$] forms a field of extremals about E_0 when E_0 is contained in the family for some value $a = a_0$ between a_1 and a_2 and when no two curves of the family have any point in common. Bliss's definition of a field, based directly on Hilbert's ideas—as we shall see in the next section—is, as mentioned earlier, different in formulation. For him a field F is a region of $x, y_1, \ldots, y_n$ space with a set of slope-functions $p_i(x, y)$ $(i = 1, 2, \ldots, n)$ such that

(1) the p_i are single-valued and have continuous partial derivatives in F;
(2) the points $x, y, p_i(x, y)$ all lie in the domain of definition of the integrand $f(x, y, p)$;
(3) the Hilbert integral

$$I^* = \int \left[f(x, y, p)\,dx + (dy_i - p_i\,dx) f_{y_i}(x, y, p) \right]$$

is independent of the path in F, where repeated indices indicate summation from 1 to n.

The proof for the existence of a field embedding a given arc in Bolza *VOR*, pp. 100–103 is interesting. Since φ_a is continuous and $\neq 0$, it cannot change sign on $[x_1 x_2]$, and Bolza fixes the sign to be > 0; thus $\varphi_a(x, a_0) > 0$ on $[x_1 x_2]$. This implies that $\phi_a(x, a) > 0$ for (x, a) on $[x_1, x_2]$, $|a - a_0| \leqslant k$ provided that k is chosen sufficiently small. Bolza now designates by S_k the map of this closed set from the xa-plane onto the xy-plane under transformation (7.4), and he chooses $P_3(x_3, y_3)$ to be any point in S_k. Then $x_1 \leqslant x_3 \leqslant x_2$, and there is a value $a = a_3$ in $|a - a_0| \leqslant k$ such that $y_3 = \varphi(x_3, a_3)$. He goes on to show that there cannot be another value $a_3' \neq a_3$ in $|a - a_0| \leqslant k$ such that $y_3 = \varphi(x_3, a_3') = \varphi(x_3, a_3)$. This is impossible since φ_a is a properly monotonic function of a. As a consequence, the curve $a = a_3$ is the unique member of family (7.4) passing through P_3 in S_k.

Bolza then remarks that the entire segment

$$\varphi(x_3, a_0 - k) \leqslant y \leqslant \varphi(x_3, a_0 + k)$$

of the line $x = x_3$ lies in S_k and that S_k is the set of the xy-plane bounded by the lines $x = x_1$ and $x = x_2$ and the two nonintersecting curves $y = \varphi(x, a_0 - k)$, $y = \varphi(x, a_0 + k)$. He can then solve equation (7.4) for $a(x, y)$ as a function of class C'. (The continuity and the differentiability follow easily from implicit-function theory.) He finally wishes to choose a neighborhood (ρ) of E_0—the set of all points $(x, \bar{y})$ such that $|y(x) - \bar{y}| < \rho$ for $x_1 \leqslant x \leqslant x_2$—which lies entirely in the interior of S_k; to do this, he remarks that the two continuous functions

$$\varphi(x, a_0 + k) - \varphi(x, a_0), \; \varphi(x, a_0) - \varphi(x, a_0 - k)$$

have positive minima on $[x, x_2]$; he chooses ρ to be smaller than either of these minima. This choice of ρ is effective, as can be seen without difficulty.

Let us see how the Hilbert–Bliss and the Schwarz definitions for a field compare. Suppose that F is a field of extremals about E_0 in Schwarz's sense. There is then a one-parameter family $y = \varphi(x, a)$ containing E_0 for $a = a_0$ such that through every point (x, y) of F there passes one and only one curve of the family. By the result of Hilbert noted in Section 7.1, there is an intimate relation between the solutions of equations (7.2) and (7.3). For the case of problems in the plane—$n = 1$ above—the relation is particularly simple. Bliss's theorem is this (Bliss, *LEC*, p. 85):

> *If a one parameter family of extremals*
>
> $$y(x, a) \qquad [a_1 \leqslant a \leqslant a_2, \; x_1(a) \leqslant x \leqslant x_2(a)]$$
>
> *is cut by a curve defined on the family by a function $x = \xi(a)$ $(a_1 \leqslant a \leqslant a_2)$ and if the family and the intersecting curve have continuity properties . . . , then every region F of the xy-plane which is simply covered by the extremals is a field with the slope function $p(x, y)$ of the family, provided that the derivative $y_a(x, a)$ is different from zero at each set of values (x, a) corresponding to a point (x, y) in F.*

The definition of a field in spaces higher than the plane is more complicated than that of Schwarz and requires the introduction of new concepts. This generalization was first carried out by Hilbert and by Mayer. We shall see below how they did this with the help of the Hilbert invariant integral.

7.3. Hilbert, Continuation

In a most elegant paper Hilbert [1906] did a number of very important things in the calculus of variations: first, he took up the problem of Mayer, so named by Kneser (*LV'*, pp. 256ff), and established the multiplier rule for the problem; second, he showed the relation between his independence theorem and the Hamilton–Jacobi theory and in the course of this gave the modern definition for a field; third, he carried out his ideas on double integral problems in the calculus of variations; and last, he gave a "general rule for the handling of variational problems and the formation of a new criterion." We shall examine briefly some of these points.

Hilbert in this paper and Kneser in *LV*, Sections 56 and 59, bridged a fundamental gap in earlier proofs of the multiplier rule. They show, in effect, that given a set of admissible variations η_i there is a family $y_i(x,\epsilon)$ of admissible comparison arcs satisfying the side- and end-conditions, containing the given arc y_i for $\epsilon = 0$, and having $\partial y_i(x,\epsilon)/\partial\epsilon|^{\epsilon=0} = \eta_i$.

Hilbert also established in his summer lectures in 1899 what is now called *Hilbert's differentiability condition* by a simple modification of du Bois-Raymond's lemma (see Whittemore [1901], p. 132). He showed that a minimizing arc $y(x)$ must satisfy the Euler equation in integral form $f_{y'} = \int_{x_0}^{x} f_y\, dx + c$; and hence near every point (x, y, y') not a corner and where $f_{y'y'} \neq 0$, the function y has continuous derivatives of order n when f has continuous partial derivatives of order $\leqslant n$ near (x, y, y') (see Bliss, *LEC*, p. 13). What Hilbert noted is that the second derivative y'' exists for all values of x when $f_{y'}$ is differentiable and $f_{y'y'} \neq 0$.

Hahn [1906], p. 254, showed that every rectifiable curve which has a defined tangent at each point must satisfy the Euler differential equation if it furnishes the integral J a minimum. Hahn [1902] also modified a proof of Kneser's and gave the differentiability condition of Hilbert. Out of this he showed it is not necessary to assume the existence of a continuous second derivative y'' for a minimizing arc. Hahn [1904] gives a nice discussion of the multiplier rule for isoperimetric problems.

Hilbert acknowledges both the prior work of Mayer [1895] and Kneser *LV*, Sections 56–58, on the multiplier rule. He considers the case of three dependent functions y, z, s of x and two conditions on these quantities of the form

$$f(y', z', s', y, z, s; x) = 0, \qquad g(y', z', s', y, z, s; x) = 0. \tag{7.5}$$

He supposes that $y(x)$, $z(x)$, $s(x)$ are three functions satisfying conditions (7.5) which have

$$\begin{vmatrix} \dfrac{\partial f}{\partial y'} & \dfrac{\partial f}{\partial z'} \\ \dfrac{\partial g}{\partial y'} & \dfrac{\partial g}{\partial z'} \end{vmatrix} \neq 0 \tag{7.6}$$

for all x between $x = a_1$ and $x = a_2$. Let $Y(x)$, $Z(x)$, $S(x)$ be any three other functions also satisfying (7.5), for which

$$Y(a_1) = y(a_1), \qquad Z(a_1) = z(a_1),$$
$$Z(a_2) = z(a_2), \qquad S(a_1) = s(a_1), \qquad S(a_2) = s(a_2).$$

It is also assumed that the functions Y, Z, S together with their derivatives differ only a little from y, z, s and their derivatives, and finally that

$$Y(a_2) \geqslant y(a_2). \tag{7.7}$$

He then states his result: "*if this minimal condition is satisfied, then there necessarily exist two functions $\lambda(x)$, $\mu(x)$ which do not both vanish identically for all x, and which together with the functions $y(x)$, $z(x)$, $s(x)$ satisfy the Lagrange differential equations*

$$\frac{d}{dx}\frac{\partial(\lambda f+\mu g)}{\partial y'}-\frac{\partial(\lambda f+\mu g)}{\partial y}=0, \tag{7.8}$$

$$\frac{d}{dx}\frac{\partial(\lambda f+\mu g)}{\partial z'}-\frac{\partial(\lambda f+\mu g)}{\partial z}=0, \tag{7.8$'$}$$

$$\frac{d}{dx}\frac{\partial(\lambda f+\mu g)}{\partial s'}-\frac{\partial(\lambda f+\mu g)}{\partial s}=0 \tag{7.8$''$}$$

which arise from setting to zero the first variation of the integral.

$$\int_{a_1}^{a_2}\{\lambda f(y',z',s',y,z,s;x)+\mu g(y',z',s',y,z,s;x)\}\,dx.\text{''}^{10}$$

To prove this theorem Hilbert considers two functions σ_1, σ_2 of x, which vanish for $x = a_1$ and $x = a_2$, and he sets

$$\begin{gathered} Y = Y(x,\epsilon_1,\epsilon_2), \qquad Z = Z(x,\epsilon_1,\epsilon_2), \\ S = s(x)+\epsilon_1\sigma_1(x)+\epsilon_2\sigma_2(x) \end{gathered} \tag{7.9}$$

for ϵ_1, ϵ_2 two parameters. These functions must satisfy the conditions

$$f(Y',Z',S',Y,Z,S;x)=0, \qquad g(Y',Z',S',Y,Z,S;x)=0, \tag{7.10}$$

which Hilbert considers as a system of two differential equations in the two functions Y, Z. By known results on differential equations, he concludes that for sufficiently small ϵ_1, ϵ_2 there are solutions $Y(x,\epsilon_1,\epsilon_2), Z(x,\epsilon_1,\epsilon_2)$ satisfying (7.10) identically in $x, \epsilon_1, \epsilon_2$ such that for $\epsilon_1 = \epsilon_2 = 0$, $Y(x,\epsilon_1,\epsilon_2) = y(x)$, $Z(x,\epsilon_1,\epsilon_2) = z(x)$ and for $x = a_1$, ϵ_1, ϵ_2 arbitrary, Y, Z take on the values $y(a_1), z(a_1)$, respectively. (He cites Picard [1891], Vol. III, Chapter VIII.)

Hilbert now considers the minimum condition (7.7). He must have $Y(a_2,\epsilon_1,\epsilon_2)$ a minimum for $\epsilon_1 = \epsilon_2 = 0$ while $Z(a_2,\epsilon_1,\epsilon_2) = z(a_2)$. By "the theory of relative minima" of a function of two variables, he knows that there must exist two constants l, m, not both zero, for which

$$\begin{aligned} \left[\frac{\partial(lY(a_2,\epsilon_1,\epsilon_2)+mZ(a_2,\epsilon_1,\epsilon_2))}{\partial\epsilon_1}\right]_0 &= 0, \\ \left[\frac{\partial(lY(a_2,\epsilon_1,\epsilon_2)+mZ(a_2,\epsilon_1,\epsilon_2))}{\partial\epsilon_2}\right]_0 &= 0, \end{aligned} \tag{7.11}$$

[10] Hilbert [1906], p. 352. It is not unreasonable to call (7.8), (7.8′), and (7.8″) the Euler–Lagrange equations since both men contributed significantly.

where the subscript zero means that the relevant expressions are evaluated for $\epsilon_1 = \epsilon_2 = 0$.

Consider now the linear, homogeneous differential equations (7.8) and (7.8′) in the functions $\lambda(x)$, $\mu(x)$ together with the end-conditions at $x = a_2$

$$\left[\frac{\partial(\lambda f + \mu g)}{\partial y'}\right]_0 = l, \qquad \left[\frac{\partial(\lambda f + \mu g)}{\partial z'}\right]_0 = m; \tag{7.12}$$

the functions f, g are evaluated for y, z, s by means of the functions defined in (7.9). From the theory of differential equations and the fact that $(l, m) \neq (0, 0)$, equations (7.8) and (7.8′) are solvable for functions $\lambda(x)$, $\mu(x)$ not both identically zero.

Hilbert now differentiates identities (7.10) as to ϵ_1, ϵ_2 and finds

$$\left[\frac{\partial Y'}{\partial \epsilon_1}\right]_0 \frac{\partial f}{\partial y'} + \left[\frac{\partial Y}{\partial \epsilon_1}\right]_0 \frac{\partial f}{\partial y} + \left[\frac{\partial Z'}{\partial \epsilon_1}\right]_0 \frac{\partial f}{\partial z'} + \left[\frac{\partial Z}{\partial \epsilon_1}\right]_0 \frac{\partial f}{\partial z} + \sigma_1' \frac{\partial f}{\partial s'} + \sigma_1 \frac{\partial f}{\partial s} = 0,$$

$$\left[\frac{\partial Y'}{\partial \epsilon_1}\right]_0 \frac{\partial g}{\partial y'} + \left[\frac{\partial Y}{\partial \epsilon_1}\right]_0 \frac{\partial g}{\partial y} + \left[\frac{\partial Z'}{\partial \epsilon_1}\right]_0 \frac{\partial g}{\partial z'} + \left[\frac{\partial Z}{\partial \epsilon_1}\right]_0 \frac{\partial g}{\partial z} + \sigma_1' \frac{\partial g}{\partial s'} + \sigma_1 \frac{\partial g}{\partial s} = 0,$$

$$\left[\frac{\partial Y'}{\partial \epsilon_2}\right]_0 \frac{\partial f}{\partial y'} + \left[\frac{\partial Y}{\partial \epsilon_2}\right]_0 \frac{\partial f}{\partial y} + \left[\frac{\partial Z'}{\partial \epsilon_2}\right]_0 \frac{\partial f}{\partial z'} + \left[\frac{\partial Z}{\partial \epsilon_2}\right]_0 \frac{\partial f}{\partial z} + \sigma_2' \frac{\partial f}{\partial s'} + \sigma_2 \frac{\partial f}{\partial s} = 0,$$

$$\left[\frac{\partial Y'}{\partial \epsilon_2}\right]_0 \frac{\partial g}{\partial y'} + \left[\frac{\partial Y}{\partial \epsilon_2}\right]_0 \frac{\partial g}{\partial y} + \left[\frac{\partial Z'}{\partial \epsilon_2}\right]_0 \frac{\partial g}{\partial z'} + \left[\frac{\partial Z}{\partial \epsilon_2}\right]_0 \frac{\partial g}{\partial z} + \sigma_2' \frac{\partial g}{\partial s'} + \sigma_2 \frac{\partial g}{\partial s} = 0.$$

He multiplies the first equation by λ, the second by μ, adds the two, and integrates the resulting relation between the limits $x = a_1$, $x = a_2$; he also does the same with the third and fouth equations. This gives him the relations

$$\int_{a_1}^{a_2} \left\{ \frac{\partial(\lambda f + \mu g)}{\partial y'} \left[\frac{\partial Y'}{\partial \epsilon_\alpha}\right]_0 + \frac{\partial(\lambda f + \mu g)}{\partial y} \left[\frac{\partial Y}{\partial \epsilon_\alpha}\right]_0 + \frac{\partial(\lambda f + \mu g)}{\partial z'} \left[\frac{\partial Z'}{\partial \epsilon_\alpha}\right]_0 + \frac{\partial(\lambda f + \mu g)}{\partial z} \left[\frac{\partial Z}{\partial \epsilon_\alpha}\right]_0 + \frac{\partial(\lambda f + \mu g)}{\partial s'} \sigma_\alpha' + \frac{\partial(\lambda f + \mu g)}{\partial s} \sigma_\alpha \right\} dx = 0$$

$$(\alpha = 1, 2). \tag{7.13}$$

Recall that $Y(a_1, \epsilon_1, \epsilon_2) = y(a_1)$, $Z(a_1, \epsilon_1, \epsilon_2) = z(a_1)$ and consequently that at $x = a_1$

$$\left[\frac{\partial Y}{\partial \epsilon_\alpha}\right]_0 = 0, \qquad \left[\frac{\partial Z}{\partial \epsilon_\alpha}\right]_0 = 0 \qquad (\alpha = 1, 2);$$

moreover, for $x = a_2$ relations (7.11) and (7.12) imply that

$$\frac{\partial(\lambda f + \mu g)}{\partial y'}\left[\frac{\partial Y}{\partial \epsilon_\alpha}\right]_0 + \frac{\partial(\lambda f + \mu g)}{\partial z'}\left[\frac{\partial Z}{\partial \epsilon_\alpha}\right]_0 = 0 \qquad (\alpha = 1, 2).$$

Hilbert now applies the usual integration by parts—he calls this *integration of a product*—to equations (7.13), taking account of relations (7.8) and (7.8′). This yields

$$\int_{a_1}^{a_2}\left\{\frac{\partial(\lambda f + \mu g)}{\partial s'}\sigma'_\alpha + \frac{\partial(\lambda f + \mu g)}{\partial s}\sigma_\alpha\right\} dx = 0 \qquad (\alpha = 1, 2).$$

He now defines the left-hand member of these equations for an arbitrary σ by the symbol $(\lambda\mu, \sigma)$ and summarizes his results in the following[11]: "*To every two functions* σ_1, σ_2 *vanishing at* $x = a_1$ *and* $x = a_2$ *there always exists a non-identically vanishing solution system* λ, μ *of the differential equations* . . . [(7.8) *and* (7.8′)], *so that there results*

$$(\lambda\mu, \sigma_1) = 0 \quad \textit{and} \quad (\lambda\mu, \sigma_2) = 0."$$

Hilbert now goes on to show that either for this pair λ, μ, $(\lambda\mu, \sigma) = 0$ for all σ or that there is another pair λ', μ' of solutions of (7.8) and (7.8′) for which $(\lambda'\mu', \sigma) = 0$ for all σ. Suppose that for the functions λ, μ there is a σ_3 for which

$$(\lambda\mu, \sigma_3) \neq 0. \tag{7.14}$$

Then by what went before there is a nonidentically zero solution system λ', μ' of (7.8) and (7.8′) for which

$$(\lambda'\mu', \sigma_3) = 0. \tag{7.14′}$$

Now if there were a σ_4 for which

$$(\lambda'\mu', \sigma_4) \neq 0, \tag{7.14″}$$

then by the lemma above there would exist a solution system λ'', μ'' of (7.8) and (7.8′) for which

$$(\lambda''\mu'', \sigma_\beta) = 0 \qquad (\beta = 3, 4). \tag{7.14‴}$$

Notice that λ, μ; λ', μ'; λ'', μ'' are solutions of a system of two homogeneous, linear differential equations of the first order. They must therefore be linearly dependent; i.e., there must be constants a, a', a'' not all zero such that

$$a\lambda + a'\lambda' + a''\lambda'' = 0, \qquad a\mu + a'\mu' + a''\mu'' = 0.$$

Now by relations (7.14), (7.14′) and (7.14‴) it would follow directly that

[11] Hilbert [1906], p. 355.

$a = 0$ and by (7.14″) and (7.14‴) that $a' = 0$; this is impossible, since $a = 0$, $a' = 0$ and λ'', μ'' not both identically zero would imply that $a'' = 0$. Hilbert then concludes, "Our assumption is consequently not correct, and we conclude therefore that either λ, μ or λ', μ' is such a system of solutions of [(7.8) and (7.8′)] that the integral relation $(\lambda\mu, \sigma) = 0$ or $(\lambda'\mu', \sigma) = 0$ is valid for every function σ. The application of product integration (integration by parts) to this relation shows that the solution system λ, μ or λ', μ' necessarily must satisfy the equation . . . [(7.8″)], and therefore the desired proof is completely finished."[12]

Hilbert then proceeds to discuss the relation between his independence theorem and the Hamilton–Jacobi theory. To do this, he drops further discussion of plane-problems and considers here the simplest problem in three-space: to find two functions $y(x)$, $z(x)$ which give the integral

$$J = \int_a^b F(y', z', y, z; x)\, dx$$

its minimal value among comparison curves through the same end-points. He then poses the problem:

> *We consider now the integral*
>
> $$J^* = \int_a^b \{ F + (y - p) F_p + (z' - q) F_q \}\, dx$$
>
> $$\left[F = F(p, q, y, z; x), \qquad F_p = \frac{\partial F(p, q, y, z; x)}{\partial p}, \qquad F_q = \frac{\partial F(p, q, y, z; x)}{\partial q} \right]$$
>
> *and ask how the functions p, q appearing therein are to be chosen so that the value of this integral J^* will be independent of the path of integration in xyz-space, i.e., independent of the functions $y(x)$, $z(x)$.*

To answer this, he considers a surface $T(x, y, z) = 0$ in xyz-space and asks what happens if he chooses paths on this surface joining two fixed points on it. He considers now the solutions of the Lagrange equations

$$\frac{d}{dx} F_{y'} = F_y, \qquad \frac{d}{dx} F_{z'} = F_z, \tag{7.15}$$

through each point P of the surface $T = 0$, for which $y' = p$, $z' = q$ at P. In this fashion he constructs a two-parameter family of integral curves which fill a field in space.[13] (This means he has assumed through each point x, y, z of the field there passes a unique extremal of the family.) He then says "*The values of the derivatives y', z' at each point x, y, z are then the functions $p(x, y, z)$; $q(x, y, z)$ with the desired property.*"

To show this, consider a point A of the surface $T = 0$ and join it to an arbitrary point Q of the spatial field by a curve w. Through each point of w

[12] Hilbert [1906], pp. 355–356.

[13] Notice how, in effect, Hilbert has defined a field in three-space with the help of his invariant integral.

there passes an integral curve of the two-parameter family. This generates a one-parameter family of integral curves

$$y = \psi(x,\alpha), \qquad z = \chi(x,\alpha); \tag{7.16}$$

moreover, the intersection of this family with the surface $T = 0$ is a curve w_T, which goes from the point A to the point P on $T = 0$, which corresponds to Q in the sense that the integral curve of (7.16) through Q cuts the surface $T = 0$ in P.

Hilbert next eliminates the parameter α from equations (7.16) and finds the surface

$$z = f(x, y). \tag{7.16'}$$

He introduces this function z into F in this way:

$$F\left(y', \frac{\partial f}{\partial x} + \frac{\partial f}{\partial y} y', y, f(x, y); x \right) = \Phi(y', y; x),$$

so that for every curve on the surface (7.16′)

$$\int_a^b F(y', z', y, z; x)\, dx = \int_a^b \Phi(y', y, x)\, dx.$$

He now has a new minimum problem, but this time it is in the xy-plane. He goes to show that the extremals for this problem are the curves $y = \psi(x,\alpha)$ and that the invariant integral corresponding to Φ is precisely the one he wishes.

It is not hard to show that the first variation of the integral in the right-hand member of this equation vanishes for every curve of the family $y = \psi(x,\alpha)$ in the xy-plane.[14]

Consider now the invariant integral for the plane problem involving Φ:

$$\int_a^b \{\Phi + (y' - p)\Phi_p\}\, dx \qquad [\Phi = \Phi(p, y; x)]. \tag{7.17}$$

Hilbert remarks that this integral is invariant since his independence theorem is known to be true in the plane. But

$$z' = f_x + f_y y', \qquad q = f_x + f_y p$$

and hence $f_y(y' - p) = z' - q$ so that

$$\begin{aligned} \Phi(p, y; x) + (y' - p)\Phi_p(p, y; x) &= F(p, q, y, z; x) + (y' - p)(F_p + F_q f_y) \\ &= F(p, q, y, z; x) + (y' - p)F_p + (z' - q)F_q. \end{aligned}$$

---

[14]To see this point of Hilbert consider the first variation of the integral of Φ from a to b. It is clearly expressible in terms of variations $\eta(x)$, with the help of the Lagrange equations (7.15), as

$$\begin{aligned} \int_a^b \{\Phi_y \cdot \eta + \Phi_{y'} \cdot \eta'\}\, dx = \int_a^b \{ & [F_{z'}(f_{xy} + f_{yy} y') + F_y + F_z \cdot f_y]\eta \\ & + [F_{y'} + F_{z'} \cdot f_y]\eta'\}\, dx = \big[(F_{y'} + F_{z'} \cdot f_y)\eta\big]_a^b = 0, \end{aligned}$$

since all varied curves pass through points A and Q and thus $\eta(a) = \eta(b) = 0$.

From this it follows that the invariance of the integral (7.17) implies the integral

$$J^* = \int_a^b \{F + (y' - p)F_p + (z' - q)F_q\}\, dx$$

has the same value for all paths lying on $T = 0$ and joining points A and P.

Hilbert, moreover, has the relation

$$\int_{(w)} \{F + (y' - p)F_p + (z' - q)F_q\}\, dx$$
$$= \int_{(w_T)} \{F + (y' - p)F_p + (z' - q)F_q\}\, dx + \int_P^Q F\, dx, \qquad (7.18)$$

since $y' = p$, $z' = q$ along members of family (7.16); he next chooses another composite path $\bar{w}$ in the pq-field joining A and Q formed of a path $\bar{w}_T$ on $T = 0$ through A and P and of the unique member of (7.16), through P and Q. He then also has

$$\int_{(\bar{w})} \{F + (y' - p)F_p + (z' - q)F_q\}\, dx$$
$$= \int_{(\bar{w}_T)} \{F + (y' - p)F_p(z' - q)F_q\}\, dx + \int_P^Q F\, dx. \qquad (7.18')$$

The first integrals in the right-hand members of (7.18) and (7.18′) have the same value since both w_T and $\bar{w}_T$ lie on the surface $T = 0$ and join the same points A and P. It therefore follows that the left-hand members of (7.18) and (7.18′) are equal, which establishes for Hilbert his independence theorem for xyz-space.

Since the integral J^* is independent of the path, he can write

$$J(x, y, z) = \int_A^{x,y,z} \{F + (y' - p)F_p + (z' - q)F_q\}\, dx;$$

moreover, he has

$$\frac{\partial J}{\partial x} = F - pF_p - qF_q, \qquad \frac{\partial J}{\partial y} = F_p, \qquad \frac{\partial J}{\partial z} = F_q. \qquad (7.19)$$

Hilbert now notes that if p, q are eliminated between these equations, there results the "Jacobi–Hamilton partial differential equation of the first order for the function J."[15] If the functions p, q of the field are so chosen that

$$F - pF_p - qF_q : F_p : F_q = \frac{\partial T}{\partial x} : \frac{\partial T}{\partial y} : \frac{\partial T}{\partial z},$$

then on the surface $T = 0$ the integral J^* will have the value 0 for all paths.[16] In this case the function $J(x, y, z)$ is clearly a solution of "the Jacobi–Hamilton equation" which vanishes on $T = 0$.

[15] Hilbert [1906], p. 361.

[16] Let us set

$$F - pF_p - qF_q = T_x, \qquad F_p = T_y, \qquad F_q = T_z$$

and solve the last two equations for $p = P(x, y, z, T_y, T_z)$ and $q = Q(x, y, z, T_y, T_z)$. If these are substituted into the first of these equations, there results $T_x + H(x, y, z, T_y, T_z) = 0$, the familar Hamilton–Jacobi equation.

Hilbert now considers a two-parameter family of surfaces containing the surface $T = 0$ and determined by parameters a, b. Then the functions p, q and $J(x, y, z)$ are functions of these parameters a, b. By differentiating $J(x, y, z)$ with respect to them, he finds

$$\frac{\partial J}{\partial a} = \int_A^{x,y,z} \left\{ (y' - p) \frac{\partial F_p}{\partial a} + (z' - q) \frac{\partial F_q}{\partial b} \right\} dx,$$

$$\frac{\partial J}{\partial b} = \int_A^{x,y,z} \left\{ (y' - p) \frac{\partial F_b}{\partial b} + (z' - q) \frac{\partial F_q}{\partial b} \right\} dx.$$

Notice that $y' = p$, $z' = q$ on each surface of the family and thus that the integrands above vanish. It follows at once that

$$\frac{\partial J}{\partial a} = c, \qquad \frac{\partial J}{\partial b} = d,$$

where a, b, c, d are constants of integration in the Lagrange equations; i.e., every curve satisfying these equations is an extremal and there is a family of solutions $y(x, a, b, c, d)$, $z(x, a, b, c, d)$ forming a four-parameter family of extremals.[17]

Hilbert then proceeds to discuss double integral and related problems in much the same way as he has just done and then concludes with a "General rule for handling variational problems and the formation of a new criterium."[18]

7.4. Mayer Families of Extremals

In a two-paper series [1903], [1905], Mayer gave a systematic treatment of the problem of Lagrange in the course of which he introduced what are now called *Mayer families of extremals*.[19] In what follows in this section we shall look briefly at his 1905 paper to see how he treated the connection between Hilbert's invariant integral and the Hamilton–Jacobi theory. This is virtually the same topic we saw Hilbert discussing in Section 7.3 above.

Mayer considers a problem of Lagrange. He assumes that there are $r < n$ given differential equations of the first order

$$f_\rho(x, y_1, \ldots, y_n, y_1', \ldots, y_n') = 0 \qquad (\rho = 1, 2, \ldots, r) \qquad (7.20)$$

which are solvable for r of the derivatives $y_1', \ldots, y_n'$. Then among all functions $y_1, \ldots, y_n$ of x which are continuous between the limits x_0 and

[17] Bliss, *LEC*, p. 73.

[18] Hilbert [1906], pp. 362–370.

[19] Bliss says in his discussion of problems in parametric form (*LEC*, p. 125) that they are families "whose slope functions form fields in every region F of y-space which they simply cover." As we shall see, such families make the Hilbert integral independent of the path.

$x_1 > x_0$, which pass through given points at x_0 and x_1, and which satisfy equations (7.20), he seeks to find a set which renders the integral

$$\int_{x_0}^{x_1} f(x, y_1, \ldots, y_n, y_1', \ldots, y_n')\,dx$$

a maximum or a minimum.

The Lagrange differential equations are

$$\frac{d}{dx}\frac{\partial\Omega}{\partial y_i'} = \frac{\partial\Omega}{\partial y_i}, \qquad f_\rho = 0, \tag{7.21}$$

in which

$$\Omega \equiv f + \sum_1^r \lambda_\rho f_\rho.$$

He introduces canonical variables

$$\frac{\partial\Omega}{\partial y_i'} = v_i, \qquad f_\rho = 0 \tag{7.22}$$

and notes that these $n + r$ equations can be solved for the $n + r$ unknowns $y_1', \ldots, y_n', \lambda_1, \ldots, \lambda_r$. He writes the solutions as

$$\begin{aligned} y_i' &= p_i(x, y_1, \ldots, y_n, v_1, \ldots, v_n), \\ \lambda_\rho &= \mu_\rho(x, y_1, \ldots, y_n, v_1, \ldots, v_n), \end{aligned} \tag{7.23}$$

and he then forms the hamiltonian function

$$\sum_1^n y_i' \frac{\partial\Omega}{\partial y_i'} - \Omega = H(x, y_1, \ldots, y_n, v_1, \ldots, v_n),$$

where the y_i', λ_ρ are replaced by the p_i, μ_ρ of (7.23). He then has the canonical equations for the extremals

$$\frac{dy_i}{dx} = \frac{\partial H}{\partial v_i}, \qquad \frac{dv_i}{dx} = -\frac{\partial H}{\partial y_i}. \tag{7.24}$$

Mayer, in effect, now observers that every extremal y_i, λ_ρ, i.e., every set y_i, λ_ρ satisfying the differential equations (7.21) defines by means of equations (7.22) a solution y_i, v_i of equations (7.24). Conversely, every solution y_i, v_i of (7.24) defines an extremal y_i, λ_ρ with the help of the last r equations (7.23).

He now writes $|\Omega|$, $|f|$, $|f_\rho|$ to mean the functions Ω, f, f_ρ with the expressions $y_i' = p_i$, $\lambda_\rho = \mu_\rho$ of (7.23) substituted for y_i', λ_ρ in those functions. Thus, e.g., $|f_\sigma| = f_\sigma(x, y_1, \ldots, y_n, p_1, \ldots, p_n)$ and

$$|\Omega| = |f| + \sum_1^r {}_\rho \mu_\rho |f_\rho|.$$

He poses what he calls Problem II: "*To determine the* p_i, μ_ρ *as functions of* $x, y_1, \ldots, y_n$ *so that the expression*

$$|\Omega| + \sum_1^n {}_i(y_i' - p_i)\frac{\partial|\Omega|}{\partial p_i}, \tag{7.25}$$

in which $y_1, \ldots, y_n$ are regarded as undetermined functions of x, is a perfect derivative and so that the r conditions $|f_\rho| = 0$ are identically satisfied."[20]

To solve the problem, he seeks a function V of x, y such that

$$|\Omega| + \sum_1^n {}_i(y_i' - p_i)\frac{\partial|\Omega|}{\partial p_i} = \frac{dV}{dx},$$

and he writes this expression as the $1 + n$ relations

$$|\Omega| - \sum_1^n {}_i p_i \frac{\partial|\Omega|}{\partial p_i} = \frac{\partial V}{\partial x}, \qquad \frac{\partial|\Omega|}{\partial p_i} = \frac{\partial V}{\partial y_i}. \tag{7.26}$$

These equations (7.26), together with $|f_\rho| = 0$, determine the p_i, μ_ρ as functions of x, y_i, $\partial V/\partial y_i$, and relations (7.23) enable him to write

$$p_i = p_i\left(x, y_i, \ldots, y_n, \frac{\partial V}{\partial y_1}, \ldots, \frac{\partial V}{\partial y_n}\right),$$

$$\mu_\rho = \mu_\rho\left(x, y_1, \ldots, y_n, \frac{\partial V}{\partial y_1}, \ldots, \frac{\partial V}{\partial y_n}\right). \tag{7.23$'$}$$

The first of equations (7.26) with these values for p_i, μ_ρ then becomes the Hamilton–Jacobi partial differential equation

$$\frac{\partial V}{\partial x} + H\left(x, y_1, \ldots, y_n, \frac{\partial V}{\partial y_1}, \ldots, \frac{\partial V}{\partial y_n}\right) = 0. \tag{7.27}$$

Mayer concludes that to each solution system p_i, μ_ρ of Problem II there corresponds, by means of a simple quadrature, a solution V of the Hamilton–Jacobi equation (7.27) so that the solution set p_i, μ_ρ satisfies equations (7.23′). Conversely, each solution V of (7.27) when substituted into the equations (7.23′) yields a solution system p_i, μ_ρ of Problem II. Moreover, expression (7.25) has the form

$$|\Omega| + \sum_1^n {}_h(y_h' - p_h)\frac{\partial|\Omega|}{\partial p_h} \equiv B + \sum_1^n {}_h B_h y_h',$$

in which B, B_h are functions of $x, y_1, \ldots, y_n$.

It is clear that if the functions p_i, μ_ρ make the expression on the left-hand side of this identity an exact derivative, then the $n + n(n-1)/2$ relations

$$\frac{\partial B_i}{\partial x} - \frac{\partial B}{\partial y_i} = 0, \qquad \frac{\partial B_i}{\partial y_h} - \frac{\partial B_h}{\partial y_i} = 0 \tag{7.28}$$

must be satisfied. It follows with the help of the relations

$$B = |\Omega| - \sum_1^n p_h \frac{\partial|\Omega|}{\partial p_h}, \qquad B_h = \frac{\partial|\Omega|}{\partial p_h}$$

[20] Mayer [1905], p. 52. See also Bliss, *LEC*, pp. 238ff.

that

$$\frac{\partial B}{\partial y_i} = \frac{\partial}{\partial y_i}\left\{|\Omega| - \sum_1^n {}_h p_h \frac{\partial|\Omega|}{\partial p_h}\right\} = \frac{\partial|\Omega|}{\partial y_i} + \sum_1^r {}_\rho \frac{\partial \mu_\rho}{\partial y_i}|f_\rho| - \sum_1^n {}_h p_h \frac{\partial}{\partial y_i}\frac{\partial|\Omega|}{\partial p_h}$$

$$= \frac{\partial|\Omega|}{\partial y_i} - \sum_1^n {}_h p_h \frac{\partial}{\partial y_i}\frac{\partial|\Omega|}{\partial p_h},$$

since $|f_\rho| = 0$. Integrability conditions (7.28) thus become

$$\frac{\partial}{\partial x}\frac{\partial|\Omega|}{\partial p_i} + \sum_1^n {}_h p_h \frac{\partial}{\partial y_h}\frac{\partial|\Omega|}{\partial p_i} = \frac{\partial|\Omega|}{\partial y_i}, \qquad \frac{\partial}{\partial y_h}\frac{\partial|\Omega|}{\partial p_i} = \frac{\partial}{\partial y_i}\frac{\partial|\Omega|}{\partial p_h},$$

and the n first of these can then be expressed as

$$\frac{\partial^2|\Omega|}{\partial p_i\,\partial x} + \sum_1^n {}_h p_h \frac{\partial^2|\Omega|}{\partial p_i\,\partial y_h} + \sum_1^n {}_k \frac{\partial^2|\Omega|}{\partial p_i\,\partial p_k}\left(\frac{\partial p_k}{\partial x} + \sum_1^n {}_h p_h \frac{\partial p_k}{\partial y_h}\right)$$

$$+ \sum_1^r {}_\rho \frac{\partial|f_\rho|}{\partial p_i}\left(\frac{\partial \mu_\rho}{\partial x} + \sum {}_h p_h \frac{\partial \mu_\rho}{\partial y_h}\right) = \frac{\partial|\Omega|}{\partial y_i}. \qquad (7.29)$$

But for the functions $y_k' = p_k$, $\lambda_\rho = \mu_\rho$ of Problem II, Mayer has

$$y_k'' = \frac{\partial p_k}{\partial x} + \sum_1^n {}_h \frac{\partial p_k}{\partial y_h} p_h,$$

$$\lambda_\rho' = \frac{\partial \mu_\rho}{\partial x} + \sum_1^n {}_h \frac{\partial \mu_\rho}{\partial y_h} p_h.$$

He substitutes these into relations (7.29) and finds that

$$\frac{\partial^2\Omega}{\partial y_i'\partial x} + \sum_1^n {}_h \frac{\partial^2\Omega}{\partial y_i'\,\partial y_h} y_h' + \sum_1^n {}_k \frac{\partial^2\Omega}{\partial y_i'\,\partial y_k'} y_k'' + \sum_1^r {}_\rho \frac{\partial f_\rho}{\partial y_i'}\lambda_\rho' = \frac{\partial\Omega}{\partial y_i}.$$

These conditions and $|f_\rho| = 0$ then reduce precisely to the Euler–Lagrange equations

$$\frac{d}{dx}\frac{\partial\Omega}{\partial y_i'} = \frac{\partial\Omega}{\partial y_i}, \qquad f_\rho = 0.$$

Mayer summarizes his result in his Theorem IV:

> *To each solution V of the partial differential equation* . . . [(7.27)] *there corresponds a system of solutions $y_1, \ldots, y_n, \lambda_1, \ldots, \lambda_r$ of the differential equations* . . . [(7.21)] *containing n arbitrary constants in respect to which the solutions $y_1, \ldots, y_n$ are independent of each other; and one obtains the solutions of this system by completely integrating the first n of the equations* . . . [(7.23′)], *formed with the help of the solution V above, that themselves constitute a system of first-order differential equations between $y_1, \ldots, y_n$ and x, and then by substituting their solutions into the last r equations* [(7.23′)].

What he is in effect saying is that given a solution V of the Hamilton–Jacobi equation (7.27), one finds an n-parameter family of extremals by integrating the system of first order differential equations

$$y_i' = p_i[x, y, V_y(x, y)] \qquad (i = 1, 2, \ldots, n);$$

when these solutions $y_i(x, a_1, \ldots, a_n)$ are substituted into the equations for the μ_ρ

$$\lambda_\rho = \mu_\rho[x, y, V_y(x, y)] \qquad (\rho = 1, 2, \ldots, r),$$

there results a family of extremals

$$y_i(x, a_1, \ldots, a_n), \qquad \lambda_\rho(x, a_1, \ldots, a_n).$$

Now there is associated with this family the slope functions

$$p_i = p_i[x, y, V_y(x, y)],$$

and multipliers 1, $\lambda_\rho = \mu_\rho[x, y, V_y(x, y)]$; moreover, the integral of function (7.25) is independent of the path in the region F of xy-space on which the p_i, λ_ρ are defined and have continuous first partial derivatives. (This is how he finds a Mayer family.)

In his Section 2 he considers a $2n$-parameter family

$$y_i = \varphi_i(x, c_1, \ldots, c_{2n}), \qquad v_i = \psi_i(x, c_1, \ldots, c_{2n}) \tag{7.30}$$

of solutions of the accessory equations (7.24), and from these he derives, with the help of the relations $v_i = \partial\Omega/\partial y_i'$, the complete system of solutions

$$y_i = \varphi_i(x, c_1, \ldots, c_{2n}), \qquad \lambda_\rho = \Theta_\rho(x, c_1, \ldots, c_{2n})$$

of the Lagrange equations (7.21). Mayer now chooses a value $x = a$ so that at this value of x the $2n$ equations (7.30) are solvable for the $c_1, \ldots, c_{2n}$. He has

$$a_i = \varphi_i(a, c_1, \ldots, c_{2n}), \qquad b_i = \psi_i(a, c_1, \ldots, c_{2n}) \tag{7.30$'$}$$

as the values of y_i, v_i at $x = a$; and he replaces in (7.30) the constants c by a, b, the solutions of (7.30′). This gives him the family

$$y_i = y_i(x, a_1, \ldots, a_n, b_1, \ldots, b_n), \qquad v_i = v_i(x, a_1, \ldots, a_n, b_1, \ldots, b_n)$$

with

$$y_i(a, a_1, \ldots, a_n, b_1, \ldots, b_n) = a_i, \qquad v_i(a, a_1, \ldots, a_n, b_1, \ldots, b_n) = b_i.$$

He now chooses an arbitrary function $A = A(a, a_1, \ldots, a_n)$ and sets $b_h = \partial A/\partial a_h$. This produces an n-parameter family of solutions of the canonical equations (7.24)

$$y_i = y_i\left(x, a_1, \ldots, a_n, \frac{\partial A}{\partial a_1}, \ldots, \frac{\partial A}{\partial a_n}\right) = \bar{y}_i(x, a_1, \ldots, a_n),$$

$$v_i = v_i\left(x, a_1, \ldots, a_n, \frac{\partial A}{\partial a_1}, \ldots, \frac{\partial A}{\partial a_n}\right) = \bar{v}_i(x, a_1, \ldots, a_n),$$

$$\lambda_\rho = \bar{\lambda}_\rho(x, a_1, \ldots, a_n)$$

with the property that at $x = a$, the $y_i = a_i$ and also that identically

$$y_i' = p_i(x, \bar{y}_i, \dots, \bar{y}_n, \bar{v}_1, \dots, \bar{v}_n),$$
$$\lambda_\rho = \mu_\rho(x, \bar{y}_1, \dots, \bar{y}_n, \bar{v}_1, \dots, \bar{v}_n), \tag{7.31}$$

as he indicated just after equations (7.24).

He defines a function V in the form

$$V = A(a, a_1, \dots, a_n) + \int_a^x \left\{ \sum_1^n {}_h v_h \frac{\partial H}{\partial v_h} - H \right\} dx,$$

where H was defined above as a function of x, y, v and $\overline{H}(x, a_1, \dots, a_n) = H(x, \bar{y}, \bar{v})$. Mayer now has

$$\frac{\partial V}{\partial a_k} = \frac{\partial A}{\partial a_k} + \int_a^x \sum_1^n {}_h \left\{ \bar{v}_h \frac{\partial}{\partial a_k} \frac{\partial \overline{H}}{\partial v_h} - \frac{\partial \overline{H}}{\partial y_h} \frac{\partial \bar{y}_h}{\partial a_k} \right\} dx;$$

but

$$\frac{\partial \overline{H}}{\partial v_h} = \frac{\partial \bar{y}_h}{\partial x}, \qquad \frac{\partial \overline{H}}{\partial y_h} = -\frac{\partial \bar{v}_h}{\partial x},$$

and consequently the sum under the integral sign above becomes

$$\sum_1^n {}_h \left\{ \bar{v}_h \frac{\partial^2 \bar{y}_h}{\partial x \, \partial a_k} + \frac{\partial \bar{v}_h}{\partial x} \frac{\partial \bar{y}_h}{\partial a_k} \right\} = \frac{\partial}{\partial x} \sum_1^n {}_h \bar{v}_h \frac{\partial \bar{y}_h}{\partial a_k}.$$

Moreover for $x = a$, $\bar{y}_h = a_h$, $\bar{v}_h = \partial A / \partial a_h$ and so

$$\left[\sum_1^n {}_h \bar{v}_h \frac{\partial \bar{y}_h}{\partial a_k} \right]_a^x = \sum_1^n {}_h \left(\bar{v}_h \frac{\partial \bar{y}_h}{\partial a_k} - \frac{\partial A}{\partial a_h} \frac{\partial a_h}{\partial a_k} \right) = \sum_1^n {}_h \bar{v}_h \frac{\partial \bar{y}_h}{\partial a_k} - \frac{\partial A}{\partial a_k}.$$

It follows from this directly that

$$\frac{\partial V}{\partial a_k} = \sum_1^n {}_h \bar{v}_h \frac{\partial \bar{y}_h}{\partial a_k}. \tag{7.32}$$

Mayer now assumes that the equations $y_i = \bar{y}_i(x, a_1, \dots, a_n)$ can be solved for the a_i and writes their solutions as

$$a_i = a_i(x, y_1, \dots, y_n) = (a_i). \tag{7.33}$$

He writes $(V) = W(x, y_1, \dots, y_n)$ to mean $V[x, a_1(x, y), \dots, a_n(x, y)]$. Recall that for $x = a$, $V = A(a, a_1, \dots, a_n)$ and hence that

$$W(a, y_1, \dots, y_n) = A(a, y_1, \dots, y_n).$$

He has therefore by (7.32)

$$\frac{\partial W}{\partial y_i} = \sum_1^n {}_k \left(\frac{\partial V}{\partial a_k} \right) \frac{\partial (a_k)}{\partial y_i} = \sum_1^n {}_h (\bar{v}_h) \sum_1^n {}_k \left(\frac{\partial \bar{y}_h}{\partial a_k} \right) \frac{\partial (a_k)}{\partial y_i}.$$

But as can be seen from (7.33), $y_i \equiv \bar{y}_i[x, a_1(x, y), \dots, a_n(x, y)]$ and so

$$\frac{\partial y_h}{\partial y_i} = \sum_1^n {}_k \left(\frac{\partial \bar{y}_h}{\partial a_k} \right) \frac{\partial (a_k)}{\partial y_i}.$$

It then follows directly that

$$\frac{\partial W}{\partial y_i} = (\bar{v}_i).$$

Mayer also has by definition

$$\frac{dV}{dx} \equiv \sum_1^n {}_h\bar{v}_h \frac{\partial \bar{H}}{\partial v_h} - \bar{H}, \tag{7.34}$$

and $V = \overline{W} = W[x, y_1(x, a_1, \ldots, a_n), \ldots, y_n(x, a_1, \ldots, a_n)]$. This gives him $\bar{v}_i = \partial \overline{W}/\partial y_i$ and

$$\frac{dV}{dx} = \frac{\partial \overline{W}}{\partial x} + \sum_1^n {}_h \frac{\partial \overline{W}}{\partial y_h} \frac{\partial \bar{y}_h}{\partial x} = \frac{\partial \overline{W}}{\partial x} + \sum_1^n {}_h\bar{v} \frac{\partial \bar{H}}{\partial v_h}.$$

It now follows from this and relation (7.34) that

$$\frac{\partial \overline{W}}{\partial x} + \bar{H} = \frac{\partial \overline{W}}{\partial x} + H\left(x, \bar{y}_1, \ldots, \bar{y}_n, \frac{\partial \overline{W}}{\partial y_1}, \ldots, \frac{\partial \overline{W}}{\partial y_n}\right) \equiv 0.$$

He replaces the a_i by $a_i(x, y_1, \ldots, y_n)$ and has

$$\frac{\partial V}{\partial x} + H\left(x, y_1, \ldots, y_n, \frac{\partial V}{\partial y_1}, \ldots, \frac{\partial V}{\partial y_n}\right) = 0.$$

His general result is then formulated in his theorem:

> V. *Let the [Euler–Lagrange] differential equations* . . . [(7.21)] *be completely integrated and then eliminate out of the complete solutions*
>
> $$y_i = \varphi_i(x_1 c_1, \ldots, c_{2n}), \qquad \lambda_\rho = \Theta_\rho(x, c_1, \ldots, c_{2n})$$
>
> *the* $2n$ *integration constants* $c_1, \ldots, c_{2n}$ *with the help of the* $2n$ *equations*
>
> $$\varphi_i(a, c_1, \ldots, c_{2n}) = a_i, \qquad \left[\frac{\partial \Omega}{\partial y_i'}\right]_{x=a} = \frac{\partial A}{\partial a_i},$$
>
> *in which* a *is a new arbitrary constant or equally a particular, chosen value of* x; *one finds a new system of solutions of these differential equations*
>
> $$y_i = \bar{y}_i(x, a_1, \ldots, a_n), \qquad \lambda_\rho = \bar{\lambda}_\rho(x, a_1, \ldots, a_n),$$
>
> *whose* n *arbitrary constants* $a_1, \ldots, a_n$ *are the initial values of the variables* $y_1, \ldots, y_n$ *for* $x = a$; *these new solutions of the differential equations* . . . [(7.22)] *belong to a special solution* $V = W$ *of the [Hamilton–Jacobi] partial differential equation* . . . [(7.27)] *and indeed a solution which for* $x = a$ *assumes the value*
>
> $$W = A(a, y_1, \ldots, y_n)$$
>
> *so that for* $V = W$ *the* $n + r$ *equations* . . . [(7.23′)] *are identically satisfied.*

This is his method for finding an n-parameter Mayer family.

In Section 4 Mayer discusses the importance of this theorem. He calculates the value of $\partial V/\partial a$ and finds

$$\frac{\partial V}{\partial a} = \frac{\partial A}{\partial a} - \left[\sum_1^n {}_h\bar{v}_h \frac{\partial \bar{H}}{\partial v_n} - \bar{H}\right]_{x=a} + \sum_1^n {}_h\bar{v}_h \frac{\partial \bar{y}_h}{\partial a} - \left[\sum_1^n {}_h\bar{v}_h \frac{\partial \bar{y}_h}{\partial a}\right]_{x=a}$$

but for $x = a$ he has

$$\bar{y}_h = a_h, \qquad \bar{v}_h = \frac{\partial A}{\partial a_h}, \qquad \frac{\partial \overline{H}}{\partial v_k} = \left[\frac{\partial \bar{y}_h}{\partial x} \right]_{x=a}.$$

This gives him the relation

$$\frac{\partial V}{\partial a} = \frac{\partial A}{\partial a} + H\left(a, a_1, \ldots, a_n, \frac{\partial A}{\partial a_1}, \ldots, \frac{\partial A}{\partial a_n} \right) + \sum_1^n {}_h \bar{v}_h \frac{\partial \bar{y}_h}{\partial a}$$
$$- \sum_1^n {}_h \frac{\partial A}{\partial a_h} \left[\frac{\partial \bar{y}_h}{\partial x} + \frac{\partial \bar{y}_h}{\partial a} \right]_{x=a}.$$

But it is clear that

$$0 = \frac{\partial a_h}{\partial a} = \left[\frac{\partial \bar{y}_h}{\partial x} + \frac{\partial \bar{y}_h}{\partial a} \right]_{x=a}$$

and consequently

$$\frac{\partial V}{\partial a} = \frac{\partial A}{\partial a} + H\left(a, a_1, \ldots, a_n, \frac{\partial A}{\partial a_1}, \ldots, \frac{\partial A}{\partial a_n} \right) + \sum_1^n {}_h \bar{v}_h \frac{\partial \bar{y}_h}{\partial a}.$$

Mayer next calculates from $V = \overline{W}$ that

$$\frac{\partial V}{\partial a} = \frac{\partial \overline{W}}{\partial a} + \sum_1^n {}_h \frac{\partial \overline{W}}{\partial y_h} \frac{\partial \bar{y}_h}{\partial a} = \frac{\partial \overline{W}}{\partial a} + \sum_1^n {}_h \bar{v}_h \frac{\partial \bar{y}_h}{\partial a}$$

and thus

$$\frac{\partial \overline{W}}{\partial a} = \frac{\partial A}{\partial a} + H\left(a, a_1, \ldots, a_n, \frac{\partial A}{\partial a_1}, \ldots, \frac{\partial A}{\partial a_n} \right).$$

If $V = A(x, y_1, \ldots, y_n)$ is a solution of the Hamilton–Jacobi equation (7.27) and is independent of a, then $\partial \overline{W}/\partial a = 0$ and thus $\partial W/\partial a = 0$. This implies that W is a solution of the Hamilton–Jacobi equation which is independent of a, and for $x = a$ it has the value

$$W = A(a, y_1, \ldots, y_n).$$

Mayer sums up his analysis in his Theorem VI. He observes that "*After a complete integration of the* [*Euler-Lagrange*] *differential equations* . . . [(7.21)] *the theorem* V *permits one in general to obtain every system of solutions of those differential equations which with some system of solutions of Problem II, satisfy the relations*

$$y_i' = p_i, \qquad \lambda_\rho = \mu_\rho.\text{"}$$

In his last section Mayer repeats in essence a part of Hilbert's paper [1906] which appeared in the same year in the *Göttinger Nachrichten*. What Mayer showed is that the total variation ΔI of the integral I is expressible as

$$\Delta I = \int_{x_0}^{x_1} E\, dx,$$

where E is given by

$$E \equiv f - |f| - \sum_{1}^{n} {}_h(y'_h - p_h)\frac{\partial|\Omega|}{\partial p_h}.$$

He did this with the help of the invariant integral just as Hilbert did for the simplest problem.

7.5. Kneser's Methods

In 1897 Kneser saw how to exploit a very elegant result due to Zermelo to give a new and simple proof for the Jacobi condition.[21] Then in 1900 he discovered as a by-product of this work two new concepts, focal points and transversality, which led, among other things, to the solution of variable end-point problems. He saw how to generalize certain important properties of geodesics on a surface to extremals in general. This forms the foundation of his work.

There is a well-known theorem of Gauss on geodesics which says this: consider a given curve on a surface and at each point of this curve, draw the geodesic through that point normal to the curve, lay off on these geodesics a fixed length. Then the end-points of the geodesics arcs form a curve which cuts the geodesics orthogonally.[22] Conversely, two curves on the surface cutting the same family of geodesics orthogonally intercept a constant length on the geodesics. This is one of the basic theorems that Kneser generalized.

Suppose that there is given a curve on a surface, consider a point P on this curve and draw through P the geodesic normal to the given curve. Suppose the curve is determined by the parameter v; further, suppose that the common length cut off on the geodesics is denoted by u and denote by Q the end-point of the geodesic arc of length $|u|$ through P. Thus given u, v the point Q is uniquely determined. Conversely, Q uniquely determines values u and v, provided that the region R of the surface being considered is chosen small enough. The curves $u = \text{const}$, $v = \text{const}$ then determine a very elegant set of curves, which can be used to define curvilinear coordinates. In fact, arc-length in this coordinate system is $ds^2 = du^2 + m^2\, dv^2$, as

[21] Kneser [1898], pp. 48–49; *LV*, pp. 93–97; and *LV'*, pp. 116–120. (The latter two are his *Lehrbuch der Variationsrechnung.*) Zermelo's so-called envelope theorem appears in his dissertation on p. 96. See Zermelo [1894]. In a related matter Osgood [1901], p. 278, says quite correctly: "Kneser has given a proof by means of power series on the assumption that all functions considered are analytic. This assumption is unfortunate, inasmuch as it introduces restrictions without being accompanied by any simplification."

[22] See, e.g., Bolza, *VOR*, pp. 333. The theorem is in Gauss, *DIS*.

Gauss showed. The integral

$$J = \int_{t_0}^{t_1} \sqrt{\left(\frac{du}{dt}\right)^2 + m^2\left(\frac{dv}{dt}\right)^2}\, dt \tag{7.35}$$

gives then the length of any arc $\mathfrak{C}$, $u = u(t)$, $v = v(t)$ on the surface, joining two points A and B with A corresponding to t_0 and B to t_1. The value of J when evaluated on the geodesic through A and B is $\bar{J} = u_1 - u_0$ and consequently

$$\Delta J = \bar{J} - J = \int_{t_0}^{t_1} \left\{ \sqrt{u'^2 + m^2 v'^2} - u' \right\} dt,$$

where the prime superscript indicates differentiation with respect to t. It is evident that ΔJ is never negative and can vanish only if $\mathfrak{C}$ is the geodesic through A and B. It is then clear that the geodesic is the shortest path between given points on a surface.

It is not difficult to see that the condition imposed on region R of the surface is in essence the Jacobi condition for the problem posed above relative to the integral (7.35). Darboux gives a theorem on geodesics which says that if a family of geodesics on a surface passing through a given point A has an envelope B_0CB as in Figure 7.1, then the value of J along AC plus the value along CB equals the value along AB. Point B is conjugate to A on the geodesic AB, and if C is near enough to B on the envelope, then the joint arc, AB plus BC, is an admissible variation for which $\Delta J = 0$; i.e., $J(AB) + J(BC) = J(AC)$ or in Kneser's notation, $(AB) + (BC) = (AC)$. But it can be shown that the envelope BC cannot be a geodesic, and hence another arc can be found joining B and C, which will give ΔJ a smaller (i.e., negative) value.[23] This means that the geodesic does not furnish a minimum in this case for arcs joining A and B and certainly would not for points beyond B on the geodesic.

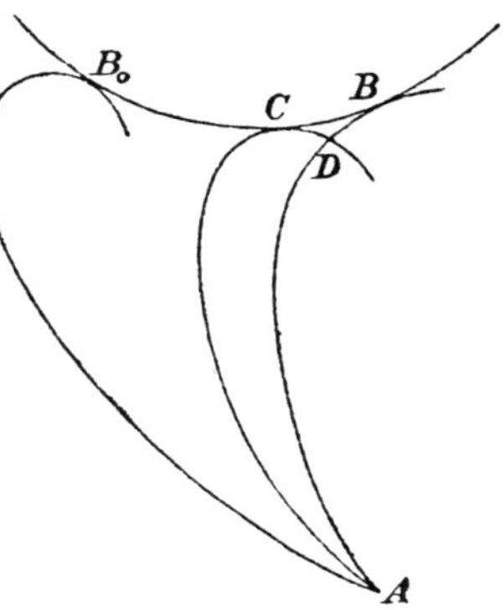

Figure 7.1

[23] Darboux, *TDS*, Vol. III, p. 88.

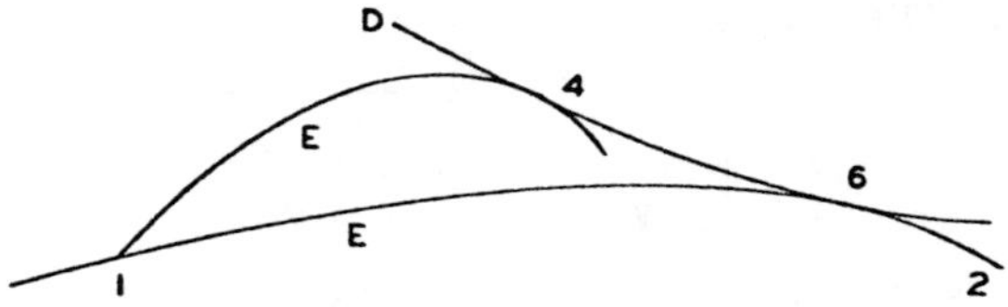

Figure 7.2

These results on geodesics were generalized by Kneser [1898] and *LV*, and the Jacobi condition follows quite elegantly and naturally from them. To see the Zermelo–Kneser envelope theorem, it is perhaps well to deviate a little from the formulation of this theorem in Kneser's paper. The envelope theorem is easily stated for the simplest problem of the calculus of variations:

> *If a one-parameter family of extremals E has an envelope D, as shown in Figure . . .* [(7.2)], *then the equation*
>
> $$I(E_{16}) = I(E_{14}) + I(D_{46})$$
>
> *holds for every position of the point* 4 *preceding the point* 6 *on D.*[24]

This is clearly the generalization of Darboux's theorem on geodesics. Kneser's generalization of the notion of orthogonality occurs in his book *LV*. We will investigate this shortly.

Kneser says of the envelope equation above[25]:

> This equation is contained in a more general one, which Zermelo deduced in his dissertation . . . by means of the $\mathcal{E}$-function introduced by Weierstrass without determining, as he himself emphasized, whether an envelope exists or whether the resulting equation can be applied to a problem of the calculus of variations. I have answered these questions in the discussion at hand; I have replaced the infinitesimal considerations mentioned above by a rigorous and direct demonstration; and I have found finally a new result concerning the termination of the maximum or minimum property of the integral J in the case that one integrates along a curve for which δJ vanishes from some point to another conjugate to it.

Actually, he is not generous enough in giving credit to Zermelo: in his dissertation Zermelo certainly proved the envelope theorem, and by a most elegant argument. He made use of Weierstrass's elegant result—often called

---

[24]Bliss, *LEC*, p. 25. See also Kneser [1898], pp. 48–49. In Figure 7.2, E_{14} and E_{16} are extremals through the fixed point 1, and D is the envelope of the family and is therefore tangent to E_{14} at 4 and to E_{16} at 6. The functional I is the value of

$$\int_{x_1}^{x_2} f(x, y, y')\, dx$$

evaluated along an appropriate arc.

[25]Kneser [1898], p. 28. The reference to Zermelo is [1894], p. 96.

Weierstrass's theorem—that if an extremal AB is contained in a field, then the variation of the integral is

$$\Delta J = \int \mathcal{E}\, dx.$$

Zermelo considered the extremal arc AB in Figure 7.1 and the comparison arc ACB made up of the extremal arc AC neighboring to AB and the arc CB of the envelope. By the Weierstrass theorem, he has

$$J(ACB) - J(AB) = \int \mathcal{E}(x, y, p, P)\, dx$$

where (x, y) is a point on ACB, p is the value of the slope function at (x, y), and P is the slope of ACB at (x, y). But $\mathcal{E}(x, y, p, P) = f(x, y, P) - f(x, y, p) - (P - p)f_{y'}(x, y, p)$; along AB and AC it is clear that $P = p$ so that $\mathcal{E} = 0$. Along CB the envelope is, by definition, tangent to the extremal through (x, y), and so $P = p$. Zermelo concluded that $J(AC) + J(CB) - J(AB) = J(ACB) - J(AB) = 0$, which is the envelope theorem. Now it is quite true that Zermelo does not discuss, as does Kneser, the question of the existence of the envelope, nor does he see how to use the result to find the Jacobi condition. Kneser does not give Zermelo credit in his *Lehrbuch* for this theorem, nor have subsequent authors such as Bliss (*LEC*).

Let us first see how Kneser established the envelope result. He considers an extremal $\mathfrak{C}$, an arbitrary point (x, y) on it, and a one-parameter family of extremals $y = \varphi(x, a)$ containing $\mathfrak{C}$ for $a = a_0$, where φ is regular in a neighborhood of (x, a_0). To establish a correspondence between points on these extremals, he considers the equations

$$\bar{y} - \varphi(\bar{x}, a) = 0, \tag{7.36}$$

$$\varphi_x(x, a_0)(\bar{y} - y) + \bar{x} - x = 0 \tag{7.36'}$$

for a near a_0. He assumes that the functional determinant of these equations with respect to $\bar{y}$ and $\bar{x}$

$$\begin{vmatrix} 1 & -\varphi_x(\bar{x}, a) \\ \varphi_x(x, a_0) & 1 \end{vmatrix}$$

is different from zero for $\bar{x} = x$, $a = a_0$. There is then the desired correspondence between x, y and $\bar{x}, \bar{y}$. In fact $(\bar{x}, \bar{y})$ is the point at which the normal at (x, y) to $\mathfrak{C}$ cuts the curve (7.36). The $\bar{x}, \bar{y}$ are regular functions of a. Kneser then forms the variations

$$\bar{y} - y = \delta y = [a - a_0]_1, \qquad \bar{x} - x = \delta x = [a - a_0]_1,$$

where the expressions in brackets are power series in $(a - a_0)$ whose lowest terms are of the first order. He indicates by $(\bar{x}_2, \bar{y}_2)$ and $(\bar{x}_3, \bar{y}_3)$ the points on $\mathfrak{C}'$ corresponding to (x_2, y_2) and (x_3, y_3) on $\mathfrak{C}$; and he writes

$$J_{23} = \int_{x_2}^{x_3} f(x, y, p)\, dx, \qquad \bar{J}_{23} = \int_{\bar{x}_2}^{\bar{x}_3} f(\bar{x}, \bar{y}, \bar{p})\, d\bar{x},$$

where $\bar{p} = \varphi_x(\bar{x}, a)$. (Thus $\bar{p} - p = [a - a_0]_1$.)

Kneser calculates the first variation of his integral and determines that

$$\delta\int_u^v f(x,y,p)\,dx=\int_u^v\left(f_y-\frac{df_p}{dx}\right)(\delta y-p\,\delta x)+(f-pf_p)_{x=v}\,\delta v$$
$$-(f-pf_p)_{x=u}\,\delta u+\left[f_p\,\delta y\right]_u^v.$$

Along an extremal the integrand in the right-hand member of this expression vanishes, and it is easy to evaluate $\Delta J=\bar{J}_{23}-J_{23}$. He finds without difficulty that

$$\bar{J}_{23}-J_{23}=(f-pf_p)^{(3)}(\bar{x}_3-x_3)+(f_p)^{(3)}(\bar{y}_3-y_3)$$
$$-(f-pf_p)^{(2)}(\bar{x}_2-x_2)-(f_p)^{(2)}(\bar{y}_2-y_2)+\left[a-a_0\right]_2,$$

where the superscripts in this result indicate that the expressions inside the parentheses are evaluated at the values $x_\nu, y_\nu, \varphi_x(x_\nu,a)$ $(\nu=2,3)$. (Kneser's analysis in 1898 is not very elegant due in large measure to the fact that he has made his argument depend unnecessarily on complex-variable theory.) By analytic continuation Kneser shows that the result above is valid even though the points 2 and 3 are not near to each other.

Suppose that $y=\varphi(x,a_0)$, $y=\varphi(x,a_1)$ are the two extremals through the points A and D in Figure 7.1. Then $\bar{J}-J=[a_1-a_0]_2$ since $(x,y)=(\bar{x},\bar{y})$ at A and D. Kneser writes the value of J along the arc AD corresponding to a_0 as $(AD)_1$ and to a_1 as $(AD)_2$. He therefore has

$$(AD)_1-(AD)_2=\left[a_1-a_0\right]_2.$$

But by the additive property of integrals,

$$(AD)_1=(AB)-(DB),\qquad (AD)_2=(AC)+(CD),$$

where, e.g., (DB) is the value of J evaluated along the extremal arc DB.

Kneser then is able to conclude that

$$(AB)+(BC)-(AC)=\left[a_1-a_0\right]_2.$$

He next considers another extremal defined by $a=a_2$ with $|a_2-a_1|$ so small that the envelope BC meets this curve in a point C' very near to C. The result above then becomes

$$(AC)+(CC')-(AC')=\left[a_2-a_1\right]_2,$$

where CC' is an arc of the envelope; and he notes that

$$(AB)+(BC)-(AC)+(AC)+(CC')-(AC')=(AB)+(BC')-(AC').$$

From this it follows that $(AC)+(CC')-(AC')$ is the change in $(AB)+(BC)-(AC)$ when C changes to C', i.e., when a_1 is replaced by a_2. Kneser concludes that $(AB)+(BC)-(AC)$ is zero.[26] Thus $(AB)+(BC)=(AC)$, which is the envelope theorem.

[26] To see this, consider $\varphi(a_1)=(AB)+(BC)-(AC)$, $\varphi(a_2)=(AB)+(BC')-(AC')$ and form $\varphi(a_2)-\varphi(a_1)$. Kneser has shown that $\varphi'(a_1)(a_2-a_1)+\ldots=\varphi(a_2)-\varphi(a_1)=\{(AC)+(CC')-(AC')\}=[a_2-a_1]_2$. He then has $\varphi'(a_1)=0$, and φ is independent of a_1.

This means directly that $(AB) = (AC) + (CB)$ so that the integral J has the same value when evaluated on the extremal joining A and B as it does along "countless other" integration paths deviating arbitrarily little from this path, i.e., along the extremal AC plus the arc CB of the envelope. "Therefore the integral J along the arc AB is surely neither a maximum nor a minimum." The points A and B are then conjugate points. Kneser goes on to comment that the points C and B can, in general, be joined by an extremal arc if they are close enough together; he designates by $(\overline{CB})$ this value of J. Then surely $(CB) > (\overline{CB})$ since CB is an arc of the envelope and cannot satisfy the Euler equation; (The reason for this is that the unique extremal with values x, y, y' at point B is the extremal through A and B itself; i.e., by the theory of differential equations, there is exactly one solution of the second-order Euler equation with the given common values x, y, y' of the extremal and the envelope at B.) It results that

$$(AB) > (AC) + (\overline{CB}),$$

and the integral (AB) is certainly not a minimum.

In case the envelope contracts down to point B, the inequality $(CB) > (\overline{CB})$ no longer holds, and Kneser remarks that J has the same value along all extremals passing through A and B and near enough to the given one. In his *LV*, Kneser has replaced this derivation by a simpler one, which depends on differentiation of an integral with respect to a parameter.[27] We will take this up in Section 7.6. We conclude Kneser's 1898 paper with his theorem on the Jacobi condition and his work on the existence of an envelope. He states (p. 50) the following result:

> "*The integral J ceases to be an extremum when it is evaluated along the curve $\mathfrak{C}$ from the initial point A not only when the interval of integration includes the point B conjugate to A but already when it is bounded by the point.*"

He goes on to say that this removes the uncertainty left by the results of Mayer and Scheeffer.[28]

To discuss the existence of an envelope Kneser ([1898], pp. 36ff) starts with a two-parameter family of extremals $y = g(x, a, b)$ containing a given arc $\mathfrak{C}$ for $a = a_0$, $b = b_0$. By simple differentiation with respect both to a

---

[27] Kneser, *LV*, pp. 89–97.

[28] In Erdmann [1877], p. 327, we find that when the second variation $\delta^2 J = 0$, then

$$\delta^3 J = -R(x')\varphi_{ax}(x', a_0)\varphi_{aa}(x', a_0),$$

where $R = f_{y'y'}$, $y = \varphi(x, a)$ is a one-parameter family of extremals through the first end-point $P_0(x_0, y_0)$, $\varphi_{ax}(x', a_0) \neq 0$, and $\varphi_{aa}(x', a_0) \neq 0$ except when the envelope of the family has a cusp at $P'(x', y')$ or when it degenerates into a point. Except in these cases, Erdmann had, in effect, already in 1877 shown Kneser's result to be correct since it is easy to show that the second variation vanishes in the situations mentioned in Kneser's result. (Recall that $\Delta J = \varepsilon\,\delta J + (\varepsilon^2/2)\,\delta^2 J + (\varepsilon^3/6)\,\delta^3 J + \ldots$.)

and b, Kneser reaches "the fundamental observation of Jacobi" that

$$\frac{d}{dx}\{f_{pp}(g_a g_{bx} - g_b g_{ax})\} = 0$$

so that $f_{pp}(g_a g_{bx} - g_b g_{ax}) = C$.

Now suppose that all the extremals of the family go through the point (x_0, y_0). He then shows that $b - b_0$ can be expressed as a power series in $(a - a_0)$ whose first term is zero, in his notation $b - b_0 = [a - a_0]_1$. He substitutes this expansion of b as a function of a into g and has

$$g(x,a,b) = \varphi(x,a). \tag{7.37}$$

When $\varphi_a(x_1,a_0) \neq 0$, Kneser observes that in a neighborhood of (x_1, y_1) the curve $\mathfrak{C}$ is not cut by any of the curves of family $y = \varphi(x,a)$ if $|a - a_0|$ is small enough. If this were not the case, the equation

$$\varphi(x,a) - \varphi(x,a_0) = 0$$

would hold for sufficiently small values of $x - x_1$ and $a - a_0$. But the left-hand member may be written as

$$\begin{aligned}(a - a_0)\varphi_a(x,a_0) &+ \frac{1}{2}(a - a_0)^2(\varphi_{aa}(x_1,a_0) + [a - a_0, x - x_1]_1)\\ &= (a - a_0)\{\varphi_a(x,a_0) + (a - a_0)[a - a_0, x - x_1]_1\},\end{aligned}$$

and this expression is not zero for $|x - x_1|$ and $|a - a_0|$ small enough. Kneser [1898] concludes (p. 39): "When therefore the curves $[y = \varphi(x,a)]$ have points in common with $\mathfrak{C}$ for arbitrarily small values of $|a - a_0|$, which accumulate without bound in the neighborhood of at least one point (x_1, y_1), one has for this [point] the equation

$$\varphi_a(x_1,a_0) = 0. \tag{7.38}$$

The accumulation point is different from (x_0, y_0) since in the vicinity of this [point] each curve . . . $[y = \varphi(x,a)]$ has no second point in common with $\mathfrak{C}$, as one can easily see."

Kneser now considers identity (7.37) and writes

$$\varphi_a = g_a + g_b\frac{db}{da};$$

since, moreover, $y_0 = g(x_0,a,b)$, he also has

$$0 = g_a(x_0,a,b) + g_b(x_0,a,b)\frac{db}{da}.$$

It follows from the equation (7.38) that

$$0 = g_a(x_1,a_0,b_0) + g_b(x_1,a_0,b_0)\left(\frac{db}{da}\right)_{a=a_0}.$$

Combining these, Kneser finds that

$$0 = \begin{vmatrix} g_a(x_0,a_0,b_0) & g_b(x_0,a_0,b_0) \\ g_a(x_1,a_0,b_0) & g_b(x_1,a_0,b_0) \end{vmatrix}, \tag{7.39}$$

which enables him to write

$$g_b(x_1, a_0, b_0)\frac{g_a(x_0, a_0, b_0)}{g_b(x_0, a_0, b_0)} = g_a(x_1, a_0, b_0).$$

He also has

$$\varphi_{xa} = g_{xa} + g_{xb}\frac{db}{da}$$

so that

$$\varphi_{xa}g_b(x_0, a, b) = g_{xa}g_b(x_0, a, b) - g_{xb}g_a(x_0, a, b).$$

From this he observes that

$$\varphi_{xa}(x_1, a_0)\, g_b(x_0, a_0, b_0) = \begin{vmatrix} g_{xa}(x_1, a_0, b_0) & g_{xb}(x_1, a_0, b_0) \\ g_a(x_0, a_0, b_0) & g_b(x_0, a_0, b_0) \end{vmatrix}$$

and thus that

$$\varphi_x(x_1, a_0)\, g_b(x_1, a_0, b_0) = \begin{vmatrix} g_{xa}(x_1, a_0, b_0) & g_{xb}(x_1, a_0, b_0) \\ g_a(x_1, a_0, b_0) & g_b(x_1, a_0, b_0) \end{vmatrix}.$$

Now the right-hand member of this equation is different from zero since by assumption the equations $y = g(x, a, b)$, $p = g_x(x, a, b)$ are such that at $x = x_0$

$$\begin{aligned} y_0 = b_0 = g(x_0, a_0, b_0), &\qquad a_0 = g_x(x_0, a_0, b_0) \\ b = g(x_0, a, b), &\qquad a = g_x(x_0, a, b). \end{aligned}$$

At least near to $x = x_0$, $a = a_0$, $b = b_0$ both the functional determinant

$$\begin{vmatrix} g_{xa} & g_{xb} \\ g_a & g_b \end{vmatrix}$$

and g_b are different from zero. Kneser then concludes that $\varphi_{xa}(x_1, a_0) \neq 0$. He also remarks that the right-hand member of (7.39) is the quantity Mayer designated by $\Delta(x_0, x_1)$.[29]

He concludes this discussion by observing that the relations

$$\varphi_a(x_1, a_0) = 0, \qquad \varphi_{xa}(x_1, a_0) \neq 0$$

are precisely the conditions which guarantee that in a neighborhood of the point (x_1, y_1) the curves $y = \varphi(x, a)$ possess an envelope. To see this, he notes that the equation $\varphi_a(\xi, a) = 0$ defines ξ as an analytic function of a near to a_0 such that

$$(\xi - x_1)\varphi_{ax}(x_1, a_0) + (a - a_0)\varphi_{aa}(x_1, a_0) + [\xi - x_1, a - a_0]_2 = 0.$$

[29] Kneser has, in effect, shown that $\Delta(x_0, x)$ is not identically zero and that its first zero x_1 is the point conjugate to x_0 on the arc $\mathfrak{E}$.

He consequently can write

$$\xi - x_1 = -\frac{\varphi_{aa}(x_1, a_0)}{\varphi_{xa}(x_1, a_0)}(a - a_0) + [a - a_0]_2;$$

furthermore, he sets

$$\eta = \varphi(\xi, a) = \varphi(x_1, a_0) + [a - a_0]_1 = y_1 + [a - a_0]_1$$

and notes that the point (ξ, η) traces out a curve, which may in a special case reduce to the point (x_1, y_1). This curve is clearly the envelope.

7.6. Kneser on Focal Points and Transversality

In his *Lehrbuch*, Section 25, pp. 93ff (LV) or Section 19, pp. 116ff (LV'), Kneser takes up a more general form of the envelope theorem with the help of his notion of transversality. He also generalizes Gauss's theorem on geodesics mentioned in Section 7.5 above, as we shall see. Kneser here considers an integral of the form

$$J = \int_{t_0}^{t_1} F(x, y, x', y')\, dt = \int_{x_0}^{x_1} f(x, y, p)\, dx,$$

where

$$F(x, y, x', y') = x' F(x, y, \pm 1, \pm y'/x') = x' f(x, y, p).^{30}$$

He goes on (LV, p. 32) to consider the problem of minimizing the integral J subject to the end-conditions

$$g_a(x_0, y_0, x_1, y_1) = 0 \qquad (a < 3),$$

and says "the most important special case is that in which x_0 and y_0 are fixed but between x_1 and y_1 there is a given relation." In this case the end-conditions become $g(x_1, y_1) = 0$, and he has the so-called transversality condition ($a = 1$ in this case)

$$F_{x'}\,\delta x + F_{y'}\,\delta y|^1 = f - pf_p|^1\,\delta x_1 + f_p|^1\,\delta y_1 = 0,$$

$$\frac{\partial g}{\partial x_1}\,\delta x_1 + \frac{\partial g}{\partial y_1}\,\delta y_1 = 0.$$

He now says that two directions $(\delta x, \delta y)$ and $(dx, p\,dx)$ with $p\,dx = dy$ are *transversal* in case

$$F_{x'}\,\delta x + F_{y'}\,\delta y = (f - pf_p)\,\delta x + f_p\,\delta y = 0.$$

---

[30]Note that F is positively homogeneous, i.e., $F(x, y, \alpha x', \alpha y') = \alpha F(x, y, x', y')$ for $\alpha > 0$ (Kneser, LV, p. 7).

He then states that an extremal 01 joining fixed point 0 with point 1 on the curve $g = 0$ can render the integral J an extremum only when the curve is cut transversally by the extremal at point 1. (Osgood [1901], p. 112, on the other hand, uses a different terminology; he says the extremal is cut transversally by the curve.)

Now consider (*LV*, p. 43) a one-parameter family of extremals

$$x = \xi(t,a), \qquad y = \eta(t,a) \tag{7.40}$$

with $\tau \leqslant t \leqslant T$, $|a - a_0| \leqslant \gamma$, with $x'^2 + y'^2 = \xi_t^2 + \eta_t^2 \neq 0$, and with F regular at every value (x, y, x', y') determined by (7.40); furthermore, suppose that for $a = a_0$ the resulting extremal $\mathfrak{C}$ defined for $\tau \leqslant t \leqslant T$ is regular and does not intersect itself. Thus along an arc $\mathfrak{B}$ of the extremal $\mathfrak{C}$ different values of t correspond to different points, and there is therefore "a uniquely defined system (t, a_0)" corresponding to each point (x, y). Finally, suppose that

$$\Delta = \frac{\partial(\xi,\eta)}{\partial(t,a)} = \xi_t\eta_a - \xi_a\eta_t \neq 0$$

for $\tau \leqslant t \leqslant T$, $|a - a_0| \leqslant \gamma$.

Kneser now says that the region of x, y space determined by the family (7.40) with $\tau \leqslant t \leqslant T$ and $|a - a_0| \leqslant \gamma$ is a field about the arc $\mathfrak{B}$ and that the curves of family (7.40) are the extremals of the field. He considers a regular curve $\mathfrak{C}_0$ inside the field whose equation is $g(x_0, y_0) = 0$. Through each point (x_0, y_0) on it there passes a unique extremal with $x_0 = \xi(t_0, a)$, $y_0 = \eta(t_0, a)$, and $g[\xi(t_0, a), \eta(t_0, a)] = 0$. He then considers the function of t and a defined by

$$u = \int_{t_0}^{t} F(\xi, \eta, \xi_t, \eta_t)\, dt,$$

and notes that it is regular in t and a. In this definition t_0 is the regular function of a discussed above. He finds (note that $F' = dF/dx$)

$$\frac{\partial u}{\partial t} = F(\xi, \eta, \xi_t, \eta_t) = \xi_t F_{x'} + \eta_t F_{y'},$$

$$\begin{aligned}\frac{\partial u}{\partial a} &= -F|^0 \frac{dt_0}{da} + \int_{t_0}^{t} \frac{\partial F(\xi, \eta, \xi_t, \eta_t)}{\partial a}\, dt \\ &= -F|^0 \frac{dt_0}{da} + \int_{t_0}^{t} (F_x\xi_a + F_y\eta_a + F_{x'}\xi_{at} + F_{y'}\eta_{at})\, dt \\ &= -F|^0 \frac{dt_0}{da} + F_{x'}\xi_a + F_{y'}\eta_a|_0^1 + \int_{t_0}^{t} \left[(F_x - F'_{x'})\xi_a + (F_y - F'_{y'})\eta_a\right] dt.\end{aligned}$$

The last integral clearly vanishes since $x = \xi$, $y = \eta$ is an extremal of the field and consequently

$$\frac{\partial u}{\partial a} = -F\Big|^0 \frac{dt_0}{da} + F_{x'}\xi_a + F_{y'}\eta_a\Big|_0^1$$

$$= F_{x'}\xi_a + F_{y'}\eta_a\Big|^1 - F_{x'}\left(\xi_a + \xi_t \frac{dt_0}{da}\right) - F_{y'}\left(\eta_a + \eta_t \frac{dt_0}{da}\right)\Big|^0.$$

To simplify notations Kneser writes

$$Dx = \xi_a + \xi_t \frac{dt_0}{da}\Big|^0 da, \qquad Dy = \eta_a + \eta_t \frac{dt_0}{da}\Big|^0 da,$$

and he has in his notation the relation

$$\frac{\partial u}{\partial a} da = F_{x'}\xi_a + F_{y'}\eta_a|^1 da - F_{x'}Dx - F_{y'}Dy|^0. \tag{7.41}$$

Suppose that $F_{x'}(\xi,\eta,\xi_t,\eta_t), F_{y'}(\xi,\eta,\xi_t,\eta_t)$ do not both vanish. Since in Kneser's fields regular functions of t and a are also regular functions of x and y, the relation at $t = t_0$,

$$F_{x'}(\xi,\eta,\xi_t,\eta_t)Dx + F_{y'}(\xi,\eta,\xi_t,\eta_t)Dy = 0, \tag{7.42}$$

defines "$Dy : Dx$ or the reciprocal value as a regular function of x and y; for that reason one obtains in the considered points a regular arc $\mathfrak{C}_0$ which will be cut transversally by the extremals of the field."[31] Recall, $F_{x'} = F_{y'} = 0$ imply $F = 0$ since $x'F_{x'} + y'F_{y} = F$ by the positive homogeneity of F.

Kneser now assumes that $F(\xi,\eta,\xi_t,\eta_t) \neq 0$ for $a = a_0$ and hence for $|a - a_0|$ sufficiently small. He then restricts his notion of a field so that this is the case throughout the field. He accordingly has through each set (t, a) corresponding to a point of the field a unique curve $\mathfrak{C}_0$ which cuts transversally the extremals of the field. It follows with the help of relation (7.41) that

$$\frac{\partial u}{\partial a} = F_{x'}\xi_a + F_{y'}\eta_a|^1.$$

and consequently that

$$du = F_{x'}(\xi_t\,dt + \xi_a\,da) + F_{y'}(\eta_t\,dt + \eta_a\,da)$$

$$= F_{x'}\,dx + F_{y'}\,dy,$$

where the arguments of $F_{x'}$, $F_{y'}$ are ξ, η, ξ_t, η_t.

Kneser notes (*LV*, p. 48) that in addition to the curves $\mathfrak{C}_0$ through each point of the given extremal arc $\mathfrak{B}$, the equations $u = \text{const}$, i.e.,

$$du = F_{x'}\,dx + F_{y'}\,dy = 0,$$

define another set of curves $\mathfrak{C}_1$ at each point in the field with the same properties as those of the $\mathfrak{C}_0$, i.e., they also cut transversally all extremals of

[31] Kneser, *LV*, p. 46. Note that $\mathfrak{C}_0$ is the curve described by the end-point t_0.

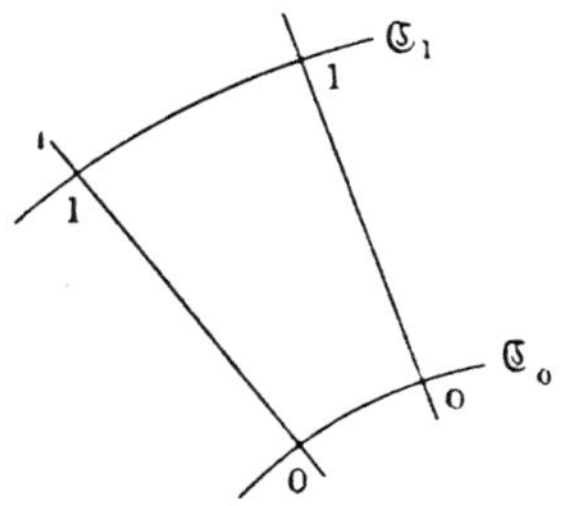

Figure 7.3

the field as in Figure 7.3. (In this figure of Kneser's the straight lines 01 are the extremals of the field.) He concludes that the value of $J, \bar{J}_{01}$, evaluated along the extremals 01 of the field between the transversal curves is a constant.[32] It is also very easy to see the converse proposition. He has thus made an elegant generalization of Gauss's result on geodesics: two transversals to the same set of extremals intercept arcs on these extremals along which the integral J is a constant. (Recall that for geodesics, the notion of transversality is the same as the notion of orthogonality.)

It is worth noticing that the envelope property is entirely analogous to the elegant property of the evolute which is sometimes called the *string property*. In Figure 7.4 we see the evolute D of a curve C—the locus of centers of curvature of curve C is called its *evolute*. It has the property that the end of a taut string wound around the evolute and then unrolled will generate the given curve. At end 3 of the string E the string is orthogonal to C. The length of the arc made up of E_{34} and D_{46} equals that of E_{56} for each position of point 4.

The results of Kneser are easy to demonstrate with the help of a scheme used by Bliss but certainly derived from Kneser (*LV*, p. 96). Bliss supposes that the problem is that of minimizing the integral

$$\int_{x_1}^{x_2} f(x, y, y')\, dx$$

in a class of arcs joining the points (x_1, y_1) and (x_2, y_2) in the plane; consider a variable arc E with end-points 3 and 4, which describe two given

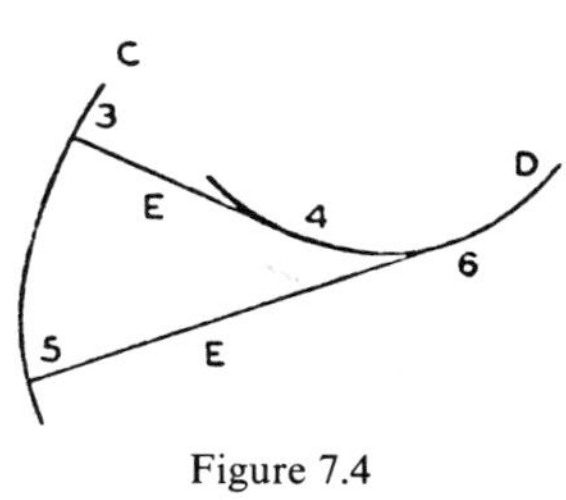

Figure 7.4

[32] Kneser, *LV*, p. 48. This follows at once since the value of $\bar{J}_{01}$ is u, which is a constant.

curves C and D. Let the arcs E be represented by $y = y(x,a)$, and let the curves C and D be represented by

$$x_3(t),\ y_3(t) = y[x_3(t), a(t)],$$
$$x_4(t),\ y_4(t) = y[x_4(t), a(t)], \qquad x_3(t) \leqslant x \leqslant x_4(t).$$

Bliss then differentiates the function of t

$$I(t) = \int_{x_3(t)}^{x_4(t)} f\{x, y[x,a(t)], y'[x,a(t)]\}\, dx,$$

finding

$$I'(t) = f\frac{dx}{dt}\Big|_3^4 + \frac{da}{dt}\int_{x_3}^{x_4}\{f_y y_a + f_{y'} y_a'\}\, dx.$$

Suppose that arc E is an extremal so that $df_{y'}/dx = f_y$; then along E

$$f_y y_a + f_{y'} y_a' = \frac{d(f_{y'} y_a)}{dx}$$

and

$$I'(t) = \left[\frac{f\,dx}{dt} + \frac{da}{dt}\cdot f_{y'} y_a\right]_3^4.$$

It is clear that

$$dy_3 = y'\,dx_3 + y_a\,da, \qquad dy_4 = y'\,dx_4 + y_a\,da$$

and consequently

$$dI = [f\,dx + (dy - y'\,dx) f_{y'}]_3^4.$$

It now follows easily that if the Hilbert invariant integral

$$I^* = \int [f\,dx + (dy - y'\,dx) f_{y'}]$$

is introduced, then (see Figure 7.5)[33]

$$I(E_{78}) - I(E_{56}) = I^*(D_{68}) - I^*(C_{57}).$$

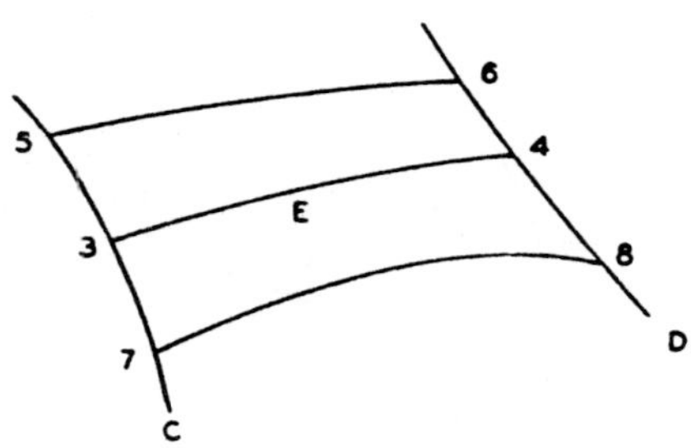

Figure 7.5

[33] Bliss, *LEC*, p. 20.

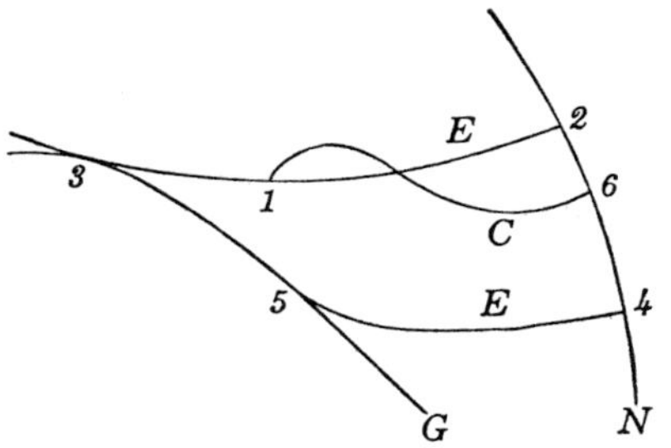

Figure 7.6

If C and D are transversal to the extremals E at their intersections, then clearly the two invariant integrals in the right-hand member of the relation vanish and $I(E_{56}) = I(E_{78})$; i.e., in Kneser's terms the integral I is constant when evaluated on extremals between two transversal curves.

In *LV* pp. 89ff Kneser generalizes another geometrical notion. Suppose that we consider a given plane curve N and the normals to that curve. Mark off on each normal the corresponding center of curvature; these centers generate a new curve, which is, as we saw, the evolute. Suppose now in Figure 7.6 that the point 1 is given as well as the curve N. Let E_{32} be the extremal through 1, which is transversal to N, and let G be the envelope of the family of extremals, such as E_{54}, transversal to N. Then point 3, where E_{32} meets the envelope, is called by Kneser the focal point of N on the extremal.

Kneser defines this point with the help of a one-parameter family $x = \xi(t, a)$, $y = \eta(t, a)$ of extremals transversal to the given curve, which he calls $\mathfrak{C}_0$. Let the given extremal $\mathfrak{C}$ be defined by $a = a_0$ and the point where $\mathfrak{C}_0$ meets $\mathfrak{C}$ by 0 as in Figure 7.7. Then $x = \xi(t, a_0)$, $y = \eta(t, a_0)$ is the extremal through the point 0 transversal to the curve $\mathfrak{C}_0$. The first point 6 at which

$$\Delta = \begin{vmatrix} \xi_t & \xi_a \\ \eta_t & \eta_a \end{vmatrix}$$

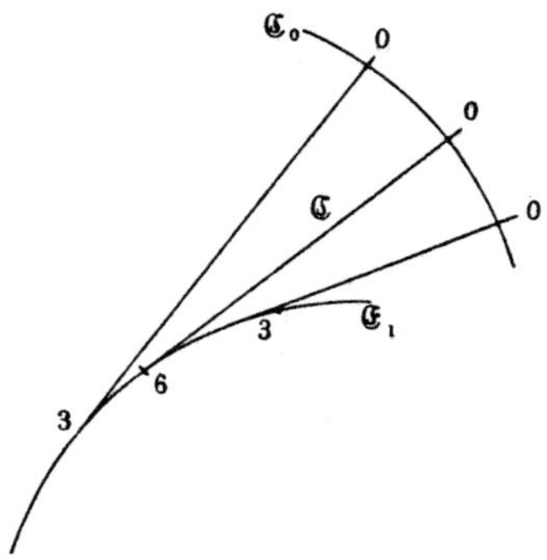

Figure 7.7

vanishes after 0 is called by him the *focal point* of the curve $\mathfrak{C}_0$ on the extremal $\mathfrak{C}$. He notes that when $\mathfrak{C}_0$ degenerates into a point the notions of focal and conjugate points coincide. With the help of the quantity

$$\omega = \frac{\Delta \cdot \delta a}{\sqrt{\xi_t^2 + \eta_t^2}},$$

Kneser shows that Δ, or rather ω, satisfies a linear differential equation of the second order in the form

$$\omega'' + M\omega' + N\omega = 0$$

along the arc 06 provided that[34]

$$F_1 = \frac{F_{x'x'}}{y'^2} = -\frac{F_{x'y'}}{x'y'} = \frac{F_{y'y'}}{x'^2} \neq 0. \tag{7.43}$$

Kneser proceeds in Section 25 to show how the envelope theorem holds for extremals transversal to the given curve. It is then not difficult to show that for an extremum for the problem of one fixed end-point 0 and one movable end-point 1 it is necessary that the focal point of the given curve $\mathfrak{C}_0$ on the extremal $\mathfrak{C}$ shall not lie between the given point 0 and the point 1 where $\mathfrak{C}$ meets $\mathfrak{C}_0$. In the case where both end-points are variable, the comparable condition is given by Bliss [1904], who uses the term *critical* point instead of focal point.[35] He also gives a sufficiency proof there.

Bliss introduced the notations I, II, III below for the following necessary conditions for a nonparametric problem

$$J = \int_{x_1}^{x_2} F(x, y, y')\,dx$$

with F regular, $F > 0$, $F_{y'y'} > 0$ in a region R of the x, y-plane, and y' arbitrary (Hilbert calls such problems *regular*); he seeks a minimum among curves in R joining two fixed and nonintersecting curves D and E.

> I. *C must be an extremal, in other words, a solution of the differential equation*
>
> $$F_y - \frac{d}{dx} F_{y'} = 0.$$
>
> II. *C must cut both D and E transversally, i.e., in its intersection points* 1 *and* 2 *with the curves D and E the condition must hold,*
>
> $$F + (\bar{y}' - y')F_{y'} = 0$$
>
> *where y, y' refer always to the curve C, and $\bar{y}$, $\bar{y}'$ to D or E.*

---

[34] It is not difficult to show that Δ satisfies Jacobi's differential equation, which may be written in the form

$$F_2 u - \frac{d}{dt}\left(F_1 \frac{du}{dt}\right) = 0,$$

where F_1 is given above and F_2 is a related function of t (see Bolza, *VOR*, pp. 224ff).

[35] Bliss [1904], p. 74.

If C is a curve joining the two given curves D and E, and satisfying the necessary conditions I *and* II, *then a third necessary condition for a minimum of the integral J is*

III. *The critical point d of D must not lie between the point* 1 . . . *and the critical point e of E.*[36]

Bliss then proceeds to prove the theorem

When the necessary conditions I, II *upon the curve C are satisfied, then a sufficient condition for a minimum of the integral J is,*

III$'$. *The critical point e of the curve E lies between E and the critical point d of D.*

In Section 16 of *LV* (pp. 49ff) or 22 of *LV*$'$ (pp. 142ff), Kneser introduced curvilinear coordinates in an elegant way. He again considers the function of t and a

$$u = \int_{t_0}^{t} F(\xi, \eta, \xi_t, \eta_t)\, dt,$$

where, as before $x = \xi(t,a)$, $y = \eta(t,a)$ are the extremals of a field. Since by hypothesis the functional determinant $F = \partial u/\partial t = \partial(u,a)/\partial(t,a) \neq 0$, Kneser has

$$\frac{\partial(x,y)}{\partial(u,a)} = \frac{\partial(\xi,\eta)}{\partial(t,a)} \cdot \frac{\partial(t,a)}{\partial(u,a)} = \Delta : \frac{\partial u}{\partial t} = \frac{\Delta}{F(\xi,\eta,\xi_t,\eta_t)} \,.$$

He now introduces *normal coordinates* of the field, $u, v = a$, and he has along the extremals of the field $v' = 0$ and along the transversals to the field, $u' = 0$.

It is not difficult to verify that the extremals $v =$ const. are cut transversally by the curves $u =$ const. Kneser sets

$$F(x, y, dx, dy) = G(u, v, du, dv) = g(u, v, s)\, du, \quad \frac{dv}{du} = s;$$

he then notes that

$$\begin{aligned} x' &= x_u u' + x_v v', & y' &= y_u u' + y_v v', \\ u' &= u_x x' + u_y y', & v' &= v_x x' + v_y y'. \end{aligned}$$

Along the extremals of the field $v = a =$ const, $s = 0$. It is easy to see that

$$F_{x'} Dx + F_{y'} Dy = G_{u'} Du + G_{v'} Dv,$$

since $Dx = x_u Du + x_v Dv$, $Dy = y_u Du + y_v Dv$ and

$$\frac{\partial u}{\partial u} = u_x x_u + u_y y_u = 1, \qquad \frac{\partial v}{\partial u} = v_x x_u + v_y y_u = 0.$$

(Note this shows that the Weierstrass E-function is an invariant.) Recall

[36] Bliss [1904], pp. 71, 72, and 74.

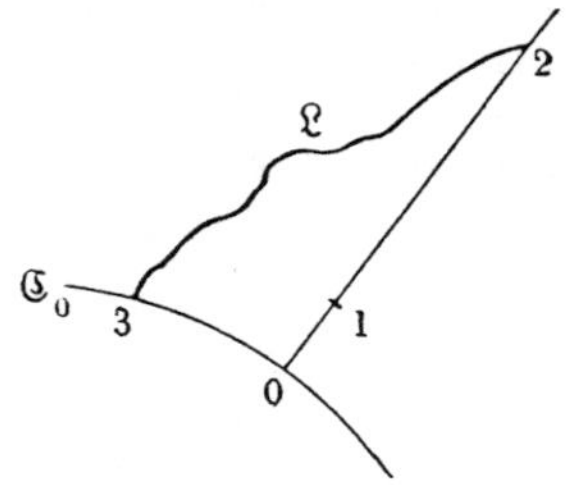

Figure 7.8

that $F_{x'}Dx + F_{y'}Dy = 0$ by (7.42) when Dx, Dy are calculated along the transversals, $u = \text{const}$. From this it is clear that

$$G_{u'}Du + G_{v'}Dv = (g - sg_s)Du + g_s Dv = 0;$$

in this the derivatives $G_{u'}, G_{v'}$ are understood to be evaluated along the extremals of the field, $v = \text{const}$. But for $s = 0$, $Du = 0$, and the last equation implies that $g_s(u, v, 0) = 0$. Kneser then has for a Taylor's expansion of g the following

$$g(u, v, s) = 1 + \frac{s^2}{2}\, g_{ss}(u, v, \theta s), \qquad |\theta| < 1. \tag{7.44}$$

Now he supposes (Figure 7.8) that his field contains the extremal $\mathfrak{C}$ through the points 0, 1, 2 with 0 the point where $\mathfrak{C}$ cuts the given curve $\mathfrak{C}_0$ transversally—he seeks to minimize the integral J among arcs joining $\mathfrak{C}_0$ to 2. He also draws an arbitrary curve $x = x(\tau)$, $y = y(\tau)$ through the point 2 cutting $\mathfrak{C}_0$ in the point 3. Then

$$J_{32} = \int_{\tau_3}^{\tau_2} F\left(x, y, \frac{dx}{d\tau}, \frac{dy}{d\tau}\right) d\tau = \int_{\tau_3}^{\tau_2} G\left(u, v, \frac{du}{d\tau}, \frac{dv}{d\tau}\right) d\tau.$$

Moreover, since $u = \text{const}$ along a transversal,

$$\int_{\tau_2}^{\tau_3} \frac{du}{d\tau}\, d\tau = u\Big|_3^2 = u\Big|_0^2$$

and consequently as long as $du/d\tau > 0$,

$$J_{32} - \bar{J}_{02} = \int_3^2 \left[G\left(u, v, \frac{du}{d\tau}, \frac{dv}{d\tau}\right) - \frac{du}{d\tau} \right] d\tau,$$

where $\bar{J}_{02}$ is the value of J along the extremal through 0 and 2.

Kneser now needs to calculate the size of G. To do this, he makes use of expansion (7.44) and can express the integrand appearing in the right-hand member of the relation above as [37]

$$\frac{s^2}{2}\, g_{ss}(u, v, \theta s)\, \frac{du}{d\tau}\,.$$

[37] Kneser, *LV*, pp. 51–52 and *LV′*, pp. 142–144.

To evaluate g_{ss} he is able to show that

$$G_{v'v'} = F_1(x_v y' - y_v x')^2,$$

where F_1 is defined by relations (7.43); it then follows that

$$\frac{g_{ss}(u,v,s)}{u'} = F_1(x_v y' - y_v x')^2 = F_1(x_v y_u - x_u y_v)^2 u'^2.$$

(Note that the expression $x_u y_v - x_v y_u \neq 0$.) From this Kneser can now infer that $J_{32} - \bar{J}_{02} > 0$ provided that $\Delta \neq 0$, $F \neq 0$, and $F_1(x, y, \cos\gamma, \sin\gamma) > 0$ along the extremal $\mathfrak{C}$ for every value of γ.

In the same issue of the *Transactions of the American Mathematical Society*, Bolza [1901] and Osgood [1901] give proofs of a suffiency theorem for a strong minimum using the Weierstrass E-function. Bolza's proof is very simple and depends on recognizing the fact that the E-function is invariant under the transformation $x = X(u,v)$, $y = Y(u,v)$. He calculates without difficulty that

$$E'(u,v;u',v';\bar{u}',\bar{v}') = E(x,y;x',y';\bar{x}',\bar{y}').$$

He supposes that

$$\Delta = \frac{\partial(\xi,\eta)}{\partial(t,a)} \neq 0 \quad \text{for } a = a_0 \qquad (t_0 - \epsilon, t_1 + \epsilon)$$

and that

$$F_1(\xi(t,a_0), \eta(t,a_0), \cos\gamma, \sin\gamma) > 0$$

for every γ. He can infer from this that in a sufficiently small neighborhood R_κ in t,a-space $t_0 \leqslant t \leqslant t_1$, $|a - a_0| \leqslant \kappa$ and hence in a neighborhood S_κ of x, y-space he has $\Delta \neq 0$, $F_1(x, y, \cos\gamma, \sin\gamma) > 0$, and a one-to-one correspondence between R_κ and S_κ.

Bolza can then embed the given extremal C in the field S_κ. He considers any other curve $\bar{C}$ in the field with continuously turning tangent and writes

$$\Delta I = I_{\bar{C}}(AB) - I_C(AB) = \int_{\tau_0}^{\tau_1} E(x,y,x',y';\bar{x}',\bar{y}')\,d\tau,$$

where x', y' are the values of the slope functions at the point x, y of the field and $\bar{x}', \bar{y}'$ are the values of x', y' for $\bar{C}$. [In his 1882 lectures Weierstrass argues that $\Delta I > 0$ since

$$E(x,y;\cos\theta,\sin\theta;\cos\bar{\theta},\sin\bar{\theta}) = (1-\cos\omega)F_1(x,y,\cos\theta^*,\sin\theta^*),$$

where $\omega = \bar{\theta} - \theta$, $|\omega| \leqslant \pi$ and θ^* lies between θ and $\bar{\theta}$.]

Osgood showed that for $0 < h < \kappa$ there is an $\epsilon_h > 0$ such that for every curve C joining the points A and B and interior to S_κ but not wholly contained in S_h, $\Delta I > \epsilon_h$. Bolza proves this by using Kneser's curvilinear

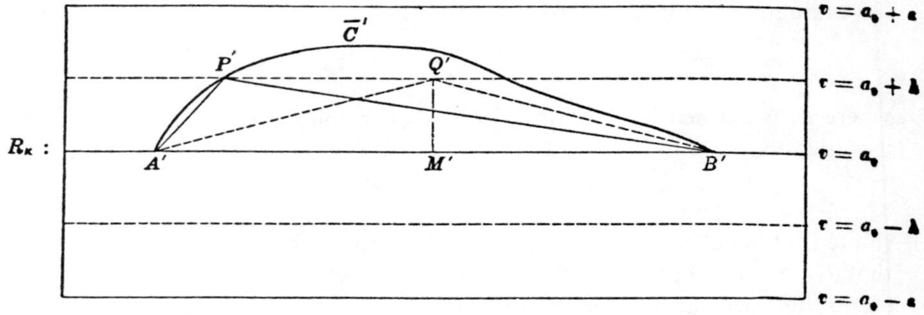

Figure 7.9

coordinates u, v with $u = t(x, y)$, $v = a(x, y)$. He accordingly has[38]

$$\Delta I = \int_{s_0}^{s_1} E'(u, v; \cos 0, \sin 0; \cos \omega, \sin \omega)\, ds$$

$$= \int_{s_0}^{s_1} (1 - \cos \omega) G_1(u, v, \cos \theta^*, \sin \theta^*)\, ds.$$

In these expressions the integrals are taken along the image $\overline{C}'$ in the u, v plane of curve $\overline{C}$; ω is the angle between the positive tangent at u, v to $\overline{C}'$ and the positive u-axis; and s is arc-length measured along $\overline{C}'$. It results that there is a constant m such that for every u, v in R_κ and every λ, $D^2 F_1 = G_1(u, v, \cos\lambda, \sin\lambda) \geqslant m > 0$; as a consequence,

$$\Delta I \geqslant m \int_{s_0}^{s_1} (1 - \cos \omega)\, ds = m(l - (t_1 - t_0)),$$

where l is the length of the curve $\overline{C}'$ and $D = \partial(X, Y)/\partial(u, v)$.

To finish the proof, Bolza now supposes that $\overline{C}$ is not wholly within S_h and therefore passes through some point P of one of the extremals defined by $a = a_0 \pm h$. Then $\overline{C}'$ passes through a point P' whose ordinate is $v = a_0 \pm h$ as in Figure 7.9. Then l is not less than the length of the broken line $A'P'B'$. He now chooses Q' on the same line $v =$ const. as P' so that $A'Q' = B'Q'$ and has $A'P'B' \geqslant A'Q'B'$. It now follows that

$$\Delta I \geqslant 2m \left[\sqrt{h^2 + \left(\frac{t_1 - t_0}{2} \right)^2} - \frac{(t_1 - t_0)}{2} \right],$$

which completes Bolza's proof.

To prove Osgood's theorem for the case of one end-point variable Bolza chooses for u, v Kneser's curvilinear coordinates and, by "a slight modifica-

[38] Bolza [1901], p. 425. He shows there that $G_1 = D^2 F_1$, where D is the functional determinant of the equations $x = X(u, v)$, $y = Y(u, v)$.

tion of the above reasoning," finds that

$$\Delta I \geqslant m\left[\sqrt{h^2 + (u_1 - u_0)^2} - (u_1 - u_0)\right],$$

provided that it is assumed that $F[\xi, \eta, \xi_t, \eta_t] > 0$ in the region.[39]

7.7. Bliss's Work on Problems in Three-Space

In Bliss and Mason [1908], [1910] there is an interesting extension of the works of Kneser and Bolza from problems in the plane to those in space. To make precise what they do consider the parametric problem of minimizing an integral

$$J = \int_{t_0}^{t_1} F(x, y, z, y', z')\, dt$$

among certain curves in a region of xyz-space

$$x = x(t), \qquad y = y(t), \qquad z = z(t) \qquad (t_0 \leqslant t \leqslant t_1)$$

with

$$F(x, y, z, \kappa y', \kappa z') = \kappa F(x, y, z, y', z') \qquad (\kappa > 0).$$

They allow the end-points to be fixed or one of them to vary on a given curve or surface. We need not repeat their derivations of the Euler or of the transversality condition. They are straightforward, but the Jacobi condition introduces new complications.

As is usual Mason and Bliss define F_1 by means of the relation

$$\begin{vmatrix} F_{x'x'} & F_{x'y'} & F_{x'z'} & x' \\ F_{y'x'} & F_{y'y'} & F_{y'z'} & y' \\ F_{z'x'} & F_{z'y'} & F_{z'z'} & z' \\ x' & y' & z' & 0 \end{vmatrix} = -F_1(x'^2 + y'^2 + z'^2)^2,$$

and they then choose arc-length as their parameter. If $F_1 \neq 0$ for all values of its arguments, there is one and only one extremal through each point x_0, y_0, z_0 in a given direction x'_0, y'_0, z'_0. These extremals are representable in the form

$$x = \varphi(s, x_0, y_0, z_0, x'_0, y'_0, z'_0),$$
$$y = \psi(s, x_0, y_0, z_0, x'_0, y'_0, z'_0),$$
$$z = \chi(s, x_0, y_0, z_0, x'_0, y'_0, z'_0),$$

---

[39] For the details, see Bolza, *LEC*, p. 191. In fact, the discussion there and on pp. 187–199 is very good and complete.

where $x'^2 + y'^2 + z'^2 = 1$ and where

$$\begin{aligned} \varphi(0,x_0,y_0,z_0,x_0',y_0',z_0') &= x_0, & \varphi_s(0,x_0,y_0,z_0,x_0',y_0',z_0') &= x_0', \\ \psi(0,x_0,y_0,z_0,x_0',y_0',z_0') &= y_0, & \psi_s(0,x_0,y_0,z_0,x_0',y_0',z_0') &= y_0', \\ \chi(0,x_0,y_0,z_0,x_0',y_0',z_0') &= z_0, & \chi_s(0,x_0,y_0,z_0,x_0',y_0',z_0') &= z_0'. \end{aligned}$$

The functions $\varphi, \psi, \chi, \varphi_s, \psi_s, \chi_s$ are of class C' for all values of their arguments, which define points in the fundamental region R of definition.

Mason and Bliss first show that the determinant

$$U = \begin{vmatrix} \varphi_s & \varphi_{x'_0} & \varphi_{y'_0} & \varphi_{z'_0} \\ \psi_s & \psi_{x'_0} & \psi_{y'_0} & \psi_{z'_0} \\ \chi_s & \chi_{x'_0} & \chi_{y'_0} & \chi_{z'_0} \\ 0 & x_0' & y_0' & z_0' \end{vmatrix}$$

is continuous even for $s = 0$ since U/s^2 approaches 1 as s approaches zero. Then if U/s^2 is different from zero along an extremal arc C_{01} joining points 0 and 1, it will remain so for a point 2 near to 0 and in the order 201 where the arguments $(x_2, y_2, z_2, x_2', y_2', z_2')$, which correspond to 2, are sufficiently near to those $(x_0, y_0, z_0, x_0', y_0', z_0')$ which correspond to 0. They now restrict the generality of their family of extremals and consider a two-parameter family of the form

$$x = \varphi(s,u,v), \qquad y = \psi(s,u,v), \qquad z = \chi(s,u,v) \tag{7.45}$$

and the associated determinant

$$\Delta(s,u,v) = \begin{vmatrix} \varphi_s & \varphi_u & \varphi_v \\ \psi_s & \psi_u & \psi_v \\ \chi_s & \chi_u & \chi_v \end{vmatrix}.$$

They then show without difficulty that "*if a surface S is transversal to an extremal C_{01} at the point* 0, *then C_{01} can be imbedded in a family of extremals* . . . [7.45)] *to each of which S is transversal. The determinant* $\Delta(s,u,v)$ *of the family does not vanish in the neighborhood of the point* 0."[40]

They now suppose that Δ vanishes for some point 2 different from 0 on the extremal C_{01} and that at least one of the determinants of order 3 of the matrix

$$\begin{vmatrix} \Delta_s & \Delta_u & \Delta_v \\ \varphi_s & \varphi_u & \varphi_v \\ \psi_s & \psi_u & \psi_v \\ \chi_s & \chi_u & \chi_v \end{vmatrix}$$

[40] Bliss [1908], pp. 448–449.

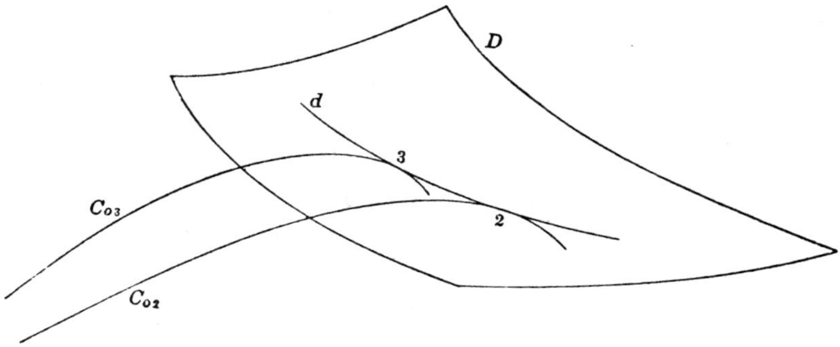

Figure 7.10

does not vanish with Δ—they suppose for definiteness that it is

$$\begin{vmatrix} \Delta_s & \Delta_u & \Delta_v \\ \varphi_s & \varphi_u & \varphi_v \\ \psi_s & \psi_u & \psi_v \end{vmatrix}$$

—then the equations

$$\Delta(s,u,v) = 0, \qquad \varphi(s,u,v) = x, \qquad \psi(s,u,v) = y$$

are solvable for s, u, v as functions of x, y

$$s = s(x, y), \qquad u = u(x, y), \qquad v = v(x, y)$$

near to x_2, y_2. If these values are substituted into $z = \chi(x,u,v)$, a surface $z = z(x, y)$ is found as in their figure (Figure 7.10). This can be shown to be the envelope of the two-parameter family (7.45).

They next consider the envelope theorem and the Jacobi condition. By a direct calculation they show the following envelope theorem[41]:

> *Suppose that an extremal C_{01} is contained in a one-parameter family of extremals, each of which passes through a fixed point 0 and which have an enveloping curve touching C_{01} at a point 2 [see Figure 7.10]. Then if 3 is a neighboring point to 2 on d, the value of J taken along C_{03} plus the value of J taken along d_{32} is always equal to the value of J taken along C_{02}. In other words*
>
> $$J(C_{03}) + J(d_{32}) = J(C_{02}).$$
>
> *When the extremals are all cut transversally by a surface S, the same equation is true if 0 is understood to denote the variable intersection of C_{03} with S.*

They show without difficulty that the envelope d cannot satisfy Euler's equation, and thus that points 2 and 3 can be joined by a curve d'_{32} such

[41] Bliss [1908], p. 453.

that

$$J(C_{03}) + J(d'_{32}) > J(C_{02}).$$

They further state: "*If therefore an extremal C_{01} minimizes or maximizes the integral J, it must not have upon it a point 2 which is conjugate to the initial point* 0."

After this Mason and Bliss establish the Legendre and Weierstrass conditions without complication and then state and prove sufficient conditions for strong and weak minima.

Hahn [1911] undertook to finish the investigation begun by Mason and Bliss on problems with variable end-point problems. In the course of his analysis Hahn proved a theorem of basic importance for establishing sufficient conditions. This result has been called by Bliss the *theorem of Hahn*. Hahn considers the Lagrange problem of making

$$\int f(x, y_1, \ldots, y_n, y'_1, \ldots, y'_n)\, dx$$

a minimum subject to the side-conditions

$$\varphi_k(x, y_1, \ldots, y_n, y'_1, \ldots, y'_n) = 0 \qquad (k = 1, 2, \ldots, m).$$

He assumes that he is dealing with a normal arc so that there are multipliers $\lambda_1, \ldots, \lambda_m$ such that

$$F_{y_i} - \frac{d}{dx} F_{y'_i} = 0 \qquad (i = 1, 2, \ldots, n),$$

where

$$F = f + \sum_{k=1}^{m} \lambda_k \varphi_k,$$

and the determinant

$$\begin{vmatrix} F_{y'_i y'_j} & \varphi_{k y'_i} \\ \varphi_{l y'_j} & 0 \end{vmatrix} \qquad \begin{matrix} (i, j = 1, 2, \ldots, n) \\ (k, l = 1, 2, \ldots, m) \end{matrix}$$

does not vanish along the arc.

The Euler–Lagrange equations above can be solved for y, λ as functions of x and of $2n$ integration constants. He expresses their solutions with the help of the variables

$$w_i = F_{y'_i}(x, y_1, \ldots, y_n, y'_1, \ldots, y'_n, \lambda_1, \ldots, \lambda_m) \qquad (i = 1, 2, \ldots, n)$$

in the form

$$y_i = y_i(x, x_0, y_1^0, \ldots, y_n^0, w_1^0, \ldots, w_n^0),$$

$$\lambda_k = \lambda_k(x, x_0, y_1^0, \ldots, y_n^0, w_1^0, \ldots, w_n^0),$$

where y_i^0, w_i^0 are the values of y_i, w_i for $x = x_0$. He writes this extremal as (x_0, y^0, w^0). In what follows he considers a given one of these extremals (x_{00}, y^{00}, w^{00}) which he writes as $y_i = y_i(x)$, $\lambda_i = \lambda_i(x)$, with

$$w_i(x) = F_{y'_i}(x, y_1(x), \ldots, y_n(x), y'_1(x), \ldots, y'_n(x), \lambda_1(x), \ldots, \lambda_m(x)).$$

He assumes two conditions hold: first, that the Weierstrass E-function is such that

$$E(x, y_1, \ldots, y_n, y'_1, \ldots, y'_n, \lambda_1, \ldots, \lambda_m; \tilde{y}'_1, \ldots, \tilde{y}'_n) \geqslant 0$$

for

$$A \leqslant x \leqslant B, \qquad |y_i - y_i(x)| \leqslant \rho, \qquad |y'_i - y'_i(x)| \leqslant \rho', \qquad |\lambda_k - \lambda_k(x)| \leqslant \rho''$$

and all values of $\tilde{y}'_1, \ldots, \tilde{y}'_n$. (The equality can hold only for $\tilde{y}'_i = y'_i$ $(i = 1, 2, \ldots, n)$.); second, that if $A < a < b < B$, the arc of the extremal (x_{00}, y^{00}, w^{00}) on $a \leqslant x \leqslant b$ has on it no point conjugate to a.

In terms of these hypotheses, Hahn's theorem (p. 129) is this:

There are three positive numbers δ, ϵ, η such that for

$$|x_0 - x_{00}| \leqslant \delta, |y_i^0 - y_i^{00}| \leqslant \delta, |w_i^0 - w_i^{00}| \leqslant \delta \qquad (i = 1, 2, \ldots, n)$$

the arc of the extremal (x_0, y^0, w^0) on $a - \epsilon \leqslant x \leqslant b + \epsilon$ furnishes a minimum relative to every admissible comparison curve which has the same initial- and end-points and remains entirely in the η-neighborhood of the extremal (x_{00}, y^{00}, w^{00}) on $(a - \epsilon, b + \epsilon)$.

(By an "η-neighborhood of an arc $y_i(x)$ on $\alpha \leqslant x \leqslant \beta$," Hahn means the set of all points $x, y_1, \ldots, y_n$ such that $\alpha \leqslant x \leqslant \beta$, $|y_1 - y_i(x)| \leqslant \eta$ and by Δ, the Mayer determinant (6.71).)

He shows directly that $|\Delta(x, x_0^*)| \neq 0$ on $a \leqslant x \leqslant a + \xi$, where x_0^* is some point on $a - \xi < x_0^* < a$ and consequently that it has a positive lower bound on that interval. Also $\Delta(x, a) \neq 0$ for $a < x \leqslant b$ since the given extremal has on it no point conjugate to a, and so $|\Delta(x, a)|$ has a positive lower bound on $a + \xi \leqslant x \leqslant b$. Notice that $\Delta(x, x_0)$ is a continuous function of x_0; by means of this fact Hahn infers that $|\Delta(x, x_0^*)| \geqslant$ some positive bound for $a + \xi \leqslant x \leqslant b$ and that x_0^* can, in fact, be so chosen that $|\Delta(x, x_0^*)|$ also has this property on $a \leqslant x \leqslant b$.[42]

Hahn now argues that the determinant

$$|y_{iw_j^0}(x, x_0^*, y_1^0, \ldots, y_n^0, w_1^0, \ldots, w_n^0)| \qquad (i, j = 1, 2, \ldots, n)$$

is a continuous function of x, y^0, w^0 and for $a \leqslant x \leqslant b$, $y^0 = y^{0*}$, $w^0 = w^{0*}$ has a positive lower bound since for these values it is $\Delta(x, x_0^*)$. He concludes that it remains different from zero for

$$a - H \leqslant x \leqslant a + H; \quad |y^0 - y^{0*}| \leqslant H, \quad |w^0 - w^{0*}| \leqslant H.$$

He can then use an implicit-function theorem to solve the equations

$$y_i - y_i(x, x_0^*, y_1^0, \ldots, y_n^0, w_1^0, \ldots, w_n^0) = 0 \qquad (i = 1, 2, \ldots, n).$$

[42] Perhaps the easiest way to see Hahn's result here is to notice that if an arc on $a \leqslant x \leqslant b$ has on it no point conjugate to a, there is always a point a' to the left of a such that the arc defined on this later interval has on it no point conjugate to a' (see Bliss, *LEC*, p. 30). With the help of this remark, one can always find a family of conjugate solutions whose determinant is nonzero on $a \leqslant x \leqslant b$ and hence, by continuity, has a nonzero lower bound.

The solutions of these equations

$$w_i^0 = w_i^0(x, y_1, \dots, y_n, y_1^0, \dots, y_n^0) \qquad (i = 1, 2, \dots, n)$$

serve Hahn to define a Mayer field which embeds the given extremal and in which the E-function is positive. Hahn's theorem then follows directly. Hahn uses that result to prove a general sufficiency theorem:

> *Let the proposed variational problem be to make the integral . . . a minimum subject to the side-conditions . . . among all sufficiently near arcs whose initial-and end-point coordinates satisfy any conditions whatsoever. Let the arc of the extremal*
>
> $$y_i = y_i(x), \qquad \lambda_k = \lambda_k(x), \qquad w_i = w_i(x) \qquad a \leqslant x \leqslant b$$
>
> *(where $A < a < b < B$) furnish a weak minimum to our variational problem. If the E-function is positive definite in the region characterized by the inequalities . . . [$A \leqslant x \leqslant B$, $|y_i - y_i(x)| \leqslant \rho$, $|y_i' - y_i'(x)| \leqslant \rho'$, $|\lambda_k - \lambda_k(x)| \leqslant \rho''$], this extremal arc also provides a strong minimum for our variational problem.*

Hahn's proof is straightforward but tedious. Bliss has given a somewhat more general formulation of the result than did Hahn (Bliss, *LEC*, pp. 160–161). But aside from this, Hahn's result made it possible to see how sufficient conditions for general problems of the calculus of variations with variable end-points could be established.

7.8. Boundary-Value Methods

Hilbert introduced into the calculus of variations a number of fundamental ideas, not the least of which was his realization that there is an intimate and fundamental relation between minimum problems and boundary-value problems.[43]

The Jacobi condition may be viewed as a necessary condition that the second variation $\delta^2 J$ be nonnegative, and when slightly strengthened will ensure that $\delta^2 J$ is actually always positive. One approach to this is the geometrical one which is related to the envelope of a family of extremals. Another one—that due to Hilbert—is an analytic one. The second variation may be viewed as a real-valued quadratic functional which possesses an infinity of eigenvalues and corresponding eigenfunctions. It can also be shown that this functional can be expanded in terms of these functions—much like a Fourier expansion—with the eigenvalues as coefficients. Thus a necessary condition is that there be no negative eigenvalue, and a sufficient condition is that all be positive.

[43] Hilbert [1912], pp. 56–66. (This reference is to Hilbert's famous *Grundzüge*.)

To develop this idea, Hilbert motivated three of his American graduate students—W. Cairns, Mason, and Richardson—to study this way of viewing the Jacobi condition.

Mason considers first the problem of minimizing the integral

$$J(y) = \int_a^b \left(\frac{dy}{dx} \right)^2 dx \tag{7.46}$$

subject to the conditions that $y(x)$ is twice differentiable on the interval (a, b) and that $y(a) = y(b) = 0$,

$$K(y) = \int_a^b Ay^2\, dx = 1, \tag{7.47}$$

where A is a positive and continuous function on (a, b).[44] He then relates this isoperimetric problem to the boundary-value problem of solving the equation

$$L(y) \equiv \frac{d^2y}{dx^2} + \lambda A(x) y = 0, \qquad A(x) > 0, \tag{7.48}$$

subject to the boundary conditions

$$y(a) = 0, \qquad y(b) = 0. \tag{7.48$'$}$$

To do this, he makes use of Fredholm's methods in the theory of integral equations as well as Burkhardt's and Bôcher's ideas of a Green's function. (It was Burkhardt who first constructed such a function for $d^2y/dx^2 = 0$ in 1894 and Bôcher who gave a general definition in 1901; see Mason [1904], p. 530n.)

Mason shows without real difficulty that the lower bound λ_0 of all values of J "is the smallest parameter value [eigenvalue] for the boundary-value problem" just posed. Let us look briefly at his proof. Since λ_0 is a lower bound, there exists a sequence of functions $y_1, y_2, y_3, \ldots$ so that

$$\lim_{h=\infty} J(y_h) = \lambda_0, \qquad K(y_h) = 1, \qquad y_h(a) = 0, \qquad y_h(b) = 0.$$

With the help of the y_h he defines a sequence of continuous functions f_h by means of the relations

$$\frac{d^2y_h}{dx^2} + \lambda_0 A y_h = f_h. \tag{7.49}$$

Now multiply each of these by y_h and integrate from a to b. This gives Mason the new equations

$$\lambda_0 - J(y_h) = \int_a^b y_h f_h\, dx,$$

from which it follows that

$$\lim_{h=\infty} \int_a^b y_h f_h\, dx = 0.$$

[44] Mason [1904], pp. 528–544; Cairns [1907]; and Richardson [1910], [1911].

Mason now selects a sequence of twice differentiable functions v_h vanishing at $x = a$ and $x = b$ and defines a corresponding sequence g_h so that

$$\frac{d^2 v_h}{dx^2} + \lambda_0 A v_h = g_h. \tag{7.49'}$$

He now gives a somewhat lengthy proof that if the v_h are such that

$$\left| \int_a^b v_h g_h \, dx \right| < B \qquad (h = 1, 2, 3, \ldots,) \tag{7.50}$$

from some B, independent of h, then

$$\lim_{h=\infty} \int_a^b y_h g_h \, dx = 0.$$ [45]

He makes an indirect proof. Suppose that infinite subsequences y_h', v_h' could be picked out of y_h, v_h so that either

$$\int_a^b y_h' g_h' \, dx \geqslant \rho \quad \text{or} \leqslant \rho$$

for all h, where in the first case $\rho > 0$ and in the second case $\rho < 0$; and ρ is, in both cases, independent of h. [The superscript $'$ does not here signify the derivative; f_h', g_h' are defined with the help of (7.49) and (7.49′).] He then could form $\eta_h = y_h' + c v_h'$ with c a constant, and write

$$\frac{d^2 \eta_h}{dx^2} + \lambda_0 A \eta_h = f_h' + c g_h'.$$

From these relations it would follow, as before, that

$$\lambda_0 K(\eta_h) - J(\eta_h) = \int_a^b y_h' f_h' \, dx + c^2 \int_a^b v_h' g_h' \, dx + c \int_a^b (v_h' f_h' + y_h' g_h') \, dx.$$

He also remarks that

$$\int_a^b (v_h' f_h' - y_h' g_h') \, dx = \int_a^b \left\{ v_h' \left(\frac{d^2 y_h'}{dx^2} + \lambda_0 A y_h' \right) - y_h' \left(\frac{d^2 v_h'}{dx^2} + \lambda_0 A v_h' \right) \right\} dx = 0,$$

and consequently that

$$\lambda_0 K(\eta_n) - J(\eta_h) = \int_a^b y_h' f_h' \, dx + c^2 \int_a^b v_h' g_h' \, dx + 2c \int_a^b y_h' g_h' \, dx.$$

The constant c could then be chosen so that $\rho c > 0$ and $2\rho c > c^2 B$; and Mason could define δ^2 to be $2c\rho - c^2 B$. Then

$$\lambda_0 K(\eta_h) - J(\eta_h) \geqslant \int_a^b y_h' f_h \, dx + \delta^2.$$

---

[45] Mason [1904], pp. 534–536. Repeated indices here do not mean summations.

Since the integral in the right-hand members of these relations has limit zero and δ^2 is independent of h, he could find a value $h = H$ so that

$$\lambda_0 K(\eta_H) - J(\eta_H) > 0.$$

It is clear from this that η_H would not be identically zero, and so he could set $\alpha = \sqrt{K(\eta_H)}$. Then the function

$$y(x) = \frac{1}{\alpha}\eta_H(x)$$

would be twice differentiable, vanish at a and b, and $K(y) = 1$. As a consequence, he would have $\lambda_0 > J(y)$, which is a contradiction of the definition of λ_0, and thus

$$\lim_{h=\infty} \int_a^b y_h g_h \, dx = 0.$$

Mason now proceeds to finish his proof by another indirect proof. He assumes that λ_0 is not an eigenvalue and notes that then the equations

$$\frac{d^2 v_h}{dx^2} + \lambda_0 A v_h = A y_h \qquad (h = 1, 2, 3, \ldots),$$

$$v_h(a) = 0, \qquad v_h(b) = 0$$

could be solved for the v_h. It remains for him to demonstrate that these v_h satisfy the conditions (7.50), which he does by an appeal to Fredholm's theory of integral equations. The v_h would satisfy an integral equation of the form

$$v_h(\xi) = F(\xi) + \lambda_0 \int_a^b G(x,\xi) A(x) v_h(x) \, dx,$$

where F has the value

$$F(\xi) = -\int_a^b G(x,\xi) A(x) y_h(x) \, dx,$$

and G is Green's function for $d^2y/dx^2 = 0$ on $[a,b]$:

$$G(x,\xi) = \frac{(b-\xi)(x-a)}{b-a} \qquad x \leqslant \xi$$

$$= \frac{(b-x)(\xi-a)}{b-a} \qquad x \geqslant \xi.$$

He can then conclude that

$$v_h(x) = -\int_a^b G(\xi, x) A(\xi) y_h(\xi) \, d\xi + \int_a^b \int_a^b \frac{D\binom{x}{y}}{D} G(\xi, y) A(\xi) y_h(\xi) \, d\xi \, dy,$$

where $D\binom{x}{y}/D$ is a finite, continuous function introduced by Fredholm. If Δ is the maximum of the absolute value of this function on $[a,b]$, then he would have

$$\left|\int_a^b v_h A y_h dx\right| < m\int_a^b \int_a^b |A(x)y_h(x)| \cdot |A(\xi)y_h(\xi)|\,dx\,d\xi,$$

where $m = (a+b)(1+\Delta(b-a))/2$ since $|G(x,\xi)| \leqslant (a+b)/2$. The argument is then completed by use of Schwarz's inequality, from which it would follow that

$$\left\{\int_a^b |A(x)y_h(x)|\,dx\right\}^2 < (b-a)\int_a^b A^2 y_h^2\,dx$$

and hence that

$$\left|\int_a^b v_h A y_h\,dx\right| < (b-a)m\int_a^b A y_h^2\,dx = m(b-a)$$

for every h. This quantity $m(b-a)$ in the right-hand member of the inequality above could now be taken as the constant B in (7.50). As a consequence,

$$\lim_{h=\infty}\int_a^b y_h A y_h\,dx = \lim_{h=\infty} K(y_h) = 0,$$

which is contradiction of the fact that $K(y_h) = 1$ for every h. The lower bound λ_0 is then an eigenvalue, and there is a $y_0 \neq 0$ such that

$$\frac{d^2 y_0}{dx^2} + \lambda_0 A y_0 = 0,$$

$$y_0(a) = 0, \qquad y_0(b) = 0, \qquad K(y_0) = 1.$$

Mason remarks that for this y_0 he has $J(y_0) = \lambda_0$, and y_0 is the solution of the variational problem, and λ_0 is the smallest eigenvalue. He then goes on to show the existence of a second eigenvalue and eigensolution by considering the new variational problem

$$J(y) = \int_a^b \left(\frac{dy}{dx}\right)^2 dx,$$

$y(a) = 0 = y(b)$, $K(y) = 1$, $K(y_0, y) = \int_a^b A y_0 y\,dx = 0$.[46] By an argument similar to his previous one he finds the lower limit λ_1 of J under these conditions on y and observes that "*if* $\lambda_1 = \lambda_0$ *there would exist for this value of* λ *a second solution of our boundary problem linearly independent of* y_0." Fredholm's theory requires in this case that $D\binom{\xi}{\eta}$ must vanish. Mason goes to show that this is impossible and concludes that $\lambda_1 > \lambda_0$. He summarizes his results in the following theorem:

> *There exists an infinite sequence of increasing eigenvalues and their corresponding solutions* y_i *of the differential equation* $d^2y/dx^2 + \lambda A(x)y = 0$ *under the boundary conditions* $y(a) = 0$, $y(b) = 0$: *The function* y_i *is the*

[46] Mason [1904], pp. 537–539.

solution of the variational problem $J(y) =$ *minimum under the conditions* $y(a) = 0, y(b) = 0, K(y) = 1, K(y_j, y) = 0$ $(j = 0, 1, 2, \dots, i - 1)$ *and gives the integral* J *the value* λ_i.

(He also briefly takes up the case in which A is periodic with period $b - a$.)

In his last section Mason generalizes his results to the case of a system

$$L_1(y,z) = \frac{d^2y}{dx^2} + \lambda(A_{11}(x)y + A_{12}(x)z) = f_1(x),$$

$$L_2(y,z) = \frac{d^2z}{dx^2} + \lambda(A_{21}(x)y + A_{22}(x)z) = f_2(x),$$

subject to the boundary conditions

$$y(a) = c_1, \qquad y(b) = c_2, \qquad z(a) = d_1, \qquad z(b) = d_2.$$

In this case his variational problem is concerned with the integral

$$J(y,z) = \int_a^b \left\{ \left(\frac{dy}{dx} \right)^2 + \left(\frac{dz}{dx} \right)^2 \right\} dx,$$

and his results are what one would expect.

Richardson [1910], following Hilbert, considers a somewhat more general problem than did Mason. He asks that the integral

$$D(u) = \int_0^1 \left\{ p(x) \left(\frac{du}{dx} \right)^2 - q(x)u^2(x) \right\} dx$$

be a minimum subject to the end-conditions $u(0) = 0$, $u(1) = 0$, and the isoperimetric condition

$$\int_0^1 k(x)u^2(x)\,dx = 1.$$

In this formulation it is to be understood that p is analytic and positive on $[0, 1]$, that q is analytic and nonpositive, and that k is analytic on the open interval $(0, 1)$. From this it is clear that $D \geqslant 0$ for all u and that the Euler–Lagrange equation for the problem is

$$L_1(u) = \frac{d(pu')}{dx} + qu + \lambda ku = 0. \tag{7.51}$$

He knows from Hilbert's results that when $k \geqslant 0$ on $[0, 1]$, there are an infinite number of positive eigenvalues $\lambda_1, \lambda_2, \dots$ and corresponding eigenfunctions $U_1, U_2, \dots$.[47] Let $K(x, \xi)$ be $k(x) \cdot k(\xi)$ times the Green's function for the self-adjoint differential expression

$$L(u) = \frac{d(pu')}{dx} + qu.$$

---

[47] Richardson notes (p. 281) that when $k \leqslant 0$, the isoperimetric condition needs to be changed to $\int_0^1 ku^2\,dx = -1$. He also notes that if k changes sign, there are infinitely many positive and infinitely many negative eigenvalues.

Then

$$k(x)u(x) = \lambda \int_0^1 K(x,\xi)u(\xi)\,d\xi.$$

His first theorem is the following:

If $u(x)$ is a continuous function satisfying the conditions

$$u(0) = u(1) = 0, \qquad \int_0^1 k(x)u^2\,dx = 1,$$

$$\int_0^1 kU_1\,dx = 0, \qquad \int_0^1 kU_2\,dx = 0, \ldots, \qquad \int_0^1 kU_{n-1}\,dx = 0,$$

the integral

$$D(u) = \int_0^1 (pu'^2 - qu^2)\,dx$$

has a smallest value λ_n; this value will be assumed for $u(x) = \pm U_n(x)$.

Richardson's proof is straightforward and need not be repeated here.

In his fourth section he takes up the Jacobi condition. He remarks that $U_1(x) = u(x) = \alpha_1 u_1(x,\lambda_1)$ is the solution of the variational problem with the quadratic isoperimetric condition $\int_0^1 k(x)u^2\,dx = 1$. Since this minimum has been shown to exist, the Jacobi condition is certainly satisfied for the function U_1. He now goes on to show that on the interval $[0,1]$ this function U_1 never oscillates. For definiteness, he supposes that $(du_1/dx)_{x=0} = \lambda > 0$. [This is no restriction since the only solution of (7.51) with $u(0) = u'(0) = 0$ is $u = 0$.]

He goes on (p. 288) to say that the Jacobi condition means geometrically that

> through each point in a given neighborhood of our extremal can be drawn one and only one extremal of the two-parameter family passing through the initial point:
>
> $$\begin{cases} u = \alpha u_1(x,\lambda), \\ v_1 = \alpha^2 \int_0^x k(x)u_1^2\,dx. \end{cases} \tag{7.52}$$
>
> This requires that for no value of $x > 0$ on the interval 0, 1 can the two corresponding equations of the family ... [(7.52)] formed by the variation of the parameters λ and α:
>
> $$u_1(x,\lambda_1)\,\delta\alpha + \alpha_1 \frac{\partial u(x,\lambda_1)}{\partial\lambda}\,\delta\lambda = 0,$$
>
> $$\left[\alpha_1 \int_0^x k(x)u_1^2\,dx\right]\delta\alpha + \left[\alpha_1^2 \int_0^x k(x)u_1 \frac{\partial u_1}{\partial\lambda}\,dx\right]\delta\lambda = 0, \tag{7.53}$$
>
> be equally satisfied, where λ_1 and α_1 signify the values of the parameters for the curve $u = U_1(x)$, $v_1 = \int_0^x kU_1^2\,dx$ and $\delta\alpha, \delta\lambda$ are constants lying under certain bounds.

If there is a point $X > 0$ where these equations (7.53) are both satisfied, then the point X is conjugate to the initial point $x = 0$ and the determinant

$$D_1(x,\lambda) = \begin{vmatrix} u_1(x,\lambda) & \dfrac{\partial u_1(x,\lambda)}{\partial \lambda} \\ \displaystyle\int_0^x k(x)u_1^2(x,\lambda)\,dx & \displaystyle\int_0^x k(x)u_1(x,\lambda)\frac{\partial u_1(x,\lambda)}{\partial \lambda}\,dx \end{vmatrix} \tag{7.54}$$

vanishes at $x = X$. He continues, "The Jacobi criterium requires that the first zero (except for the point $x = 0$) of the determinant $D_1(x,\lambda_1)$ not lie in the interval 0, 1. From this we can conclude that the function $u_1(x)$ and hence the function $U_1(x) = \alpha_1 u_1(x)$ do not oscillate in the interval." He summarizes his comparison results in the theorems (pp. 239–290):

Theorem 2. *If a is a zero of the solution $u(x)$ of the differential equation*

$$(pu'(x))' + qu(x) + \lambda k(x)u(x) = 0 \qquad (p(x) > 0,\ q(x) \leqslant 0,\ \lambda > 0)$$

and $a_1 > a$ is a second zero of $u(x)$ or a zero of $u'(x)$, then

$$\int_a^{a_1} k(x)u^2(x)\,dx > 0.$$

The proof is very simple. Multiply both members of the differential equation above by u and integrate from a to a_1; there results

$$\begin{aligned} \lambda\int_a^{a_1} ku^2\,dx &= -\int_a^{a_1}\{(pu')'u + qu^2\}\,dx \\ &= -\left[puu'\right]_a^{a_1} + \int_a^{a_1}\left[pu'^2 - qu^2\right]dx > 0 \end{aligned}$$

since $pu'^2 - qu^2 > 0$. He further proposes:

Theorem 3. *Let $u(x,\lambda)$, $u^*(x,\lambda^*)$ be two solutions of the differential equation*

$$(pu')' + qu + \lambda k(x)u = 0 \qquad (p(x) > 0,\ q(x) \leqslant 0),$$

which satisfy the initial conditions $u(0) = 0$, $u^*(0) = 0$ and correspond to the parameter values λ, λ^* respectively; if $\lambda^* > \lambda > 0$, then the second, third, . . . zero of u^* precedes the second, third, . . . zero of u.

Richardson says that for his purposes it suffices to prove this last result when $\lambda^* = \lambda + \epsilon$ with $\epsilon > 0$ arbitrarily small. He considers the equations

$$\begin{gathered} (pu')' + qu + \lambda ku = 0, \\ (pu^{*\prime})' + qu^* + [\lambda + \epsilon]ku^* = 0, \end{gathered}$$

multiplies both members of the first by u^* and of the second by u, subtracts, and integrates from 0 to a zero α of u. This gives

$$p(\alpha)u'(\alpha)u^*(\alpha) = \epsilon\int_0^\alpha kuu^*\,dx.$$

He then argues that the right-hand member of this can be replaced by

$$\epsilon\left[\int_0^\alpha ku^2\,dx + \epsilon'\right],$$

"where ϵ' is an infinitesimal quantity of the same order as ϵ." [48] But by Richardson's Theorem 2 above, the integral in this last expression is positive and hence $u'(\alpha)u^*(\alpha) > 0$. If now $\alpha = a_1$ is the first zero of u, then $u'(a_1) < 0$ and consequently $u^*(a_1) < 0$. But he assumes that at $x = 0$ his solutions have $u = 0$ and $u' > 0$, and so $u^*(x) > 0$ near to $x = 0$. This implies at once that there is a zero of u^* between $x = 0$ and $x = a_1$. If $\alpha = a_2$ is the second zero of u, then $u'(a_2) > 0$ and consequently $u^*(a_2) > 0$. Since $u^*(a_1) < 0$, there must be a second root of u^* between a_1 and a_2. The remainder of the argument is similar.

Richardson now arrays the zeros $a_1 > 0, a_2, a_3, \ldots$ of u_1 on the positive real axis. He has $u_1'(a_1) < 0$, $u_1'(a_2) > 0$, $u_1'(a_3) < 0, \ldots$. Moreover, for $x = a_1$, he finds that the function D_1 of (7.54) is expressible as

$$D_1(a_1,\lambda) = -\left(\frac{\partial u_1}{\partial \lambda}\right)_{x=a_1}\cdot\int_0^{a_1} ku_1^2\,dx,$$

since u_1 vanishes at $x = a_1$. In the right-hand member of this equation the first factor is negative, as he concludes from Theorem 3 by letting λ^* approach λ; and the second factor is positive by Theorem 2. This implies that $D_1(a_1,\lambda) > 0$. By a similar argument he finds $D_1(a_2,\lambda) < 0$, $D_1(a_3,\lambda) > 0, \ldots$. Thus "between a_1 and a_2, between a_2 and a_3, etc. the continuous function $D_1(x,\lambda)$ must have zeros. Since however $D_1(x,\lambda)$ has no zero in the interval $0, 1, a_1$ must $= 1$, and the function U_1 cannot oscillate in the interval." [49]

His fourth section concludes with the result:

Theorem 4. *The zeros of the functions $u_1(x,\lambda)$, $D_1(x,\lambda)$, which vanish at the initial point, separate one another; and the first one of $u_1(x,\lambda)$ precedes the first of $D_1(x,\lambda)$.*

In his fifth section Richardson considers the Jacobi condition for the case of two conditions

$$\int_0^1 k(x)u^2\,dx = 1, \qquad \int_0^1 k(x)U_1u\,dx = 0.$$

In this case he shows "that the Jacobi condition requires the minimal function U_2 be a one-time oscillating function on the interval 0, 1." Then in his sixth section he takes up the general case. He finds[50]:

[48] Richardson [1910], p. 290.
[49] Richardson [1910], p. 291.
[50] Richardson [1910], p. 300.

Principal Theorem. *The Jacobi condition of the calculus of variations signifies that the solution* $u(x) = U_1(x)$ *of the minimum problem . . .*

$$\int_0^1 (pu'^2 - qu^2)\,dx = \min, \quad [p(x) > 0, q(x) \leqslant 0, u(0) = u(1) = 0]$$

with the quadratic side-condition $\int_0^1 k(x)u^2\,dx = 1$ *does not oscillate in the interval* 0, 1; *the solution* $u(x) = U_2(x)$ *of this problem with the quadratic and a linear side-condition* $\int_0^1 U_1(x)u(x)\,dx = 0$ *oscillates once in the interval; and that, in general, the solution* $u(x) = U_{n+1}(x)$ *of the problem with the quadratic and the n linear side-conditions*

$$\int_0^1 k(x)U_1(x)u(x)\,dx = 0, \qquad \int_0^1 k(x)U_2(x)u(x)\,dx = 0, \ldots,$$

$$\int_0^1 k(x)U_n(x)u(x)\,dx = 0$$

oscillates exactly n times in the interval.

His paper closes with expansions of the determinants $D_1, D_2, \ldots$.

In his second paper on this same topic Richardson [1912] generalizes somewhat the problem he treats and finds quite analogous results. The problem he considers is to minimize $D(u)$ subject to the same end-conditions as before, but with the side-conditions

$$\int_0^1 l(x)u^2(x)\,dx = \pm 1, \qquad \int_0^1 r(x)u^2(x)\,dx = 0.$$

This leads to the Euler–Lagrange equation

$$(pu')' + qu + (\lambda l + \mu r)u = 0,$$

where λ and μ are now the Lagrange multipliers. He then proceeds to still more general sorts of problems.

The elegant work of Morse on the calculus of variations in the large was influenced in some small measure by these ideas. Morse was one of the first to appreciate the deep significance of comparison, separation, and oscillation theorems. Much more importantly, he was also the first to see how to bring the tools of topology to bear on the classical calculus of variations in order to develop a macro-analysis. By this means Morse changed the entire subject into a very deep and elegant branch of modern mathematics. The work of Hilbert on existence theory, discussed in the next section, is the first example of the calculus of variations in the large. The ideas of Morse were carried out by Morse himself and a vigorous school that helped develop his ideas.

7.9. Hilbert's Existence Theorem

It was believed by many that at least for nonnegative functionals J, there was always an extremal which rendered the integral an absolute minimum. It was first shown by Weierstrass *VOR* in Volume II of his *Collected Works*

(p. 49) that this is not necessarily true. Consider the problem of minimizing the integral

$$J = \int_{-1}^{1} x^2 y'^2 \, dx$$

among all suitable curves of class C' joining the point $(-1, a)$ to $(+1, b)$ when $a \neq b$. It can be shown that the lower bound of the integral is zero. To see this note that $J \geqslant 0$ and that the curves

$$y = \frac{a+b}{2} + \frac{b-a}{2} \frac{\arctan x/\epsilon}{\arctan 1/\epsilon}$$

pass through the given end-points and for them

$$J < \int_{-1}^{+1} (x^2 + \epsilon^2) y'^2 \, dx = \frac{\epsilon (b-a)^2}{2 \arctan 1/\epsilon}.$$

(This last expression goes to zero with ϵ.) There is however no curve with y' continuous which gives the integral the value zero since this would require the integrand $x^2 y'^2$ to vanish and this would mean that $y' = 0$; but y = const. cannot satisfy the end-conditions.[51]

Hilbert [1899] investigated this question of whether a given integral of the calculus of variations actually attains its lower bound calculated over the class of admissible arcs. He indicated how it is possible to restrict the integrand F or the class of admissible arcs so that there is an *a priori* assurance of the existence of a solution. Bolza[52] notes that Hilbert "illustrates the gist of his method by the example of the shortest line upon a surface and by Dirichlet's problem. In a subsequent course of lectures (Göttingen, summer, 1900) he gave the details of his method for the shortest line on a surface and some indications . . . concerning its extension to the problem of minimizing the integral

$$J = \int_{x_0}^{x_1} F(x, y, y') \, dx."$$

Shortly after Hilbert's original result was published, H. Lebesgue [1902], Carathéodory [1906], and Hadamard [1906] made substantial simplifications and generalizations of Hilbert's results, and since then other authors have dealt with this and similar problems. Lebesgue's paper is concerned both with the single integral $\int F \, dx = \int f(x, y)(x'^2 + y'^2)^{1/2} \, dx$ and the double integral $\int\int (\mathrm{EG} - \mathrm{F}^2)^{1/2} \, du \, dv$.

However, let us consider only Hilbert's original paper [1900], where he

[51] Bolza, *VOR*, pp. 419–420.

[52] Bolza, *LEC*, pp. 245–246. The details of Hilbert's method are contained in Noble's thesis, "Eine neue Methode in der Variationsrechnung," *Dissertation*, Göttingen, 1901. This presentation contains a number of inadequacies, and it is better to read Bolza's proof.

says (p. 186):

> How this principle can serve as a pole star for discovering rigorous and simple existence proofs is shown in the following two examples:
> "I. *To draw on a given surface $z = f(x, y)$ the shortest curve joining two given points P and P'.*"

To do this, he calls l the lower bound of the lengths of all curves on the surface, joining the points given above, and chooses a sequence of these curves $C_1, C_2, C_3, \ldots$ whose lengths $L_1, L_2, L_3, \ldots$ approach the bound l. He then marks off on C_1 from the first point P the length $L_1/2$ and calls that end-point $P_1^{(1/2)}$; similarly, he marks off from P on C_2 the length $L_2/2$ and finds $P_2^{(1/2)}$, etc. The points $P_1^{(1/2)}, P_2^{(1/2)}, P_3^{(1/2)}, \ldots$ have an accumulation point $P^{(1/2)}$, which is again on the surface $z = f(x, y)$. By extension he can then produce $P^{(1/4)}, P^{(3/4)}$ as well as $P^{(1/8)}, P^{(3/8)}, P^{(5/8)}, P^{(7/8)}; P^{(1/16)}, \ldots$. Hilbert concludes that "all these points and their accumulation points form on the surface $z = f(x, y)$ a continuous curve, which is the desired shortest line."[53]

Hilbert remarks that the proof for this statement is easy to make if one views the length of a curve as the limit of the lengths of inscribed polygons. He then states his next problem:

> II. *To find a potential function $z = f(x, y)$, which takes on preassigned boundary values along a given boundary curve.*

We will not pursue this further. The subject of existence theorems has been carried forward steadily by Carathéodory, Tonelli, McShane, and others.

7.10. Bolza and the Problem of Bolza

In a very elegant paper, Bolza [1913] first formulated what Bliss in 1932 was to call the *problem of Bolza*. Bolza says in his 1913 paper that in Bolza [1907] he has generalized Hilbert's method for the simplest case of the Mayer problem (see Hilbert [1906]). He continues in [1913] by remarking on Hadamard's presentation of the Mayer problem.

---

[53] Hilbert [1900], p. 186. Bolza, (*LEC*, pp. 253ff) gives a detailed proof of Hilbert's assertions. To carry out the analysis for a more general integrand, Hilbert introduced in his lectures a generalized integral. This appears in Bolza *LEC* on p. 247n; it is of interest to note in Section 31, pp. 156ff how Weierstrass and later Osgood also introduced generalized integrals for curves having no tangents.

The problem Bolza proposes to treat is one which is sufficiently general to include both the Lagrange and Mayer problems as special cases.[54] It is this: Given the functional

$$U = \int_{t_0}^{t_1} f(y_1, \ldots, y_n, y_1', \ldots, y_n')\, dt + G(y_{10}, \ldots, y_{n0}; y_{11}, \ldots, y_{n1})$$

to find in a class of arcs satisfying p differential and q finite equations

$$\varphi_\alpha(y_1, \ldots, y_n, y_1', \ldots, y_n') = 0 \qquad (\alpha = 1, 2, \ldots, p), \quad (7.55)$$

$$\psi_\beta(y_1, \ldots, y_n) = 0 \qquad (\beta = 1, 2, \ldots, q), \quad (7.56)$$

as well as r equations on the end-points

$$\chi_\gamma(y_{10}, \ldots, y_{n0}; y_{11}, \ldots, y_{n1}) = 0 \qquad (\gamma = 1, 2, \ldots, r) \quad (7.57)$$

one which renders U a minimum.[55] He further assumes that the determinant

$$\begin{vmatrix} \dfrac{\partial \varphi_\alpha}{\partial y_1'}, & \dfrac{\partial \varphi_\alpha}{\partial y_2'}, & \ldots, & \dfrac{\partial \varphi_\alpha}{\partial y_m'} \\ \dfrac{\partial \psi_\beta}{\partial y_1}, & \dfrac{\partial \psi_\beta}{\partial y_2}, & \ldots, & \dfrac{\partial \psi_\beta}{\partial y_m} \end{vmatrix} \neq 0 \qquad \begin{array}{l} (\alpha = 1, 2, \ldots, p; \beta = 1, 2, \ldots, q; \\ m = p + q) \end{array}$$

along the given arc $\mathfrak{E}_0$.

Bolza now says that he needs to reduce the end-conditions to an independent subset in the sense to be explained below. To do this, he supposes the finite equations (7.56) to be decomposable into two sets

$$\overset{0}{\psi}_\beta \equiv \psi_\beta(y_{10}, \ldots, y_{n0}) = 0, \qquad \overset{1}{\psi}_\beta \equiv \psi_\beta(y_{11}, \ldots, y_{n1}) = 0$$
$$(\beta = 1, 2, \ldots, q), \qquad (7.56')$$

and that the given arc is representable as

$$\mathfrak{E}_0 : y_i = \overset{0}{y}_i(t) \quad t_0 \leqslant t \leqslant t_1 \qquad (i = 1, 2, \ldots, n),$$

whose end-coordinates $(\overset{0}{y}_{10}, \ldots, \overset{0}{y}_{n0}), (\overset{0}{y}_{11}, \ldots, \overset{0}{y}_{n1})$ satisfy (7.56′) and (7.57). He then says that an admissible arc must satisfy the end-conditions (7.56′) and (7.57), and he writes them in the generic form

$$\omega_g(y_{10}, \ldots, y_{n0}; y_{11}, \ldots, y_{n1}) = 0 \qquad (g = 1, 2, \ldots, 2q + r).$$

He further assumes that in a neighborhood of the place

$$A_0 : \left(\overset{0}{y}_{10}, \ldots, \overset{0}{y}_{n0}; \overset{0}{y}_{11}, \ldots, \overset{0}{y}_{n1} \right)$$

---

[54] Bliss later showed that this new problem which he called the problem of Bolza is actually equivalent to the other two problems. (Bliss, *LEC*, pp. 189ff).

[55] Although he does not make the point explicitly, the functions f and φ_α are positively homogeneous of dimension 1 in the $y_1', \ldots, y_n'$.

there is an integer R, so that all determinants of order $R+1$ of the matrix

$$\left\| \frac{\partial \omega_g}{\partial y_{10}}, \ldots, \frac{\partial \omega_g}{\partial y_{n0}}; \frac{\partial \omega_g}{\partial y_{11}}, \ldots, \frac{\partial \omega_g}{\partial y_{n1}} \right\|$$

vanish identically while at least one of order R does not vanish identically in that neighborhood. This implies that $R \leqslant 2q + r$ and $R \leqslant 2n$. He explicitly does not permit A_0 to be a place where all determinants of order R vanish by supposing that at least one determinant of order R is different from zero at A_0. In the contrary case, Bolza says that the end-points are *singular*; whereas if $R = 2q + r$, he says that the end-conditions are *reduced*. If $R < 2q + r$, some of the $\omega_1, \ldots, \omega_{2q+r}$ can be expressed as functions of the others. In this case the relations

$$\omega_{R+1} = \mathfrak{F}_{R+l}(\omega_1, \ldots, \omega_R), \ldots, \omega_{2q+r} = \mathfrak{F}_{2q+r}(\omega_1, \ldots, \omega_R)$$

must hold identically at each place $(y_{10}, \ldots, y_{n0}; y_{11}, \ldots, y_{n1})$. In particular, when this place is A_0, then

$$\mathfrak{F}_{R+1}(0, \ldots, 0) = 0, \ldots, \mathfrak{F}_{2q+r}(0, \ldots, 0) = 0.$$

It follows "that every place $(y_{10}, \ldots, y_{n0}; y_{11}, \ldots, y_{n1})$ in some given neighborhood of A_0, which satisfies the R equations

$$\omega_1 = 0, \ldots, \omega_R = 0, \tag{7.58}$$

must also satisfy the remaining equations

$$\omega_{R+1} = 0, \ldots, \omega_{2q+r} = 0."$$

Bolza proceeds to say that "For our purposes it is now important . . . one can arrange that among the equations . . . [(7.58)] exactly the $2q$ equations . . . [(7.56′)] appear." He knows then that at least one determinant of order $2q$ of the matrix

$$\left\| \begin{matrix} \frac{\partial \overset{0}{\psi}_\beta}{\partial y_{10}} & \cdots & \frac{\partial \overset{0}{\psi}_\beta}{\partial y_{n0}} & 0 & \cdots & 0 \\ 0 & \cdots & 0 & \frac{\partial \overset{1}{\psi}_\beta}{\partial y_{11}} & \cdots & \frac{\partial \overset{1}{\psi}_\beta}{\partial y_{n1}} \end{matrix} \right\|$$

is different from zero at the place A_0 and $R \geqslant 2q$. He also concludes that at least one determinant of order $2q$ of the matrix

$$\left\| \begin{matrix} \frac{\partial \overset{0}{\psi}_\beta}{\partial \omega_1} & \cdots & \frac{\partial \overset{0}{\psi}_\beta}{\partial \omega_R} \\ \frac{\partial \overset{1}{\psi}_\beta}{\partial \omega_1} & \cdots & \frac{\partial \overset{1}{\psi}_\beta}{\partial \omega_R} \end{matrix} \right\|$$

at the place $\omega_1 = 0, \ldots, \omega_R = 0$ is different from zero, where the $2q$

functions $\overset{0}{\psi}_\beta, \overset{1}{\psi}_\beta$ have been expressed in terms of the independent functions $\omega_1, \ldots, \omega_R$. He summarizes:

Theorem I. If we exclude the case of singular end-points, we are permitted to assume without restriction of generality that the system ... [(7.57), (7.56′)] *of initial conditions is reduced, i.e., that* $2q + r \leqslant 2n$, *and that at least one determinant of order* $(2q + r)$ *of the matrix*

$$\left\| \begin{matrix} \dfrac{\partial \overset{0}{\psi}_\beta}{\partial y_{10}} & \cdots & \dfrac{\partial \overset{0}{\psi}_\beta}{\partial y_{n0}}, & 0 & \cdots & 0 \\ 0 & \cdots & 0 & \dfrac{\partial \overset{1}{\psi}_\beta}{\partial y_{11}} & \cdots & \dfrac{\partial \overset{1}{\psi}_\beta}{\partial y_{n1}} \\ \dfrac{\partial \chi_\gamma}{\partial y_{10}} & \cdots & \dfrac{\partial \chi_\gamma}{\partial y_{n0}} & \dfrac{\partial \chi_\gamma}{\partial y_{11}} & \cdots & \dfrac{\partial \chi_\gamma}{\partial y_{n1}} \end{matrix} \right\| \qquad \begin{matrix} \beta = 1, 2, \ldots, q; \\ \gamma = 1, 2, \ldots, r \end{matrix}$$

is different from zero at the place $A_0(\overset{0}{y}_{10}, \ldots, \overset{0}{y}_{n0}; \overset{0}{y}_{11}, \ldots, \overset{0}{y}_{n1})$.

There is then no system of $2q + r$ *multipliers* $c_1, c_2, \ldots, c_{2q+r}$, *not all zero, for which at the place* A_0

$$\begin{aligned} &\sum_\beta c_\beta \frac{\partial \overset{0}{\psi}_\beta}{\partial y_{i0}} + \sum_\gamma c_{2q+\gamma} \frac{\partial \chi_\gamma}{\partial y_{i0}} = 0, \\ &\sum_\beta c_{q+\beta} \frac{\partial \overset{1}{\psi}_\beta}{\partial y_{i1}} + \sum_\gamma c_{2q+\gamma} \frac{\partial \chi_\gamma}{\partial y_{i1}} = 0, \end{aligned} \qquad (i = 1, 2, \ldots, n). \tag{7.59}$$

From here on Bolza assumes that the end-points are not singular and that the initial conditions are reduced.[56]

In his second section (p. 435) Bolza takes up the problem of finding the Euler–Lagrange equations for this case. To do this, he first establishes an embedding theorem which shows in effect that the given admissible arc $\mathfrak{E}_0$ is not an isolated point.

Lemma. Let $q + r + 1$ *systems of functions*

$$\overset{\sigma}{\eta}_1(t), \overset{\sigma}{\eta}_2(t), \ldots, \overset{\sigma}{\eta}_n(t) \qquad \sigma = 1, 2, \ldots, q + r + 1$$

be given, which are twice continuously differentiable on the interval $[t_0, t_1]$ *and satisfy the* $m = p + q$ *differential equations*

$$\sum_i \left(\frac{\partial \varphi_k}{\partial y_i} \overset{\sigma}{\eta}_i + \frac{\partial \varphi_k}{\partial y_i'} \overset{\sigma}{\eta}{}_i' \right) = 0 \qquad k = 1, 2, \ldots, m \tag{7.60}$$

[56]Notice that Bolza assumes in this theorem that the finite conditions are of the form (7.56′).

where . . .

$$\varphi_{p+\beta} = \sum_i \frac{\partial \psi_\beta}{\partial y_i} y_i';^{57} \tag{7.60'}$$

one can always construct a $(q + r + 1)$*-parameter family of curves*

$$y_i = \mathfrak{Y}_i(t, \epsilon_1, \epsilon_2, \ldots, \epsilon_{q+r+1}), \tag{7.61}$$

which possess the necessary continuity properties, satisfy the m differential equations

$$\varphi_k(\mathfrak{Y}_1, \ldots, \mathfrak{Y}_n, \mathfrak{Y}_1', \ldots, \mathfrak{Y}_n') = 0, \qquad k = 1, 2, \ldots, m$$

and the equations

$$\mathfrak{Y}_i(t, 0, 0, \ldots, 0) = \overset{0}{y}_i(t), \qquad (i = 1, 2, \ldots, n)$$

and for which

$$\left(\frac{\partial \mathfrak{Y}_i}{\partial \epsilon_\sigma}\right)_0 = \overset{\sigma}{\eta}_i(t),$$

where the index 0 *signifies the substitution* $\epsilon_1 = 0, \epsilon_2 = 0, \ldots, \epsilon_{q+r+1} = 0.$

Bolza's proof is straightforward. He sets

$$\mathfrak{Y}_{m+\rho}(t, \epsilon_1, \ldots, \epsilon_{q+r+1}) = \overset{0}{y}_{m+\rho}(t) + \sum_\sigma \epsilon_\sigma \overset{\sigma}{\eta}_{m+\rho}(t)$$
$$(\rho = 1, 2, \ldots, n - m),$$

and solves the differential equations

$$\varphi_k(y_1, \ldots, y_m, \mathfrak{Y}_{m+1}, \ldots, \mathfrak{Y}_n, y_1', \ldots, y_m', \mathfrak{Y}_{m+1}', \ldots, \mathfrak{Y}_n') = 0$$
$$(k = 1, 2, \ldots, m)$$

for the quantities $y_1, \ldots, y_m$ given the initial conditions

$$y_k|^{t_0} = \overset{0}{y}_{k0} + \sum_\sigma \epsilon_\sigma \overset{\sigma}{\eta}_k(t_0).$$

If the $q + r + 1$ quantities ϵ_σ are such that the $q + r$ conditions

$$\begin{aligned} \psi_\beta(\mathfrak{Y}_1(t_0), \ldots, \mathfrak{Y}_n(t_0)) &= 0, \\ \chi_\gamma(\mathfrak{Y}_1(t_0), \ldots, \mathfrak{Y}_n(t_0); \mathfrak{Y}_1(t_1), \ldots, \mathfrak{Y}_n(t_0)) &= 0 \end{aligned} \tag{7.62}$$

are satisfied, then the relations (7.61) define an admissible variation of the curve $\mathfrak{E}_0$.

---

[57] What Bolza has done here is to modify his sets of side- and end-conditions to eliminate the finite equations $\psi_\beta(y_1, \ldots, y_n) = 0$; he enlarges the set of conditions $\varphi_\alpha = 0$ $(\alpha = 1, 2, \ldots, p)$ by appending the additional new conditions

$$\varphi_{p+\beta} = \frac{d}{dt}\psi_\beta = \sum_i \frac{\partial \psi_\beta}{\partial y_i} y_i' = 0$$

and the set of conditions $\chi_\gamma = 0$ $(\gamma = 1, \ldots, r)$ by the additional new ones

$$\overset{0}{\psi}_\beta(y_{10}, \ldots, y_{n0}) = 0, \qquad \overset{1}{\psi}_\beta(y_{11}, \ldots, y_{n1}) = 0.$$

He now writes

$$U(\epsilon_1,\epsilon_2,\ldots,\epsilon_{q+r+1}) = \int_{t_0}^{t_1} f(\mathfrak{Y}_1,\ldots,\mathfrak{Y}_n,\mathfrak{Y}_1',\ldots,\mathfrak{Y}_n')\,dt$$
$$+ G(\mathfrak{Y}_1(t_0),\ldots,\mathfrak{Y}_n(t_0);\mathfrak{Y}_1(t_1),\ldots,\mathfrak{Y}_n(t_1))$$

and observes that at $\epsilon_1 = 0, \ldots, \epsilon_{q+r+1} = 0$ this function must be a minimum, subject to the side conditions (7.62). By the theory of extrema of ordinary functions, there must then exist constant multipliers l_0, $l_1, \ldots, l_{q+r}$, not all zero, such that

$$l_0 V(\overset{\sigma}{\eta}) + \sum_\beta l_\beta \overset{0}{\Psi}_\beta(\overset{\sigma}{\eta}) + \sum_\gamma l_{q+\gamma} \mathrm{X}_\gamma(\overset{\sigma}{\eta}) = 0 \qquad (\sigma = 1,2,\ldots,q+r+1) \tag{7.63}$$

where

$$V(\eta) = \int_{t_0}^{t_1}\left\{\sum_i \frac{\partial f}{\partial y_i}\eta_i + \frac{\partial f}{\partial y_i'}\eta_i'\right\}dt + \sum_i\left\{\frac{\partial G}{\partial y_{i0}}\eta_i(t_0) + \frac{\partial G}{\partial y_{i1}}\eta_i(t_1)\right\},$$

$$\overset{0}{\Psi}_\beta(\eta) = \sum_i \left.\frac{\partial \psi_\beta}{\partial y_i}\right|^{t_0} \eta_i(t_0),$$

$$\mathrm{X}_\gamma(\eta) = \sum_i\left\{\frac{\partial \chi_\gamma}{\partial y_{i0}}\eta_i(t_0) + \frac{\partial \chi_\sigma}{\partial y_{i1}}\eta_i(t_1)\right\} \tag{7.64}$$

and where the arguments of the functions that enter are those of the curve $\mathfrak{E}_0$.

Bolza then asserts that, as he showed in his 1907 paper, the multipliers l are independent of the choice of at least one of the $q + r + 1$ systems of functions $\overset{\sigma}{\eta}_1, \ldots, \overset{\sigma}{\eta}_n$. It then follows that the equation

$$l_0 V(\eta) + \sum_\beta l_\beta \overset{0}{\Psi}_\beta(\eta) + \sum_\gamma l_{q+\gamma}\mathrm{X}_\gamma(\eta) = 0 \tag{7.63'}$$

holds with the same multipliers for all $\eta_1, \ldots, \eta_n$ which are of class C'' and satisfy the m differential equations

$$\Phi_k(\eta) \equiv \sum_i\left(\frac{\partial \varphi_k}{\partial y_i}\eta_i + \frac{\partial \varphi_k}{\partial y_i'}\eta_i'\right) = 0 \qquad (k = 1,2,\ldots,m). \tag{7.65}$$

The proof given in Bolza [1907] is concerned with the rank of the matrix[58]

$$\|V(\overset{\sigma}{\eta}), \overset{0}{\Psi}_\beta(\overset{\sigma}{\eta}), \mathrm{X}_\gamma(\overset{\sigma}{\eta})\| \qquad (\sigma = 1,2,\ldots,q+r+1).$$

[58] See Bolza [1907], p. 372 or Bliss, *LEC*, p. 201. The matrix above must have a maximal rank ρ less than $q + r + 1$ for some system $\overset{\sigma}{\eta}$; otherwise, the equation $U(\epsilon) = U(0) + v$ together with (7.62) would have their functional determinant different from zero at an initial solution $(\epsilon, v) = (0,0)$; in this case there would be a solution $\epsilon_\sigma(v)$ near $v = 0$ and $U(\epsilon) < U(0)$ for v small and negative, which is a contradiction. If now there were a set η for which the relation (7.63′) were not valid, the rank of the matrix could be made $\rho + 1$, which is a contradiction.

The Φ_k have a peculiar form for $k = p + 1, \ldots, m$: they are expressible as

$$\Phi_{p+\beta}(\eta) = \sum \frac{d}{dt}\left(\frac{\partial \psi_\beta}{\partial y_i}\right)\eta_i + \frac{\partial \psi_\beta}{\partial y_i}\eta_i'$$

so that

$$\int_{t_0}^{t} \Phi_{p+\beta}\, dt = \sum_i \left\{ \frac{\partial \psi_\beta}{\partial y_i}\eta_i\bigg|^{t} - \frac{\partial \psi_\beta}{\partial y_i}\eta_i\bigg|^{t_0} \right\} = 0. \tag{7.66}$$

Bolza now multiplies the p first equations (7.65)

$$\sum_i \left(\frac{\partial \varphi_\alpha}{\partial y_i}\eta_i + \frac{\partial \varphi_\alpha}{\partial y_i'}\eta_i' \right) = 0 \qquad (\alpha = 1, 2, \ldots, p) \tag{7.67}$$

by undetermined multipliers $\lambda_\alpha(t)$ and integrates the resulting equations from t_0 to t_1. He next multiplies equations (7.60) above by undetermined multipliers $\mu_\beta(t)$ and again integrates from t_0 to t_1. He lastly evaluates equations (7.65) at $t = t_1$

$$\sum_i \frac{\partial \psi_\beta}{\partial y_i}\bigg|^{t_1} \eta_i(t_1) - \sum_i \frac{\partial \psi_\beta}{\partial y_i}\bigg|^{t_0} \eta_i(t_0) = 0 \tag{7.68}$$

and multiplies them by undetermined constant multipliers l_β^1. He then sums up these three sets as to α and β and adds the result to equation (7.63). With the help of an integration by parts, he finds

$$\int_{t_0}^{t_1} \sum_i \left(\frac{\partial \Omega}{\partial y_i} - \frac{d}{dt}\frac{\partial \Omega}{\partial y_i'} \right) dt + \sum_i H_i^0 \eta_i(t_0) + \sum_i H_i^1 \eta_i(t_1) = 0 \tag{7.69}$$

holding for all η of class C'' and satisfying the m differential equations (7.65). In this relation

$$\Omega = l_0 f + \sum_\alpha \lambda_\alpha \varphi_\alpha + \sum_\beta \mu_\beta \psi_\beta,$$

$$H_i^0 = -\frac{\partial \Omega}{\partial y_i'}\bigg|^{t_0} + l_0 \frac{\partial G}{\partial y_{i0}} + \sum_\beta l_\beta^0 \frac{\partial \psi_\beta}{\partial y_i}\bigg|^{t_0} + \sum_\gamma l_{q+\gamma} \frac{\partial \chi_\gamma}{\partial y_{i0}},$$

$$H_i^1 = \frac{\partial \Omega}{\partial y_i'}\bigg|^{t_1} + l_0 \frac{\partial G}{\partial y_{i1}} + \sum_\beta l_\beta^1 \frac{\partial \psi_\beta}{\partial y_i'}\bigg|^{t_1} + \sum_\gamma l_{q+\gamma} \frac{\partial \chi_\gamma}{\partial y_{i1}},$$

$$l_\beta^0 = l_\beta - l_\beta^1 - \int_{t_0}^{t_1} \mu_\beta(t)\, dt. \tag{7.70}$$

To complete the argument, the functions $\lambda_\alpha(t)$, $\mu_\beta(t)$ and the constants l_β^1 are so determined that the equations

$$\frac{\partial \Omega}{\partial y_k} - \frac{d}{dt}\frac{\partial \Omega}{\partial y_k'} = 0, \qquad H_k^1 = 0 \qquad (k = 1, 2, \ldots, m)$$

are valid. It follows as a consequence that

$$\int_{t_0}^{t_1} \sum_{\rho} \left(\frac{\partial \Omega}{\partial y_{m+\rho}} - \frac{d}{dt} \frac{\partial \Omega}{\partial y'_{m+p}} \right) \eta_{m+\rho} \, dt + \sum_i H_i^0 \eta_i(t_0)$$

$$+ \sum_{\rho} H^1_{m+\rho} \eta_{m+\rho}(t_1) = 0 \qquad (\rho = 1, 2, \ldots, n-m),$$

for all η of the sort described above. Bolza sums up in

Theorem II. If the curve $\mathfrak{E}_0$ furnishes an extremum for the expression U subject to the given conditions, there are $p+q$ multipliers $\lambda_\alpha(t)$, $\mu_\beta(t)$ depending upon t and equally $2q+r+1$ constant multipliers: $l_0, l_\beta^0, l_\beta^1, l_{q+\gamma}$ of such a kind that the n differential equations

$$\frac{\partial \Omega}{\partial y_i} - \frac{d}{dt} \frac{\partial \Omega}{\partial y'_i} = 0 \qquad i = 1, 2, \ldots, n \tag{7.71}$$

and the $2n$ boundary conditions

$$H_i^0 = 0, \qquad H_i^1 = 0 \tag{7.72}$$

are satisfied and besides the $q+r+1$ multipliers $l_0, l_\beta, l_{q+\gamma}$ are not all equal to zero. At the same time the quantities $\Omega, H_i^0, H_i^1, l_\beta$ are defined by the equations . . . [(7.70)].

He also has a partial converse:

Theorem IIa. If there is a system of multipliers

$$\lambda_\alpha(t),\ \mu_\beta(t);\ l_0,\ l_\beta^0,\ l_\beta^1,\ l_{q+\gamma}$$

for which $l_0 \neq 0$, and for which the differential equations [(7.71)] *and the boundary conditions* [(7.72)] *hold, then $V(\eta) = 0$ for every system of admissible variations, i.e., for every two times continuously differentiable system of functions $\eta_1, \ldots, \eta_n$ which satisfy the q finite equations*

$$\Psi_\beta(\eta) = \sum_i \frac{\partial \psi_\beta}{\partial y_i} \eta_i = 0 \tag{7.73}$$

and the r conditions

$$X_\gamma(\eta) = \sum_i \left\{ \frac{\partial \chi_\gamma}{\partial y_{i0}} \eta_i(t_0) + \frac{\partial \chi_\gamma}{\partial y_{i1}} \eta_i(t_1) \right\} = 0 \tag{7.74}$$

[*as well*].

To see this latter result, notice that the expression

$$V(\eta) + \sum_\alpha \int_{t_0}^{t_1} \lambda_\alpha \Phi_\alpha(\eta)\, dt + \sum_\beta \int_{t_0}^{t_1} \mu_\beta \Psi_\beta(\eta)\, dt + \sum_\beta l_\beta^0 \overset{0}{\Psi}_\beta(\eta)$$

$$+ \sum_\beta l_\beta^1 \overset{1}{\Psi}(\eta) + \sum_\gamma l_{q+\gamma} X_\gamma(\eta)$$

is equal to $V(\eta)$ for all η_i satisfying equations (7.65), (7.73), and (7.74).

Further, by an integration by parts, it reduces to the left-hand member of (7.69) and as a consequence of (7.71) and (7.72), to zero.[59]

In his third section Bolza discusses abnormality. To do so, he considers what he calls[60];

> Case I: It is possible to select $q + r + 1$ systems of functions $\overset{\sigma}{\eta}_1, \ldots, \overset{\sigma}{\eta}_n$ from $\mathfrak{H}$ so that at least one determinant of degree $q + r$ of the matrix
>
> $$\left\| \overset{0}{\Psi}_1(\overset{\sigma}{\eta}), \ldots, \overset{0}{\Psi}_q(\overset{\sigma}{\eta}), X_1(\overset{\sigma}{\eta}), \ldots, X_r(\overset{\sigma}{\eta}) \right\|,$$
>
> $$\sigma = 1, 2, \ldots, q + r + 1 \tag{7.75}$$
>
> is different from zero.

Then he can set $l_0 = 1$ in his previous discussion; he calls this the *principal case*, and the curve $\mathfrak{E}_0$ is said to be *normal*. In this case it is possible to embed $\mathfrak{E}_0$ in a one-parameter family of admissible curves

$$y_i = \bar{y}_i(t, \epsilon) \qquad (i = 1, 2, \ldots, n),$$

which contains $\mathfrak{E}_0$ for $\epsilon = 0$ and for which

$$\left. \frac{\partial \bar{y}_i}{\partial \epsilon} \right|^{\epsilon = 0} = \eta_i.$$

(This shows that $\mathfrak{E}_0$ is not an isolated point in the set of admissible arcs.) He goes on to what he terms

> Case II. All determinants of the matrix . . . [(7.75)] vanish however the $q + r + 1$ systems of functions $\overset{\sigma}{\eta}_1, \ldots, \overset{\sigma}{\eta}_n$ of $\mathfrak{H}$ may be chosen.

Then the quantity $l_0 = 0$ and, following Hahn's terminology, Bolza calls this the *abnormal* case.[61] In this case it follows directly from (7.63) that there are $q + r$ multipliers, not all zero, so that

$$\sum_\beta l_\beta \overset{0}{\Psi}(\eta) + \sum_\gamma l_{q+\gamma} X_\gamma(\eta) = 0 \qquad (\mathfrak{H}). \tag{7.76}$$

Bolza then states:

> *Theorem III. In order that the curve $\mathfrak{E}_0$ be abnormal it is necessary and sufficient that there exist a system of multipliers*
>
> $$\lambda_\alpha(t), \ \mu_\beta(t); \ l_\beta^0, \ l_\beta^1, \ l_{q+\gamma}$$

---

[59] Bolza [1913], p. 439. Bolza says the result is a generalization of one of Hadamard's in his *Leçons*, Section 204.

[60] Bolza [1913], p. 440. Bolza uses the expression $\mathfrak{H}$ to mean the set of all η of class C'', which satisfy conditions (7.65).

[61] Hahn [1904], p. 152.

for which 1. *the n differential equations*

$$\frac{\partial \Omega_0}{\partial y_i} - \frac{d}{dt}\frac{\partial \Omega_0}{\partial y_i'} = 0 \qquad i = 1, 2, \ldots, n$$

hold with

$$\Omega_0 = \sum_\alpha \lambda_\alpha \varphi_\alpha + \sum_\beta \mu_\beta \psi_\beta;$$

2. *the* $2n$ *boundary conditions*

$$\begin{aligned} -\frac{\partial \Omega_0}{\partial y_i'}\bigg|^{t_0} + \sum_\beta l_\beta^0 \frac{\partial \psi_\beta}{\partial y_i}\bigg|^{t_0} + \sum_\gamma l_{q+\gamma} \frac{\partial \chi_\gamma}{\partial y_{i0}} &= 0, \\ \frac{\partial \Omega_0}{\partial y_i'}\bigg|^{t_1} + \sum_\beta l_\beta \frac{\partial \Psi_\beta}{\partial y_i'}\bigg|^{t_1} + \sum_\gamma l_{q+\gamma} \frac{\partial \chi_\gamma}{\partial y_{i1}} &= 0 \end{aligned} \tag{7.77}$$

are valid; 3. *the* $q + r$ *constants* l_β, l_{q+r} *are not all zero, where*

$$l_\beta = l_\beta^0 + l_\beta^1 + \int_{t_0}^{t_1} \mu_\beta(t)\,dt.$$

The proof is not difficult. The necessity follows at once from his Theorem II with $l_0 = 0$. To show the sufficiency he proceeds as follows: the systems of equations (7.66), (7.67), and (7.68) are valid for every η in $\mathfrak{H}$; multiply these equations with the multipliers λ_α, μ_β, l_β^1 whose existence is guaranteed by the theorem, integrate the first two sets from t_0 to t_1, sum them as to α and β, and add up the results. He next applies the usual integration by parts and uses the equations (7.77). It follows directly that (7.76) holds.

For the normal case Bolza states and proves:

> *Theorem IV. If the curve* $\mathfrak{E}_0$ *is normal and there is a set of multipliers*
>
> $$\lambda_\alpha(t),\ \mu_\beta(t),\ l_\beta^0,\ l_\beta^1,\ l_{q+\gamma}$$
>
> *for which the differential equations* . . . [(7.71)] *and the boundary conditions* . . . [(7.72)] *are satisfied with* $l_0 = 1$, *then there is exactly one such set of multipliers.*

The proof is indirect but straightforward. Suppose that there were another set $\bar{\lambda}_\alpha(t)$, $\bar{\mu}_\beta(t)$; $\bar{l}_\beta^0, \bar{l}_\beta, \bar{l}_{q+\gamma}$ for which (7.71) and (7.72) are satisfied. By forming $\Omega_0 = \Omega - \bar{\Omega}$, etc., Bolza remarks that $\mathfrak{E}_0$ would then be abnormal unless all the differences $\bar{l}_\beta - l_\beta$, $\bar{l}_{q+\gamma} - l_{q+\gamma}$ vanish. By the assumption that the determinant

$$\left| \frac{\partial \varphi_\alpha}{\partial y_h}, \frac{\partial \chi_\beta}{\partial y_h'} \right| \neq 0 \qquad (\alpha = 1, 2, \ldots, p;\ \beta = 1, 2, \ldots, q;\ h = 1, 2, \ldots, m)$$

along the arc $\mathfrak{E}_0$ the equations $H_i^0 - \overline{H}_i^0 = 0$, $H_i^1 - \overline{H}_i^1 = 0$ imply that

$\lambda_\alpha(t_0) = \bar{\lambda}_\alpha(t_0)$, $\lambda_\alpha(t_1) = \bar{\lambda}_\alpha(t_1)$, $l_\beta^0 = \bar{l}_\beta^0$, $l_\beta^1 = \bar{l}_\beta^1$; and the equations

$$\frac{\partial(\Omega - \bar{\Omega})}{\partial y_i} - \frac{d}{dt}\frac{\partial(\Omega - \bar{\Omega})}{\partial y_i'} = 0$$

then imply that $\lambda_\alpha = \bar{\lambda}_\alpha$, $\mu_\beta = \bar{\mu}_\beta$.

Bolza next defines two special problems. To do this, he calls $\mathfrak{M}$ the set of all admissible curves which satisfy the conditions (7.55), (7.56), and (7.57) and $\mathfrak{M}'$ the set of admissible curves which satisfy all these conditions with the exception of $\chi_{\gamma_0}(y_{10}, \ldots, y_{n0}; y_{11}, \ldots, y_{n1}) = 0$ for some particular γ_0 out of the numbers $1, 2, \ldots, r$. He then considers Problem I to be

$$U = \text{extremum in } \mathfrak{M}$$

and Problem II

$$\chi_{\gamma_0}(y_{10}, \ldots, y_{n0}; y_{11}, \ldots, y_{n1}) = \text{extremum in } \mathfrak{M}'.$$

For these problems, he notes:

> *Theorem V. If the curve $\mathfrak{E}_0$ satisfies the differential equations and boundary conditions for Problem II, then $\mathfrak{E}_0$ is abnormal for Problem I and conversely.*

Bolza closes his paper with a discussion of some special cases and the problems that arise in applying the general theory to them.

7.11. Carathéodory's Method

In an appendix to his Göttingen dissertation, Carathéodory [1904] (pp. 71–79) made an elegant generalization of John Bernoulli's method for treating the brachystochrone problem (see p. 58n above). It is worthy of note.

He considers an integral of the form

$$J = \int_{x_1}^{x_2} f(x, y, p)\,dx, \qquad p = \frac{dy}{dx}$$

and a one-parameter family of curves $\varphi(x, y) = \lambda$. Let β be the angle between the normal to these curves and the x-axis as in Figure 7.11 so that

$$\cos\beta = \frac{\varphi_x}{\sqrt{\varphi_x^2 + \varphi_y^2}}, \qquad \sin\beta = \frac{\varphi_y}{\sqrt{\varphi_x^2 + \varphi_y^2}}.$$

The distance dn between two curves defined by λ and $\lambda + d\lambda$ is given by $d\lambda = \varphi_x dx + \varphi_y dy = dn(\varphi_x^2 + \varphi_y^2)^{1/2}$. Carathéodory now draws a curve segment, meeting the x-axis in the angle α as in Figure 7.11, and asks that it be

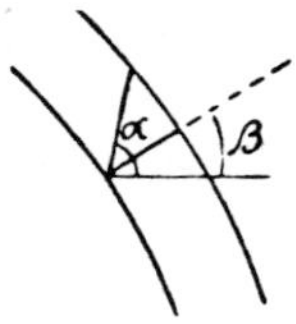

Figure 7.11

such that the expression $f\,dx$ is an extremum. He clearly has

$$ds = \frac{dn}{\cos(\alpha - \beta)}, \qquad dx = \frac{dn \cos\alpha}{\cos(\alpha - \beta)},$$

where ds is the length of the segment. It follows that

$$dx = \frac{d\lambda}{\varphi_x + p\varphi_y}, \qquad f\,dx = \frac{f(x, y, p)\,d\lambda}{\varphi_x + p\varphi_y},$$

since $p = \tan\alpha$. This implies that for an extremum,

$$\frac{\partial}{\partial p}\left(\frac{f}{\varphi_x + p\varphi_y}\right) = 0$$

and hence that

$$(\varphi_x + p\varphi_y) f_p - f\varphi_y = 0 \tag{7.78}$$

since $\varphi_x + p\varphi_y \neq 0$. This equation can be rewritten as

$$\frac{\varphi_x}{f - pf_p} = \frac{\varphi_y}{f_p}. \tag{7.78$'$}$$

The sign of

$$\frac{\partial^2}{\partial p^2}\left(\frac{f}{\varphi_x + p\varphi_y}\right)$$

determines whether an extremum is a maximum or minimum; and, as a consequence of (7.78), the sign of

$$\frac{f_{pp}}{\varphi_x + p\varphi_y} = \frac{f_{pp}\cos\alpha}{\sqrt{\varphi_x^2 + \varphi_y^2}\cdot\cos(\alpha - \beta)}$$

determines whether it is a maximum or a minimum. But the denominator of the second member is positive, and the sign of $f_{pp}\cos\alpha$ is then the relevant quantity to examine. This gives directly the Legendre condition. The condition implied by relation (7.78) on a curve is the transversality one; i.e., it implies that the curve is cut transversally by the family $\varphi = \lambda$. Relation (7.78) may be viewed as a first-order differential equation whose solutions form a one-parameter family cut transversally by the family $\varphi(x, y) = \lambda$.[62]

Carathéodory next defines what it means for the family $\varphi = \lambda$ to be *geodesically equidistant*. He says that the family has this property provided that the expression $f/(\varphi_x + p\varphi_y)$ has a constant value along each curve $\varphi = \lambda$. In this case f is expressible as

$$f = (\varphi_x + p\varphi_y)\psi(\varphi), \tag{7.79}$$

where ψ is an arbitrary function and p is calculated by means of relation (7.78) as a function of $x, y, \varphi_x, \varphi_y$.[63] Carathéodory chooses ψ to be unity and finds that (7.78), and (7.79) reduce to

$$f = \varphi_x + p\varphi_y, \qquad f_p = \varphi_y. \tag{7.80}$$

He differentiates the first of these along an arbitrary curve and obtains the relation

$$f_x\, dx + f_y\, dy + f_p\, dp = (\varphi_{xx} + p\varphi_{xy})\, dx + (\varphi_{xy} + p\varphi_{yy})\, dy + \varphi_y\, dp,$$

which reduces to

$$f_x\, dx + f_y\, dy = (\varphi_{xx} + p\varphi_{xy})\, dx + (\varphi_{xy} + p\varphi_{yy})\, dy$$

because of the second of equations (7.80). This yields the two equations

$$f_x = \varphi_{xx} + p\varphi_{xy}, \qquad f_y = \varphi_{xy} + p\varphi_{yy}\,. \tag{7.81}$$

He next differentiates the second of the equations (7.80) along a curve for which $dy/dx = p$ and finds that

$$\varphi_{xy} + p\varphi_{yy} = \frac{d}{dx} f_p,$$

which combines with the second of equations (7.81) to yield the Euler condition

$$f_y = \frac{d}{dx} f_p. \tag{7.82}$$

---

[62] Relation (7.78) may be considered as follows: let $y(x)$ $x_1 \leqslant x \leqslant x_2$ be an admissible arc nowhere tangent to $\varphi(x, y) = \lambda$. The equation $\varphi[x, y(x)] = \lambda$ then determines x as a function of λ, $x(\lambda)$, so that $x'(\lambda) = 1/(\varphi_x + p\varphi_y)$. The integral J taken along $y(x)$ from x_1 to $x_2 = x(\lambda)$ is such that $dJ/d\lambda = f/(\varphi_x + p\varphi_y)$. At each point (x, y) this expression may be viewed as a function of $y' = p$; the direction $1 : p$ in which it has its least value is *the direction of quickest descent* (see Bliss *LEC*, pp. 77ff). Then the arc $y(x)$ has the direction of quickest descent at a point (x, y) where it intersects the curve $\varphi(x, y) = \lambda$ if and only if the curve $\varphi = \lambda$ is transversal to the arc at the point. For the direction $\varphi_y : -\varphi_x$ to be transversal to the direction $1 : y' = p$, the integrand $f\,dx + (dy - p\,dx)f_p$ of the Hilbert invariant integral must vanish; but this is $f\varphi_y - f_p(\varphi_x + p\varphi_y) = 0$, which is an immediate consequence of (7.78).

[63] Two curves of the family are geodesically equidistant in case the values of the integral J calculated along the segments of curves of quickest descent, bounded by the two curves of the family, are all equal.

He concludes that "*Each curve which is cut transversally by a family of equidistant curves must satisfy the equation* . . . [(7.82)]."[64]

He continues by examining the converse problem. Let

$$y = g(x, a), \qquad g_a(x, a) \neq 0$$

be a one-parameter family of integrals of the differential equation (7.82), the curves of which are cut transversally by $\varphi = \lambda$. The solution of the equation of $y = g(x, a)$ is $a = a(x, y)$, and $p = g_x(x, a)$.

Carathéodory wishes to show that the curves $\varphi = \lambda$, which these extremals—the solutions of (7.82)—cut transversally, are geodesically equidistant. To do this he wishes to show that $\varphi_x = f - pf_p$, $\varphi_y = f_p$ or equivalently that the equation

$$\frac{\partial}{\partial y}\{f[x, y, p(x, y)] - p(x, y)f_p[x, y, p(x, y)]\} = \frac{\partial}{\partial x} f_p[x, y, p(x, y)]$$

holds at every point (x, y) of a region of the plane. Carathéodory uses a bar over a function, such as p, to indicate that it is to be viewed as a function of x, y alone. He thus has $\bar{f} = f[x, y, p(x, y)]$. With the help of this notation, he writes this relation as

$$\begin{aligned} \frac{\partial}{\partial y}(\bar{f} - \overline{pf_p}) &= \bar{f}_y - \bar{p}\bar{f}_{py} - \bar{p}\bar{f}_{pp}\frac{\partial \bar{p}}{\partial y} \\ &= \bar{f}_y - \bar{p}\bar{f}_{py} - \bar{p}\bar{f}_{pp}\frac{\bar{g}_{xa}}{\bar{g}_a} \end{aligned} \tag{7.83}$$

since $\bar{p} = p(x, y) = g_x[x, a(x, y)]$ and $y = g[x, a(x, y)]$. He further calculates for the same reason that

$$\frac{\partial}{\partial x}\bar{f}_p = \bar{f}_{px} + \bar{f}_{pp}\left\{\bar{g}_{xx} - \frac{\bar{g}_x\bar{g}_{xa}}{\bar{g}_a}\right\}.$$

Since $y = g(x, a)$ satisfies (7.82), it is true that

$$f_y[x, g(x, a), g_x(x, a)] = \frac{d}{dx} f_p[x, g(x, a), g_x(x, a)]$$

and hence that $f_y = f_{px} + f_{py}g_x + f_{pp}g_{xx}$ holds identically in x, a. If the values $a(x, y)$, $p(x, y)$ are put in place of a and p, then this relation holds identically in x, y. This enables Carathéodory to write

$$\frac{\partial}{\partial x}\bar{f}_p = \bar{f}_y - \overline{pf}_{py} - \overline{pf}_{pp}\frac{\bar{g}_{xa}}{\bar{g}_a},$$

from which it is evident that (7.83) is valid. This establishes the converse he desired.

Later Carathéodory noticed that the curves $\varphi = \lambda$ satisfy a Hamilton–

[64] Carathéodory [1904], p. 74.

Jacobi partial differential equation of the form

$$\varphi_x + H(x, y, \varphi_y) = 0 \tag{7.84}$$

if and only if these curves form a geodesically equidistant family. Even though this result is timewise outside the purview of this book, let us see how it comes about. If the second of equations (7.80) is solved for $p = P(x, y, \varphi_y)$ and the result substituted into the first one, there results a relation of the form (7.84), the Hamilton–Jacobi equation. Conversely, if φ satisfies equation (7.84), the equation $u = f_p(x, y, p)$ can be solved for p as a function of x, y, u and equations (7.80) will be satisfied. This implies that the function f is expressible in the form (7.79) with $\psi \equiv 1$, and the curves $\varphi = \lambda$ are geodesically equivalent. This serves to show how Carathéodory's ideas are related to fields and to the Hamilton–Jacobi theory.

It is perhaps germane to notice that in the case of the brachystochrone problem the family $\varphi = \lambda$ of curves is the family of synchrones referred to on p. 42 above. Had John Bernoulli used those curves in his 1718 paper instead of the normal at each point (see Section 1.11), he would have arrived at a general procedure for solving problems of the calculus of variations. His method, as we saw, lacked the generality necessary. In fact, Carathéodory says of Bernoulli's method: "The method of Bernoulli is now to pick out on each curve φ = const., those points for which the expression $f/(\varphi_x + p\varphi_y)$ is a maximum or a minimum. In the general case his procedure would lead to a disaster since the curve joining such points will not be cut transversally by the family $\varphi = \lambda$." [65]

7.12. Hahn on Abnormality

In the study of more general problems such as those of Lagrange or Mayer a peculiar phenomenon may occur, which was first noted by Mayer ([1886], p. 79), then emphasized by von Escherich ([1899], p. 1290), and studied in considerable detail by Hahn ([1904], pp. 151–152). (In fact, there are several important papers by Hahn in this period concerned with the calculus of variations.) To understand what happens, consider the Lagrange problem of minimizing the integral

$$J = \int_{t_0}^{t_1} f(y_1, \ldots, y_n; y_1', \ldots, y_n')\, dt$$

subject to the side-conditions

$$\varphi_k(y_1, \ldots, y_n; y_1', \ldots, y_n') = 0 \qquad (k = 1, 2, \ldots, m).$$

[65] Carathéodory [1904], p. 73.

Hahn supposes that along the arc in question, not all subdeterminants of the matrix

$$\left| \frac{\partial \varphi_i}{\partial y_k'} \right| \qquad (i = 1, 2, \ldots, m;\ k = 1, 2, \ldots, n)$$

vanish identically, and that f and the φ_i are positively homogeneous in the y'.

Hahn first shows that this problem is a special case of a problem of Mayer. To do this, he sets

$$y_0 = \int_{t_0}^{t} f(y_1, \ldots, y_n; y_1', \ldots, y_n')\, dt$$

and[66]

$$\varphi_0 = f(y_1, \ldots, y_n; y_1', \ldots, y_n') - y_0' = 0.$$

Now Hahn finds that there are multipliers $\lambda_0, \lambda_1, \ldots, \lambda_m$, which are functions of t, such that the equations

$$\sum_{k=0}^{m} \left(\lambda_k \frac{\partial \varphi_k}{\partial y_i} - \frac{d}{dt} \lambda_k \frac{\partial \varphi_k}{\partial y_i'} \right) = 0 \qquad (i = 0, 1, \ldots, n)$$

are satisfied. He goes on to note that for the problem of Lagrange the first of these equations reduces to $d\lambda_0 / dt = 0$, and the remaining equations are then expressible as

$$\lambda_0 \left(\frac{\partial f}{\partial y_i} - \frac{d}{dt} \frac{\partial f}{\partial y_i'} \right) + \sum_{k=1}^{m} \left(\lambda_k \frac{\partial \varphi_k}{\partial y_i} - \frac{d}{dt} \lambda_k \frac{\partial \varphi_k}{\partial y_i'} \right) = 0 \qquad (i = 1, 2, \ldots, n). \tag{7.85}$$

If $\lambda_0 \neq 0$, it can clearly be taken to be unity; whereas if it is zero, this is not possible, and the Euler–Lagrange equations (7.85) become

$$\sum_{k=1}^{m} \left(\lambda_k \frac{\partial \varphi_k}{\partial y_i'} - \frac{d}{dt} \lambda_k \frac{\partial \varphi_k}{\partial y_i'} \right) = 0 \qquad (i = 1, 2, \ldots, n). \tag{7.86}$$

Hahn now says: "We wish to say that our extremal shows itself in an *abnormal* condition on each interval in which the equations . . . [(7.86)] can be satisfied with functions λ, which are not all identically zero." He goes on to assume that the extremals are not abnormal on the interval $(t_0 t_1)$. He calls this the *principal* case. His result is then:

"*Each extremal of the Lagrange problem, which does not behave abnormally on any sub-interval of* $(t_0 t_1)$, *must satisfy the equations* . . . [(7.85) *with* $\lambda_0 = 1$]."

The reader will recall that Mayer [1886] was the first to observe the difference between normal and abnormal arcs and that von Escherich ([1899], p. 1290) discussed the distinction in some detail. After that various

[66] For a full discussion of the equivalence of problems, see Bliss, *LEC*, pp. 189–193.

papers were written on the problem of Lagrange, assuming that the arc in question was normal on every subinterval or on even stronger assumptions. In recent times Bliss, Morse and Myers, Graves, and McShane have shown that the multiplier rule as well as the conditions of Weierstrass and Clebsch can be established with no assumptions of normality. It was Morse in a series of papers in 1930–31 who "was the first to avoid the use of a hypothesis on an extension of the interval of the minimizing arc. Hestenes . . . , Morse . . . , and Reid . . . have proved sufficiency theorems applicable to arcs which are normal but not necessarily normal on every subinterval, and to abnormal arcs of a special type."[67]

Since the work of Hahn there has been a great deal of research on general problems, mainly the problem of Bolza, so that by the middle of this century what one might well call the "classical theory of the calculus of variations" was completed. In very large measure this work was accomplished by Bliss and his Chicago school of students and colleagues. This group included Graves, Hestenes, McShane, Reid, and many others. Their work is admirably summarized in Bliss, *LEC*.

---

[67] Bliss, *LEC*, p. 219. An arc is said to be normal or abnormal on an interval for the problem of Bolza if it is normal or abnormal for the corresponding problem with fixed end-points on that interval.

Bibliography

Adams *CAT*:
John Couch Adams, *A Catalogue of the Portsmouth Collection of Books and Papers Written by or Belonging to Sir Isaac Newton*, Cambridge, 1888.

Andrade [1958]:
Edward N. da C. Andrade, *Sir Isaac Newton, His Life and Work*, London, 1954. Reprinted Garden City, N. Y., 1958.

Armanini [1900]:
Egidio Armanini, Sulla superficie di minima resistenza, *Ann. di Mat.*, Ser. IIIa, Vol. IV, 1900, pp. 131–149.

Beltrami

OPERE: Eugenio Beltrami, *Opere Matematiche*, 4 Vols., Milan, 1902–1920.

[1868]:———Sulla teoria delle linee geodetiche, *Rend. d. R. Ist. Lombardo* (2), Vol. I, 1868, pp. 708–718 = *OPERE*, Vol. I., pp. 366–373.

Bernoulli, Ja.

O: James Bernoulli, *Jacobi Bernoulli, Basileensis, Opera*, 2 vols., Geneva, 1744. Reprinted Brussels, 1967.

CL:———Curvatura laminae elasticae, *Act. Erud.*, Leipzig, June 1694, p. 262 = *O*, Vol. I, pp. 576–600.

JB:———Jacobi Bernoulli solutio problematum fraternorum, *Act. Erud.*, Leipzig, May 1697, p. 214 = *O*, Vol. II, pp. 768–778. German translation (partial) by P. Stäckel in No. 46, *Ostwald's Klassiker der exacten Wissenschaften*, Leipzig, 1894, pp. 14–20. English translation (partial) in D. J. Struik, ed., *A Source Book in Mathematics*, 1200–1800, Cambridge, 1969, pp. 396–399.

SO:———Jacobi Bernoulli solutio propria problematis isoperimetrici, *Act. Erud.* , Leipzig, June 1700, p. 261 = *O*, Vol. II, pp. 874–887.

MP:———Analysis magni problematis isoperimetrici, *Act. Erud.*, Leipzig, May 1701, p. 213 = *O*, Vol. II, pp. 895–920.

Bernoulli, Jo.

OO: John Bernoulli, *Johannis Bernoulli, Opera Omnia*, J. E. Hofmann, ed., 4 Vols., Lausanne and Geneva, 1742, Reprinted Hildesheim, 1968.

LB:———"Lectori benevolo," *Act. Erud.*, Leipzig, December 1696, p. 560 = *OO*, Vol. I, p. 165.

PN:———Problema novum ad cujus solutionem mathematici invitantur, *Act. Erud.*, Leipzig, June 1696, p. 269 = *OO*, Vol. I, p. 161. German translation by P. Stäckel in No. 46 of Ostwald's *KLASSIKER*, p. 3.

PR:———Programma, *OO*, Vol. I, pp. 166–169. German translation by Stäckel, *KLASSIKER*, No. 46, pp. 3–6.

CR:———Curvatura radii in diaphanis non uniformibus solutioque problematis a se in Actis 1696, p. 269, propositi, de invenienda linea brachystochrona, id est, in qua grave a dato puncto ad datum punctum brevissimo tempore decurrit; & de curva synchrona, seu radiorum unda, construenda, *Act. Erud.*, Leipzig, May 1697, pp. 206ff = *OO*, Vol. I, pp. 187–193. German translation by Stäckel, *KLASSIKER*, No. 46, pp. 6–13. English translation in D. J. Struik, *A Source Book in Mathematics, 1200–1800*, pp. 391–396.

LE:———Lettre de Mr. Jean Bernoulli à Monsieur Basnage, Docteur en Droit, Auteur de l'Histoire des Ouvrages des Savans. Sur le probléme des isopérimétres, *OO*, Vol. 1, pp. 194–204 = *Histoire des Ouvrages des Sçavans*, June 1697, pp. 452ff.

PA:———Problémes a resoudre, *Jour. d. Savans*, 26 August 1697, p. 394, Paris ed. = *OO*, Vol. I, pp. 204–205.

SI:———Lettre de Mr. Bernoulli, Professeur de Groningue, à Mr. Varignon, du 15, Octobre 1697. Sur le probléme des isopérimétres, *Jour. d. Savans*, 2 December 1697, p. 458, Paris ed. = *OO*, Vol. I, pp. 206–213. (An extract of this paper appears in *Act. Erud.*, January 1698, p. 52.)

PD:———Reponse, *Jour. d. Savans*, 21 April 1698, p. 172, Paris ed. = *OO*, Vol. I, pp. 215–220.

SD:———Solution du probléme proposé par M. Jacques Bernoulli, dans les Actes de Leipsic du mois de May de l'année 1697; trouvée en deux manieres par M. Jean Bernoulli son Frère, & communiquée à M. Leibnitz au mois de Juin 1698. *Mém. Acad. Roy. Sci. Paris*, 1706, p. 235 = *OO*, Vol. I, pp. 424–435.

RE:———Remarques sur ce qu'on a donné jusqu'ici de solutions des problémes sur des isopérimetrés; avec une nouvelle méthode courte & facile de les resoudre sans calcul, laquelle s'étend aussi à d'autres problêmes qui ont raport à ceux-là, *Mém. Acad. Roy. Sci. Paris*, 1718, p. 100 = *OO*, Vol. II, pp. 235–269.

Bertrand

[1841]: J. L. F. Bertrand, Démonstration d'un théoreme de M. Jacobi, *Jour. l'École Poly.*, Vol. XXVIII, 1841, pp. 276–283.

[1842]:———Sur un point du calcul des variations, *Jour. d. Math.*, Vol. VII, 1842, pp. 55–58.

Bliss

[1904]: Gilbert A. Bliss, Jacobi's criterion when both end-points are variable, *Math. Ann.*, Vol. LVIII, 1904, pp. 70–80.

[1904′]:———An existence theorem for a differential equation of the second order, with an application to the calculus of variations, *Trans. Am. Math. Soc.*, Vol. V, 1904, pp. 113–125.

[1908]:———and Max Mason, The properties of curves in space which minimize a definite integral, *Trans. Am. Math. Soc.*, Vol. IX, 1908, pp. 440–466.

[1910]:———and Max Mason, Fields of extremals in space, *Trans. Am. Math. Soc.*, Vol. XI, 1910, pp. 325–340.

COV:———*Calculus of Variations*, Chicago, 1925.

LEC:———*Lectures on the Calculus of Variations*, Chicago, 1946.

PRINCE:———*The Princeton Colloquium Lectures on Mathematics, Fundamental Existence Theorems*, New York, 1913. (This is pp. 1–107; the remainder of the volume is a lecture by E. Kasner on dynamics.)

Bolza

[1901]: Oskar Bolza, New proof of a theorem of Osgood's in the calculus of variations, *Trans. Am. Math. Soc.*, Vol. II, 1901, pp. 422–427.

[1902]:———Some instructive examples of the calculus of variations, *Bull. Am. Math. Soc.*, Vol. IX, 1902/03, pp. 1–10.

LEC:———*Lectures on the Calculus of Variations*, Chicago, 1904.

[1906]:———Ein Satz über eindeutige Abbildung und seine Anwendung in der Variationsrechnung, *Math. Ann.*, Vol. LXIII, 1906, pp. 246–252.

[1907]:———Die Lagrangesche Multiplikatorenregel in der Variationsrechnung für den Fall von gemischten Bedingungen und die zugehörigen Grenzgleichungen bei variablen Endpunkten, *Math. Ann.*, Vol. LXIV, 1907, pp. 370–386.

VOR:———*Vorlesungen über Variationsrechnung*, Leipzig u. Berlin, 1909.

[1912/13]:———Bemerkungen zu Newtons Beweis seines Satzes über den Rotationskörper kleinsten Widerstandes, *Bibl. Math.*, Ser. 3, Vol. XIII, 1912–1913, pp. 146–149.

[1913]:———Über den "Anormalen Fall" beim Lagrangeschen und Mayerschen Problem mit gemischten Bedingungen und variablen Endpunkten, *Math. Ann.*, Vol. LXXIV, 1913, pp. 430–446.

Borda [1767]:

Jean-Charles, Chevalier de Borda, Éclaircissement sur les méthodes de trouver les courbes qui jouissent de quelque propriéte du *maximum* ou du *minimum*, *Mém. Acad. Sci.*, 1767, pp. 551–563.

Brunet *ACT*:

Pierre Brunet, *Étude historique sur le Principe de la Moindre Action*, Paris, 1938.

Cairns [1907]:

W. Cairns, *Die Anwendung der Integralgleichungen auf die zweite Variation bei isoperimetischen Problemen*, Dissertation, Göttingen, 1907.

Carathéodory

GMS: Constantin Carathéodory, *Gesammelte Mathematische Schriften*, 5 vols., Münich 1954–1957.

[1904]:———*Über die diskontinuierlichen Lösungen in der Variationsrechnung*, Dissertation, Göttingen, 1904 = *GMS*, Vol. I, pp. 3–79.

[1906]:———Über die starken Maxima and Minima bei einfachen Integralen, *Math. Ann.*, Vol. LXII, 1906, pp. 449–503 = *GMS*, Vol. I, (Habilitationschrift, Göttingen, 1905), pp. 80–142.

[1930]:———Untersuchungen über das Delaunaysche Problem der Variationsrechnung, *Abh. Hamburg*, Vol. VIII, 1930, pp. 32–55 = *GMS*, Vol. II, pp. 12–39.

[1937]:———The beginning of research in the calculus of variations," *Osiris*, Vol. III, 1937, Part I, pp. 224–240 = *GMS*, Vol. II, pp. 93–107.

[1945]:———Basel und der Beginn der Variationsrechnung, *Festschrift zum 60. Geburtstag von Prof. Dr. Andreas Speiser*, Zürich, 1945, pp. 1–18 = *GMS*, Vol. II, pp. 108–128.

Clebsch

[1858]: Rudolff F. A. Clebsch, Ueber die Reduktion der zweiten Variation auf ihre einfachste Form, *Jour. für Math.*, Vol. LV, 1858, pp. 254–273.

[1858′]:———Ueber diejenigen Probleme der Variationsrechnung, welche nur eine unäbhangige Variable enthalten, *Jour. für Math.*, Vol. LV, 1858, pp. 335–355.

Cramer [1752]:

Gabriel Cramer, Mémoire posthume de géométrie, *Abh. Akad. Berlin*, 1752, pp. 283–290.

Darboux *TDS*:

Gaston Darboux, *Leçons sur la Théorie générale des Surfaces*, 4 vols., Paris, 1887–1896.

Delaunay

[1841]: Charles Delaunay, Sur la distinction des maxima et des minima dans les questions qui dépendant de la méthode des variations, *Jour. d. Math.*, Vol. VI, 1841, pp. 209–237.

[1843]:———Mémoire sur le calcul des variations, *Jour. l'École Poly.*, Vol. XVII, 1843, pp. 37–120.

du Bois-Raymond

[1879]: Paul du Bois-Raymond, Erläuterungen zu den Anfangsgründen der Variationsrechnung, *Math. Ann.*, Vol. XV, 1879, pp. 283–314.

[1879′]:———Fortsetzung der Erläuterungen zu den Anfangsgründen der Variationsrechnung, *Math Ann.*, Vol. XV, 1879, pp. 564–578.

Egorow [1906]:

D. Egorow, Die hinreichenden Bedingungen des Extremums in der Theorie des Mayerschen Problems, *Math. Ann.*, Vol. LXII, 1906, pp. 371–380.

Erdmann

[1877]: G. Erdmann, Untersuchung der höheren Variationen einfacher Integrale, *Zeit. f. Math. u. Phy.*, Vol. XXII, 1877, pp. 327.

[1877′]:———Über unstetige Lösungen in der Variationsrechnung, *Jour. für Math.*, Vol. LXXXII, 1877, pp. 21–30.

[1878]:———Zur Untersuchung der zweiten Variation einfacher Integrale, *Zeit. f. Math. u. Phy.*, Vol. XXIII, 1878, pp. 362–379.

Euler

OPERA: Leonard Euler, *Leonhardi Euleri Opera Omnia*, 72 vols., Bern, 1911–1975. Series I, *Opera Mathematica*.

I, XXIV:———*Methodus Inveniendi Lineas Curvas Maximi Minimive Proprietate Gaudentes sive Solutio Problematis Isoperimetrici Latissimo Sensu Accepti*, Lausanne and Geneva, 1744 = *OPERA*, I, Vol. XXIV, C. Carathéodory, ed., Bern, 1952.

I, XXV:———Commentationes analyticae ad calculum variationum pertinentes, *OPERA*, I, Vol. XXV, C. Carathéodory, ed., Bern, 1952.

[1755]:———Euler à Lagrange, Berolini, die 6 Sept. 1755, *Oeuvres de Lagrange*, Vol. XIV, pp. 144–146. This is Euler's response to Lagrange's first letter to him.

[1764]:———Elementa calculi variationum, I, XXV, pp. 141–176.

[1764′]:———Analytica explicatio methodi maximorum et minimorum, I, XXV, pp. 177–207.

[1771]:———Methodus nova et facilis calculum variationum tractandi, I, XXV, pp. 208–235.

Fatio [1699]:

Fatio de Duillier, *Investigatio Geometrica Solidi Rotundi, in quod Minima fiat Resistentia*. This is an appendix to his *Lineae Brevissimi Descensus Investigatio Geometrica Duplex*, London, 1699.

Fermat *OP*:

Pierre de Fermat, *Oeuvres de Fermat*, P. Tannery and C. Henry, eds., 4 vols., Paris, 1891–1912; *Supplément*, C. de Waard, ed., Paris, 1922.

Galileo *TWO*:

Galileo Galilei, *Discourses and Mathematical Demonstrations Concerning Two New Sciences, Pertaining to Mechanics and Local Motions, with an Appendix on Centers of Gravity of Solids*, Leyden, 1638, (English translation by S. Drake, Madison, 1974).

Gauss

WERKE: Carl Friedrich Gauss, *WERKE*, 12 vols., Göttingen, 1870–1933.

DIS:———Disquisitiones generales circa superficies curvas, *WERKE*, Vol. IV, p. 241 = Allgemeine Flächentheorie, *Klassiker*, Vol. 4, pp. 1–51.

Gregory *NMF*:

David Gregory, Newtoni methodus fluxionum. This appears as an appendix to Andrew Motte's 1729 translation of the *Principia, The Mathematical Principles of Natural Philosophy*, by Sir Isaac Newton, Vol. II, *Appendix*, pp. v–viii.

Hadamard

LEC: J. Hadamard, *Leçons sur le Calcul des Variations*, Paris, 1910.

[1906]:———Sur une méthode de calcul des variations, *Comptes Rendus*, Vol. CXLIII, 1906, pp. 1127–1129.

Hahn

[1902]: Hans Hahn, Über die Lagrangesche Multiplikatorenmethode in der Variationsrechnung, *Monat. f. Math. u. Phy.*, Vol. XIV, 1903, pp. 325–342.

[1904]:———Bemerkung zur Variationsrechnung, *Math. Ann.*, Vol. LVIII, 1904, pp. 148–168.

[1906]:———Über das allgemeine Problem der Variationsrechnung, *Monat. f. Math. u. Phy.*, Vol. XVII, 1906, pp. 295–304.

[1906′]:———Über einen Satz von Osgood in der Variationsrechnung, *Monat. f. Math. u. Phy.*, Vol. XVII, 1906, pp. 63–77.

[1907]:———Über die Herleitung der Differentialgleichungen der Variationsrechnung, *Math. Ann.*, Vol. LXIII, 1907, pp. 253–272.

[1911]:———Über Variationsprobleme mit variablen Endpunkten, *Monat. f. Math. u. Phy.*, Vol. XXII, 1911, pp. 127–136.

Hamilton

MATH: W. R. Hamilton, *The Mathematical Papers of Sir William Rowan Hamilton*, 2 vols., Cambridge, 1931–1940, A. W. Conway and J. L. Synge, eds.

[1834]:———On a general method in dynamics by which the study of the Motions of all free systems of attracting or repelling points is reduced to the search and differentiation of one central relation, or characteristic function, *Phil. Trans. Roy. Soc.*, Vol. 124, 1834, pp. 247–308 = *MATH*, Vol. II, pp. 103–167.

[1835]:———Second essay on a general method in dynamics, *Phil. Trans. Roy. Soc.*, Vol. 125, 1835, pp. 95–144 = *MATH*, Vol. II, pp. 162–216.

Hancock *LEC*:

Hancock, *Lectures on the Calculus of Variations*, Cincinnati, 1904.

Heath *ARCH*:

T. L. Heath, *The Works of Archimedes*, Cambridge, 1897, plus *Supplement* (reprinted, New York).

Hedrick [1902]:

E. R. Hedrick, On the sufficient conditions in the calculus of variations, *Bull. Am. Math. Soc.*, Vol. IX, pp. 11–24.

Heine

[1857]: Eduard Heine, Bemerkungen zu *Jacobi's* Abhandlung über Variationsrechnung, *Jour. für Math.*, Vol. LIV, 1857, pp. 68–71.

[1870]:———Aus brieflichen Mittheilungen (namentlich über Variationsrechnung), *Math. Ann.*, Vol. II, 1870, pp. 188—191.

Hesse [1857]:

Otto Hesse, Ueber die Criterien des Maximums und Minimums der einfachen Integrale, *Jour. für Math.*, Vol. LIV, 1857, pp. 227–273.

Hiebert *HR*:

Erwin N. Hiebert, *Historical Roots of the Principle of Conservation of Energy*, Madison, 1962.

Hilbert

[1899]: David Hilbert, Ueber das Dirichlet'sche Princip, *Jahr. Ber.*, Vol. VIII, 1899, pp. 184–188.

[1900]:———Mathematische Probleme. Vortrag, gehalten auf dem internationalen Mathematiker-Kongress zu Paris, 1900, *Gött. Nach.*, 1900, pp. 253–297 = *Arch. Math. u. Phy.*, Vol. I, pp. 44–63, 213–237. English translation by M. W. Newson in *Bull. Am. Math. Soc.*, Vol. VIII, 1902, pp. 437–479.

[1901]:———Über das Dirichletsche Prinzip, *Math. Ann.*, Vol. LIX, 1901, pp. 161–186.

[1905]:———Über das Dirichletsche Prinzip, *Jour. für Math.*, Vol. CXXIX, 1905, pp. 63–67.

[1906]:———Zur Variationsrechnung, *Math. Ann.*, Vol. LXII, 1906, pp. 351–370 = *Gött. Nach.* 1905, pp. 159–180.

[1912]:———*Grundzüge einer allegemeinen Theorie der linearen Integralgleichungen*, Leipzig u. Berlin, 1912.

Huygens *OC*:

Christiaan Huygens, *Oeuvres complétes*, 22 vols., The Hague, 1888–1950.

Jacobi

WERKE: C. G. J. Jacobi, *C. G. J. Jacobi's Gesammelte Werke*, K. Weierstrass, ed., 8 vols., Berlin, 1881–1891.

[1838]:———Zur Theorie der Variations-Rechnung und der Differential-Gleichungen, *Jour. für Math.*, Vol. XVII, 1937, pp. 68–82 = *WERKE*, Vol. IV, pp. 39–55 = Sur le calcul des variations et sur la théorie des équations differentielles, *Jour. de Math.*, Vol. III, 1838, pp. 44–59. This paper is also reprinted by Stäckel in *Ostwald's Klassiker*, Vol. 47, pp. 87–98.

[1838′]:———Über die Reduction der Integration der partiellen Differentialgleichungen erster Ordnung zwischen irgend einer Zahl Variabeln auf die Integration eines einzigen Systemes gewöhnlicher Differentialgleichungen, *Jour. für Math.*, Vol. XVII, 1838, pp. 97–162 = *WERKE*, Vol. IV, pp. 57–127.

NACH:———Über diejenigen Probleme der Mechanik in welchen eine Kräftefunction existirt und über die Theorie der Störungen, *WERKE*, Vol. V, pp. 217–395.

NACH′:———Über die vollständigen Lösungen einer partiellen Differentialgleichung erster Ordnung, *WERKE*, Vol. V, pp. 399–438.

Jordan *COURS*:

C. Jordan, *Cours d'Analyse de l'École Polytechnique*. 3 vols., Paris, 1882.

Kneser

[1898]: Adolph Kneser, Zur Variationsrechnung, *Math. Ann.*, Vol. L, 1898, pp. 27–50.

[1899]:———Ableitung hinreichender Bedingungen des Maximums oder Minimums einfacher Integrale aus der Theorie der zweiten Variation, *Math. Ann.*, Vol. LI, 1899, pp. 321–345.

LV:———*Lehrbuch der Variationsrechnung*, Braunschweig, 1900.

LV′:———*Lehrbuch der Variationsrechnung*, Braunschweig, 1925.

Kobb [1892/93]:

Gustaf Kobb, Sur les maxima et les minima des intégrales doubles, *Acta Math.*, Vol. XVI, 1892/93, pp. 65 – 140; and Vol. XVII, 1893, pp. 321–344.

Lagrange

OEUVRES: Joseph Louis Lagrange, *OEUVRES*, J. – A. Serret and G. Darboux, eds., 14 vols., Paris, 1867–1892.

MA:———*Mécanique Analytique*, 2 vols., Paris, 1788 and 2nd aug. ed. 1811 – 15 = *OEUVRES*, Vols. XI and XII.

[1755]:———Lagrange à Euler. Die 12 Augusti [1755], *OEUVRES*, Vol. XIV, pp. 138–144. In fact, pp. 135–245 contain the correspondence between the two men.

[1759]:———Recherches sur la méthode de maximis et minimis, *Misc. Soc. Tur.*, Vol. I, 1759 = *OEUVRES*, Vol. I, pp. 3–20.

[1760]:———Essai sur une nouvelle méthode pour déterminer les maxima et les minima des formules intégrales indéfinies, *OEUVRES*, Vol. I, pp.

333–362 = *Misc. Soc. Tur.*, Vol. II, 1760–1761, pp. 173–195. German translation by P. Stäckel, *Klassiker*, Vol. 47, pp. 3–56 and English translation (in part) by D. J. Struik, *A Source Book in Mathematics*, pp. 407–418.

[1760′]:———Application de la méthode exposée dans le mémoire précedént à la solution de differents problèmes de dynamique, *OEUVRES*, Vol. I, pp. 363–468 = *Misc. Soc., Tur.*, Vol. II.

[1766/69]:———Sur la méthodes des variations, *Misc. Soc. Tur.*, Vol. IV, 1766 = *OEUVRES*, Vol. II, pp. 37–63.

TAF:———*Théorie des Fonctions analytiques, contenant les Principes du Calcul différentiel, dégagés de toute Considération d'infiniment Petits, d'Evanouissans, de Limites et de Fluxions, et réduits a l'Analyse algébrique des Quantités finies*, Paris 1797, 2nd ed. 1813 = *OEUVRES*, Vol. IX.

LCF:———*Leçons sur le Calcul des Fonctions*, Paris, 1806 = *OEUVRES*, Vol. X.

Lebesgue, H. [1902]:
Henri Lebesque, Intégrale, longueur, aire, *Ann. di Mat.*, (3), Vol. VII, 1902, pp. 342–359.

Lebesgue, V.-A. [1841]:
V.-A. Lebesgue, Mémoire sur une formule de Vandermonde, et son application a la démonstration d'un théorème de M. Jacobi, *Jour. de Math.*, Vol. VI, 1841, pp. 17–35.

Lecat
BIB: Maurice M. A. Lecat, *Bibliographie du Calcul des Variations 1850–1913*, Ghent and Paris, 1913.

BIB':———*Bibliographie du Calcul des Variations, depuis les Origines jusqu'à 1850 comprenant la Liste des Travaux qui ont Prépare ce Calcul*, Ghent and Paris, 1916.

Legendre [1786]:
Adrien-Marie Legendre, Sur la mainère de distinguer les maxima des minima dans le calcul des variations, *Mém. Acad. Sci.*, (1786) 1788, pp. 7–37. German translation by P. Stäckel, *Klassiker*, Vol. 47, pp. 57–86.

Leibniz *LMS*:
G. W. Leibniz, *Leibnizens mathematische Schriften*, C. I. Gerhardt, ed., Halle, Part I, Vol. III, 1855.

Lindelöf–Moigno *LEC*:
Ernst L. Lindelöf and François N. M. Moigno, *Leçons de Calcul des Variations*, Paris, 1861.

Lundström [1869]:
O. Lundström, Distinction des maxima et des minima dans un problème isopérimétrique, *Nova Act. Upsal.*, Ser. 3, Vol. VII, 1869.

Mach *MEC*:
Ernst Mach, *Die Mechanik in ihrer Entwickelung historisch-kritisch dargestellt*, Leipzig, 1933.

Maclaurin *FLUX*:
Colin Maclaurin, *A Treatise of Fluxions*, 2 vols., Edinburgh, 1742; 2nd end. London, 1801. French translation. Paris, 1749.

MacMillan *SAD*:
William D. MacMillan, *Statics and the Dynamics of a Particle*, New York, 1927.

Mainardi [1852]:
G. Mainardi, Ricerche sul calcolo delle variazioni, *Ann. Sc. Math. e. Fis.*, Vol. III, 1852, pp. 149–192, 379–383.

Mason [1904]:
Max Mason, Zur Theorie der Randwertaufgaben, *Math. Ann.*, Vol. LVIII, 1904, pp. 528–544. See also Bliss.

Maupertuis

[1740]: Pierre Louis Moreau de Maupertuis, Loi du repos des corps, *Mém. Acad. Sci.*, Paris, 1740, pp. 170–176.

[1744]:———Accord de différentes lois de la nature qui avaient jusqu'ici paru incompatibles, *Mém. Acad. Sci.*, 1744, pp. 417–426.

Mayer

[1866]: Adolph Mayer, *Beiträge zur Theorie der Maxima und Minima der einfachen Integrale,* Leipzig, 1866.

[1868]:———Ueber die Kriterien des Maximums und Minimums der einfachen Integrale, *Jour. für Math.*, Vol. LXIX, 1868, pp. 238–263.

[1870]:———Der Satz der Variationsrechnung, welcher dem Princip der kleinsten Wirkung in der Mechanik entspricht, *Math. Ann.*, Vol. II, 1870, pp. 143–149.

GP:———*Geschichte des Princips der kleinsten Action*, Leipzig, 1877.

[1877]:———Die Kriterien des Maximums und Minimums der einfachen Integrale in den isoperimetrischen Problemen, *Math. Ann.*, Vol. XIII, 1878, pp. 53–68 = *Leip. Ber.*, Vol. XXIX, 1877, pp. 114–132.

[1878]:———Ueber das allgemeinste Problem der Variationsrechnung bei einer einzigen unabhängigen Variablen, *Leip. Ber.*, Vol. XXX, 1878, pp. 16–32.

[1884]:———Zur Aufstellung der Kriterien des Maximums und Minimums der einfachen Integrale bei variablen Grenzwerthen, *Leip. Ber.*, Vol. XXXVI, 1884, pp. 99–128.

[1886]:———Begründung der Lagrange'schen Multiplicatorenmethode in der Variationsrechnung, *Math. Ann.*, Vol. XXVI, 1886, pp. 74–82.

[1886′]:———Die beiden allgemeinen Sätze der Variationsrechnung, welche den beiden Formen des Princips der kleinsten Action in der Dynamik entsprechen, *Leip. Ber.*, Vol. XXXVIII, 1886, pp. 343–355.

[1895]:———Die Lagrang'sche Multiplicatorenmethode und das allgemeinste Problem der Variationsrechnung bei einer unabhängigen Variabeln, *Leip. Ber.*, Vol. XLVII, 1895, pp. 129–144.

[1896]:———Die Kriterien des Minimums einfacher Integrale bei variablen Grenzwerthen, *Leip. Ber.*, Vol. XLVIII, 1896, pp. 436–465.

[1903], [1905]:———Über den Hilbertschen Unäbhangigkeitsatz in der Theorie des Maximums und Minimums der einfachen Integrale, *Math. Ann.*, Vol. LVIII, 1904, pp. 235–248 = *Leip. Ber.*, Vol. LV, 1903, pp. 131–145. II Mitteilung, *Math. Ann.*, Vol. LXII, 1906, pp. 335–350 = *Leip. Ber.*, Vol. LXII, pp. 49–67, 313–314.

Morse [1934]:

Marston Morse, *The Calculus of Variations in the Large*, New York, 1934. (Here there is an excellent bibliography of his earlier papers. The interested reader should certainly consult these.)

Moulton *BAL*:

F. R. Moulton, *New Methods in Exterior Ballistics*, Chicago, 1925.

Newton

PAPERS: Isaac Newton, *The Mathematical Papers of Isaac Newton*, 7 vols., D. T. Whiteside, ed., Cambridge, 1967–1976.

SOL:———The solid of revolution of least resistance to motion in a uniform fluid, *PAPERS*, Vol. VI, late 1685, pp. 456–465.

APP 1:———Appendix 1. The resistance of a sphere to 'rapid' rectilinear motion, *PAPERS*, Vol. VI, pp. 466–469.

APP 2:———Appendix 2. Recomputation of surfaces of least resistance (1694), *PAPERS*, Vol. VI, pp. 470–480.

PRIN:———*Philosophiae Naturalis Principia Mathematica*, 1st ed. London, 1687, 2nd ed. 1713, 3rd ed., London 1725/26. English translation by A.

Motte, *Sir Isaac Newton's Mathematical Principles of Natural Philosophy and his System of the World*, London, 1729 and English translation by F. Cajori, *Mathematical Principles of Natural Philosophy and his System of the World*, Berkeley, Calif., 1934.

PT:———Epistola missa ad praenobilem virum D. Carolum Mountague Armigerum, Scaccarii Regii apud Anglos Cancellarium & Societatis Regiae Praesidem, in qua solvantur duo problemata Mathematica à Johanne Barnoullo Mathematico celeberrimo proposita, *Phil. Trans.*, Vol. XIX, 1695–1697, pp. 384–389 (dated Jan. 30. $169\frac{6}{7}$).

Noble [1901]:

Ch. A. Noble, *Eine neue Methode in der Variationsrechnung*, Dissertation, Göttingen, 1901.

Osgood

[1901]: William F. Osgood, On the existence of a minimum of the integral $\int_{x_0}^{x_1} F(x, y, y')dx$ when x_0 and x_1 are conjugate points, and the geodesics on an ellipsoid of revolution: a revision of a theorem of Kneser's, *Trans. Am. Math. Soc.*, Vol. II, 1901, pp. 166–182.

[1901′]:———On a fundamental property of a minimum in the calculus of variations and the proof of a theorem of Weierstrass's, *Trans. Am. Math. Soc.*, Vol. II, 1901, pp. 273–295.

[1901″]:———Sufficient conditions in the calculus of variations, *Ann. Math.*, Ser. 2, Vol. II, 1901, pp. 105–129.

LEHR:———*Lehrbuch der Funktionentheorie*, 2 vols., Leipzig u. Berlin, 1907.

Ostwald *KLASSIKER*:

Wilhelm Ostwald, *Klassiker der exacten Wissenschaften*, 239 Nos., Leipzig, 1889–1937. The numbers 46 and 47 are translations by P. Stäckel of various papers in the calculus of variations, and they are entitled *Abhandlungen über Variations-Rechnung*.

Pascal *VR*:

Ernesto Pascal, *Calcolo delle Variazioni*, Milan, 1897. German translation by A. Schepp, *Variationsrechnung*, Leipzig, 1899. All references are to this translation.

Peano-Genocchi [1884]:

Giusseppe Peano-Angelo Genocchi, *Calcolo Differenziale e Principii di Calcolo Integrale*, Turin, 1884. German translation *Differentialrechnung und Grundzüge der Integralrechnung*, Leipzig, 1899.

Picard [1891]:

Émil Picard, *Traité d'Analyse*, 3 vols., Paris, 1891, 4th ed. 1942.

Poisson *TM*:

Siméon-Denis Poission, *Traité de Mecanique*, 2 vols., Paris, 1811 and 2nd ed., 1833.

Reiff [1879]:

R. Reiff, *Inauguraldissertation über den Einfluss der Capillarkräfte auf die Form der Oberfläche einer bewegten Flüssigkeit*, Tubingen, 1879.

Richardson [1910], [1911]:

R. G. D. Richardson, Das Jacobische Kriterium der Variationsrechnung und die Oszillationseigenschaften linearer Differentialgleichungen 2. Ordnung, *Math. Ann.*, Vol. LXVIII, 1910, pp. 279–304. This is Part I. Part II appeared in *Math. Ann.*, LXXI, 1912, pp. 214–232.

Sabra *TOL*:

A. I. Sabra, *Theories of Light from Descartes to Newton*, New York, 1967.

Scheeffer

[1885]: Ludwig Scheeffer, Die Maxima und Minima der einfachen Integrale zwischen festen Grenzen, *Math. Ann.*, Vol. XXV, 1885, pp. 522–594.

[1885′]:———Bemerkungen zu den vorstehende Aufsatze, *Math. Ann.*, Vol. XXV, 1885, pp. 594–595.

[1886]:———Ueber die Bedeutung der Begriffe 'Maximum und Minimum' in der Variationsrechnung, *Math. Ann.*, Vol. XXVI, 1886, pp. 197–208 = *Leip. Ber.*, Vol. XXXVII, 1885, pp. 92–105.

Schwarz

GMA: Karl H. A. Schwarz, *Gesammelte mathematische Abhandlungen*, 2 vols., Berlin, 1890.

[1885]:———Ueber ein die Flächen kleinsten Inhalts betreffendes Problem der Variationsrechnung, *Acta. Soc. Sci. Fennica.*, Vol. XV, 1885, pp. 315–362 = *GMA*, Vol. I, pp. 223–269. (This is a paper in a *Festschrift* on Weierstrass's 70th birthday.)

Sommerfeld [1900]:

A. Sommerfeld, Bemerkungen zur Variationsrechnung, *Jahr. Ber.*, Vol. VIII, 1900, pp. 188–193.

Spitzer [1854]:

Simon Spitzer, Über die Kriterien des Grössten und Kleinsten bei den Problemen der Variationsrechnung, *Wien. Ber..*, Vol. XII, 1854, pp. 1014–1071; Vol. XIV, 1855, pp. 41–120.

Stäckel. See Ostwald's *Klassiker*.

Stegmann *LV*:

Friedrich L. Stegmann, *Lehrbuch der Variationsrechnung*, Kassel, 1854.

Struik *SOURCE*:

A Source Book in Mathematics, 1200–1800, Cambridge, Mass., 1969 D. J. Struik, ed.

Sturm [1900]:

Rudolf Sturm, Mémoire sur les equations différentielles du second ordre, *Jour. de Math.*, Vol. I, 1836, p. 131.

Taylor *MET*:

Brook Taylor, *Methodus Incrementorum Directa and Inversa*, London, 1715.

Todhunter *HIS*:

Isaac Todhunter, *A History of the Progress of the Calculus of Variations during the Nineteenth Century*, Cambridge, 1861 = *A History of the Calculus of Variations* . . . , Reprinted New York.

Tonelli [1923]:

Tonelli, *Fondamenti di Calcolo delle Variazioni*, 2 vols., Bologna, 1923–1924.

Turksma [1896]:

B. Turksma, Begründung der Lagrange'schen Multiplicatorenmethode in der Variationsrechnung durch Vergleich derselben mit einer neuen Methode, welche zu den nämlichen Lösungen führt, *Math. Ann.*, Vol. XLVII, 1896, pp. 33–46.

von Escherich

[1898]: G. von Escherich, "Die zweite Variation der einfachen Integrale," *Wien. Ber,* Abt. IIa, Vol. CVII, 1898, pp. 1191–1250, 1267–1326, 1383–1430.

[1899]:———"Die zweite Variation der einfachen Integrale," *Wien.Ber.* Abt. IIa, Vol. CVIII, 1899, pp. 1269–1340.

Weierstrass

WERKE: Karl Weierstrass, *Mathematische Werke*, 7 vols., Berlin and Leipzig, 1894–1927.

VOR:———*Vorlesungen über Variationsrechnung* (1865 – 1890), *WERKE*, Vol. VII, R. Rothe, ed.

Whittemore [1901]:

J. K. Whittemore, Lagrange's equation in the calculus of variations, and the extension of a theorem of Erdmann, *Ann. Math.*, Ser. 2, Vol. II, 1899–1901, pp. 130–136.

Williamson *DC*:

B. Williamson, *An Elementary Treatise on the Differential Calculus*, New York, 1889.

Woodhouse *COV*:

Robert Woodhouse, *A Treatise on isoperimetrical Problems and the Calculus of Variations*, Cambridge, 1810 = *A History of the Calculus of Variations in the eighteenth Century*, Reprinted New York.

Zermelo

[1894]: Ernst Zermelo, *Untersuchungen zur Variationsrechnung*, Dissertation, Berlin, 1894.

ENC:———u. H. Hahn, Weiterentwicklung der Variationsrechnung in den letzten Jahren, *Encyklopädie der mathematischen Wissenschaften*, Vol. IIA, pp. 628–641.

Index

Studies in the History of Mathematics and Physical Sciences

Edited by **G.J. Toomer**

Volume 1
A History of Ancient Mathematical Astronomy
By **O. Neugebauer**
ISBN 0-387-**06995**-X

Volume 2
A History of Numerical Analysis from the 16th through the 19th Century
By **H. H. Goldstine**
ISBN 0-387-**90277**-5

Volume 3
I. J. Bienaymé: Statistical Theory Anticipated
By **C. C. Heyde** and **E. Seneta**
ISBN 0-387-**90261**-9

Volume 4
The Tragicomical History of Thermodynamics, 1822-1854
By **C. Truesdell**
ISBN 0-387-**90403**-3

Volume 5
A History of the Calculus of Variations from the 17th through the 19th Century
By **H. H. Goldstine**
ISBN 0-387-**90521**-9

